NEW FRONTIERS
IN ASYMMETRIC
CATALYSIS

THE WILEY BICENTENNIAL–KNOWLEDGE FOR GENERATIONS

 ach generation has its unique needs and aspirations. When Charles Wiley first opened his small printing shop in lower Manhattan in 1807, it was a generation of boundless potential searching for an identity. And we were there, helping to define a new American literary tradition. Over half a century later, in the midst of the Second Industrial Revolution, it was a generation focused on building the future. Once again, we were there, supplying the critical scientific, technical, and engineering knowledge that helped frame the world. Throughout the 20th Century, and into the new millennium, nations began to reach out beyond their own borders and a new international community was born. Wiley was there, expanding its operations around the world to enable a global exchange of ideas, opinions, and know-how.

For 200 years, Wiley has been an integral part of each generation's journey, enabling the flow of information and understanding necessary to meet their needs and fulfill their aspirations. Today, bold new technologies are changing the way we live and learn. Wiley will be there, providing you the must-have knowledge you need to imagine new worlds, new possibilities, and new opportunities.

Generations come and go, but you can always count on Wiley to provide you the knowledge you need, when and where you need it!

WILLIAM J. PESCE
PRESIDENT AND CHIEF EXECUTIVE OFFICER

PETER BOOTH WILEY
CHAIRMAN OF THE BOARD

NEW FRONTIERS IN ASYMMETRIC CATALYSIS

Edited by

KOICHI MIKAMI
Department of Applied Chemistry, Tokyo Institute of Technology

MARK LAUTENS
Department of Chemistry, University of Toronto

WILEY-INTERSCIENCE

A JOHN WILEY & SONS, INC., PUBLICATION

Published by John Wiley & Sons, Inc., Hoboken, New Jersey
Published simultaneously in Canada

Wiley Bicentennial Logo: Richard J. Pacifico

Library of Congress Cataloging-in-Publication Data:

New frontiers in asymmetric catalysis / edited by Koichi Mikami and Mark Lautens.
p. cm.
Includes bibliographical references and index.
978-0-471-68026-0
1. Catalysis–Research. 2. Asymmetry (Chemistry)–Research. I. Mikami, Koichi. II. Lautens, M. (Mark)
QD505.N474 2007
541′.395- -dc22 2006020555

Printed in the United States of America

10 9 8 7 6 5 4 3 2 1

CONTENTS

11   History and Perspective of Chiral Organic Catalysts   313

Gérald Lelais and David W. C. MacMillan

12   Chiral Brønsted/Lewis Acid Catalysts   359

Kazuaki Ishihara and Hisashi Yamamoto

13   Chiral Bifunctional Acid/Base Catalysts   383

Masakatsu Shibasaki and Motomu Kanai

PREFACE

New Frontiers in Asymmetric Catalysis provides readers with a comprehensive perspective on understanding the concepts and applications of asymmetric catalysis reactions. Despite the availability of excellent comprehensive multivolume treatises in this field, we felt that researchers in pharmaceutical and chemical companies as well as university faculty and graduate students would benefit from a selection of some of the most important recent advances in this ever-growing area.

The key to efficient asymmetric catalysis lies in the creation of robust chiral catalysts by a suitable combination of chiral organic compounds and metal centers to which they are ligated. Many chiral organic ligands are *atropisomeric* (*a + tropos* in Greek) compounds with C_2 symmetry, such as BINOL and BINAP. The use of C_2 symmetric ligands originally introduced by Kagan had a strong impact on subsequent ligands design for asymmetric catalysis. In recent years new nonsymmetric ligands that are more effective than their C_2 counterparts have been reported. The first chapters of this book are dedicated to "rational" ligand design, which is critically dependent on the reaction type (reduction, oxidation, and C–C bond formation). The concept of C_2 symmetry for bidentate ligands can be extended to the design of C_n symmetric multidentate ligands bearing phosphorous, nitrogen, and other coordinating elements.

Catalyst systems can be described as biomimetic assemblies of multifunctional or bimetallic catalysts. Ideally, their design can be based on quantitative analysis of the transition state for a given reaction. Alternatively, a combinatorial screening of metal centers and chiral ligands can also lead to new catalyst systems. The development of efficient high-throughput screening methods for finding a good lead or an optimized catalytic system is still in its infancy.

In asymmetric catalysis, Sharpless emphasized the importance of "ligand-accelerated catalysis" through the construction of an asymmetric catalyst from an achiral precatalyst via ligand exchange with a chiral ligand. By contrast, a dynamic combinatorial approach, where an achiral precatalyst combined with several multicomponent chiral ligands (L^1*, — —) and several chiral activator ligands (A^1*, — —) may selectively assemble into the most active and highest enantioselective activated catalyst ($ML^{m}* A^{n}*$).

In Chapters 4–6, recent findings on activation of C–H bonds, C–C bonds and small molecules (C=O, HCN, RN=C, and CO_2) are covered. The latest developments on C–C bond reorganization such as metathesis (which earned the Nobel prize in chemistry, 2005) are also described.

Studies on the origin of chirality generated from achiral or racemic "primitive earth" provide the basis for asymmetric catalysis starting from racemic or achiral catalysts. Asymmetric catalysis through enantiomeric fluctuation or discrimination by an external chiral bias and subsequent amplification of chirality can be developed through autocatalysis with nonlinear effects. One strategy for achieving this enantiomeric discrimination is the addition of a chiral source, which selectively transforms one catalyst enantiomer into a highly activated or deactivated catalyst enantiomer. Recent progress on "chirally economical" nonlinear phenomena, racemic catalysis, and autocatalysis are highlighted in Chapters 7–9.

Asymmetric catalysis in target- or diversity-oriented synthesis becomes an increasingly important tool. Desymmetrization of symmetric intermediates (asymmetric or enantioselective desymmetrization) is one important synthetic strategy reviewed in Chapter 10. Use of naturally occurring enzymes is one of the oldest and most important approaches employed in asymmetric desymmetrization, the so-called classic mesotrick process. Generally effective methods for highly enantioselective aziridination of olefins, reduction and C–C bond formation of aliphatic ketones, are also expected to become practical (often via a pseudodesymmetrization process) in the very near future. Asymmetric catalytic tendem (domino) reaction sequences are likely to remain at the forefront of future research efforts.

Finally in Chapters 11–13, some of the more recent discoveries that have led to a renaissance in the field of organocatalysis are described. Included in this section are the development of chiral Brönsted acids and Lewis acidic metals bearing the conjugate base of the Brönsted acids as the ligands and the chiral bifunctional acid-base catalysts.

Although tremendous progress has been made in the field of asymmetric catalysis, very few systems have become widely applicable on an industrial scale because of challenges of catalyst efficiency (turnover number (TON) and frequency (TOF), catalyst loading, applicability to a wide range of systems and with feedstocks of varying purity and the levels of enantioselectivity). The best known are the Takasago menthol process, the Novartis imine hydrogenation for metolachlor, the Sumitomo cyclopropanation for cilastatine, and the Firmenich process of fragrant paradisone. For many asymmetric reactions, the recovery and recycling of the catalysts are a serious concern for both industry and society in order to limit the amount of waste, and impurities, that affect the overall costs of the processes.

Thus, the use of catalysts in new "green" reaction media such as ionic liquids, fluorous solvents, and supercritical carbon dioxide has become a viable alternative to those discussed within the chapters.

We hope that readers will find helpful and thought-provoking information in this book written by frontrunners in their respective fields, including the areas recognized by recent Nobel prizes in chemistry.

KOICHI MIKAMI

Department of Applied Chemistry Tokyo Institute of Technology

MARK LAUTENS

Department of Chemistry University of Toronto

December 25, 2006

CONTRIBUTORS

Kohsuke Aikawa, Department of Applied Chemistry, Tokyo Institute of Technology, Meguro-ku, Tokyo 152-8552, Japan

Tamio Hayashi, Department of Chemistry, Graduate School of Science, Kyoto University, Sakyo, Kyoto 606-8502, Japan

Kazuaki Ishihara, Graduate School of Engineering, Nagoya University, Furo-cho, Chikusa, Nagoya 464-8603, Japan

Henri B. Kagan, Laboratoire de Catalyse Moléculaire, Institut de Chimie Moléculaire et des Matériaux d'Orsay (CNRS UMR 8182), Université Paris-Sud, 91405-Orsay, France

Motomu Kanai, Graduate School of Pharmaceutical Sciences, The University of Tokyo, Hongo 7-3-1, Bunkyo-ku, Tokyo 113-0033, Japan

Tsuneomi Kawasaki, Department of Applied Chemistry, Tokyo University of Science, Kagurazaka, Shinjuku-ku, Tokyo 162-8601, Japan

Masato Kitamura, Department of Chemistry and Research Center for Materials Science, Nagoya University, Chikusa, Nagoya 464-8602, Japan

Gérald Lelais, Department of Chemistry, MC 164-30, California Institute of Technology, Pasadena, CA 91125, USA

Chao-Jun Li, Department of Chemistry, McGill University, 801 Sherbrooke Street West, Montreal, Quebec H3A 2K6, Canada

Zhiping Li, Department of Chemistry, McGill University, 801 Sherbrooke Street West, Montreal, Quebec H3A 2K6, Canada

David W. C. MacMillan, Department of Chemistry, MC 164-30, California Institute of Technology, Pasadena, CA 91125, USA

Koichi Mikami, Department of Applied Chemistry, Tokyo Institute of Technology, Meguro-ku, Tokyo 152-8552, Japan

Miwako Mori, Health Science University of Hokkaido, Ishikari-Toubetsu, Hokkaido 061-0293, Japan

Ryoji Noyori, Department of Chemistry and Research Center for Materials Science, Nagoya University, Chikusa, Nagoya 464-8602, Japan

Kyoko Nozaki, Department of Chemistry and Biotechnology, Graduate School of Engineering, The University of Tokyo, Hongo 7-3-1, Bunkyo-ku, Tokyo 113-8656, Japan

Takeshi Ohkuma, Division of Chemical Process Engineering, Graduate School of Engineering, Hokkaido University, Sapporo 060-8628, Japan

Tomislav Rovis, Department of Chemistry, Colorado State University, Fort Collins, CO 80523, USA

Itaru Sato, Department of Applied Chemistry, Tokyo University of Science, Kagurazaka, Shinjuku-ku, Tokyo 162-8601, Japan

Masakatsu Shibasaki, Graduate School of Pharmaceutical Sciences, The University of Tokyo, Hongo 7-3-1, Bunkyo-ku, Tokyo 113-0033, Japan

Ryo Shintani, Department of Chemistry, Graduate School of Science, Kyoto University, Sakyo, Kyoto 606-8502, Japan

Kenso Soai, Department of Applied Chemistry, Tokyo University of Science, Kagurazaka, Shinjuku-ku, Tokyo 162-8601, Japan

Tohru Yamada, Department of Chemistry, Keio University, Hiyoshi, Yokohama 223-8522, Japan (email: yamada@chem.keio.ac.jp)

Hisashi Yamamoto, Department of Chemistry, University of Chicago, 5735 South Ellis Avenue (SCL 317), Chicago, IL 60637, USA

1

LIGAND DESIGN FOR CATALYTIC ASYMMETRIC REDUCTION

TAKESHI OHKUMA

Division of Chemical Process Engineering, Graduate School of Engineering, Hokkaido University, Sapporo, Japan

MASATO KITAMURA AND RYOJI NOYORI

Department of Chemistry and Research Center for Materials Science, Nagoya University, Chikusa, Nagoya, Japan

1.1 INTRODUCTION

Molecular catalysts consisting of a metal or metal ion and a chiral organic ligand are widely used for asymmetric synthesis. Figure 1.1 illustrates a typical (but not general) scheme of asymmetric catalytic reaction.[1] The initially used chiral precatalyst **1A** is converted to the real catalyst **1B** through an induction process. An achiral reactant A and substrate B are activated by **1B** to form reversibly an intermediate **1C**. The chiral environment of **1C** induces asymmetric transformation of A and B to the chiral product A–B (*R* or *S*) through an intermediate **1D** with reproduction of catalyst **1B**. The absolute configuration of A–B is kinetically determined at the first irreversible step, **1C→1D**. The efficiency of catalysis depends on several kinetic and thermodynamic parameters, because most catalytic reactions proceed through such multistep transformation.

Catalytic asymmetric reduction of unsaturated compounds is one of the most reliable methods used to synthsize the corresponding chiral saturated products.[2-4] Chiral transition metal complexes repeatedly activate an organic or inorganic hydride source, and transfer the hydride to olefins, ketones, or imines from one

New Frontiers in Asymmetric Catalysis, Edited by Koichi Mikami and Mark Lautens
Copyright © 2007 John Wiley & Sons, Inc.

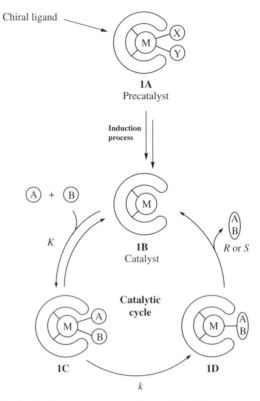

Figure 1.1. The principal of asymmetric catalysis with chiral organometallic molecular catalysts (M = metal; A, B = reactant and substrate; X, Y = neutral or anionic ligand).

of two enantiofaces selectively, resulting in the enantio-enriches alkanes, alcohols, or amines, respectively. The three-dimensional (3D) structure and functionality of the chiral ligand, among other factors, are the obvious key for efficient asymmetric reduction. Rational design of chiral ligand can be done on the basis of full understanding of the corresponding catalytic reaction. This chapter presents successful examples of catalytic asymmetric reduction and the concepts of the ligand design. The description is brought to focus on the BINAP–transition metal chemistry.[2]

1.2   HYDROGENATION OF OLEFINS

1.2.1   Enamide Hydrogenation with Rhodium Catalysts

The discovery of Wilkinson complex, $RhCl[P(C_6H_5)_3]_3$,[5] acting as an effective catalyst for hydrogenation of olefins opened the door for developing asymmetric reaction catalyzed by rhodium complexes with a chiral phosphine ligand.[1,6–10] The enantioselective ability of chiral ligands has often been evaluated by hydrogenation of α-hydroxycarbonyl- or α-alkoxycarbonyl-substituted enamides. Figure 1.2

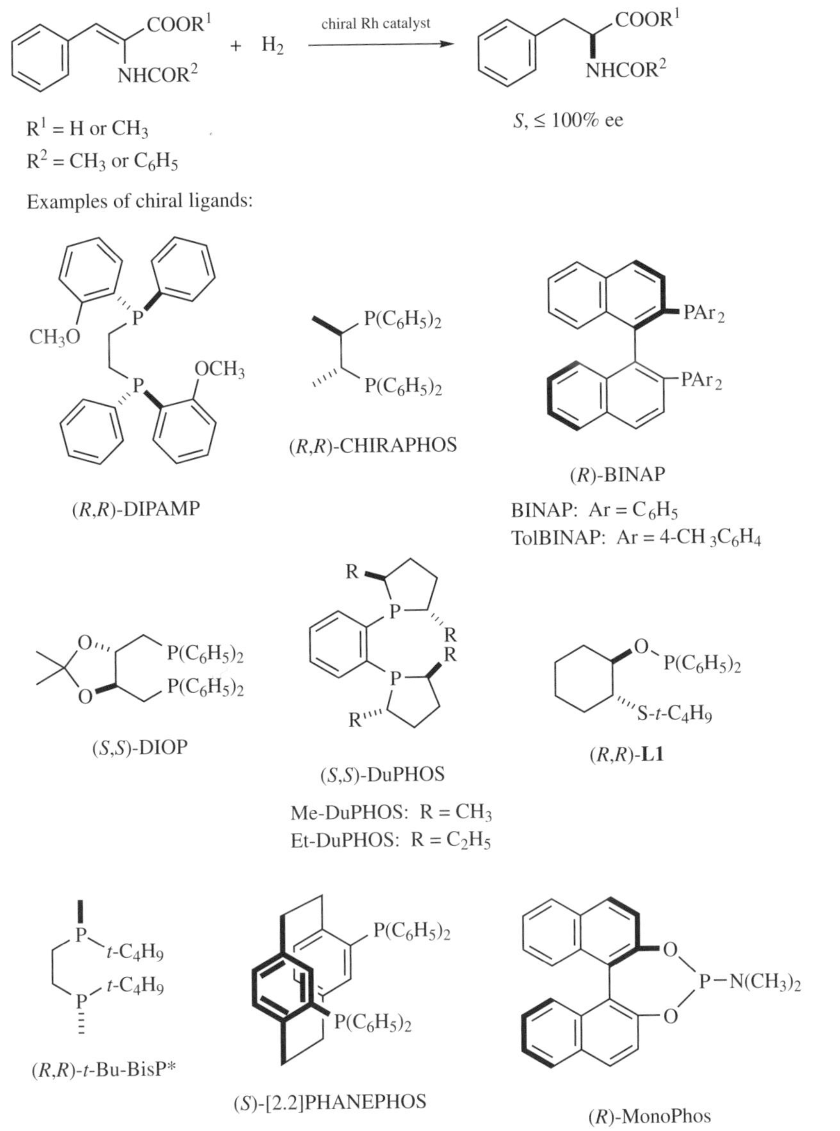
COOR1
NHCOR2
+ H2
chiral Rh catalyst
COOR1
NHCOR2
R1 = H or CH3
R2 = CH3 or C6H5
S, ≤ 100% ee
Examples of chiral ligands:
CH3O
P
P
OCH3
(R,R)-DIPAMP
P(C6H5)2
P(C6H5)2
(R,R)-CHIRAPHOS
PAr2
PAr2
(R)-BINAP
BINAP: Ar = C6H5
TolBINAP: Ar = 4-CH3C6H4
O
O
P(C6H5)2
P(C6H5)2
(S,S)-DIOP
R
P
R
R
P
R
(S,S)-DuPHOS
Me-DuPHOS: R = CH3
Et-DuPHOS: R = C2H5
O
P(C6H5)2
S-t-C4H9
(R,R)-L1
P
t-C4H9
t-C4H9
P
(R,R)-t-Bu-BisP*
P(C6H5)2
P(C6H5)2
(S)-[2.2]PHANEPHOS
O
P−N(CH3)2
O
(R)-MonoPhos

Figure 1.2. Asymmetric hydrogenation of *N*-acylated dehydroamino acids and esters.

illustrates typical examples of phosphorus-based chiral ligands, with which Rh(I) catalyst selectively afforded (*S*)-amino acid derivatives in hydrogenation of (*Z*)-2-(acylamido)cinnamic acids and the methyl esters. Key factors for designing of these ligands are: (1) monodentate or bidentate, (2) steric effects (bulkiness, conformational flexibility, space coordinate, etc.), (3) electronic effects (alkylphosphine, arylphosphine, phosphite, phosphoramidite, etc.), (4) bite angle for bidentate ligands, (5) C_1 or C_2 symmetry for bidentate ligands, and (6) chirality on the back-bone or on phosphorus atoms. A DIPAMP–Rh-catalyzed hydrogenation of an enamide substrate is industrially used in the synthesis of L-dopa, a drug for the parkinsonian disease.[8]

The mechanism of hydrogenation of methyl (Z)-2-(acetamido)cinnamate catalyzed by a CHIRAPHOS–[11] or DIPAMP–Rh[8] complex have been exhaustively studied by Halpern[12–14] and Brown.[7,15,16] They proposed the "unsaturate/dihydride mechanism" as illustrated in Figure 1.3. The Rh complex with the R,R ligand [(R)-3A] (solvate) and an enamide reversibly form the substrate complex 3B, which undergoes irreversible oxidative addition of molecular H_2 to the Rh center, affording Rh(III) dihydride species 3C. Both hydrides on Rh migrate onto the C—C double bond of the coordinated substrate. The first hydride migration to the C3 position forms a five-membered alkyl–hydride complex 3D, and then reductive elimination of the hydrogenation product (second hydride migration) completes the cycle with regeneration of 3A. The stereochemistry of product is determined at the first irreversible step, 3B → 3C, although a detailed theoretical investigation suggests the possibility that the process 3B → 3C is reversible and the step 3C → 3D constitutes the turnover-limiting step.[17] The BINAP–Rh-catalyzed hydrogenation of enamides is proposed to proceed with the same Halpern–Brown mechanism.[18–21]

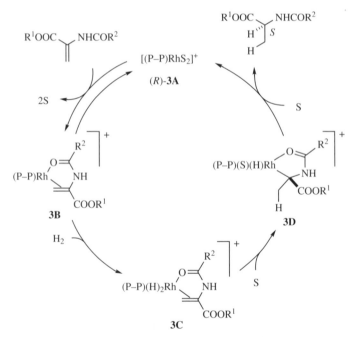

Figure 1.3. Catalytic hydrogenation of N-acylated dehydroamino esters via an unsaturated/dihydride mechanism; the β substituents in the substrates are omitted for clarity [P–P = (R,R)-DIPAMP, (R,R)-CHIRAPHOS, or (R)-BINAP; S = solvent or a weak ligand].

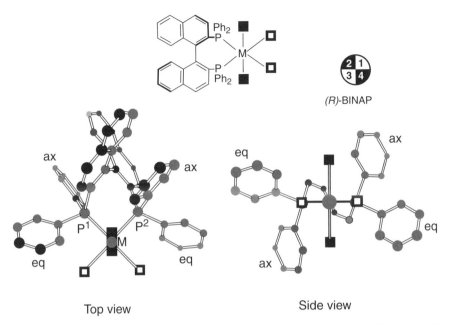

Figure 1.4. Chiral environment of an (*R*)-BINAP–transition metal complex (M = metallic element, ax = axial, eq = equatorial; □ = coordination site in the P^1–M–P^2 plane; ■ = coordination site out of the P^1–M–P^2 plane).

CHIRAPHOS, DIPAMP, and BINAP are all chiral diphosphines with a C_2 symmetry (Figure 1.2) forming chelate complexes with transition metallic elements. DIOP developed by Kagan is the origin of this type of chiral ligand.[22] Figure 1.4 illustrates the chiral template created by an (*R*)-BINAP–transition metal complex.[18–21] The naphthalene rings are omitted in the side view for clarity. In this template, the chiral information of binaphthyl backbone is transmitted through the *P*-phenyl rings to the four coordination sites shown by □ and ■. The in-plane coordination sites, □, are sterically affected by the "equatorial" phenyl rings, whereas the out-of-plane coordination sites, ■, are influenced by the "axial" phenyl groups. Consequently, the two kinds of quadrant of the chiral template (first and third vs. second and fourth) are clearly differentiated spatially, where the second and fourth quadrants are sterically congested, while the first and third ones are relatively uncrowded. (*R,R*)-CHIRAPHOS[23] and (*R,R*)-DIPAMP[24] form a similar chiral environment with metals.

As shown in Figure 1.3, the Rh catalyst (*R*)-**3A** and a bidentate enamide substrate reversibly form the substrate complex **3B**. Figure 1.5 illustrates two possible diastereomeric structures of **3B**, depending on the *Si/Re*-face selection at C2, which leads to the *R* or *S* hydrogenation product. Therefore, the enantioselectivity is determined by the relative equilibrium ratio and reactivity of *Si*-**3B** and *Re*-**3B**. A ^{31}P NMR spectrum of the Rh complex and an enamide substrate in CH_3OH showed a single signal for thermodynamically more favored *Si*-**3B**.[20] Most importantly,

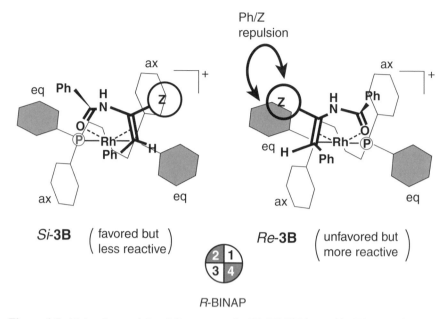

Figure 1.5. Molecular models of diastereomeric (R)-BINAP/enamide Rh complexes **3B** (not transition state) ($Z = CO_2R^1$; ax = axial, eq = equatorial).

the *Re*-**3B** which is less favored because of the nonbonded repulsion between an equatorial phenyl ring of the (R)-BINAP ligand and a carboxylate function of substrate reacts with H_2 much faster than the more stable *Si*-**3B**, leading to the S isomer as a major product. The observed enantioselectivity is a result of the delicate balance of the stability and reactivity of the diastereomeric **3B**. This inherent mechanistic problem requires careful choice of reaction parameters. For instance, the hydrogenation should be conducted under a low substrate concentration and low H_2 pressure to minimize reaction via the major diastereomeric intermediate *Si*-**3B**. Therefore, the hydrogenation of enamides catalyzed by BINAP–, CHIRAPHOS–, or DIPAMP–Rh complex, though giving amino acids in high enantiomeric excess (ee), is not ideal from the mechanistic standpoint. A Rh complex bearing Et-DuPHOS,[25] a C_2-chiral diphosphines (see Figure 1.2), catalyzes the hydrogenation basically with the same mechanism.[26,27]

This mechanistic problem can be solved when the more stable diastereomer of **3B** gives the major enantiomeric product. A Rh complex with a C_1-chiral P/S mixed ligand, (R,R)-**L1** (see Figure 1.2), catalyzes hydrogenation of methyl (Z)-2-(acetamido)cinnamate to afford the S product in excellent ee.[28] The enamide substrate is reduced along with the catalytic cycle illustrated in Figure 1.3. However, unlike traditional catalyst systems, the stereochemistry of hydrogenation product suggests that the major S product is obtained via the most stable diastereomer of **3B**. A substrate complex, *Re*-**3B**-**L1**, is the only visible species among four possible diastereomers. Figure 1.6 illustrates the structure of an

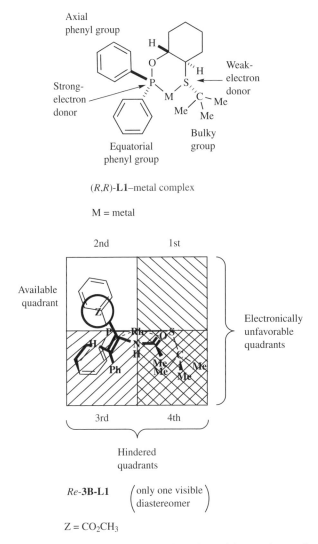

Figure 1.6. Molecular models of (R,R)-**L1**/enamide metal complexes.

(R,R)-**L1**–metal complex. The bulky t-butyl group on sulfur plays a crucial role in achieving high enantioface selectivity. This group is placed at the axial position to avoid steric hindrance with the ligand backbone. The two phenyl groups on phosphorus atom occupy the axial and equatorial positions. The high enantiodiscriminatory ability of the catalyst is rationalized by means of the quadrant model of Re-**3B-L1**.[28] The electron-donating olefin function of the enamide substrate preferably binds to Rh at the trans position to the less electron-donating sulfur atom instead of phosphorus, that is, the first and fourth quadrants are unfavorable for the olefinic function for electronic reasons. The third

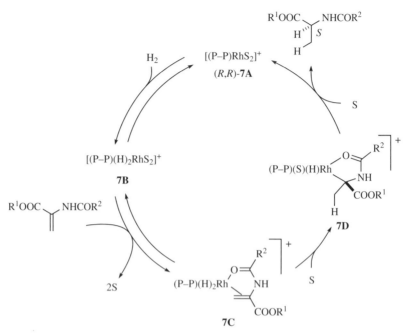

Figure 1.7. Catalytic hydrogenation of *N*-acylated dehydroamino esters via dihydride/ unsaturate mechanism; the β substituents in the substrates are omitted for clarity [P–P = (*R,R*)-*t*-Bu-BisP*; S = solvent or a weak ligand].

and fourth quadrants are blocked by equatorial *P*-phenyl and bulky *S*-*t*-butyl group, respectively. Therefore, only the second quadrant is available for approach of methoxycarbonyl group (Z).

The unsaturate/dihydride is not a sole mechanism for enamide hydrogenation. Its mechanistic problem can be resolved by a total change in catalytic cycle. *t*-Bu-BisP* is a C_2-symmetric, fully alkylated diphosphine with chiral centers at phosphorus (see Figure 1.2).[29,30] Hydrogenation of enamides catalyzed by an (*R,R*)-*t*-Bu-BisP*–Rh complex gives the *S* product in excellent ee. The hydrogenation is revealed to proceed through the "dihydride/unsaturate mechanism" as shown in Figure 1.7.[30–32] The major difference of this cycle from the unsaturate/ dihydride cycle in Figure 1.3 is the order of reaction of the substrate and H_2. Now the catalyst (*R,R*)-**7A** first reversibly reacts with H_2, giving **7B**, followed by inter- action with an enamide substrate to form a substrate–RhH₂ complex **7C**. The stereochemistry of product is determined at the first irreversible step, **7C** → **7D**. Because of the C_2-symmetric structure of (*R,R*)-*t*-Bu-BisP*, the quadrants of the chiral template are spatially differentiated into two kinds. The first and third quad- rants are crowded by the location of bulky *P*-*t*-butyl groups, whereas the second and fourth ones are open for substrate approach owing to the presence of only small methyl groups. Therefore, two diastereomers of bidentate substrate–Rh(III)H₂ complex, *Re*-**7C** and *Si*-**7C**, are possible (Figure 1.8). Formation of *Si*-**7C** is

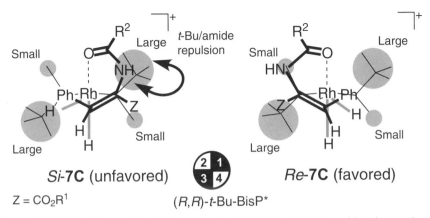

Figure 1.8. Molecular models of diastereomeric (R,R)-t-Bu-BisP*/enamide Rh complexes **7C** (not transition state).

unfavored because it suffers from serious steric repulsion between bulky P-t-butyl group and substrate amide function. On the other hand, only small methyl/amide repulsive interaction exists in Re-**7C**. The major S enantiomeric product is derived from the more stable diastereomeric species, Re-**7C**. The hydrogenation catalyzed by a [2.2]PHANEPHOS–Rh complex[33] (see Figure 1.2) is also suggested to proceed through the dihydride/unsaturate mechanism.[34]

Chiral monodentate phosphites[35,36] and phosphoramidites[37,38] are also effective ligands for Rh-catalyzed asymmetric hydrogenation of enamide substrates. As seen in the structure of MonoPhos[37,38] illustrated in Figure 1.2, combination of the modified BINOL backbone and the amine part gives a structural variety to this type of ligand.[39] Combinatorial methods are effective for optimization of the chiral structures.[40,41] Elucidation of the hydrogenation mechanism catalyzed by the MonoPhos–Rh complex is in progress.[42–44]

1.2.2 Hydrogenation of Functionalized Olefins with Ruthenium Catalysts

The BINAP–Rh catalyzed hydrogenation of functionalized olefins has a mechanistic drawback as described in Section 1.2.1. This problem was solved by the exploitation of BINAP–Ru(II) complexes.[1,2] Ru(OCOCH$_3$)$_2$(binap)[18] catalyzes highly enantioselective hydrogenation of a variety of olefinic substrates such as enamides, α,β- and β,γ-unsaturated carboxylic acids, and allylic and homoallylic alcohols (Figure 1.9).[6,7,45–48] Chiral citronellol is produced in 300 ton quantity in year by this reaction.[9]

It is worth noting that an opposite sense of enantioface selection is observed in going from the BINAP–Rh complex to the Ru catalyst. Hydrogenation of methyl (Z)-2-(acetamido)cinnamate with the (R)-BINAP–Ru catalyst in CH$_3$OH gives the R (not S) product selectively (Figure 1.9).[1,18,21,45] Figure 1.10 illustrates the

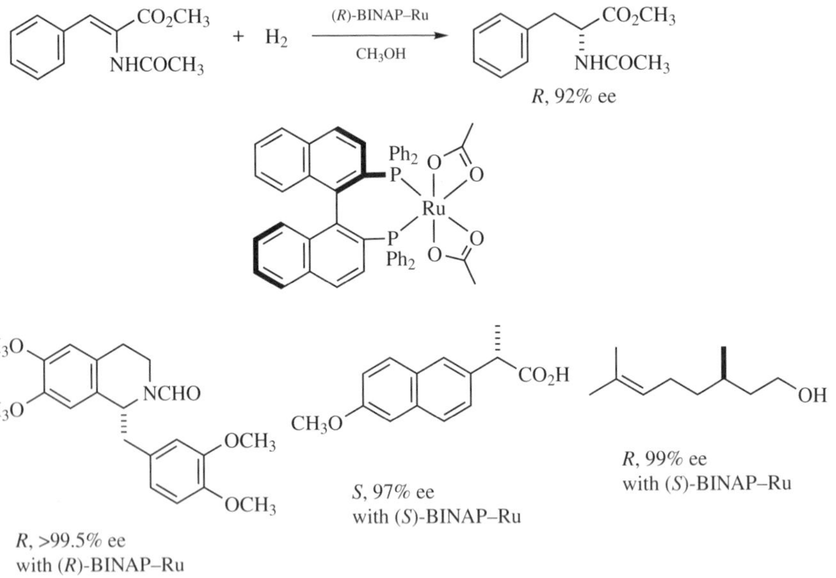

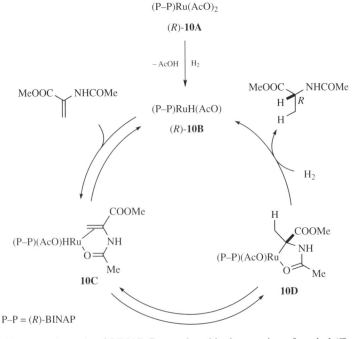

Figure 1.9. Asymmetric hydrogenation of functionalized olefins catalyzed by BINAP–Ru complexes.

Figure 1.10. Catalytic cycle of BINAP–Ru-catalyzed hydrogenation of methyl (Z)-α-acetamidocinnamate involving a monohydride/unsaturated mechanism. The β substituents in the substrates are omitted for clarity.

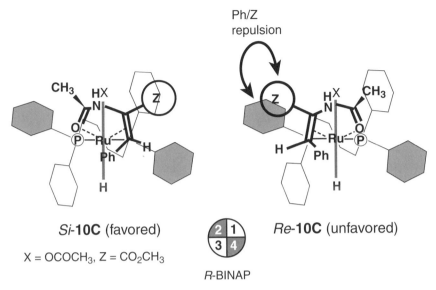

Figure 1.11. Molecular models of diastereomeric (R)-BINAP/enamide Ru complexes **10C** (not transition state).

"monohydride/unsaturate mechanism," in which the RuH(AcO) species **10B**, formed by the heterolytic cleavage of H_2 by the precatalyst **10A**, acts as a real catalyst.[21,49-53] Thus, the Ru hydride species is generated before the substrate coordination forming **10C**. The migratory insertion giving **10D**, in which the Ru—C bond is cleaved mainly by H_2, but also by CH_3OH solvent to some extent. The irreversible step determines the absolute configuration of the product. Because the diastereomers of **10D** would have a similar reactivity, the enantioselectivity well corresponds to the relative stability of the diastereomeric substrate–RuH(AcO) complexes, *Re*-**10C** and *Si*-**10C** (Figure 1.11). In order for **10C** to undergo migratory insertion, the Ru—H and C2—C3 double bond must have a *syn*-parallel alignment. As discussed above, the intermediate *Re*-**10C** is unfavored relative to *Si*-**10C** because of existence of *P*-Ph/COOCH₃ repulsion. Therefore, the major *Si*-**10C** is converted to the R hydrogenation product through **10D**. The two hydrogen atoms incorporated in the product are from two different H_2 molecules, or H_2 and protic CH_3OH.

1.2.3 Hydrogenation of Simple Olefins with Iridium Catalysts

Phosphinodihydroxazole (PHOX) compounds, **L2–4**, act as P/N bidentate ligands showing excellent enantioselectivity in Ir-catalyzed hydrogenation of simple α,α-disubstituted and trisubstituted olefins (Figure 1.12).[54-58] The use of tetrakis[3,5-bis(trifluoromethyl)phenyl]borate (BArF) as a counter anion achieves high catalytic efficiency due to avoidance of an inert Ir trimer

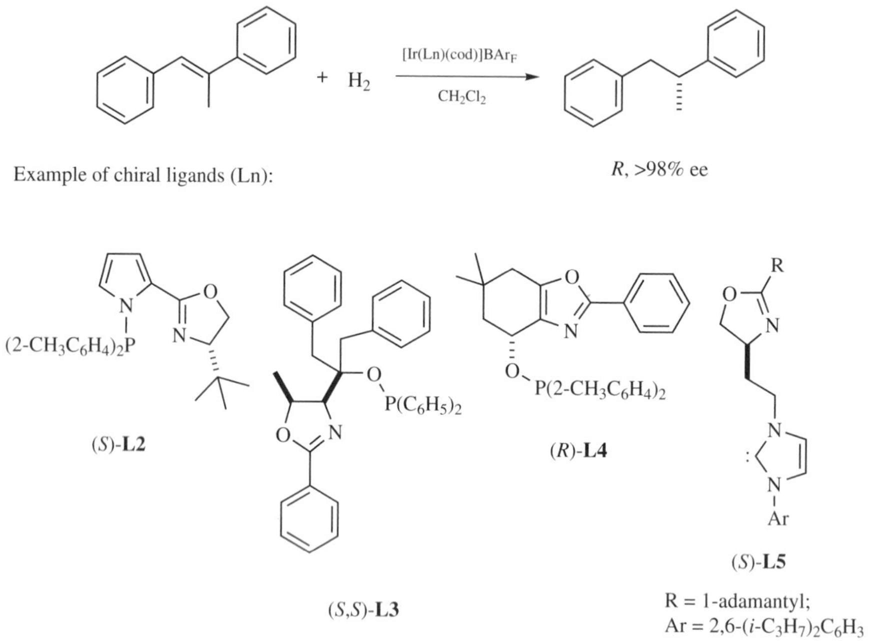

Example of chiral ligands (Ln):

R, >98% ee

(S)-L2

(S,S)-L3

(R)-L4

(S)-L5

R = 1-adamantyl;
Ar = 2,6-$(i$-C$_3$H$_7$)$_2$C$_6$H$_3$

Figure 1.12. Asymmetric hydrogenation of simple olefins catalyzed by chiral Ir complexes.

formation. A chiral carbene–oxazoline ligand **L5** is also useful for this purpose.[59] The mechanism of this reaction is to be elucidated by experimental[60–63] and theoretical[64,65] studies. Chiral titanocene catalysts also show high enantioselectivity for hydrogenation of simple olefins.[66] This subject is discussed in Section 1.4.

1.3 REDUCTION OF KETONES

1.3.1 Hydrogenation of Functionalized Ketones

Although Ru(OCOCH$_3$)$_2$(binap) exhibits excellent catalytic performance on asymmetric hydrogenation of functionalized olefins, it is feebly active for reaction of ketones. This failure is due to the property of the anionic ligands. Simple replacement of the carboxylate ligand by halides achieves high catalytic activity for reaction of functionalized ketones.[1,18,21,67] Thus, chiral precatalysts including RuCl$_2$[(R)-binap] (polymeric form),[67] RuCl$_2$[(R)-binap](dmf)$_n$ (oligomeric form),[68] [RuCl{(R)-binap}(arene)]Cl,[69] [NH$_2$(C$_2$H$_5$)$_2$][{RuCl[(R)-binap]}$_2$-(μ-Cl)$_3$],[67] and other in situ formed (R)-BINAP–Ru complexes[70] are successfully used for hydrogenation of β-keto esters, resulting in the R β-hydroxy esters in >99% ee (Figure 1.13). An intermediate for the synthesis of carbapenem antibiotics is produced industrially by this method.[18]

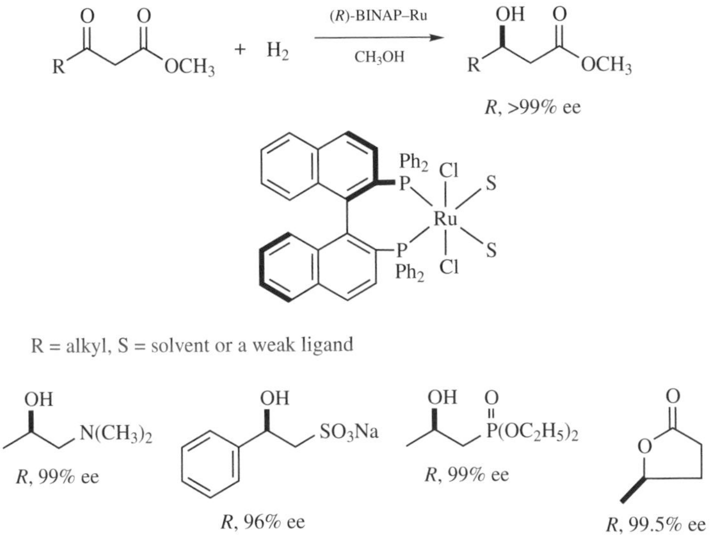

R = alkyl, S = solvent or a weak ligand

Figure 1.13. Asymmetric hydrogenation of functionalized ketones catalyzed by BINAP–Ru complexes.

Figure 1.14 illustrates a mechanistic model of this hydrogenation.[21,71] The true catalytic RuHCl species **14B** is generated by the reaction of the RuCl$_2$ precatalyst **14A** and H$_2$ by releasing HCl. The catalyst **14B** reversibly interacts with the β-keto ester to form the σ-type chelate complex **14C**. Protonation of **14C** at the carbonyl oxygen increases electrophilicity of the carbonyl carbon, inducing conversion of the geometry from σ to π. Consequently, the hydride on Ru smoothly migrates to the substrate carbonyl carbon. The hydroxy ester ligand in the resulting complex **14D** is replaced by solvent molecule. The Ru cationic species **14E** cleaves H$_2$, reproducing the catalyst **14B**. Enantioface selection of the reaction occurs in the hydride transfer step, **14C** → **14D**. The carbonyl protonation with HCl in **14C** is crucial for the transformation to **14D**.[72–74] When Ru(OCOCH$_3$)$_2$(binap) is used, acetic acid is produced instead of HCl. Such a weak acid does not sufficiently protonate the carbonyl oxygen. Thus, even achiral anionic ligands in the precatalyst are important to achieve high catalytic efficiency.

Figure 1.15 illustrates two diastereomeric transition states *Re*-**15A** and *Si*-**15A** in the hydride migration step in the (*R*)-BINAP–Ru-catalyzed hydrogenation.[21] The protonated carbonyl group, C=O$^+$H, becomes parallel to the H–Ru linkage, bringing the "R" group close to the *P*-phenyl ring of the (*R*)-BINAP ligand. Thus, transition state *Re*-**15A** producing the *S* alcohol is unfavored because of the R/Ph repulsive interaction in the fourth quadrant. Therefore, the (*R*)-β-hydroxy ester is obtained selectively through *Si*-**15A**.

Figure 1.16 lists other chiral ligands useful for asymmetric hydrogenation of α- and/or β-keto esters.[6,75–77] A Ru complex with BPE, a fully alkylated diphosphine,

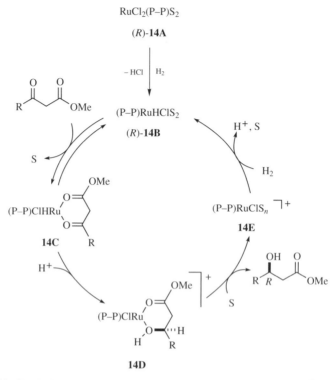

Figure 1.14. Catalytic cycle of BINAP–Ru-catalyzed hydrogenation of β-keto esters involving a monohydride mechanism [P–P=(*R*)-BINAP; S = solvent or a weak ligand].

shows very high activity for reaction of β-keto esters.[78,79] Its electron-donating character may accelerate the hydride transfer in **14C** → **14D** (see Figure 1.14). An axially chiral SEGPHOS efficiently discriminates two enantiofaces of α- and β-keto esters in the Ru-catalyzed hydrogenation.[80] The small angle, 65°, between

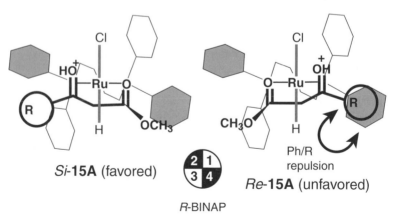

Figure 1.15. Molecular models of diastereomeric transition states in (*R*)-BINAP–Ru-catalyzed hydrogenation of β-keto esters.

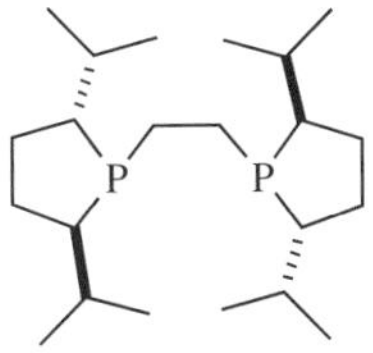

(*R,R*)-*i*-Pr-BPE

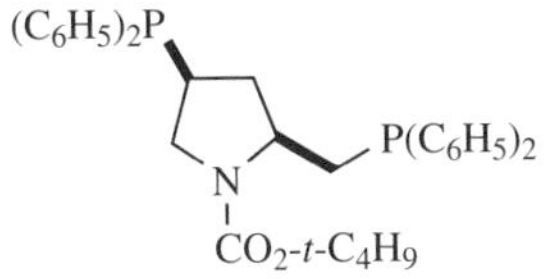

(2*S*,4*S*)-BPPM

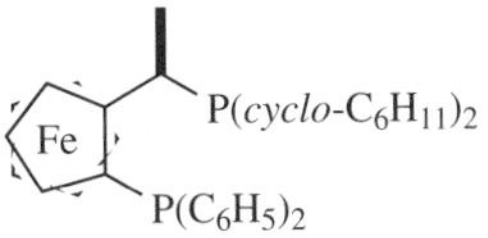

(*R*)-(*S*)-JOSIPHOS

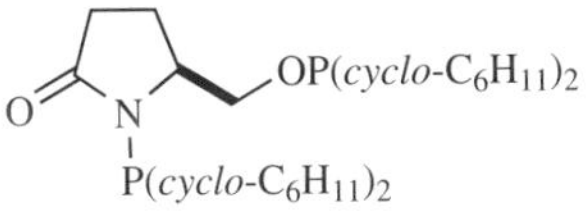

(*S*)-Cy,Cy-oxoProNOP

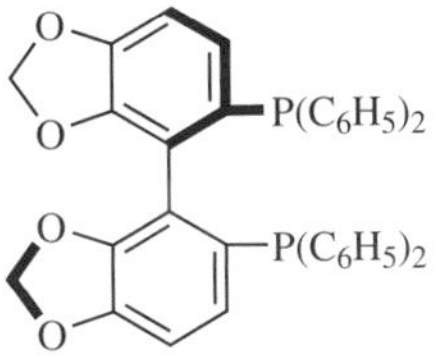

(*R*)-SEGPHOS

Figure 1.16. Chiral ligands useful for hydrogenation of keto esters.

the two aryl planes (dihedral angle) of the SEGPHOS–Ru complex may cause the high stereoselectivity. BPPM[81] and JOSIPHOS[82] are original for the design of C_1-chiral diphosphines, whereas the origin of their efficiency is hard to rationalize. Rh complexes with oxoProNOP ligands achieve high catalytic activity and enantios-electivity in hydrogenation of α-keto esters.[83]

1.3.2 Hydrogenation of Simple Ketones

β-Keto esters are highly reactive for the BINAP–$RuCl_2$-catalyzed hydrogenation (Figure 1.13).[18,67,68] However, simple unfunctionalized ketones are totally inert to this catalyst system, because the substrates are unable to stabilize the transition state by forming a chelate structure (Figure 1.15).[21] A combined system of *trans*-$RuCl_2$(binap)(dpen) and alkaline base[84–86] or *trans*-RuH(η^1-BH_4)-(binap)(dpen)[87] with or without a base exhibits excellent catalytic performance in asymmetric hydrogenation of a variety of simple ketones.[21,86] For example, acetophenone is hydrogenated in the presence of an (*S*)-XylBINAP/(*S*,*S*)-DPEN–Ru catalyst (ketone:Ru = 100,000:1) in 2-propanol to give (*R*)-1-pheny-lethanol in 99% ee (Figure 1.17).[85,87] Notably, now the reaction conditions are slightly basic.

The excellent catalytic activity is rationalized by a nonclassical metal–ligand bifunctional mechanism using an "NH effect."[86,88–90] As shown in Figure 1.18, *trans*-RuH(η^1-BH_4)(tolbinap)(dpen) (**18A**) (TolBINAP; see Figure 1.2), a precata-lyst, is converted to 16-electron cationic species **18B** in 2-propanol.[88,91,92] It accepts an H_2 molecule to form **18C**, which undergoes deprotonation with an alcoholic solvent giving the Ru dihydride **18D**. This step is accelerated by a base. Ketonic substrate is rapidly reduced by **18D** to give the alcohol and 16-electron Ru amide **18E**. This complex is easily protonated by alcoholic media to recover the amino

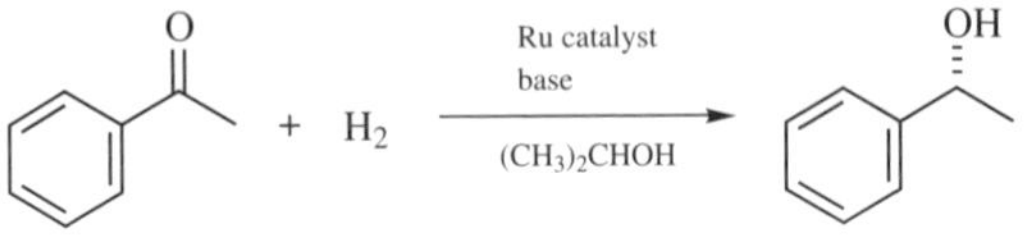

Ru catalyst:

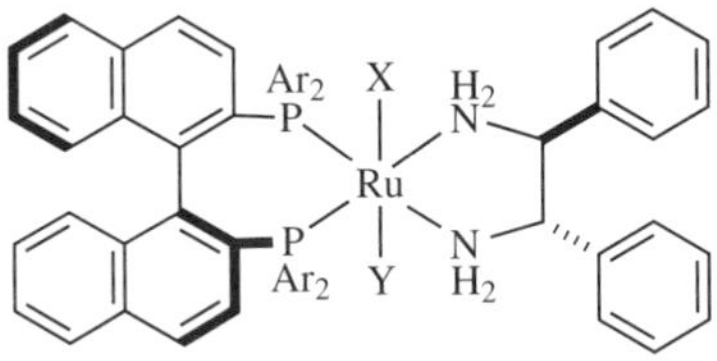

Ar = 3,5-$(CH_3)_2C_6H_3$

trans-RuCl$_2$[(S)-xylbinap][(S,S)-dpen]: X = Y = Cl

trans-RuH(η^1-BH$_4$)[(S)-xylbinap][(S,S)-dpen]: X = H, Y = η^1-BH$_4$

Figure 1.17. Asymmetric hydrogenation of simple ketones catalyzed by BINAP/1,2-diamine–Ru complex.

complex **18B**, while it partially gives **18D** by reaction with H$_2$. The reducing species **18D** has a *fac* structure for the hydride and two nitrogen atoms, allowing reaction with a ketone via the six-membered pericyclic transition state **18F**. Ketone is reduced in the outer coordination sphere of **18D**, where neither ketone/Ru nor alkoxy/Ru interaction is involved. This hydrogenation is carbonyl-selective. Alkenyl and alkynyl groups, which are normally reduced through a substrate/metal complex, are left intact under the hydrogenation conditions.[85,87,93]

The combination of (S)-TolBINAP and (S,S)-DPEN (or R/R,R combination) is crucial to achieve high enantioselectivity.[94,95] Figure 1.19 illustrates a transition-state model in hydrogenation of acetophenone using an (S)-TolBINAP/(S,S)-DPEN–RuH$_2$ catalyst.[88,89] Both (S)-TolBINAP and (S,S)-DPEN bind to a Ru center resulting in a C_2-symmetric RuH$_2$ complex. The skewed five-membered DPEN chelate ring has two kinds of diastereotopic hydrogen at the nitrogen atoms. The axially arranged hydrogens, H$_{ax}$, are more reactive than the equatorial ones for stereoelectronic reasons. The H$^{\delta-}$—Ru$^{\delta+}$—N$^{\delta-}$—H$_{ax}$$^{\delta+}$ moiety with a small dihedral angle fits well with the C$^{\delta+}$=O$^{\delta-}$ function. Then a hydride on Ru smoothly migrates to the electrophilic carbonyl carbon, while the amine proton is transferred to the oxygen atom. Acetophenone approaches the reaction site in a way to minimize nonbonded repulsion and to maximize electronic attraction. The *Si*-**19A** is favored over the diastereomeric *Re*-**19A**, which suffers from significant nonbonded repulsion between the *P*-tolyl group of TolBINAP and the acetophenone phenyl ring. The *Si*-**19A** could further be stabilized by the secondary attractive interaction between an NH$_{eq}$ and the phenyl ring of the substrate. This view is consistent with the fact that sterically less demanding alkenyl alkyl ketones are hydrogenated with equally high enantioselectivity.[85,87,93]

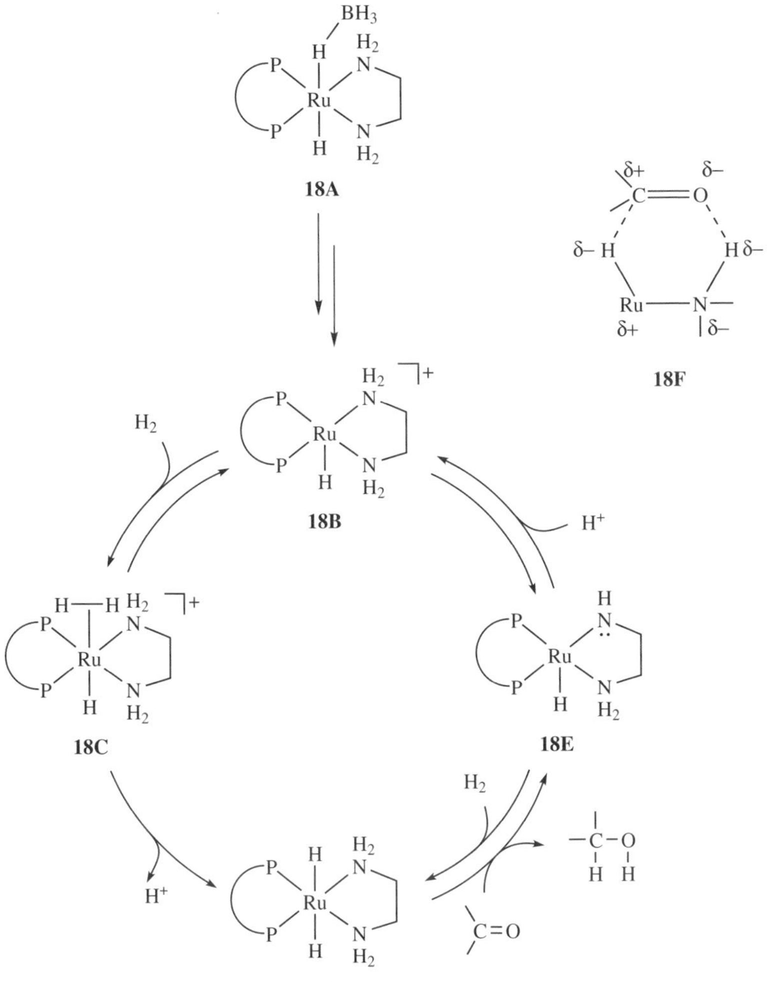

Figure 1.18. Catalytic cycle of TolBINAP/1,2-diamine–Ru-catalyzed hydrogenation of simple ketones.

The chiral environment of this catalyst system is easily modified by changing the combination of diphosphine and diamine ligands.[77,96–98] Although the BINAP/1,2-diamine–Ru catalysts are feebly active for hydrogenation of 1-tetralones, this problem is solved simply by use of chiral 1,4-diamines instead of conventional 1,2-diamine ligands.[99] For example, hydrogenation of 5-methoxy-1-tetralone in the presence of an (S)-TolBINAP/(R)-IPHAN–Ru catalyst in 2-propanol results in the R alcohol in 98% ee (Figure 1.20). Hydrogenation of *tert*-alkyl ketones with the BINAP/1,2-diamine–Ru catalysts is also difficult obviously because of steric hindrance of the substrates. Ru catalysts wearing BINAP and α-picolylamine

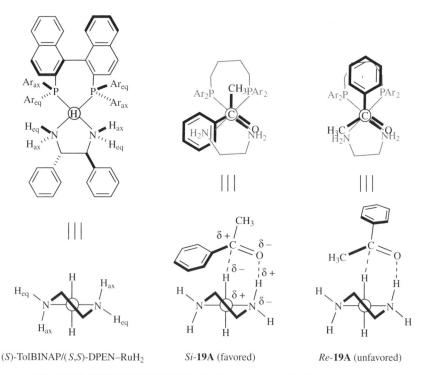

(S)-TolBINAP/(S,S)-DPEN–RuH$_2$ Si-**19A** (favored) Re-**19A** (unfavored)

Figure 1.19. (S)-TolBINAP/(S,S)-DPEN–RuH$_2$ species and diastereomeric transition states in the metal–ligand bifunctional catalysis; the equatorially oriented phenyl substituents in the DPEN ligands are omitted in the transition states Si-**19A** and Re-**19A** (Ar = 4-CH$_3$C$_6$H$_4$; ○ = Ru; ax = axial, eq = equatorial).

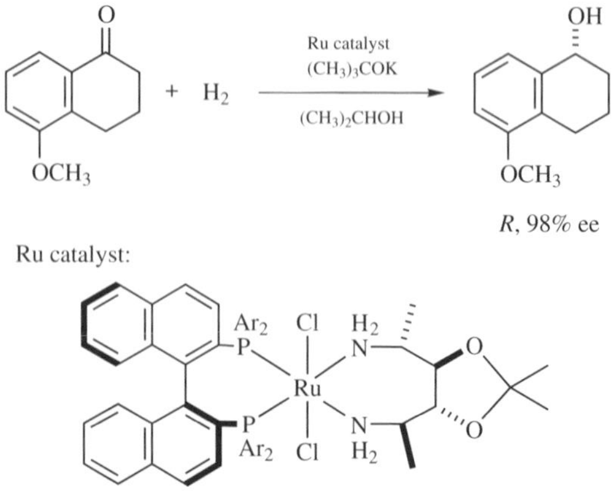

Ar = 4-CH$_3$C$_6$H$_4$

$trans$-RuCl$_2$[(S)-tolbinap][(R)-iphan]

Figure 1.20. Asymmetric hydrogenation of 1-tetralones catalyzed by BINAP/1,4-diamine–Ru complex.

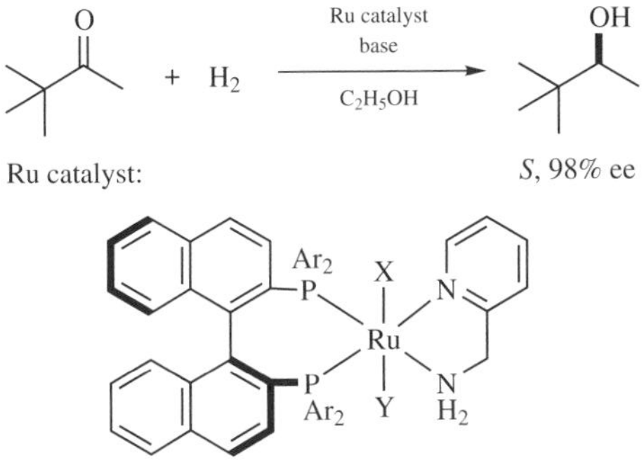

Ar = 4-CH$_3$C$_6$H$_4$

RuCl$_2$[(S)-tolbinap](pica): X = Y = Cl

RuH(η^1-BH$_4$)[(S)-tolbinap](pica): X = H, Y = η^1-BH$_4$

diastereomeric mixture

Figure 1.21. Asymmetric hydrogenation of *tert*-alkyl ketones catalyzed by BINAP/PICA–Ru complex.

(PICA) show excellent activity and enantioselectivity for reaction of such bulky ketones.[100] Selection of alcoholic solvent is important to achieve high catalytic performance. Thus, hydrogenation of pinacolone with the (S)-TolBINAP/PICA–Ru catalyst (S/C = 100,000) in C$_2$H$_5$OH quantitatively gives (S)-3,3-dimethyl-2-butanol in 98% ee (Figure 1.21). The reaction in conventional 2-propanol with the same catalyst results in the S alcohol in only 36% ee.

Some chiral amino phosphine–Ru catalysts are also effective for asymmetric hydrogenation of simple ketones.[101]

A Rh complex with (R,S,R,S)-Me-PennPhos efficiently catalyzes asymmetric hydrogenation of simple ketones (Figure 1.22).[102] Addition of catalytic amounts of 2,6-lutidine is crucial to achieve high enantioselectivity. This catalyst is also

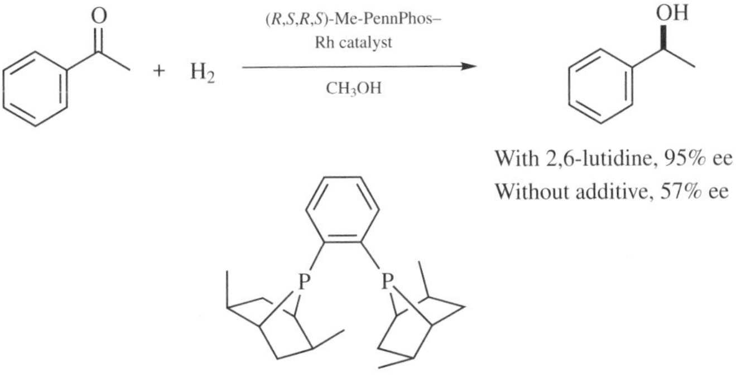

With 2,6-lutidine, 95% ee
Without additive, 57% ee

(R,S,R,S)-Me-PennPhos

Figure 1.22. Asymmetric hydrogenation of simple ketones catalyzed by a PennPhos–Rh complex.

effective for reaction of some aliphatic ketones. The origin of high stereoselectivity has not been elucidated yet.

1.3.3 Transfer Hydrogenation of Ketones

As illustrated in Figure 1.23, chiral arene–Ru catalysts achieve high enantioselectivity in transfer hydrogenation of aryl, alkenyl, and alkynyl ketones.[1,103] Various secondary alcohols are obtainable in >95% ee. The combination of TsDPEN (Ts = p-toluenesulfonyl) and arene ligands controls enantio-face selection. Importantly, selection of an achiral arene ligand is crucial for high stereoselection. In place of TsDPEN, some chiral β-amino alcohols are also usable.[6,49,75,76,104] 2-Propanol or formic acid is selected as a hydride source. The reduction of ketones with 2-propanol is reversible, because the products are also secondary alcohols. In many cases, therefore, formic acid, an irreversible reducing agent, with

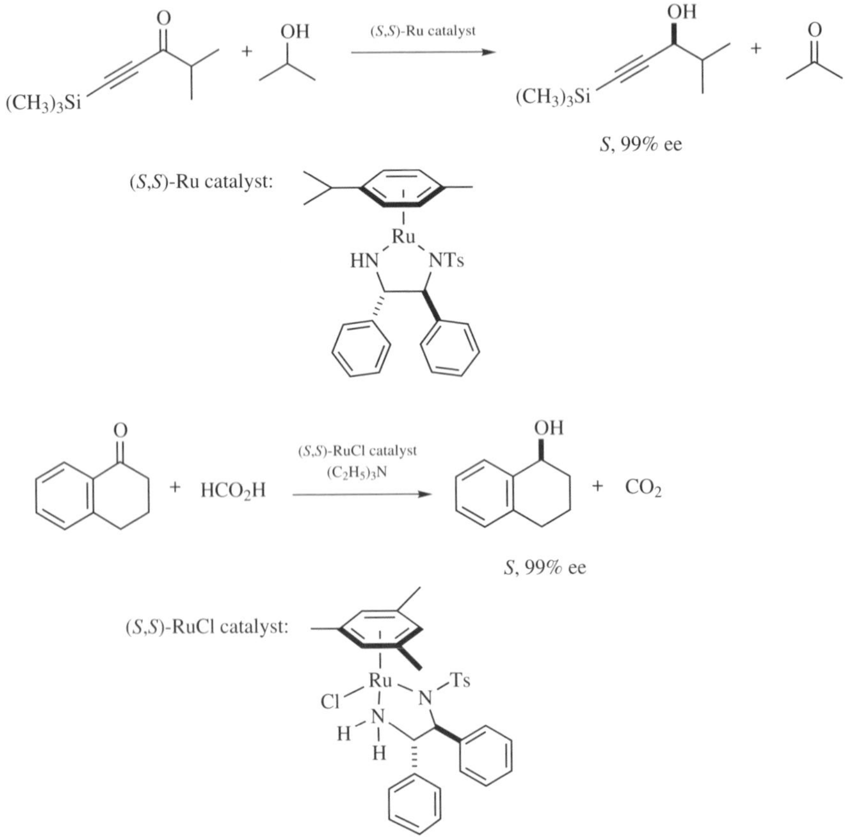

Figure 1.23. Asymmetric transfer hydrogenation of ketones catalyzed by chiral arene–Ru complexes.

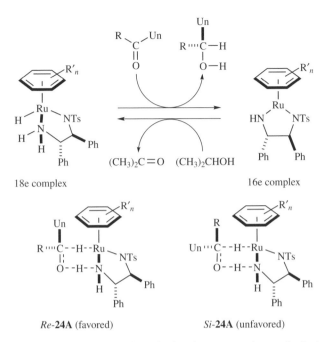

Figure 1.24. Metal–ligand bifunctional mechanism in asymmetric transfer hydrogenation of ketones (R = alkyl; Un = alkenyl, alkynyl, or aryl).

$(C_2H_5)_3N$ for tuning the acidity of reaction media gives higher conversion and better enantioselectivity.[105]

Detailed experimental[106] as well as theoretical[107–110] studies revealed the mechanism of the asymmetric transfer hydrogenation in 2-propanol. As summarized in Figure 1.24, the NH effect is evident. The 18-electron Ru complex, $RuH[(S,S)$-$TsNCH(C_6H_5)CH(C_6H_5)NH_2](\eta^6$-arene), smoothly reduces ketonic substrate through a six-membered pericyclic transition state, in which Ru—H and N—H are simultaneously delivered onto the C=O function, giving an S alcohol and 16-electron $Ru[(S,S)$-$TsNCH(C_6H_5)CH(C_6H_5)NH](\eta^6$-arene). The stereochemical outcome indicates that the *Re*-**24A** is much more favored than the diastereomeric *Si*-**24A**, due to stabilization caused by the CH–π interaction between the η^6-arene ligand and the aromatic, olefinic, or acetylenic group in the substrates.[109]

Figure 1.25 exemplifies the structures of certain efficient precatalysts for asymmetric transfer hydrogenation of ketones. Precatalysts **C1**–**C3** use the "NH effect" described above.[111–113] A turnover frequency, defined as moles of product per mol of catalyst per hour, of 30,000 h^{-1} is achieved by using of **C2** and an alkaline base in 2-propanol.[112] A Rh complex **C3** is an isolobal to the corresponding arene–Ru complex (see Figure 1.23).[113] The Ru complexes **C4**[114] and **C5**[115] without NH group in ligand catalyze the reaction by different mechanisms. A higher than 90% optical yield is achieved by using **C5** in reduction of certain aliphatic ketones.[115]

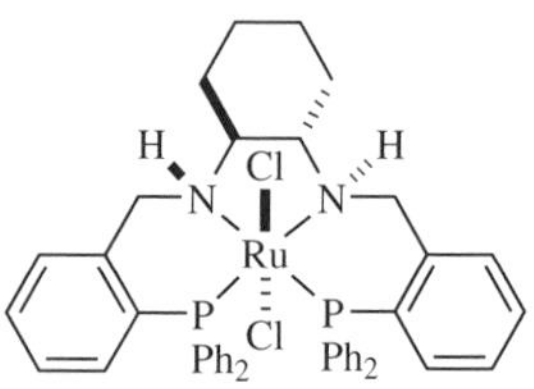

(S,S)-**C1**

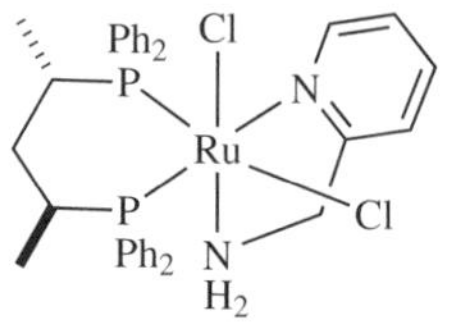

(S,S)-**C2**

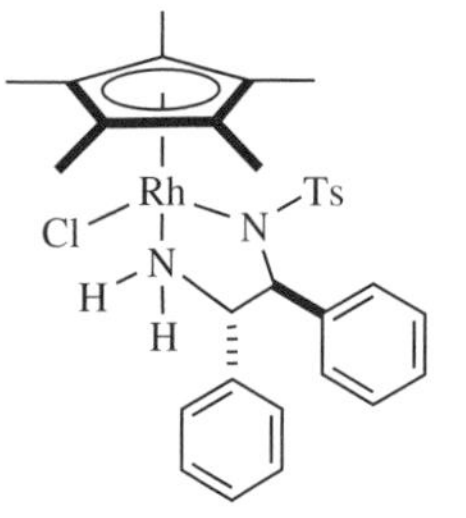

(S,S)-**C3**

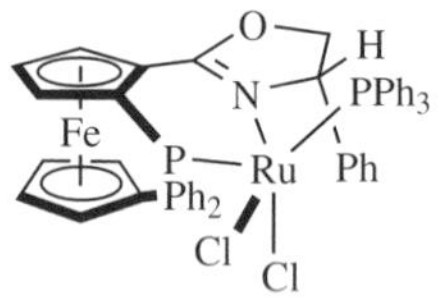

(S)-**C4**

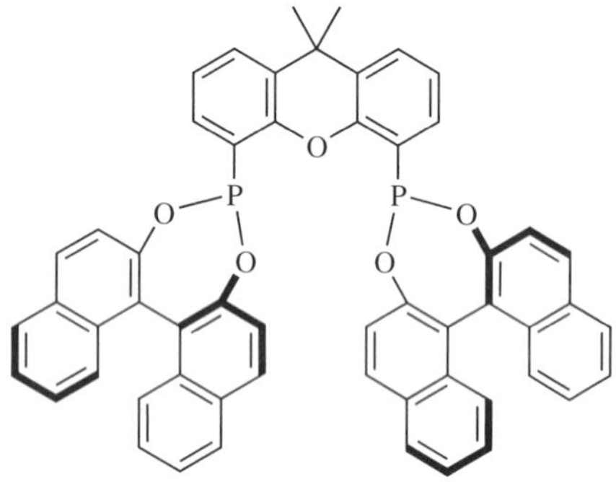

with

[RuCl$_2$(p-cymene)]$_2$

(S,S)-**C5**

Figure 1.25. Examples of chiral precatalyst for asymmetric transfer hydrogenation of ketones.

As shown in Figure 1.26, a chiral Sm(III) complex catalyzes asymmetric reduction of aromatic ketones in 2-propanol with high enantioselectivity.[116] Unlike other late-transition-metal catalysis, the hydrogen at C2 of 2-propanol directly migrates onto the carbonyl carbon of substrate via a six-membered transition state **26A**, as seen in the Meerwein–Ponndorf–Verley reduction.[6]

1.3.4 Hydroboration of Ketones

Borane reduction catalyzed by chiral oxazaborolidines (CBS reduction, CBS = Corey, Bakshi, and Shibata) exhibits excellent enantio- and chemoselectivity for a wide variety of ketonic substrates (Figure 1.27).[117–119] This reaction was originally developed as a stoichiometric system consisting of diphenylvalinol and borane,[120] but was later extended to a useful catalytic method.[121] Because of the high efficiency of this reaction, many chiral oxazaborolidines have been synthesized from β-amino alcohols.[117–119] Among them the prolinol-derived oxazaborolidine is one of the most widely used catalysts.[117,122]

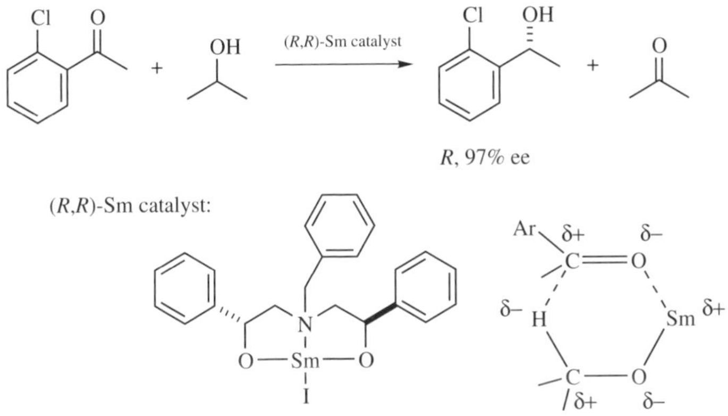

Figure 1.26. Asymmetric Meerwein–Ponndorf–Verley-type reduction of ketones catalyzed by a Sm complex.

The proposed catalytic cycle for reduction of acetophenone is illustrated in Figure 1.28.[117] The (S)-oxazaborolidine catalyst (S)-**28A** has both Lewis acidic and basic sites, and its borane adduct **28B** acts as a chiral Lewis acid. The B center in the borolidine ring selectively interacts with a sterically more accessible electron

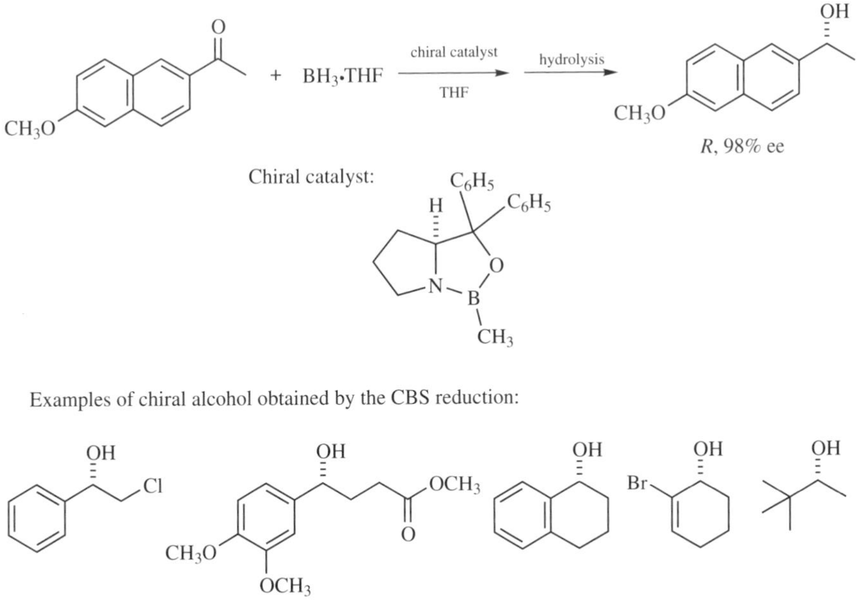

Examples of chiral alcohol obtained by the CBS reduction:

Figure 1.27. Asymmetric reduction of ketones with borane catalyzed by oxazaborolidines.

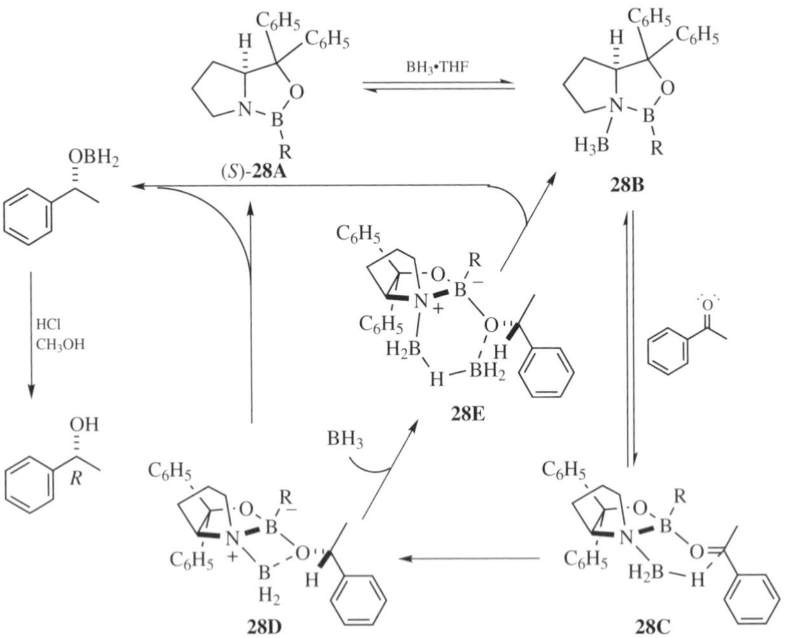

Figure 1.28. Catalytic cycle of oxazaborolidine-catalyzed asymmetric reduction of ketones.

pair of the carbonyl oxygen to avoid steric repulsion between the R substituent on B and the acetophenone phenyl, leading to a favored six-membered transition state **28C**. Migration of the borane hydride to the carbonyl carbon gives **28D**, which is converted to **28A** directly or **28B** through a borane adduct **28E** with releasing of the R alkoxide product. For this reaction, borane–THF complex is most commonly used, whereas catecholborane gives better results for reaction of α,β-unsaturated ketones[123,124] and alkynyl ketones.[124]

The skeletally fixed prolinol ring gives the best performance for enantioface selection of substrates in **28C**.[117] Furthermore, the substitution by *gem*-phenyl or -2-naphthyl groups at the carbinol center is recommended (see structure in Figure 1.27). More hindered groups and heteroaryl rings reduce the enantioselectivity. Methyl and *n*-butyl groups are commonly used as a substituent at the boron atom. The use of (trimethylsilyl)methyl group instead of simple alkyls gives better enantioselection in the reaction of (triisopropylsilyl)acetylenic ketones, which requires recognition of the bulky group located far from the reaction site.[122]

A chiral β-keto iminato Co complex in the presence of tetrahydrofuryl alcohol (THFA) and ethanol (or methanol) results in high enantioselectivity in reduction of aromatic ketones using $NaBH_4$ as a hydride source (Figure 1.29).[125,126] The in situ generated $NaBH_2(OR)(OC_2H_5)$ (ROH = THFA) reduces the Co complex to form a true catalytic CoH species.

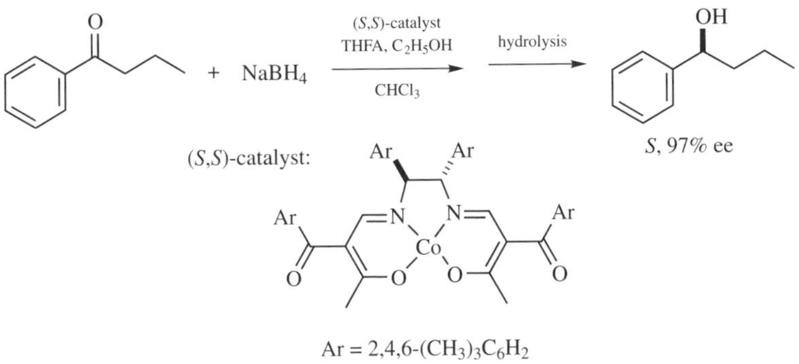

$Ar = 2,4,6\text{-}(CH_3)_3C_6H_2$

Figure 1.29. Asymmetric reduction of ketones with $NaBH_4$ catalyzed by a β-keto iminato Co complex.

1.4   REDUCTION OF IMINES

The Brintzinger-type C_2-chiral titanocene catalysts[127] efficiently promote asymmetric hydrogenation of imines (Figure 1.30).[128,129] A variety of cyclic and acyclic imines are reduced with excellent enantioselectivity by using these catalysts. The active hydrogenation species **30B** is produced by treatment of the titanocene binaphtholate derivative **30A** with *n*-butyllithium followed by phenylsilane.

Figure 1.31 illustrates a mechanism proposed for this hydrogenation. The titanocene hydride **31A** is expected to be a catalytic species. The imine substrate is inserted into the Ti—H bond of **31A** with a 1,2-fashion to form a titanocene amide complex **31B**. Then the hydrogenolysis of **31B** through a σ-bond metathesis produces the amine product with regeneration of **31A**. The enantioface selection

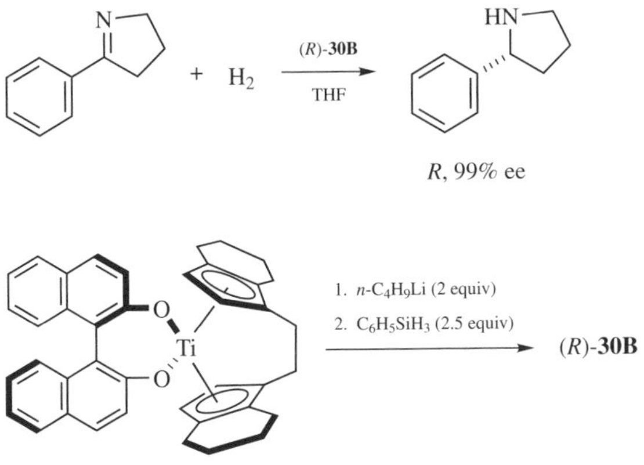

Figure 1.30. Asymmetric hydrogenation of imines with a chiral titanocene catalyst.

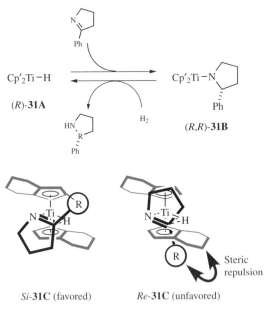

Figure 1.31. Reaction mechanism and diastereomeric transition states of titanocene-catalyzed hydrogenation of imines.

occurs at the first 1,2-insertion step **31A** → **31B**, in which two diastereomeric four-membered transition states *Re*-**31C** and *Si*-**31C** are possible.[129] *Re*-**31C** suffers from significant steric repulsion between the R substituent of the cyclic imine and the tetrahydroindenyl ligand. Thus the *R* amine product is predominantly produced via the favored transition state *Si*-**31C**.

Substituents on imino nitrogen influence both reactivity and enantioselectivity in hydrogenation of imino compounds.[6,9,130] Figure 1.32 shows two successful examples. An f-BINAPHANE–Ir complex effects asymmetric hydrogenation of *N*-aryl aromatic imines.[131] On the other hand, an Et-DuPHOS–Rh complex[25] (see Figure 1.2) is effective for hydrogenation of *N*-acylhydrazones.[132]

Figure 1.33 illustrates asymmetric hydrogenation of a functionalized imine with a XYLIPHOS–Ir catalyst, occurring with a catalyst turnover number of 2,000,000.[9,133] The presence of I⁻ under acidic conditions is crucial to achieve high catalytic performance. (*S*)-Metolachlor, a herbicide, is industrially produced in a >10,000-ton quantity per year by this reaction.

Asymmetric transfer hydrogenation of imines catalyzed by chiral arene–Ru complexes achieves high enantioselectivity (Figure 1.34).[103,134] Formic acid in aprotic dipolar solvent should be used as a hydride source. The reaction proceeds through the "metal–ligand bifunctional mechanism" as shown in the carbonyl reduction (Figure 1.24).

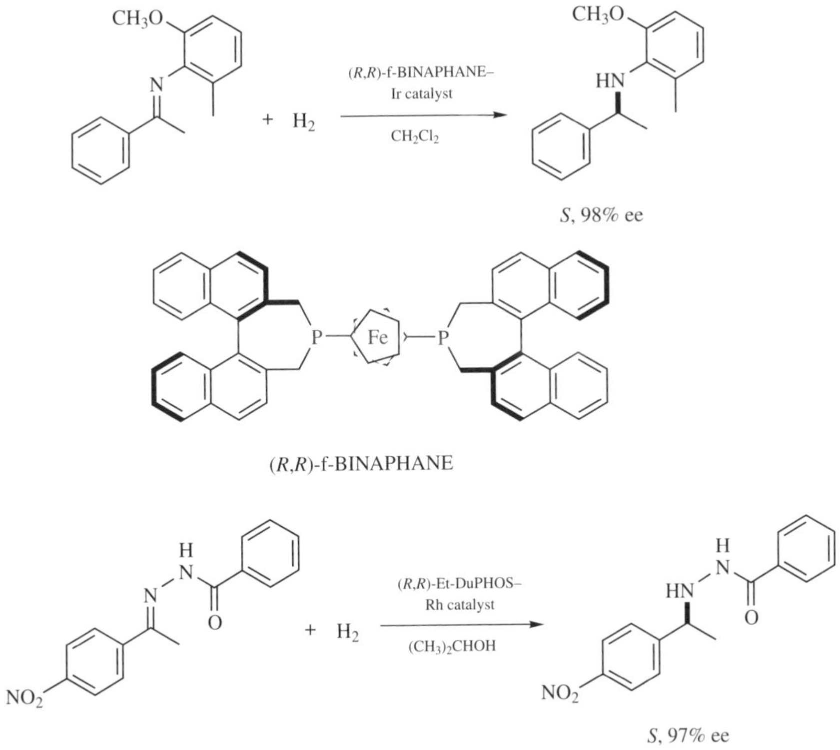

(R,R)-f-BINAPHANE

Figure 1.32. Asymmetric hydrogenation of N-arylimines and N-acylhydrazones catalyzed by chiral Ir and Rh complexes.

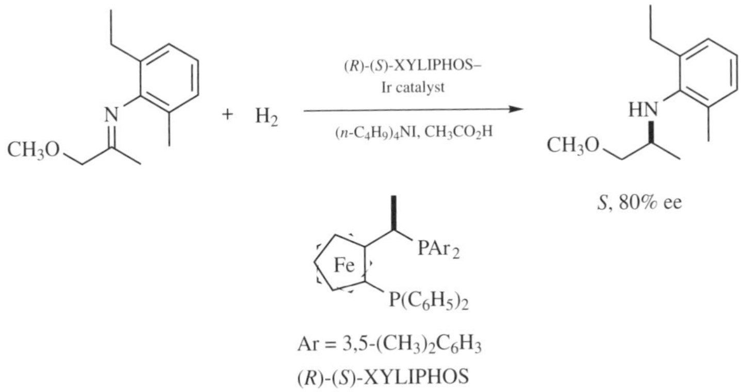

Ar = 3,5-(CH$_3$)$_2$C$_6$H$_3$

(R)-(S)-XYLIPHOS

Figure 1.33. Industiral asymmetric hydrogenation of a functionalized imine catalyzed by a XYLIPHOS–Ir complex.

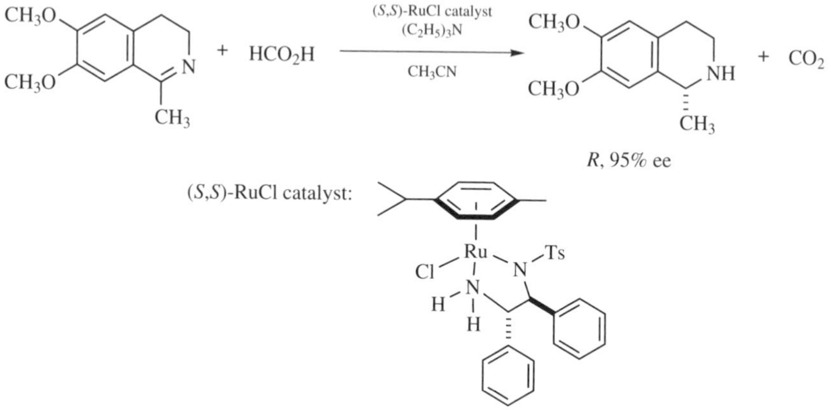

Figure 1.34. Asymmetric transfer hydrogenation of imines catalyzed by chiral Ru complexes.

REFERENCES

1. Noyori, R. *Angew. Chem. Int. Ed.* **2002**, *41*, 2008.
2. Noyori, R. In *Asymmetric Catalysis in Organic Synthesis*, Wiley, New York, **1994**.
3. *Comprehensive Asymmetric Catalysis*, Jacobsen, E. N.; Pfaltz, A.; Yamamoto, H. (Eds.), Springer, Berlin–Heidelberg, **1999**, Vol. 1.
4. *Catalytic Asymmetric Synthesis*, 2nd ed., Ojima, I. (Ed.), Wiley-VCH, New York, **2000**.
5. Evans, D.; Osborn, J. A.; Jardine, F. H.; Wilkinson, G. *Nature* **1965**, *208*, 1203.
6. Ohkuma, T.; Kitamura, M.; Noyori, R. In *Catalytic Asymmetric Synthesis*, 2nd ed., Ojima, I. (Ed.), Wiley-VCH, New York, **2000**, p. 1.
7. Brown, J. M. In *Comprehensive Asymmetric Catalysis*, Jacobsen, E. N.; Pfaltz, A.; Yamamoto, H. (Eds.), Springer; Berlin–Heidelberg, **1999**, Vol. 1, p. 121.
8. Knowles, W. S. *Angew. Chem. Int. Ed.* **2002**, *41*, 1998.
9. Blaser, H.-U.; Malan, C.; Pugin, B.; Spindler, F.; Steiner, H.; Studer, M. *Adv. Synth. Catal.* **2003**, *345*, 103.
10. Au-Yeung, T. T.-L.; Chan, S.-S.; Chan, A. S. C. In *Transition Metals for Organic Synthesis*, 2nd ed., Beller, M.; Bolm, C. (Eds.), Wiley-VCH, Weinheim, **2004**, Vol. 2, p 14.
11. Fryzuk, M. D.; Bosnich, B. *J. Am. Chem. Soc.* **1977**, *99*, 6262.
12. Chan, A. S. C.; Halpern, J. *J. Am. Chem. Soc.* **1980**, *102*, 838.
13. Landis, C. R.; Halpern, J. *J. Am. Chem. Soc.* **1987**, *109*, 1746.
14. Halpern, J. *Precious Met.* **1995**, *19*, 411.
15. Brown, J. M.; Chaloner, P. A. *Tetrahedron Lett.* **1978**, 1877.
16. Brown, J. M.; Chaloner, P. A. *J. Am. Chem. Soc.* **1980**, *102*, 3040.
17. Landis, C. R.; Hilfenhaus, P.; Feldgus, S. *J. Am. Chem. Soc.* **1999**, *121*, 8741.
18. Noyori, R. In *Asymmetric Catalysis in Organic Synthesis*, Wiley, New York, **1994**, p. 16.
19. Miyashita, A.; Yasuda, A.; Takaya, H.; Toriumi, K.; Ito, T.; Souchi, T.; Noyori, R. *J. Am. Chem. Soc.* **1980**, *102*, 7932.

20. Miyashita, A.; Takaya, H.; Souchi, T.; Noyori, R. *Tetrahedron* **1984**, *40*, 1245.

21. Noyori, R.; Kitamura, M.; Ohkuma, T. *Proc. Natl. Acad. Sci. USA* **2004**, *101*, 5356.

22. Dang, T. P.; Kagan, H. B. *J. Chem. Soc. Chem. Commun.* **1971**, 481.

23. Ball, R. G.; Payne, N. C. *Inorg. Chem.* **1977**, *16*, 1187.

24. Vineyard, B. D.; Knowles, W. S.; Sabacky, M. J.; Bachman, G. L.; Weinkauff, D. J. *J. Am. Chem. Soc.* **1977**, *99*, 5946.

25. Burk, M. J. *J. Am. Chem. Soc.* **1991**, *113*, 8518.

26. Landis, C. R.; Feldgus, S. *Angew. Chem. Int. Ed.* **2000**, *39*, 2863.

27. Armstrong, S. K.; Brown, M. J.; Burk, M. K. *Tetrahedron Lett.* **1993**, *34*, 879.

28. Evans, D. A.; Michael, F. E.; Tedrow, J. S.; Campos, K. R. *J. Am. Chem. Soc.* **2003**, *125*, 3534.

29. Imamoto, T.; Watanabe, J.; Wada, Y.; Masuda, H.; Yamada, H.; Tsuruta, H.; Matsukawa, S.; Yamaguchi, K. *J. Am. Chem. Soc.* **1998**, *120*, 1635.

30. Crépy, K. V. L.; Imamoto, T. *Adv. Synth. Catal.* **2003**, *345*, 79.

31. Gridnev, I. D.; Higashi, N.; Asakura, K.; Imamoto, T. *J. Am. Chem. Soc.* **2000**, *122*, 7183.

32. Gridnev, I. D.; Imamoto, T. *Acc. Chem. Res.* **2004**, *37*, 633.

33. Pye, P. J.; Rossen, K.; Reamer, R. A.; Tsou, N. N.; Volante, R. P.; Reider, P. J. *J. Am. Chem. Soc.* **1997**, *119*, 6207.

34. Giernoth, R.; Heinrich, H.; Adams, N. J.; Deeth, R. J.; Bargon, J.; Brown, J. M. *Chem. Commun.* **2001**, 1296.

35. Claver, C.; Fernandez, E.; Gillon, A.; Heslop, K.; Hyett, D. J.; Martorell, A.; Orpen, A. G.; Pringle, P. G. *Chem. Commun.* **2000**, 961.

36. Reetz, M. T.; Mehler, G. *Angew. Chem. Int. Ed.* **2000**, *39*, 3889.

37. van den Berg, M.; Minnaard, A. J.; Schudde, E. P.; van Esch, J.; de Vries, A. H. M.; de Vries, J. G.; Feringa, B. L. *J. Am. Chem. Soc.* **2000**, *122*, 11539.

38. Feringa, B. L. *Acc. Chem. Res.* **2000**, *33*, 346.

39. Jerphagnon, T.; Renaud, J.-L.; Bruneau, C. *Tetrahedron: Asymmetry* **2004**, *15*, 2101.

40. Huttenloch, O.; Laxman, E.; Waldmann, H. *Chem. Commun.* **2002**, 673.

41. Bernsmann, H.; van den Berg, M.; Hoen, R.; Minnaard, A. J.; Mehler, G.; Reetz, M. T.; de Vries, J. G.; Feringa, B. L. *J. Org. Chem.* **2005**, *70*, 943.

42. van den Berg, M.; Minnaard, A. J.; Haak, R. M.; Leeman, M.; Schuddle, E. P.; Meetsma, A.; Feringa, B. L.; de Vries, A. H. M.; Maljaars, C. E. P.; Willans, C. E.; Hyett, D.; Boogers, J. A. F.; Henderickx, H. J. W.; de Vries, J. G. *Adv. Synth. Catal.* **2003**, *345*, 308.

43. Pena, D.; Minnaard, A. J.; de Vries, A. H. M.; de Vries, J. G.; Feringa, B. L. *Org. Lett.* **2003**, *5*, 475.

44. Monti, C.; Gennari, C.; Piarulli, U.; de Vries, J. G.; de Vries, A. H. M.; Lefort, L. *Chem. Eur. J.* **2005**, *11*, 6701.

45. Kawano, H.; Ikariya, T.; Ishii, Y.; Saburi, M.; Yoshikawa, S.; Uchida, Y.; Kumobayashi, H. *J. Chem. Soc. Perkin Trans. I* **1989**, 1571.

46. Kitamura, M.; Hsiao, Y.; Ohta, M.; Tsukamoto, M.; Ohta, T.; Takaya, H.; Noyori, R. *J. Org. Chem.* **1994**, *59*, 297.

47. Ohta, T.; Takaya, H.; Kitamura, M.; Nagai, K.; Noyori, R. *J. Org. Chem.* **1987**, 52, 3174.

48. Takaya, H.; Ohta, T.; Sayo, N.; Kumobayashi, H.; Akutagawa, S.; Inoue, S.; Kasahara, I.; Noyori, R. *J. Am. Chem. Soc.* **1987**, *109*, 1596, 4129.

49. Kitamura, M.; Noyori, R. In *Ruthenium in Organic Synthesis*, Murahashi, S.-I. (Ed.); Wiley-VCH, Weinheim, **2004**, p. 3.

50. Tsukamoto, M.; Kitamura, M. *J. Synth. Org. Chem. Jpn.* **2005**, *63*, 899.

51. Wiles, J. A.; Bergens, S. H. *Organometallics* **1998**, *17*, 2228.

52. Wiles, J. A.; Bergens, S. H. *Organometallics* **1999**, *18*, 3709.

53. Ishibashi, Y.; Bessho, Y.; Yoshimura, M.; Tsukamoto, M.; Kitamura, M. *Angew. Chem. Int. Ed.* **2005**, *44*, 7287.

54. Crabtree, R. *Acc. Chem. Res.* **1979**, *12*, 331.

55. Pfaltz, A.; Blankenstein, J.; Hilgraf, R.; Hörmann, E.; McIntyre, S.; Menges, F.; Schönleber, M.; Smidt, S. P.; Wüstenberg, B.; Zimmermann, N. *Adv. Synth. Catal.* **2003**, *345*, 33.

56. Källström, K.; Hedberg, C.; Brandt, P.; Bayer, A.; Andersson, P. G. *J. Am. Chem. Soc.* **2004**, *126*, 14308.

57. Cui, X.; Burgess, K. *Chem. Rev.* **2005**, *105*, 3272.

58. Halterman, R. L. In *Comprehensive Asymmetric Catalysis*, Supplement 2, Jacobsen, E. N.; Pfaltz, A.; Yamamoto, H. (Eds.), Springer, Berlin–Heidelberg, **2004**, p. 1.

59. Perry, M. C.; Cui, X.; Powell, M. T.; Hou, D.-R.; Reibenspies, J. H.; Burgess, K. *J. Am. Chem. Soc.* **2003**, *125*, 113.

60. Drago, D.; Pregosin, P. S.; Pfaltz, A. *Chem. Commun.* **2002**, 286.

61. Smidt, S. P.; Zimmermann, N.; Studer, M.; Pfaltz, A. *Chem. Eur. J.* **2004**, *10*, 4685.

62. Cui, X.; Fan, Y.; Hall, M. B.; Burgess, K. *Chem. Eur. J.* **2005**, *11*, 6859.

63. Dietiker, R.; Chen, P. *Angew. Chem. Int. Ed.* **2004**, *43*, 5513.

64. Brandt, P.; Hedberg, C.; Andersson, P. G. *Chem. Eur. J.* **2003**, *9*, 339.

65. Fan, Y.; Cui, X.; Burgess, K.; Hall, M. B. *J. Am. Chem. Soc.* **2004**, *126*, 16688.

66. Broene, R. D.; Buchwald, S. L. *J. Am. Chem. Soc.* **1993**, *115*, 12569.

67. Noyori, R.; Ohkuma, T.; Kitamura, M.; Takaya, H.; Sayo, N.; Kumobayashi, H.; Akuta-gawa, S. *J. Am. Chem. Soc.* **1987**, *109*, 5856.

68. Kitamura, M.; Tokunaga, M.; Ohkuma, T.; Noyori, R. *Org. Synth.* **1993**, *71*, 1.

69. Mashima, K.; Kusano, K.; Sato, N.; Matsumura, Y.; Nozaki, K.; Kumobayashi, H.; Sayo, N.; Hori, Y.; Ishizaki, T.; Akutagawa, S.; Takaya, H. *J. Org. Chem.* **1994**, *59*, 3064.

70. Genêt, J. P.; Ratovelomanana-Vidal, V.; Cano de Andrade, M. C.; Pfister, X.; Guerreiro, P.; Lenoir, J. Y. *Tetrahedron Lett.* **1995**, *36*, 4801.

71. Daley, C. J. A.; Bergens, S. H. *J. Am. Chem. Soc.* **2002**, *124*, 3680.

72. Taber, D. F.; Silverberg, L. J. *Tetrahedron Lett.* **1991**, *32*, 4227.

73. King, S. A.; Thompson, A. S.; King, A. O.; Verhoeven, T. R. *J. Org. Chem.* **1992**, *57*, 6689.

74. Kitamura, M.; Yoshimura, M.; Kanda, N.; Noyori, R. *Tetrahedron* **1999**. *55*, 8769.

75. Ohkuma, T.; Noyori, R. In *Comprehensive Asymmetric Catalysis*, Jacobsen, E. N.; Pfaltz, A.; Yamamoto, H. (Eds.), Springer, Berlin–Heidelberg, **1999**, Vol. 1, p. 199.

76. Ohkuma, T.; Noyori, R. In *Comprehensive Asymmetric Catalysis*, Supplement 1, Jacob-sen, E. N.; Pfaltz, A., Yamamoto, H. (Eds.), Springer, Berlin–Heidelberg, **2003**, p. 6.

77. Ohkuma, T.; Noyori, R. In *Transition Metals for Organic Synthesis*, 2nd ed., Beller, M.; Bolm, C. (Eds.), Wiley-VCH, Weinheim, **2004**, Vol. 2, p. 29.

78. Burk, M. J.; Harper, T. G. P.; Kalberg, C. S. *J. Am. Chem. Soc.* **1995**, *117*, 4423.

79. Burk, M. J.; Gross, M. F.; Harper, T. G. P.; Kalberg, C. S.; Lee, J. R.; Martinez, J. P. *Pure Appl. Chem.* **1996**, *68*, 37.

80. Saito, T.; Yokozawa, T.; Ishizaki, T.; Moroi, T.; Sayo, N.; Miura, T.; Kumobayashi, H. *Adv. Synth. Catal.* **2001**, *343*, 264.

81. Ojima, I.; Kogure, T.; Yoda, Y. *Org. Synth.* **1985**, *63*, 18.

82. Togni, A.; Breutel, C.; Schnyder, A.; Spindler, F.; Landert, H.; Tijani, A. *J. Am. Chem. Soc.* **1994**, *116*, 4062.

83. Roucoux, A.; Thieffry, L.; Carpentier, J.-F.; Devocelle, M.; Meliet, C.; Agbossou, F.; Mortreux, A. *Organometallics* **1996**, *15*, 2440.

84. Doucet, H.; Ohkuma, T.; Murata, K.; Yokozawa, T.; Kozawa, M.; Katayama, E.; England, A. F.; Ikariya, T.; Noyori, R. *Angew. Chem. Int. Ed.* **1998**, *37*, 1703.

85. Ohkuma, T.; Koizumi, M.; Doucet, H.; Pham, T.; Kozawa, M.; Murata, K.; Katayama, E.; Yokozawa, T.; Ikariya, T.; Noyori, R. *J. Am. Chem. Soc.* **1998**, *120*, 13529.

86. Noyori, R.; Ohkuma, T. *Angew. Chem. Int. Ed.* **2001**, *40*, 40.

87. Ohkuma, T.; Koizumi, M.; Muñiz, K.; Hilt, G.; Kabuto, C.; Noyori, R. *J. Am. Chem. Soc.* **2002**, *124*, 6508.

88. Sandoval, C. A.; Ohkuma, T.; Muñiz, K.; Noyori, R. *J. Am. Chem. Soc.* **2003**, *125*, 13490.

89. Noyori, R.; Sandoval, C. A.; Muñiz, K.; Ohkuma, T. *Phil. Trans. Roy. Soc. A* **2005**, *363*, 901.

90. Abdur-Rashid, K.; Clapham, S. E.; Hadzovic, A.; Harvey, J. N.; Lough, A. J.; Morris, R. H. *J. Am. Chem. Soc.* **2002**, *124*, 15104.

91. Sandoval, C. A.; Yamaguchi, Y.; Ohkuma, T.; Kato, K.; Noyori, R. *Magn. Resol. Chem.* **2006**, *44*, 66.

92. Hamilton, R. J.; Leong, C. G.; Bigam, G.; Miskoizie, M.; Bergens, S. H. *J. Am. Chem. Soc.* **2005**, *127*, 4152.

93. Ohkuma, T.; Ooka, H.; Ikariya, T.; Noyori, R. *J. Am. Chem. Soc.* **1995**, *117*, 10417.

94. Ohkuma, T.; Ooka, H.; Hashiguchi, S.; Ikariya, T.; Noyori, R. *J. Am. Chem. Soc.* **1995**, *117*, 2675.

95. Ohkuma, T.; Doucet, H.; Pham, T.; Mikami, K.; Korenaga, T.; Terada, M.; Noyori, R. *J. Am. Chem. Soc.* **1998**, *120*, 1086.

96. Ohkuma, T.; Noyori, R. In *Handbook of Homogeneous Hydrogenation*, Vol. 3, de Vries, J. G.; Elsevier, C. J. (Eds.), Wiley-VCH, Weinheim, **2007**, p. 1105.

97. Genov, D. G.; Ager, D. J. *Angew. Chem. Int. Ed.* **2004**, *43*, 2816.

98. Mikami, K.; Wakabayashi, K.; Aikawa, K. *Org. Lett.* **2006**, *8*, 1517.

99. Ohkuma, T.; Hattori, T.; Ooka, H.; Inoue, T.; Noyori, R. *Org. Lett.* **2004**, *6*, 2681.

100. Ohkuma, T.; Sandoval, C. A.; Srinivasan, R.; Lin, Q.; Wei, Y.; Muñiz, K.; Noyori, R. *J. Am. Chem. Soc.* **2005**, *127*, 8288.

101. Abdur-Rashid, K.; Guo, R.; Lough, A. J.; Morris, R. H.; Song, D. *Adv. Synth. Catal.* **2005**, *347*, 571.

102. Jiang, Q.; Jiang, Y.; Xiao, D.; Cao, P.; Zhang, X. *Angew. Chem. Int. Ed.* **1998**, *37*, 1100.

103. Noyori, R.; Hashiguchi, S. *Acc. Chem. Res.* **1997**, *30*, 97.

104. Gladiali, S.; Alberico, E. In *Transition Metals for Organic Synthesis*, 2nd ed., Beller, M.; Bolm, C. (Eds.), Wiley-VCH, Weinheim, **2004**, Vol. 2, p. 145.

105. Fujii, A.; Hashiguchi, S.; Uematsu, N.; Ikariya, T.; Noyori, R. *J. Am. Chem. Soc.* **1996**, *118*, 2521.

106. Haack, K.-J.; Hashiguchi, S.; Fujii, A.; Ikariya, T.; Noyori, R. *Angew. Chem. Int. Ed. Engl.* **1997**, *36*, 285.

107. Yamakawa, M.; Ito, H.; Noyori, R. *J. Am. Chem. Soc.* **2000**, *122*, 1466.

108. Alonso, D. A.; Brandt, P.; Nordin, S. J. M.; Andersson, P. G. *J. Am. Chem. Soc.* **1999**, *121*, 9580.

109. Yamakawa, M.; Yamada, I.; Noyori, R. *Angew. Chem. Int. Ed.* **2001**, *40*, 2818.

110. Noyori, R.; Yamakawa, M.; Hashiguchi, S. *J. Org. Chem.* **2001**, *66*, 7931.

111. Gao, J.-X.; Ikariya, T.; Noyori, R. *Organometallics* **1996**, *15*, 1087.

112. Baratta, W.; Herdtweck, E.; Siega, K.; Toniutti, M.; Rigo, P. *Organometallics* **2005**, *24*, 1660.

113. Mashima, K.; Abe, T.; Tani, K. *Chem. Lett.* **1998**, 1201.

114. Nishibayashi, Y.; Takei, I.; Uemura, S.; Hidai, M. *Organometallics* **1999**, *18*, 2291.

115. Reetz, M. T.; Li, X. *J. Am. Chem. Soc.* **2006**, *128*, 1044.

116. Evans, D. A.; Nelson, S. G.; Gagné, M. R.; Muci, A. R. *J. Am. Chem. Soc.* **1993**, *115*, 9800.

117. Corey, E. J.; Helal, C. J. *Angew. Chem. Int. Ed.* **1998**, *37*, 1986.

118. Itsuno, S. *Org. React.* **1998**, *52*, 395.

119. Itsuno, S. In *Comprehensive Asymmetric Catalysis*, Jacobsen, E. N.; Pfaltz, A.; Yamamoto, H. (Eds.), Springer, Berlin–Heidelberg, **1999**, Vol. 1, p. 289.

120. Itsuno, S.; Nakano, M.; Miyazaki, K.; Masuda, H.; Ito, K. *J. Chem. Soc. Perkin Trans. 1* **1985**, 2039.

121. Corey, E. J.; Bakshi, R. K. Shibata, S.; Chen, C.-P.; Singh, V. K. *J. Am. Chem. Soc.* **1987**, *109*, 7925.

122. Helal, C. J.; Magriotis, P. A.; Corey, E. J. *J. Am. Chem. Soc.* **1996**, *118*, 10938.

123. Corey, E. J.; Bakshi, R. K. *Tetrahedron Lett.* **1990**, *31*, 611.

124. Corey, E. J.; Helal, C. J. *Tetrahedron Lett.* **1995**, *36*, 9153.

125. Nagata, T.; Sugi, K. D.; Yorozu, K.; Yamada, T.; Mukaiyama, T. *Catal. Surv. Jpn.* **1998**, *2*, 47.

126. Yamada, T.; Nagata, T.; Ikeno, T.; Ohtsuka, Y.; Sagara, A.; Mukaiyama, T. *Inorg. Chim. Acta* **1999**, *296*, 86.

127. Wild, F. R. W. P.; Zsolnai, L.; Huttner, G.; Brintzinger, H. H. *J. Organomet. Chem.* **1982**, *232*, 233.

128. Willoughby, C. A.; Buchwald, S. L. *J. Am. Chem. Soc.* **1992**, *114*, 7562.

129. Willoughby, C. A.; Buchwald, S. L. *J. Org. Chem.* **1993**, *58*, 7627.

130. Spindler, F.; Blaser, H.-U. In *Transition Metals for Organic Synthesis*, 2nd ed., Beller, M.; Bolm, C. (Eds.), Wiley-VCH, Weinheim, **2004**, Vol. 2, p. 113.

131. Dengming, X.; Zhang, X. *Angew. Chem. Int. Ed.* **2001**, *40*, 3425.

132. Burk, M. J.; Martinez, J. P.; Feaster, J. E.; Cosford, N. *Tetrahedron* **1994**, *50*, 4399.

133. Bader, R. R.; Blaser, H.-U. In *Heterogeneous Catalysis and Fine Chemicals IV*, Blaser, H. U.; Baiker, A.; Prins, R. (Eds.), Elsevier, Amsterdam, **1997**, p. 17.

134. Uematsu, N.; Fujii, A.; Hashiguchi, S.; Ikariya, T.; Noyori, R. *J. Am. Chem. Soc.* **1996**, *118*, 4916.

2

LIGAND DESIGN FOR OXIDATION

Tohru Yamada

Department of Chemistry, Keio University, Hiyoshi, Yokohama, Japan

2.1 INTRODUCTION

The oxidation reaction is one of the most fundamental transformations in organic chemistry. Its catalytic enantioselective versions, compared with the reductions of unsaturated compounds, have more recently appeared as the common procedures for organic syntheses.[1] In 1970, molybdenum or vanadium complexes[2] were reported to catalyze the epoxidation of alkenes with peroxides as the terminal oxidants, and it was observed that the epoxidation of allylic alcohols was specifically enhanced by a vanadium catalyst to suggest the interaction between the substrates and the catalyst. Since the enantioselective catalysis was demonstrated in the cyclopropanation with diazo compounds and chiral copper complexes,[3] a wide variety of transition metal complexes have been developed as catalysts for asymmetric syntheses. In 1977, two pioneering reports appeared on the catalytic enantioselective epoxidation of allylic alcohols using molybdenum[4] and vanadium[5] complexes. Sharpless et al. examined various metal complexes, such as aluminum, vanadium, molybdenum, and titanium. They found that the titanium complexes with optically active tartaric ester effectively catalyzed the enantioselective epoxidation with hydroperoxide as the terminal oxidant.[6] In the presence of molecular sieves, a truly catalytic system was realized for the enantioselective epoxidation of a wide variety of allylic alcohols[7]. In addition, the absolute configuration of the produced epoxides is predictable considering the tartaric acid employed as the chiral ligand. The Katsuki–Sharpless epoxidation has been established as an indispensable tool for

New Frontiers in Asymmetric Catalysis, Edited by Koichi Mikami and Mark Lautens
Copyright © 2007 John Wiley & Sons, Inc.

both the total synthesis and determination of the absolute configurations in the natural product chemistry:

$$(2.1)$$

As well as epoxidation, dihydroxylation of alkenes by osmium tetraoxide is reliable method for obtaining oxygen-functionalized compounds. On the basis of the report that the formation of the osmate ester from an alkene and osmium tetraoxide was enhanced by 100–10000 times through the coordination of tertiary amines,[8] various optically active amines **1–7** were desgined as the chiral ligands for the catalytic enantioselective dihydroxylation of alkenes, but in stoichiometric reactions.[9] In order to realize the truly catalytic oxidation, Sharpless examined the structure of amine ligands. Finally, he designed the (DHQ)$_2$-PHAL **8** and (DHQD)$_2$-PHAL **9** ligands for the osmium-catalyzed asymmetric dihydroxylation (AD) of various alkenes.[10] The absolute configuration of the AD product could also be predicted. In addition, he provided the simple procedure for the elegant enantioselective transformation for all organic chemists. The AD-mix-α and AD-mix-β are all-in-one reagents including the osmium source, chiral ligand, and terminal oxidant:[11]

(DHQ)₂-PHAL **8**   (DHQD)₂-PHAL **9**

In this chapter, the ligand design for the catalytic enantioselective oxidations developed after the Katsuki–Sharpless epoxidation and the Sharpless AD will be discussed.

2.2   CATALYTIC ENANTIOSELECTIVE EPOXIDATION OF UNFUNCTIONALIZED OLEFINS

Kochi first reported that cationic manganese(III) complexes with salen ligands were effective catalysts for the epoxidation of various olefins with iodosylbenzene as the terminal oxidant.[12] Jacobsen and Katsuki then independently developed chiral salen–manganese complexes for the enantioselective epoxidation of unfunctionalized olefins. Jacobsen proposed[13] the cationic manganese(III) complex **10a**, and the X-ray crystallography of the bis(acetone) adduct of (*S,S*)-**10b** complex reveals that the tetradentate ligand adopts a near-planar geometry with the phenyl groups of the diphenylethylene unit occupying a pseudoequatorial position (see Figure 2.1). Whereas Katsuki examined similar complexes **11a-11c** bearing another chirality on the C3 and C3′ positions of salicylaldehydes.[14] In both cases, it was confirmed that the presence of bulky alkyl groups on C3(3′) in the salen ligand is crucial to the selectivity and stability of these catalysts. The salen–Mn(III) complex **10b**

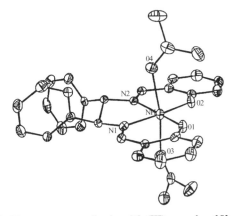

Figure 2.1. X-ray structure of salen–Mn(III) complex **10b**-[acetone]₂.

bearing no substituents on C3(3′) positions catalyzes the epoxidation of β-methylstyrene with only 0–3% ee and about five catalyst turnovers. Alkyl substituents smaller than *tert*-butyl lead to poor results. Increasing the size of the alkyl group as in **10c** leads to minor improvements in selectivity, but more hindered systems such as **10d** tend to be less selective.[15] It was found that asymmetric induction by salen complexes bearing four asymmetric centers was affected by the relative configuration of the stereogenic carbons at the ethylenediamine and salicylaldehyde parts and by the conformation of the C3 and C3′ substituents:[16]

The electronic properties of chiral catalysts were examined. Condensation of the optically active 1,2-diphenylethylenediamine with appropriate C5(5′)-substituted *tert*-butyl salicylaldehyde derivatives followed by complexation with manganese(III) center led to the corresponding catalysts **12a–12e**. Then three model substrates, 2,2-dimethylchromene, *cis*-β-methylstyrene, and *cis*-2,2-dimethyl-3-hexene, were subjected to enantioselective expoxidation catalyzed by 5-substituted

X = OMe **12a**
Me **12b**
H **12c**
Cl **12d**
NO$_2$ **12e**

12f

salen manganese complexes **12a–12e** with NaClO as the terminal oxidant. The Hammett plots of these reactions (see Figure 2.2) revealed that the electron-donating groups on the catalysts led to higher enantioselectivities.[17] While the electron-withdrawing group on the catalysts increase the rate of epoxidation. The electron-donating substituents attenuate the reactivity of the catalyst, resulting in a milder oxidant, which is expected to transfer oxygen to the alkene via a more product-like transition state to achieve high enantioselectivities. On the basis of these observations, a new catalyst **12f** bearing sterically hindered and electron-donating OSi(^{i}Pr)$_3$ substituents at the C5 and C5$'$ positions was designed.[18] By using the catalyst **12f**, the enantioselectivity for almost all olefin classes was improved higher than the *tert*-butyl substituted analog **10a**.

The high enantioselectivities were observed in the epoxidations for a wide range of substrates with diaminocyclohexane-derived salen–Mn(III) catalysts **13a–13c**. The crystal structure of a series of these complexes were determined by X-ray analysis[19] and suggested that these Mn complexes might be quite flexible conformationally. The *tert*-butyl groups on C3(3$'$) positions were large enough to block side-on approaches other than those over the diimine backbone. The addition of electron-donor compounds, such as pyridine *N*-oxide derivatives, to the reaction was examined. In the reaction catalyzed by the complex **13a**, a measurable increase in the epoxide yield and reaction rate was observed, but a negligible influence on the

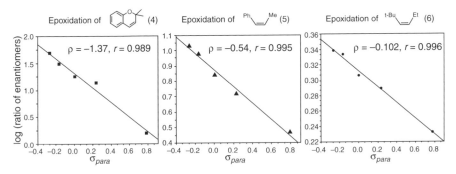

Figure 2.2. Hammett plots for the oxidation by catalysts **12a–12e**.

stereoselectivity of the epoxidation.[20] It was demonstrated that pyridine N-oxide acts as an axial ligand in epoxidation mediated by cationic chromium(salen) complexes.[21] Jacobsen proposed the effect of the axial ligand as follows. The active (salen)Mn^V=O complex undergoes reversible coupling with the Mn^{III} complex to generate an inactive μ-oxo dimer. In the presence of pyridine N-oxide derivatives, the equilibrium is shifted toward the Mn^V oxo intermediate as a result of the additive binding to the coordinatively unsaturated Mn^{III} complex. Acceleration in the reaction rate is then expected because of the increased concentration of the active Mn^V oxo species in the solution. In some examples, the addition of 2-methylimidazole as a donor ligand to the reaction improved the enantioselectivity.[22] Katsuki proposed that the conformational change of the optically active oxo(salen)metal complexes due to the coordination of donor ligands seemed to affect the refinement of their asymmetric induction ability as well as the enhancement of the reaction rate:

13a   13b

13c

$$(2.2)$$

Katsuki et al. developed new types of salen–manganese(III) complexes bearing binaphthyl groups of axial chirality instead of the chiral C3 and C3′ substituents in complex **14**.[23] By the addition of donative ligands, such as 4-dimethylaminopyridine N-oxide or pyridine N-oxide, the enantioselectivities were remarkably improved. They considered that the ligand effect was attributed to the conformational change of the skeleton of the salen complex and of its binaphthyl moieties on coordination of the ligand. They also demonstrated that two sets of stereogenic centers at C8(8′) and C9(9′) had a high potential to discriminate the facial selectivity of unfunctionalized olefins.[24] In the presence of 2.3 mol% of Mn(III) complex **15**, *trans*-stilbene was epoxidized with iodosylbenzene as a terminal oxidant into the stilbene oxide with 49% ee. The effect of the axial ligands was then reexamined. In the presence of the optically active (−)-sparteine and achiral

14

15

Mn(III) catalyst **16**, the chromene derivative was oxidized with iodosylbenzene as the terminal oxidant. The corresponding epoxide was obtained in poor yield, but the enantioselectivity was up to 70% ee.[25] In case the ethylenediamine bridge does not have a chiral element, the Mn–salen complex is considered to exist as an equilibrium mixture of enantiomeric conformational isomers in solution (see Eq. (2.3)). Both isomers, of course, equally catalyze the epoxidation to afford a racemic epoxide. If this equilibrium could be shifted to one side even though the Mn–salen complex is achiral, the enantioselective epoxidation should be possible. As mentioned above, the oxo Mn–salen complexes could accept the donor ligand on the axial position, and the axial ligand was replaced with an optically active one that caused the equilibrium to shift toward one conformer. Therefore, it can be expected that achiral Mn–salen complexes can be employed as catalysts for the enantioselective epoxidation by conformational control with an optically active axial ligand. On the basis of these hypotheses, the optically active bipyridine N-oxide derivatives were employed as additives to achieve the high enantioselectivity up to 86% ee using the achiral Mn–salen complex catalyst **17**:[26]

16

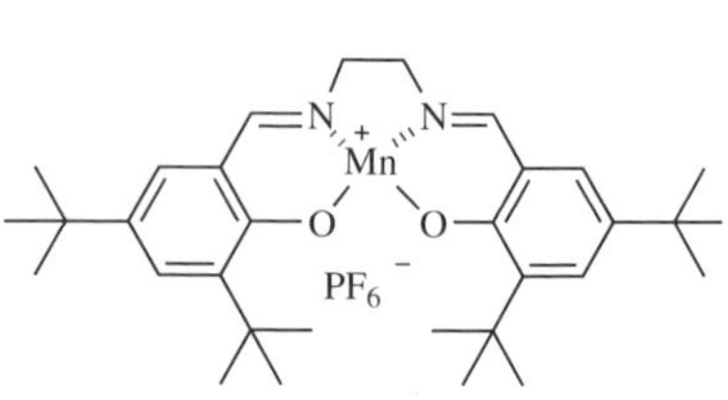

17

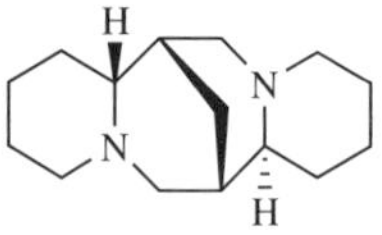

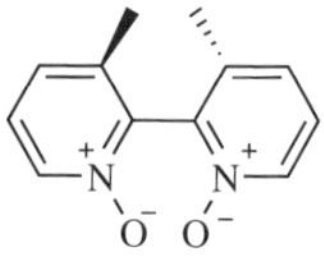

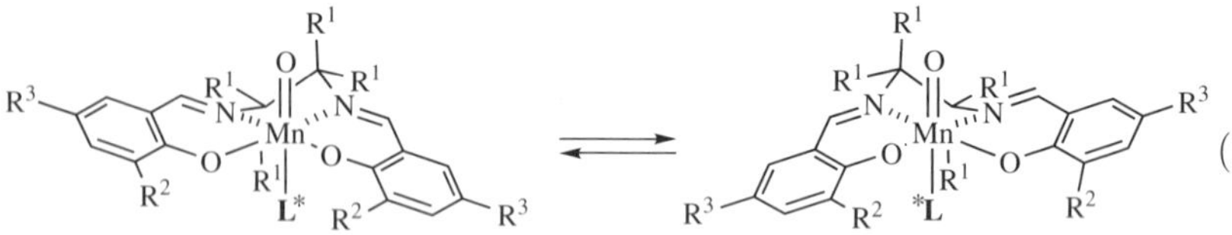

$$(2.3)$$

18   **19**

The ruthenium complexes are well known to serve as catalysts for oxidation. Katsuki synthesized the (ON^+)(salen)ruthenium(II) complex and examined the enantioselective epoxidation of chromene derivatives. After the screening of various terminal oxidants, 2,6-dichloropyridine N-oxide[27] was found to be an excellent terminal oxidant as well as for the ruthenium porphyrin-catalyzed epoxidation. In addition, salen-Ru complex-catalyzed epoxidation was found to be effectively accelerated by exposure to sunlight around 450 nm. This observation suggested that the electron transfer from ruthenium ion to a ligand and the subsequent ligand dissociation were responsible for this photoacceleration. The conjugated olefins were converted to the corresponding epoxides with higher selectivities than 80% ee.[28]

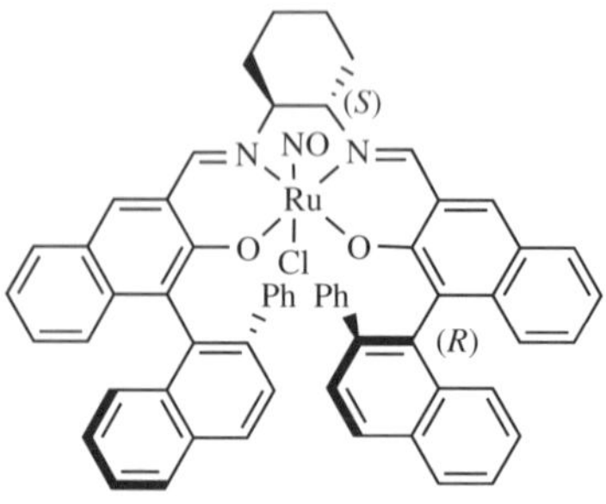

20

In the presence of a catalytic amount of the optically active Mn(III)–salen complexes **23** and **24**, the enantioselective epoxidation of unfunctionalized olefins with the combined use of molecular oxygen and pivalaldehyde was demonstrated.[29] This report revealed that the *tert*-butyl group on the C3 position of salicylaldehyde in the chiral ligand was essential to realize a high enantioselection, and that pivalaldehyde was the most effective reductant for both the enantioselectivity and chemical yield. It should be noted that the sence of enantioselectivity to form the epoxide from 1,2-

$$(2.4)$$

21

cat·chiral Mn(III)complex

RT
1 atm O_2

CHO
pivalaldehyde

COOH

22
optically active

dihydronaphthalene using the (*S*,*S*)-catalyst is opposite to the results reported by Jacobsen or Katsuki. Furthermore, in the presence of a catalytic amount of *N*-methylimidazole as the axial ligand, the absolute configuration of the epoxide catalyzed by the (*S*,*S*)-complex is completely reversed to give the epoxide with a (1*S*,2*R*) configuration and the enantiomeric excess is also improved. The influence of the position of an alkyl group attached to an imidazole ring and a pyridine ring on the optical yield of the obtained epoxides was also examined.[30] The results decreased sharply when 2-methylimidazole or 4-methylimidazole was added, and similar effects were observed in the case of using pyridine or 2,6-lutidine. These results suggested that the alkyl groups attached to the carbon next to the nitrogen would prevent the coordination to the central manganese atom because of their steric hindrance, and that Mn(III) complex molecules could not always be coordinated with imidazole derivatives. Reactive catalysts for the epoxidation would include both the manganese complex coordinated by imidazole derivatives and the imidazole-free complex, which afforded the epoxide with the reversed absolute configuration. Therefore, the total optical yield decreased. In the case of porphyrinato–iron complexes, the acylperoxo species were converted into the corresponding oxo–iron complexes by coordination to the imidazole derivatives.[31] Reversal of the absolute configuration of the epoxides suggested that a similar phenomenon is observed in the aerobic epoxidation by the combined use of molecular oxygen and aldehyde; that is, the acylperoxo–Mn complex **29** would be formed from molecular

23

24

oxygen, pivalaldehyde, and the original Mn(III) complex in the first step. In the absence of N-methylimidazole, the acylperoxomanganese (S,S)-salen complex **29** itself would react with an olefin to afford the ($1R,2S$)-(+)-epoxide. In the presence of N-alkylimidazole, the acylperoxo–Mn complex **29** was converted into the oxo-Mn complex **30** by the coordination of the donor axial ligand. The oxo–Mn complex **30** has been widely accepted as a reactive intermediate for epoxidations by using terminal oxidants such as iodosylbenzene and sodium hypochlorite.[32] New and effective kinds of optically active manganese(III) complexes **23** and **24** with N,N'-bis(3-oxo-butylidene) diamine ligands were designed for aerobic enantioselective epoxidation

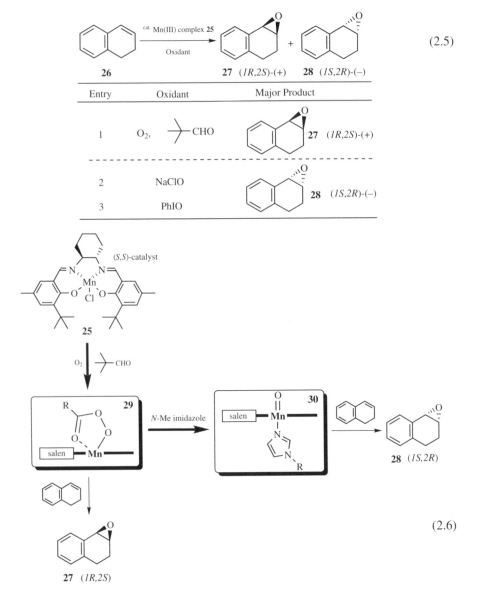

Entry	Oxidant	Major Product
1	O$_2$, ⤴CHO	**27** (*1R,2S*)-(+)
2	NaClO	**28** (*1S,2R*)-(−)
3	PhIO	

(2.5)

(2.6)

in the presence of pivalaldehyde.[33] After several experiments using these chiral catalysts having bulkier substituents proved to catalyze the aerobic enantioselective epoxidation of unfunctionalized olefins effectively. The present aerobic enantioselective epoxidation proceeded with the opposite enantioface selection from those obtained terminal oxidants, such as sodium hypochlorite and iodosylbenzene.

Although the chiral ketoiminatomanganese(III) complexes were reported to catalyze the asymmetric aerobic alkene epoxidations, an aldehyde such as pivalaldehyde is required as a sacrificial reducing agent. Groves reported that the dioxo(porphyrinato)ruthenium complexes **31**, prepared with *m*-chloroperoxybenzoic acid, catalyzed the aerobic epoxidation without any reductant.[34] On the basis of these reports, Che synthesized the optically active D_4-porphyrin **35** and applied it to the truly aerobic enantioselective epoxidation of alkenes catalyzed by the chiral *trans*-dioxo (D_4-porphyrinato)ruthenium(VI) complex.[35] The dioxoruthenium complex catalyzed the enantioselective aerobic epoxidation of alkenes with moderate to good enantiomeric excess without any reductant. In the toluene solvent, the turnovers for the epoxidation of *cis*-β-methylstyrene reached 20 and the ee of the epoxide was increased to 73% ee.

$$(2.7)$$

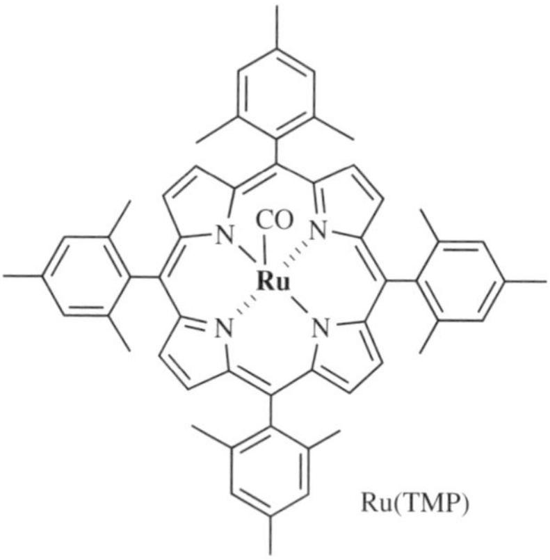

Ru(TMP)

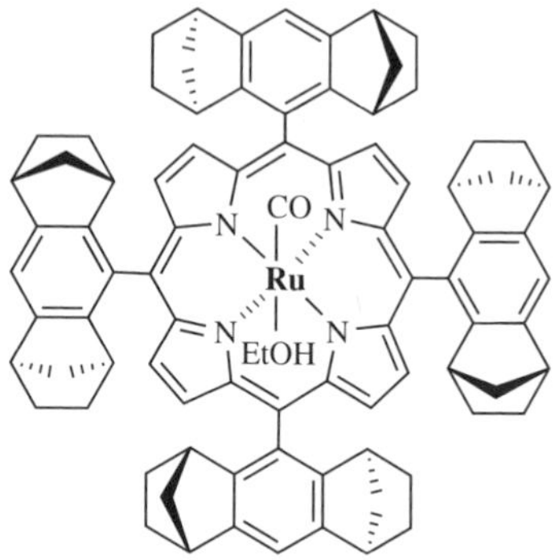

D_4-porphyrinatoRu complex

34   35

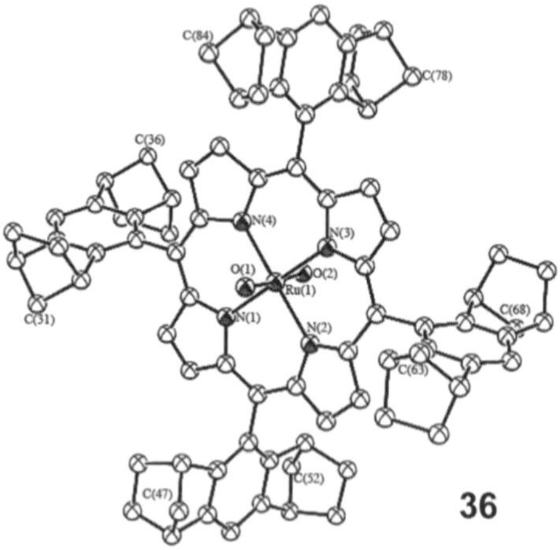

36

Figure 2.3. X-ray structure of dioxo D_4-porphyrinatoRu complex.

2.3 ENANTIOSELECTIVE METAL-CATALYZED BAEYER–VILLIGER OXIDATION

The Baeyer–Villiger oxidation is one of the most reliable reactions employed to convert cyclic ketones into the corresponding lactones and has been widely used both in degradation and synthesis, although peroxides, such as hydrogen peroxide, peracetic acid, or m-chloroperbenzoic acid, are generally used for this purpose. Mukaiyama and Yamada reported that molecular oxygen could also be used as a simple oxidizing agent for the asymmetric oxidation of olefins in the presence of metal complex catalysts and an oxygen acceptor, such as an aldehyde.[36] For the Baeyer–Villiger reaction, although several procedures for the preparation of ε-caprolactone from cyclohexanone using molecular oxygen and an aldehyde catalyzed by transition-metal salts or complexes were found in patents,[37] the selectivities were not satisfactory and the conversion of cyclohexanone was generally low. With β-diketonatonickel complex catalysts the truly aerobic oxidation system was applied to the Baeyer-Villiger reaction. It was found that cyclic and acyclic ketones are smoothly converted into the corresponding lactones or esters in good to high yields by the combined use of isovaleraldehyde **38** or benzaldehyde as a reductant under the atmospheric pressure of oxygen at room temperature:[38]

$$(2.8)$$

On the basis of their oxidation system, Bolm developed an enantioselective, metal-catalyzed version of the Baeyer–Villiger oxidation.[39] In the presence of a chiral copper complex and pivalaldehyde, the racemic 2-phenylcyclohexanone **42** was treated with molecular oxygen to obtain the corresponding optically active lactone **43** along with the starting cyclic ketone **44**, as a result of kinetic resolution. After optimization of the ligand structure, the best results were obtained with a *p*-nitro-substituted copper complex **41** in water-saturated benzene. In the presence of 1 mol% of a chiral copper catalyst, the lactone with 69% ee was obtained in 47% yield at 6°C. They applied this oxidation system to chiral cyclobutanones.[40] The copper-complex-catalyzed oxidation of *rac*-cyclobutanone **45** with the (*S,S*)-complex, led to the formation of the two regioisomeric lactones **46** and **47** in an appropriate 1 : 1 ratio. The "normal" Baeyer–Villiger product **46**, in which the oxygen was incorporated between the more substituted carbon atom and the carbonyl group, was formed with 67% ee. Its regioisomer **47** was obtained with 92% ee. The "normal" Baeyer–Villiger product has the *R* configuration at the bridgehead carbon atom attached to the oxygen, whereas the other one, the "abnormal" product, has the *S* configuration at the α-carbon atom to the carbonyl group. As a result, the (*S,S*)-copper-catalyst transformed the enantiomeric ketones into different regioisomeric lactones. The highest enantiomeric induction of normal and abnormal lactones reached 76% ee and 95% ee, respectively, for the oxidation of the tricyclic ketone. The present system could be applied to the prochiral cyclobutanones for the corresponding lactones with moderate enantioselectivity:[41]

$$(2.9)$$

$$(2.10)$$

46
"normal" lactone
67% ee / 45% yield

47
"abnormal" lactone
92% ee / 15% yield

The stereochemistry of the Baeyer–Villiger reaction is reported to be controlled by two factors: (1) face selectivity during oxidant addition and (2) enantiotopos selectivity during migration. As the Criegee adduct formation is a reversible step and its migration to a lactone is an irreversible and rate-determining one, topos selection in the second step is considered to influence the stereoselection of the Baeyer–Villiger oxidation. With these considerations, the salen–cobalt complex with a square-planar geometry **50a** was employed as the catalyst for the Baeyer–Villiger oxidation of 3-substituted cyclobutanone, but no enantioselection was observed, whereas the salen–cobalt complex possessing a cis-β-structure **50b** realized a good enantioselectivity up to 78% ee.[42] A peroxy zirconium-salen complex of the cis-β-structure **51** was also effective as the catalyst of the enantio-selective Baeyer–Villiger oxidation. The reaction of 3-phenylcyclobutanone **48** using the urea•hydrogen peroxide adduct (UHP) as an oxidant afforded the corresponding lactone **49** in 68% yield with 87% ee.[43] They examined the chiral cationic palladium(II) 2-(phosphinophenyl)pyridine complex **52** for the enantioselective Baeyer–Villiger oxidation of 3-phenylcyclobutanone.[44] In a THF solution at −60°C, the lactone was obtained in 91% yield with 80% ee and the excellent enantioselectivity of >99% ee was observed during the reaction of the tricyclic cyclobutanone:

$$(2.11)$$

chelated Criegee adduct

$$(2.12)$$

48

49

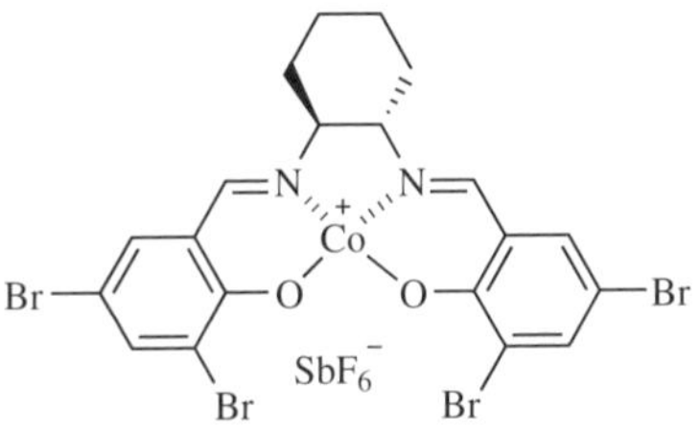

29% yield / 0% ee
50a

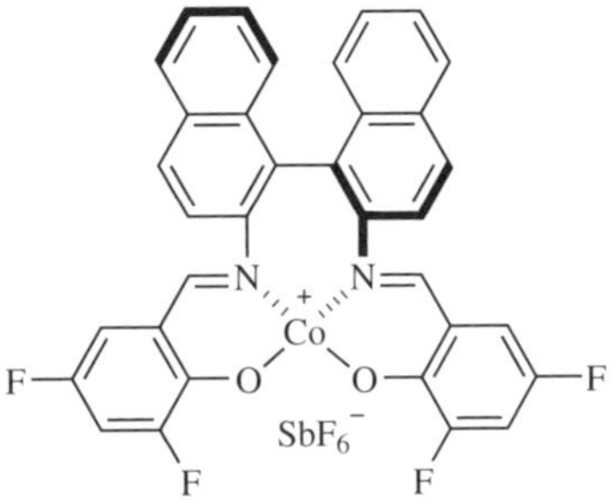

72% yield / 77% ee
50b

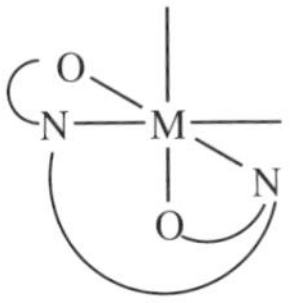

cis-β structure 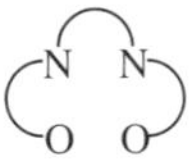   salen ligand

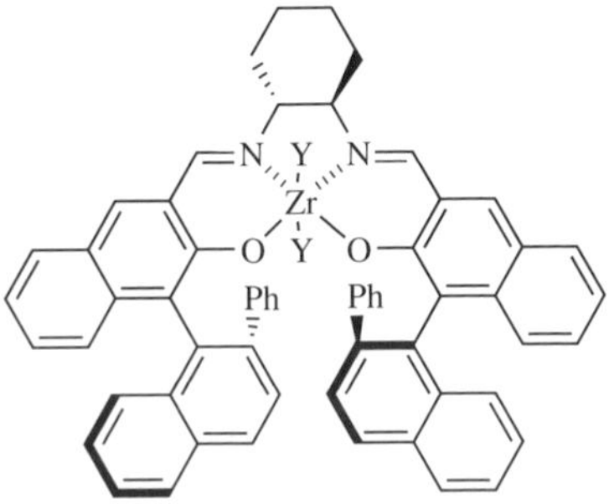

Y = PhO
68% yield / 87% ee
51

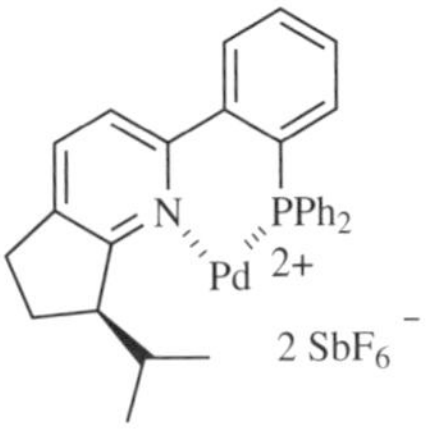

(-60°C, in THF)

91% yield / 80% ee
52

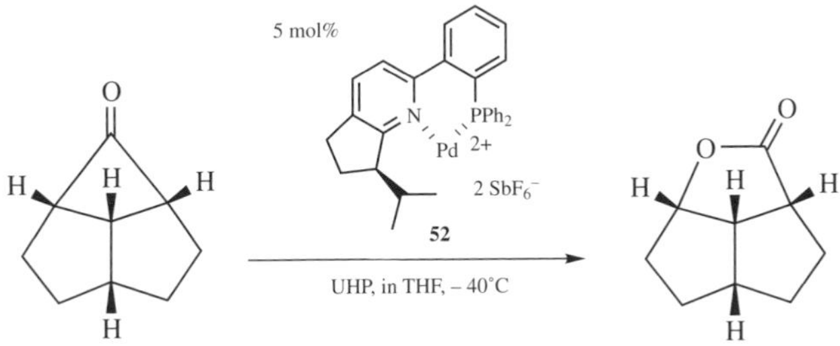

5 mol%

52

UHP, in THF, – 40°C

53

(2.13)

54

(1*S*, 4*R*, 7*R*, 10*S*)
89% yield / >99% ee

2.4   OPTICAL RESOLUTION DURING OXIDATION OF ALCOHOLS

Although the oxidation of a secondary alcohol is one of the most common and well-studied reactions in organic chemistry, there are relatively few catalytic enantioselective examples of the reliable alcohol oxidation. Stoltz developed a catalytic oxidative kinetic resolution of secondary alcohols that uses molecular oxygen as the terminal oxidant. Using 1-phenylethanol **56** as the test substrate, they surveyed a number of variations in the catalytic reaction and found that the conditions developed by Uemura[45] were particularly suited to the rapid screening of a variety of chiral ligands. From the structurally diverse set of ligands explored for the oxidation reaction, (−)-sparteine **55** quickly emerged as the most selective one. Also, the nature of the palladium source was found to be critical. Among the various palladium sources, such as $PdCl_2$, $Pd(OAc)_2$, $Pd(CH_3CN)_2Cl_2$, and $Pd(PhCN)_2Cl_2$, it was revealed that $Pd(nbd)Cl_2$ provided the most selective catalytic system.[46] The resolution was found to have a selectivity factor of 23.1, thereby providing acetophenone **57** in 59.9% conversion and unreacted alcohol **58** of 98.7% ee:

$$(2.14)$$

The empirical observation that (−)-sparteine **55** is necessary for catalysis implicates a base-promoted pathway in the mechanism. In the first step, a palladium alkoxide is formed after alcohol binding, followed by β-hydride elimination of the alkoxide to yield a ketone product. On the basis of a kinetic study of the enantioselective oxidation of 1-phenylethanol, it was revealed that (−)-sparteine plays a dual role in the oxidative kinetic resolution of alcohols, as a ligand on palladium and an exogeneous base:[47]

$$(2.15)$$

Katsuki found that the optically active (nitroso)(salen)ruthenium(II) chloride **20** could be employed as an effective catalyst for the aerobic oxidation of racemic secondary alcohols under irradiation with a halogen or fluorescent lamp, and that if proceeded with good enantiomer differentiation.[48] For example, 4-phenyl-3-butyn-2-ol **59** was subjected to the present oxidation system to afford the corresponding ketone **60** in 64.6% yield along with an unreacted alcohol **61** of >99.5% ee, and the selectivity factor reached up to 20. Their catalysis system could be applied to *meso*-diols to give the corresponding optically active lactones with moderate enantioselectivity.[49] The selectivity was dependent on the nature of their apical ligand.[50] The aerobic oxidation of the *meso*-diol was catalyzed by the ruthenium chloride complex **62a** to obtain the lactone in low yield with moderate enantioselectivity. In contrast, the oxidation reaction catalyzed by the ruthenium hydroxo complex **62b** improved the enantioselectivity. They also tried to apply the ruthenium hydroxo complex catalyst **62b** to the aerobic chemoselective oxidation of diols **63** to the corresponding lactols **64**.[51] Remarkably, the reactions proceeded with a good to high level of enantiotopos differentiation to give enantiomerically enriched lactols up to 82% ee as the sole products:

$$(2.16)$$

$$(2.17)$$

2.5 CATALYTIC ENANTIOSELECTIVE OXIDATIVE COUPLING OF 2-NAPHTHOLS

The optically active 1,1'-bi-2-naphthol (BINOL) and its derivatives possessing an axial chirality have been widely employed as versatile chiral ligands in asymmetric transformations and catalyses. The enantioselective reduction of prochiral carbonyl compounds was achieved by using aluminum hydride modified by the optically active 1,1'-bi-2-naphthol and applied to various syntheses of biologically active compounds and natural products.[52] The titanium and zirconium complexes with BINOL and their derivaives were developed as effective chiral Lewis acids for a wide variety of asymmetric syntheses.[53] Their conversion into the corresponding 2,2'-bis(diphenylphosphino)-1,1'-binaphthyls (BINAP) was established by Noyori to expand their synthetic utility.[54] A variety of enantioselective catalyses employing BINAP as the chiral ligand were developed. The oxidative coupling of 2-naphthols is a well-known method to afford 1,1'-bi-2-naphthols, and various stoichiometric oxidants including molecular oxygen have been examined for this purpose. In addition, the optical resolution of the racemic BINOL obtained was also examined to provide enantio-enriched BINOL.

The enantioselective oxidative coupling of 2-naphthols with molecular oxygen was achieved by using chiral catalysts prepared from proline-derived diamines **65** and cuprous chloride.[55] In this reaction, the enantioselectivity of the oxidative coupling of 2-naphthol was not efficient, whereas the ester moiety on the 3-position is essential to realize the high enantioselection. The mechanism of the aerobic oxidative coupling was not clear though it was postulated that this reaction consisted of three successive processes: (1) exchange of the hydroxy group on the copper complex for a phenolic hydroxy group followed by the additional coordination of the ester carbonyl to a copper atom, (2) oxidative coupling affording a diketone with central chirality, and (3) transfer of the central chirality to axial chirality through keto–enol isomerization along with dissociation of the copper/amine complex (see Eq. (2.18)). The chiral 1,5-diaza-*cis*-decalins **68** could be employed as the ligand for the copper-catalyzed oxidative coupling of 2-naphthols.[56] In this enantioselective catalysis, the ester group on the 3-position of 2-naphthol was essential and the stereochemical model described was tentatively proposed:

17% ee / 78% yield (*S*)
10 mol%, CH₂Cl₂, RT, 24 h

78% ee / 85% yield (*S*)

65   **66**   **67**

13% ee / 80% yield (R)
10 mol%, CH$_2$Cl$_2$, RT, 5 days   CuI / CH$_3$CN•CH$_2$Cl$_2$ 90% ee / 79% yield (R)

68   **66**   **69**

(2.18)

The enantioselective oxidative coupling of 2-naphthol itself was achieved by the aerobic oxidative reaction catalyzed by the photoactivated chiral ruthenium(II)-salen complex **73**.[57] It was reported that the (R,R)-chloronitrosyl(salen)ruthenium complex [(R,R)-(NO)Ru(II)salen complex] effectively catalyzed the aerobic oxidation of racemic secondary alcohols in a kinetic resolution manner under visible-light irradiation. The reaction mechanism is not fully understood although the electron transfer process should be involved. The solution of 2-naphthol was stirred in air under irradiation by a halogen lamp at 25°C for 24 h to afford BINOL **66** as the sole product. The screening of various chiral diamines and binaphthyl chirality revealed that the binaphthyl unit influences the enantioselection in this coupling reaction. The combination of (R,R)-cyclohexanediamine and the (R)-binaphthyl unit was found to construct the most matched ligand to obtain the optically active BINOL **66** in 65% ee.

73

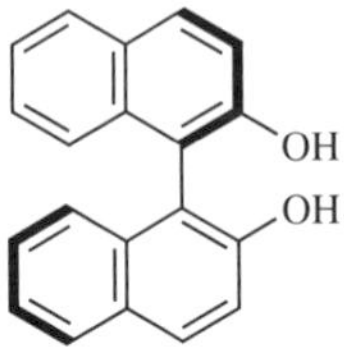

65% ee / 72% yield (*R*)

2 mol%, toluene, RT, 24 h
under irradiation of visible light

66

Numerous synthetic reactions using oxovanadium(V) reagents have been reported for the oxidative coupling of phenolic compounds. Because these oxovanadium(V) compounds, such as VOCl$_3$ and VOF$_3$, are moisture-sensitive and hazardous materials, their practical use is considered to be difficult. Meanwhile, VO(acac)$_2$ was reported to be a stable and available catalyst in the aerobic oxidative coupling of 2-naphthol.[58] On the basis of these observations, various ligands have been designed for the enantioselective version of the aerobic oxidative coupling of 2-naphthols. The tridentate Schiff bases derived from functionalized salicylaldehydes and α-amino acids was found to exhibit unique catalytic activities toward the aerobic oxidative coupling of 2-naphthol with the combined use of vanadyl salts.[59] Various salicylaldehydes and α-amino acids were examined for the chiral ligands, and it was found that 5,6-benzosalicylaldehyde with valine **74** was the best combination. The Schiff bases from (+)-ketopinic acid and α-amino acids were effective ligands for the vanadium complex catalysts **75** of the aerobic oxidative coupling of 2-naphthol.[60] The corresponding BINOL was obtained in quantitative yield with 84% ee:

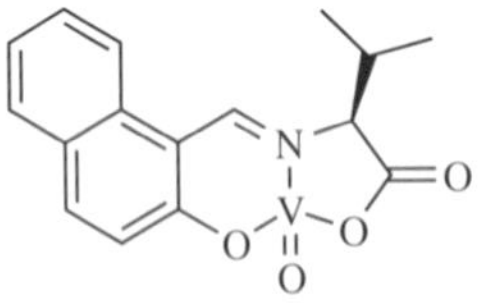

74

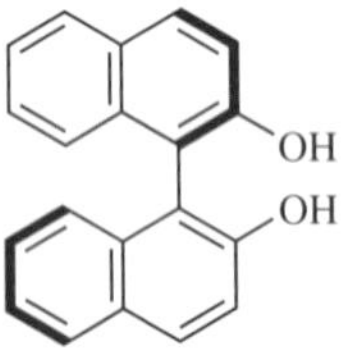

62% ee / 94% yield (*R*)

10 mol%, CCl$_4$, RT, 9 days

66

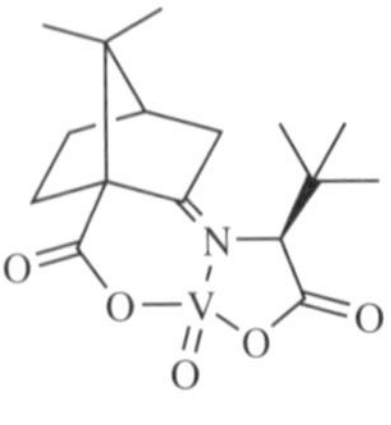

75

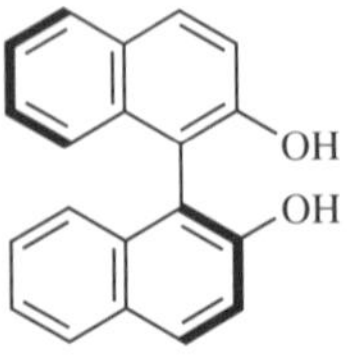

84% ee / 99% yield (*S*)

3 mol%, CCl$_4$, RT, 7 days

66

The optically active binaphthol itself was an effective frame for chiral ligands.[61] The condensation of (S)-3-formyl-1,1′-bi-2-naphthol with chiral amino acids and vanadyl sulfate provided the chiral oxovanadium complex catalysts **76** for oxidative coupling. It was noted that the (R)-product was isolated in the reaction catalyzed by the combination of (S)-binaphthol and (R)-amino acid within 12 h. However, the complex derived from (S)-binaphthol and (S)-amino acids produced an inferior enantioselection. The combination of the (S)-BINOL-derived aldehyde and the (R)-amino acid represents a matched pair for the enantioselective oxidative coupling of 2-naphthol, whereas the complex derived from (S)-BINOL and the (S)-amino acids is mismatched. On the basis of a similar concept, the condensation of chiral 3,3′-diformyl-2,2′-dihydroxy-1,1′-bi-2-naphthol with (S)-amino acids and vanadyl sulfate afforded the dimetallic complex catalysts **78**, whose structures were confirmed by high-resolution mass spectrometry (HRMS) and IR.[62] The oxidative coupling catalyzed by the complex derived from the (S)-BINOL derivative and (S)-phenylalanine afforded the (R)-product with 39% ee, while (R)-BINOL and (S)-phenylalanine produced the (R)-product with 6% ee. This result indicated that the double chiralities (R,S) were suited for the oxidative coupling of 2-naphthol. When the achiral salicylaldehyde, such as biphenyl derivatives, was used for the ligand, diastereomeric complexes will in principle be generated during their condensation with (S)-amino acids because of its higher rotation barrier.[63] The matched chirality in these diastereomers should provide a high enantioselectivity. Therefore, 3,3′-diformyl-2,2′-dihydroxy-1,1′-biphenyl was treated with (S)-amino acids and vanadyl sulfate to obtain the bimetallic, diastereometric complex **79**. It was actually subjected to the aerobic oxidative coupling of 2-naphthol to afford the BINOL in good yield and up to 91% ee:

54% ee / 93% yield (R)
5 mol%, CHCl₃, RT, 16 h

73% ee / 90% yield (R)
5 mol%, CHCl₃, 44°C, 8 h

76   **66**   **77**

81% ee / 86% yield (R)
10 mol%, CCl₄, 0°C, 8 days

78a   **66**

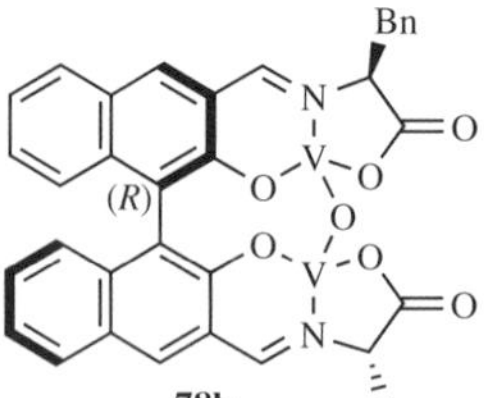

78b

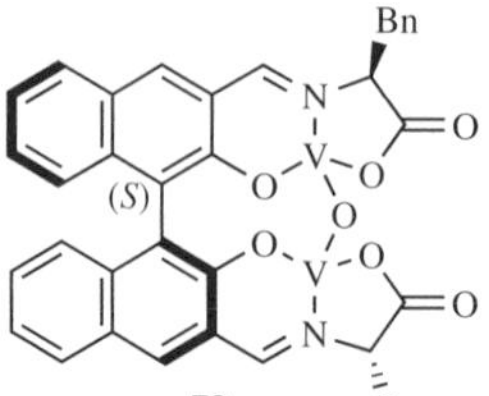

78c

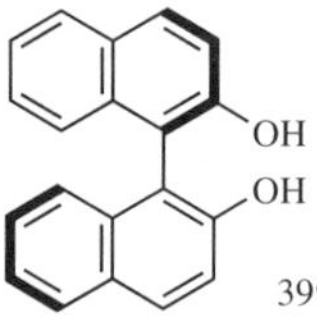

66

39% ee / 75% yield (*R*)

6% ee / 70% yield (*R*)

10 mol%, CCl₄, 20°C, 5 days

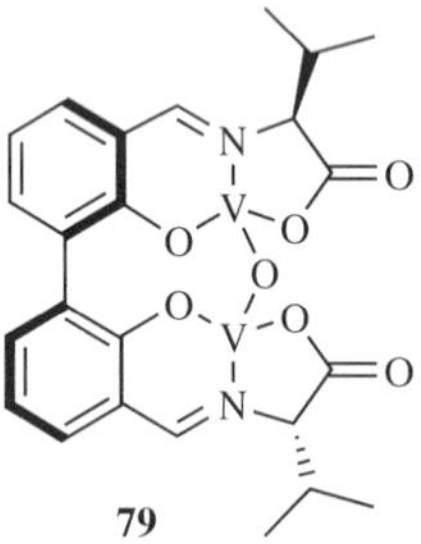

79

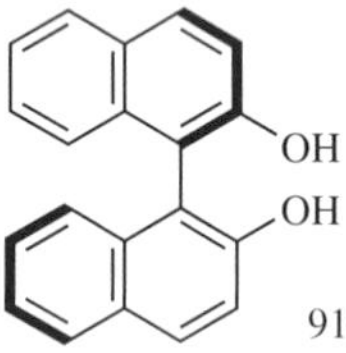

66

91% ee / 60% yield (*R*)

5 mol%, CCl₄, 0°C, 7 days

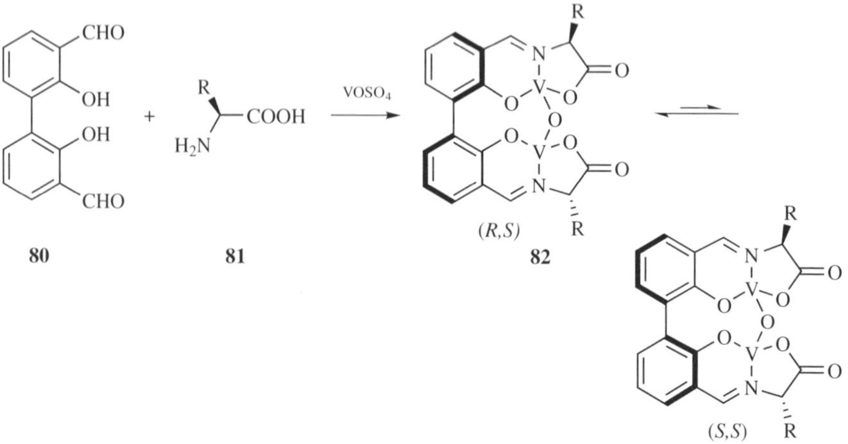

80   **81**   (*R,S*)   **82**

(*S,S*)   **83**

$$(2.19)$$

Newer types of the dinuclear vanadium(IV) complex catalysts **84** have been developed.[64] The abovementioned dinuclear vanadium complexes possess a V—O—V linkage whereas the ESR study on the catalyst **84** revealed no V—O—V linkage. The sense of enantioselection by the catalyst **84** of the (R,S,S)-structure is opposite to that of the binuclear complex **78a** of the same (R,S,S)-structure. These results suggested two active sites attached to the binaphthyl skeleton in the catalyst **84** performed the "dual activation" of 2-naphthols in the oxidative coupling to achieve high enantioselectivity:

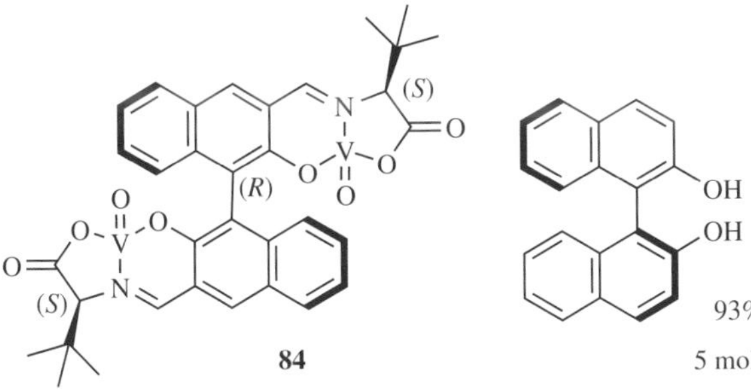

84

66
93% ee / 91% yield (S)
5 mol%, CH₂Cl₂, 0°C, 72 h

There are no reports of chiral complex catalysts that could catalyze the aerobic oxidative coupling of 2-naphthol at a reasonable reaction rate with high enantioselectivity. The reactions mentioned above are accomplished in 24 h or in 7–10 days. These drawbacks should be overcome in the near future.

2.6 CONCLUDING REMARKS

The enantioselective catalyses reviewed in this chapter have provided useful procedures for syntheses of chiral compounds, although the aerobic oxidation reaction has been avoided because the side reactions involving radical reactions should be often considered. The transition metal complexes can capture and activate molecular oxygen to generate oxo–metal intermediates, which react with substrates to afford the oxidation products. For example, ruthenium–porphyrin complexes were employed as reliable catalysts for the aerobic epoxidation of styrene derivatives. Although the enantioselective level and the applicabilities for substrates could not be satisfied, the reaction control of the aerobic and enantioselective epoxidation was realized using the ruthenium complexes with the chiral ligands. For the catalytic enantioselective reductions, hydrogen gas has been practically employed, and also in the asymmetric oxidations, reliable complex catalysts and/or reaction systems should be developed for the utilization of the ultimate oxidizing agent, "molecular oxygen."

REFERENCES

1. *Asymmetric Oxidation Reactions*, Katsuki, T. (Ed.) Oxford Univ. Press, **2004**.
2. Sheng, M. N.; Zajacek, J. G. *J. Org. Chem.* **1970**, *35*, 1839–1843.

3. Nozaki, H.; Moriuti, S.; Takaya, H.; Noyori, R. *Tetrahedron Lett.* **1966**, 5239–5244.

4. Yamada, S.; Mashiko, T.; Terashima, S. *J. Am. Chem. Soc.* **1977**, *99*, 1988–1990.

5. Michaelson, R. C.; Palermo, R. E.; Sharpless, K. B. *J. Am. Chem. Soc.* **1977**, *99*, 1990–1992.

6. Katsuki, T.; Sharpless, K. B. *J. Am. Chem. Soc.* **1980**, *102*, 5974–5976.

7. Hanson, R. M.; Sharpless, K. B. *J. Org. Chem.* **1986**, *51*, 1922.–1925

8. Marzilli, L. G. *Prog. Inorg. Chem.* **1977**, *23*, 255–378.

9. (a) Hentges, S. G.; Sharpless, K. B. *J. Am. Chem. Soc.* **1980**, *102*, 4263–4265; (b) Narasaka, K.; Yamada, T. *Chem. Lett.* **1986**, 131–134; (c) Tomioka, K.; Nakajima, M.; Koga, K. *J. Am. Chem. Soc.* **1987**, *109*, 6213–6215; (d) Corey, E. J.; Jardine, P. D.; Virgil, S.; Yuen, P.-W.; Connell, R. D. *J. Am. Chem. Soc.* **1989**, *111*, 9243–9244; (e) Hirama, M.; Oishi, T.; Itô, S. *J. Chem. Soc., Chem. Commun.* **1989**, 665–666; (f) Hanessian, S.; Meffre, P.; Girard, M.; Beaudoin, S.; Sancéau, J.-Y.; Bennani, Y. L. *J. Org. Chem.* **1993**, *58*, 1991–1993.

10. Sharpless, K. B.; Amberg, W.; Bennani, Y. L.; Crispino, G. A.; Hartung, J.; Jeong, K.-S.; Kwong, H.-L.; Morikawa, K.; Wang, Z.-M.; Xu, D.; Zhang, X.-L. *J. Org. Chem.* **1992**, *57*, 2768–2771.

11. Kolb, H. C.; VanNieuwenhze, M. S.; Sharpless, K. B. *Chem. Rev.* **1994**, *94*, 2483–2547.

12. Srinivasan, K.; Michaud, P.; Kochi, J. K. *J. Am. Chem. Soc.* **1986**, *108*, 2309–2320.

13. Zhang, W.; Loebach, J. L.; Wilson, S. R.; Jacobsen, E. N. *J. Am. Chem. Soc.* **1990**, *112*, 2801–2803.

14. Irie, R.; Noda, K.; Ito, Y.; Matsumoto, N.; Katsuki, T. *Tetrahedron Lett.* **1990**, *31*, 7345–7348.

15. Zhang, W.; Jacobsen, E. N. *J. Org. Chem.* **1991**, *56*, 2296–2298.

16. Hatayama, A.; Hosoya, N.; Irie, R.; Ito, Y.; Katsuki, T. *Synlett* **1992**, 407–409.

17. Jacobsen, E. N.; Zhang, W.; Güler, M. L. *J. Am. Chem. Soc.* **1991**, *113*, 6703–6704.

18. Chang, S.; Heid, R. M.; Jacobsen, E. N. *Tetrahedron Lett.* **1994**, *35*, 669–672.

19. Pospisil, P. J.; Carsten, D. H.; Jacobsen, E. N. *Chem. Eur. J.* **1996**, *2*, 974–980.

20. Jacobsen, E. N.; Deng, L.; Furukawa, Y.; Martínez, L. E. *Tetrahedron* **1994**, *50*, 4323–4334.

21. Samsel, E. G.; Srinivasan, K.; Kochi, J. K. *J. Am. Chem. Soc.* **1985**, *107*, 7606–7617.

22. Irie, R.; Ito, Y.; Katsuki, T. *Synlett* **1991**, 265–266.

23. Sasaki, H.; Irie, R.; Katsuki, T. *Synlett* **1993**, 300–302.

24. Hosoya, N.; Irie, R.; Ito, Y.; Katsuki, T. *Synlett* **1991**, 691–692.

25. Hashihayama, T.; Ito, Y.; Katsuki, T. *Tetrahedron* **1997**, *53*, 9541–9552.

26. (a) Hashihayama, T.; Ito, Y.; Katsuki, T. *Synlett* **1996**, 1079–1081; (b) Miura, K.; Katsuki, T. *Synlett* **1999**, 783–785.

27. Higuchi, T.; Ohtake, H.; Hirobe, M. *Tetrahedron Lett.* **1989**, *30*, 6545–6548.

28. Takeda, T.; Irie, R.; Shinoda, Y.; Katsuki, T. *Synlett* **1999**, 1157–1159.

29. Yamada, T.; Imagawa, K.; Nagata, T.; Mukaiyama, T. *Chem. Lett.* **1992**, 2231–2234.

30. Imagawa, K.; Nagata, T.; Yamada, T.; Mukaiyama, T. *Chem. Lett.* **1994**, 527–530.

31. Yamaguchi, K.; Watanabe, Y.; Morishima, I. *J. Am. Chem. Soc.* **1993**, *115*, 4058–4065.

32. Yamada, T.; Imagawa, K.; Nagata, T.; Mukaiyama, T. *Bull. Chem. Soc. Jpn.* **1994**, *67*, 2248–2256.

33. Nagata, T.; Imagawa, K.; Yamada, T.; Mukaiyama, T. *Bull. Chem. Soc. Jpn.* **1995**, *68*, 1455–1465.

34. Groves, J. T.; Quinn, R. *J. Am. Chem. Soc.* **1985**, *107*, 5790–5792.

35. Lai, T.-S.; Zhang, R.; Cheung, K.-K.; Kwong, H.-L.; Che, C.-M. *Chem. Commun.* **1998**, 1583–1584.

36. Yamada, T.; Takai, T.; Rhode, O.; Mukaiyama, T. *Bull. Chem. Soc. Jpn.* **1991**, *64*, 2109–2117.

37. (a) Stamicarbon N. V., JP Patent Tokkaisho 46–12456 (**1971**); (b) Union Carbide Corp., U. S. Patent 3,025,306 (**1962**); (c) Knapsack A. G., U.S. Patent 3,483,222 (**1962**); (d) Mitsubishi Kasei Co., JP Patent Tokkaisho 47–47896 (**1972**) and Tokkaisho 56–14095 (**1981**).

38. Yamada, T.; Takahashi, K.; Kato, K.; Takai, T.; Inoki, S.; Mukaiyama, T. *Chem. Lett.* **1991**, 641–644.

39. Bolm, C.; Schlingloff, G.; Weickhardt, K. *Angew. Chem. Int. Ed. Engl.* **1994**, *33*, 1848–1849.

40. Bolm, C.; Schlingloff, G. *J. Chem. Soc., Chem. Commun.* **1995**, 1247–1248.

41. Bolm, C.; Luong, T. K. K.; Schlingloff, G. *Synlett* **1997**, 1151–1152.

42. (a) Uchida, T.; Katsuki, T. *Tetrahedron Lett.* **2001**, *42*, 6911–6914; (b) Uchida, T.; Katsuki, T.; Ito, K.; Akashi, S.; Ishii, A.; Kuroda, T. *Helv. Chim. Acta* **2002**, *85*, 3078–3089.

43. Watanabe, A.; Uchida, T.; Ito, K.; Katsuki, T. *Tetrahedron Lett.* **2002**, *43*, 4481–4485.

44. Ito, K.; Ishii, A.; Kuroda, T.; Katsuki, T. *Synlett* **2003**, 643–646.

45. Nishimura, T.; Onoue, T.; Ohe, K.; Uemura, S. *J. Org. Chem.* **1999**, *64*, 6750–6755.

46. (a) Ferreira, E. M.; Stoltz, B. M. *J. Am. Chem. Soc.* **2001**, *123*, 7725–7726; (b) Caspi, D. D.; Ebner, D. C.; Bagdanoff, J. T.; Stoltz, B. M. *Adv. Synth. Catal.* **2004**, *346*, 185–189.

47. (a) Mueller, J. A.; Jensen, D. R.; Sigman, M. S. *J. Am. Chem. Soc.* **2002**, *124*, 8202–8203; (b) Mueller, J. A.; Sigman, M. S. *J. Am. Chem. Soc.* **2003**, *125*, 7005–7013.

48. Masutani, K.; Uchida, T.; Irie, R.; Katsuki, T. *Tetrahedron Lett.* **2000**, *41*, 5119–5123.

49. Shimizu, H.; Nakata, K.; Katsuki, T. *Chem. Lett.* **2002**, 1080–1081.

50. Shimizu, H.; Katsuki, T. *Chem. Lett.* **2003**, *32*, 480–481.

51. Irie, R.; Katsuki, T. *Chem. Rec.* **2004**, *4*, 96–109.

52. Noyori, R.; Tomino, I.; Tanimoto, Y.; Nishizawa, M. *J. Am. Chem. Soc.* **1984**, *106*, 6709–6716.

53. Mikami, K.; Terada, M.; Nakai, T. *J. Am. Chem. Soc.* **1990**, *112*, 3949–3954.

54. Miyashita, A.; Takaya, H.; Souchi, T.; Noyori, R. *Tetrahedron* **1984**, *40*, 1245–1253.

55. Nakajima, M.; Miyoshi, I.; Kanayama, K.; Hashimoto, S.-i.; Noji, M.; Koga, K. *J. Org. Chem.* **1999**, *64*, 2264–2271.

56. Li, X.; Yang, J.; Kozlowski, M. *Org. Lett.* **2001**, *3*, 1137–1140.

57. Irie, R.; Masutani, K.; Katsuki, T. *Synlett*, **2000**, 1433–1436.

58. Hwang, D.-R.; Chen, C.-P.; Uang, B.-J. *Chem. Commun.* **1999**, 1207–1208.

59. Hon, S.-W.; Li, C.-H.; Kuo, J.-H.; Barhate, N. B.; Liu, Y.-H.; Wang, Y.; Chen, C.-T. *Org. Lett.* **2001**, *3*, 869–872.

60. Barhate, N. B.; Chen, C.-T. *Org. Lett.* **2002**, *4*, 2529–2532.

61. Chu, C.-Y.; Uang, B.-J. *Tetrahedron Asym.*, **2003**, *14*, 53–55.

62. Luo, Z.; Liu, Q.; Gong, L.; Cui, X.; Mi, A.; Jiang, Y. *Angew. Chem. Int. Ed.* **2002**, *41*, 4532–4535.

63. Luo, Z.; Liu, Q.; Gong, L.; Cui, X.; Mi, A.; Jiang, Y. *Chem. Commun.* **2002**, 914–915.

64. Somei, H.; Asano, Y.; Yoshida, T.; Takizawa, S.; Yamataka, H.; Sasai, H. *Tetrahedron Lett.* **2004**, *45*, 1841–1844.

3

LIGAND DESIGN FOR C—C BOND FORMATION

Ryo Shintani and Tamio Hayashi

Department of Chemistry, Graduate School of Science, Kyoto University, Sakyo, Kyoto, Japan

3.1 INTRODUCTION

Carbon–carbon bond-forming reactions are one of the most basic, but important, transformations in organic chemistry. In addition to conventional organic reactions, the use of transition metal-catalyzed reactions to construct new carbon–carbon bonds has also been a topic of great interest. Such transformations to create chiral molecules enantioselectively is therefore very valuable. While various carbon–carbon bond-forming asymmetric catalyses have been described in the literature, this chapter focuses mainly on the asymmetric 1,4-addition reactions under copper or rhodium catalysis and on the asymmetric cross-coupling reactions catalyzed by nickel or palladium complexes.

3.2 1,4-ADDITION AND RELATED REACTIONS

1,4-Addition of organometallic reagents to α,β-unsaturated compounds is a useful strategy for the construction of carbon–carbon bonds.[1] The development of chiral catalysts that can achieve this transformation enantioselectively has therefore been investigated extensively.[2] Although various transition metals have been utilized to effect asymmetric 1,4-additions, copper- and rhodium-catalyzed reactions have been most extensively studied and are worth mentioning in detail. In general,

New Frontiers in Asymmetric Catalysis, Edited by Koichi Mikami and Mark Lautens
Copyright © 2007 John Wiley & Sons, Inc.

copper catalysis[3] is used for the introduction of alkyl groups in combination with alkylzinc reagents as the nucleophile, whereas rhodium catalysis[4] provides an introduction of alkenyl or aryl groups by the use of various organometallic reagents such as organoboronic acids. This section deals with recent progress of the copper- and rhodium-catalyzed asymmetric 1,4-addition and related reactions.

3.2.1 Copper Catalysis

Cyclic Enones as Substrates Copper-catalyzed 1,4-addition of organometallic reagents to α,β-unsaturated compounds was first described by Kharasch in 1941, using isophorone (3,5,5-trimethyl-2-cyclohexen-1-one) and methylmagnesium bromide in the presence of 1 mol% CuCl,[5] and the first asymmetric variant was reported by Lippard in 1988.[6] Since then, early efforts were mostly devoted to achieving high catalytic activity and enantioselectivity. The use of cyclic enones (e.g., 2-cyclohexen-1-one) was focused mainly in combination with organozinc reagents (e.g., diethylzinc). The breakthrough came when Feringa utilized a chiral phosphoramidite ligand in 1997 [Eq. (3.1)],[7] and a variety of effective ligands have been developed on the basis of this framework.[8–10] In addition to these phosphoramidite-based ligands, several other chiral ligands have also been found highly effective for asymmetric addition of diethylzinc to 2-cyclohexen-1-one, and ligands that provide 3-ethylcyclohexanone in $\geq 95\%$ ee are shown in Figure 3.1.[11–14]

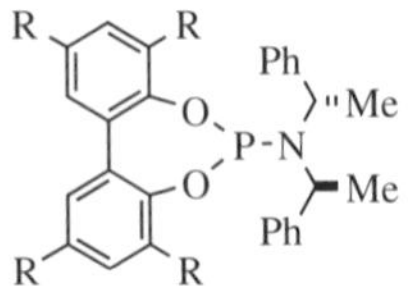

2: R = H; 96% ee [8]
3: R = Me; 99% ee [8]

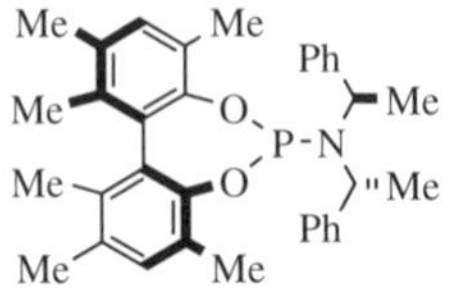

4: 97–99% ee [9]

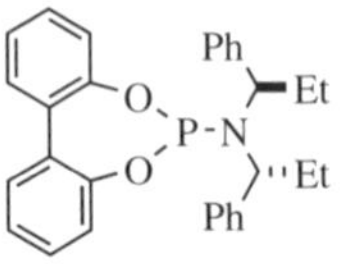

5: 99% ee [8]

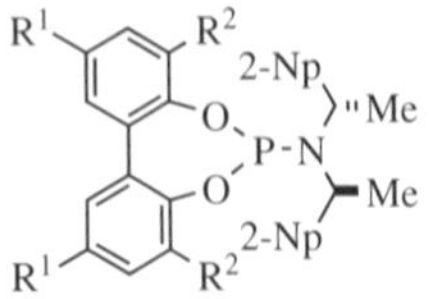

11: 98% ee [11]

12: >97% ee [12]

6: R^1 = R^2 = Me; >99% ee [10]
7: R^1 = R^2 = Cl; 95% ee [10]
8: R^1 = Me, R^2 = OMe; 98% ee [10]
9: R^1 = H, R^2 = allyl; 99% ee [10]
10: R^1 = H, R^2 =methallyl; 98% ee [10]

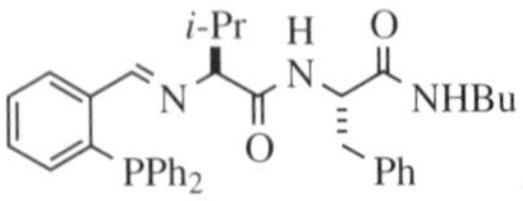

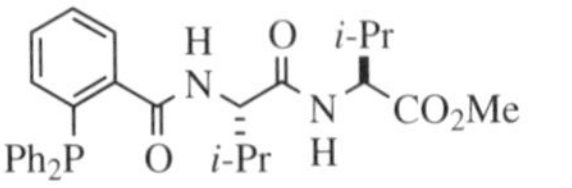

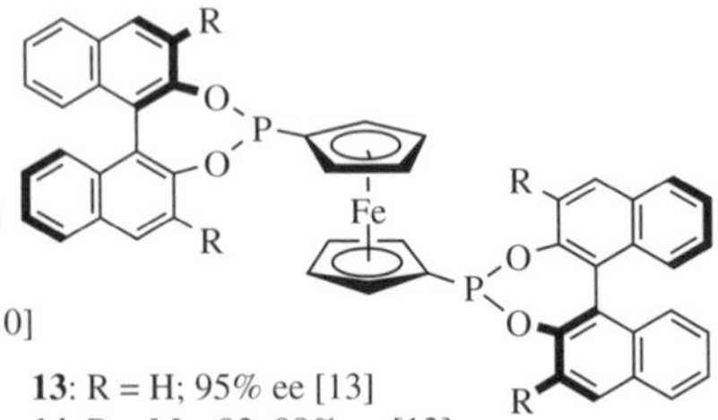

13: R = H; 95% ee [13]
14: R = Me; 93–99% ee [13]

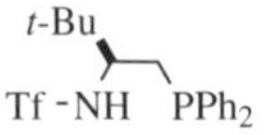

15: 97% ee [14]

Figure 3.1. Chiral ligands that exhibit $\geq 95\%$ ee in the reaction between 2-cyclohexen-1-one and diethylzinc under copper catalysis.

$$(3.1)$$

For the copper-catalyzed 1,4-addition to 2-cyclohexen-1-one, other alkylmetal reagents have also been employed, achieving high enantioselectivity in some cases [Eqs. (3.2)–(3.4)].[15–17] Recently, one example appeared that utilized diphenylzinc as the nucleophile in the presence of phosphoramidite ligand **1** to produce highly enantio-enriched 3-phenylcyclohexanone [Eq. (3.5); 94% ee].[18]

$$(3.2)$$

$$(3.3)$$

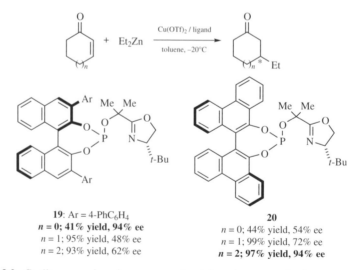

(3.4)

(3.5)

In constrast to many successful results with 2-cyclohexen-1-one, considerably fewer ligands to date are effective for other cyclic enones with various ring sizes to induce high enantioselectivity. It is often the case that the ligand structure has to be finely tuned depending on the substrate structure (e.g., Figure 3.2).[19] Among various chiral ligands, ligand **1** has relatively broad scope in the copper-catalyzed

19: Ar = 4-PhC$_6$H$_4$
$n = 0$; **41% yield, 94% ee**
$n = 1$; 95% yield, 48% ee
$n = 2$; 93% yield, 62% ee

20
$n = 0$; 44% yield, 54% ee
$n = 1$; 99% yield, 72% ee
$n = 2$; **97% yield, 94% ee**

Figure 3.2. Cyclic enone ring-size versus optimal ligand structure in the copper-catalyzed asymmetric 1,4-addition of diethylzinc.

entry	n	R	yield (%)	ee (%)
1	0	Et	75	10
2	1	Et	94	>98
3	2	Et	95	>98
4	3	Et	95	97
5	1	n-Hept	95	95
6	1	i-Pr	95	94

1

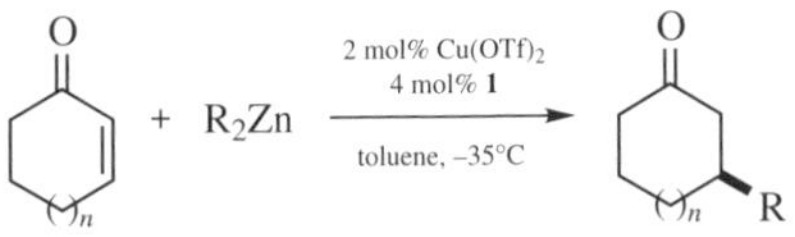

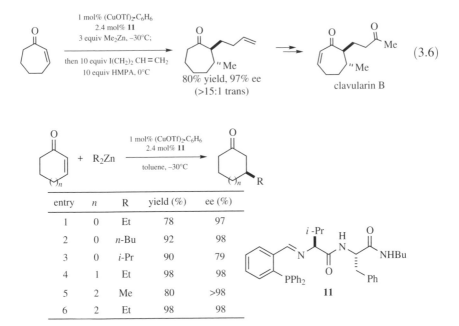

Figure 3.3. Scope of Cu/**1**-catalyzed asymmetric 1,4-addition of diorganozinc reagents to cyclic enones.

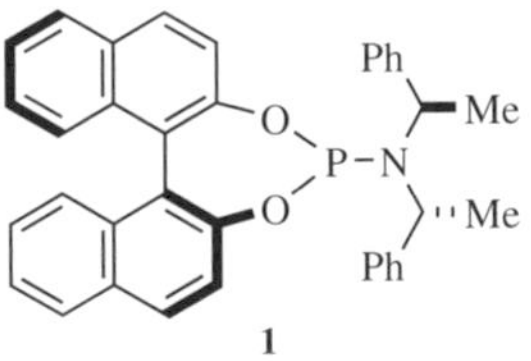

asymmetric 1,4-addition of diorganozinc reagents to various cyclic enones except for 2-cyclopenten-1-one (Figure 3.3).[3a] Furthermore, easily tunable, dipeptide-based ligand **11**, developed by Hoveyda, can provide high stereoselection to a wide range of cyclic substrate/nucleophile combination (Figure 3.4).[11] This process in the presence of ligand **11** was successfully applied to the enantioselective synthesis of anticancer agent clavularin B by quenching the 1,4-addition reaction with 4-iodo-1-butene, and the subsequent three-step transformations [Eq. (3.6)].

(3.6)

entry	n	R	yield (%)	ee (%)
1	0	Et	78	97
2	0	n-Bu	92	98
3	0	i-Pr	90	79
4	1	Et	98	98
5	2	Me	80	>98
6	2	Et	98	98

11

Figure 3.4. Scope of Cu/**11**-catalyzed asymmetric 1,4-addition of diorganozinc reagents to cyclic enones.

Acyclic Enones as Substrates Compared to cyclic enones, acyclic enones are generally more challenging substrates for the copper-catalyzed asymmetric 1,4-addition reactions. Several ligands have been reported that can achieve high ee when β-aryl acyclic enones are used as substrates in the 1,4-addition of diethylzinc (Figure 3.5).[20–23]

However, very few catalyst systems reported to date are highly effective for both β- aryl and β-alkyl acyclic enone substrates in the copper-catalyzed asymmetric 1,4-addition. Ligand **27**, developed by Hoveyda, shows high enantioselectivity in the 1,4-addition of dialkylzinc reagents to various acyclic enones (Figure 3.6).[24]

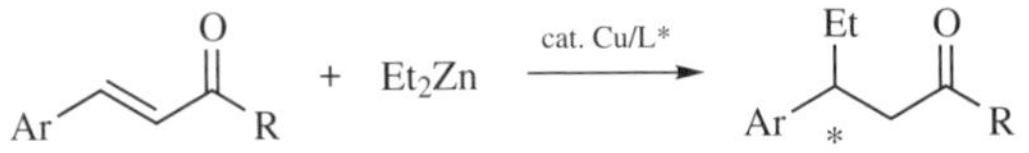

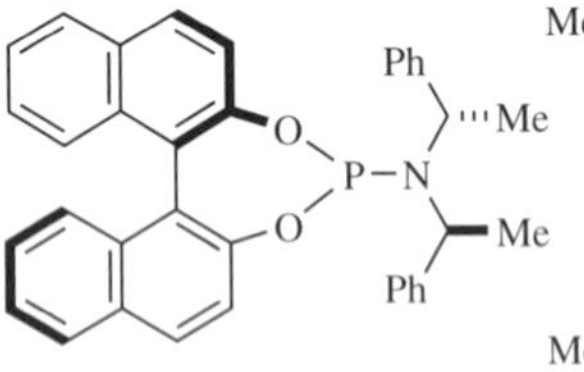

ent-**1** [8]
93% ee
(1 example)

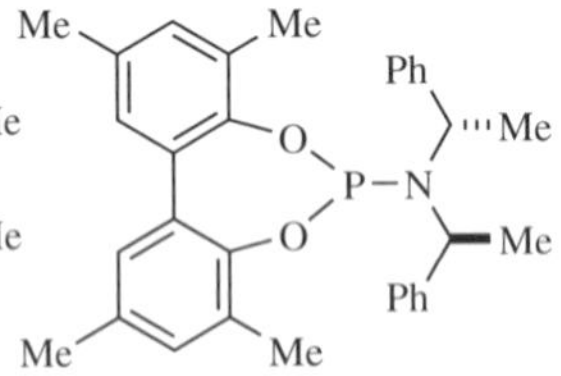

3 [8]
93% ee
(1 example)

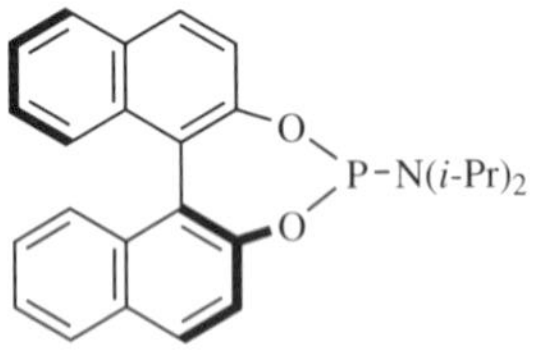

21 [3a]
90% ee
(1 example)

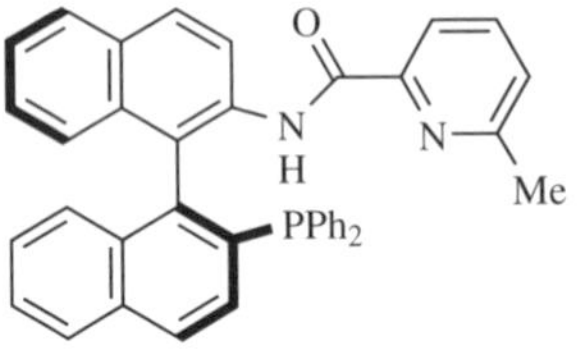

22 [20]
90–98% ee
(6 examples)

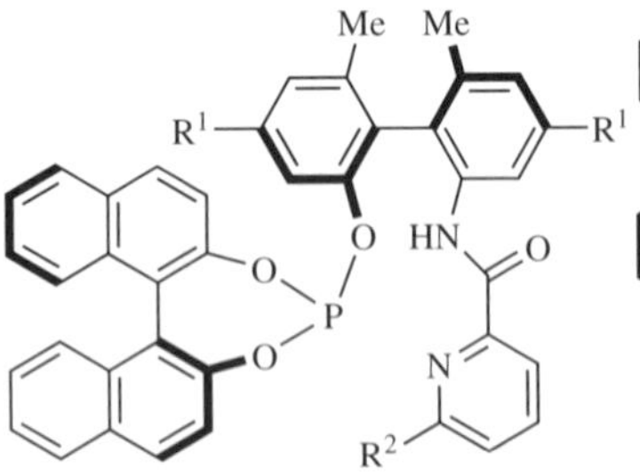

23: R¹, R² = H or Me [21]
90–98% ee
(11 examples)

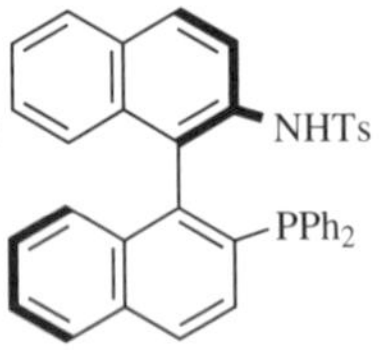

24 [22]
95–98% ee
(5 examples)

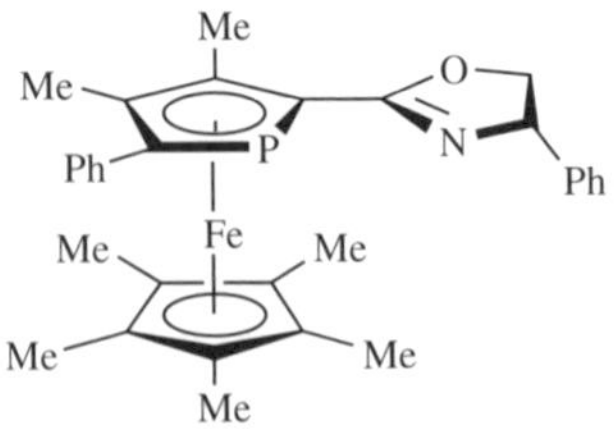

25 [23]
90% ee
(1 example)

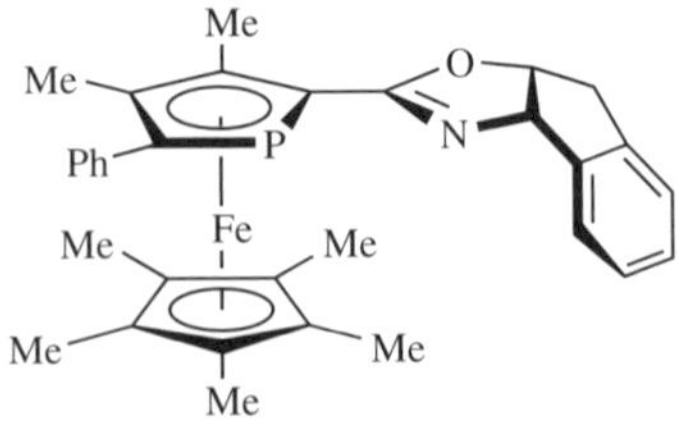

26 [23]
91% ee
(1 example)

Figure 3.5. Chiral ligands that exhibit ≥90% ee in the reaction between β-aryl acyclic enones and diethylzinc under copper catalysis.

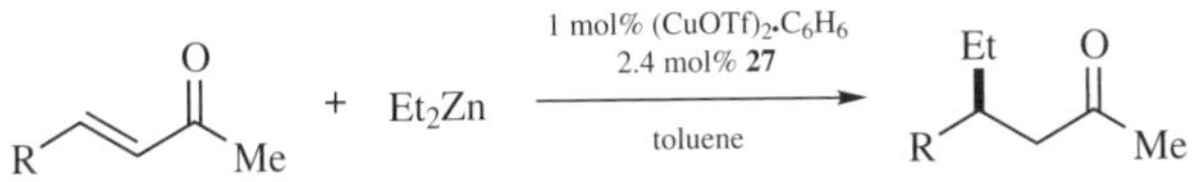

entry	R	temp (°C)	yield (%)	ee (%)
1	Ph	−20	90	93
2	4-MeOC$_6$H$_4$	−20	93	94
3	4-CF$_3$C$_6$H$_4$	−20	87	90
4	n-Pent	22	85	95
5^a	n-Pent	22	71	94
6	i-Pr	22	69	91

^{a} Me$_2$Zn was used as a nucleophile.

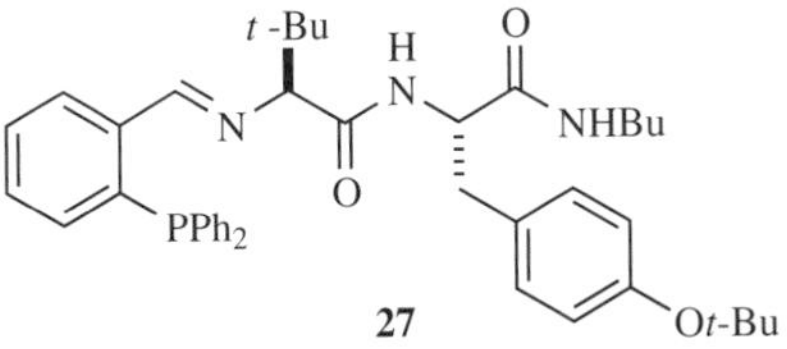

Figure 3.6. Scope of Cu/**27**-catalyzed asymmetric 1,4-addition of diorganozinc reagents to acyclic enones.

The utility of this system is further enhanced by coupling with a ruthenium-catalyzed olefin cross-metathesis. Thus, as shown in Eq. (3.7), a β-alkyl acyclic enone can be prepared by the metathesis reaction using catalyst **28**, and the subsequent Cu/**27**-catalyzed 1,4-addition of diethylzinc furnishes the cyclohexyl ketone product in good yield with high diastereo- and enantioselectivity. Hoveyda also reported that ligand **29** is suitable for the copper-catalyzed asymmetric 1,4-addition to trisubstituted cyclic enones (Figure 3.7).[25] When Grignard reagents are used as the nucleophilic component in combination with acyclic enones, ligand **30** is the ligand of choice to furnish the 1,4-adducts in high ee, as recently described by Feringa (Figure 3.8).[26]

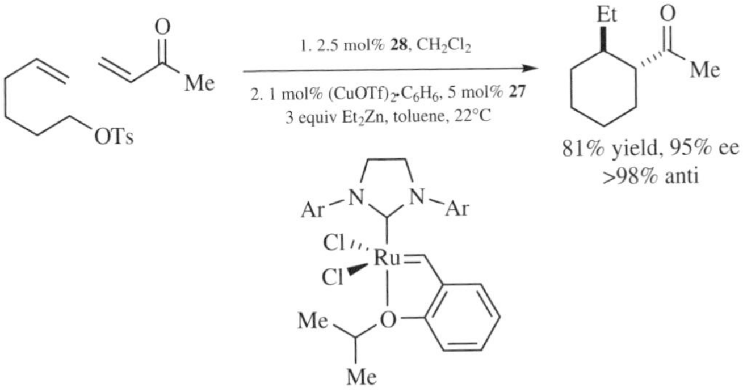

$$(3.7)$$

$$\text{substrate} + R_2Zn \xrightarrow[\text{toluene, 0°C}]{\begin{array}{c}\text{1–5 mol\% (CuOTf)}_2\cdot\text{C}_6\text{H}_6\\ \text{2.4–12 mol\% 29}\end{array}} \text{product}$$

entry	substrate	R	product	yield (%)	anti:syn	ee (%)
1	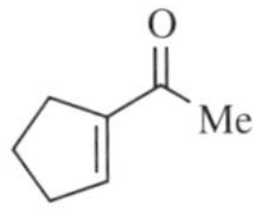	Me	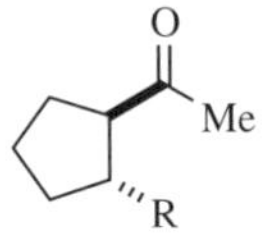	42	16:1	97
2		Et		66	>25:1	97
3		i-Pr		69	>25:1	98
4	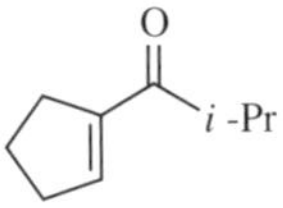	Et	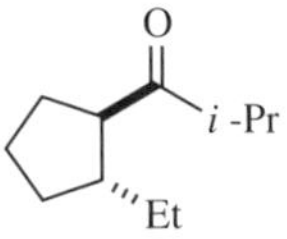	77	16:1	96
5	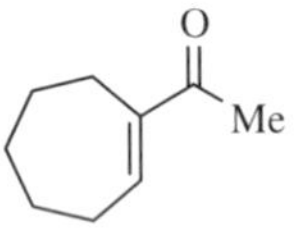	Et	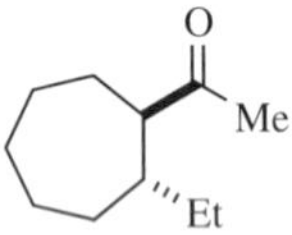	85	4:1	97

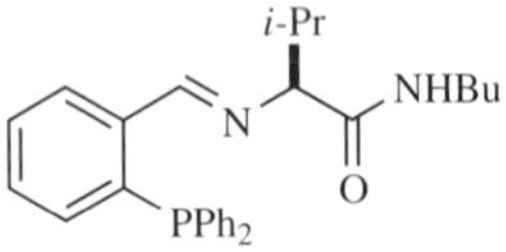

29

Figure 3.7. Scope of Cu/**29**-catalyzed asymmetric 1,4-addition of diorganozinc reagents to trisubstituted cyclic enones.

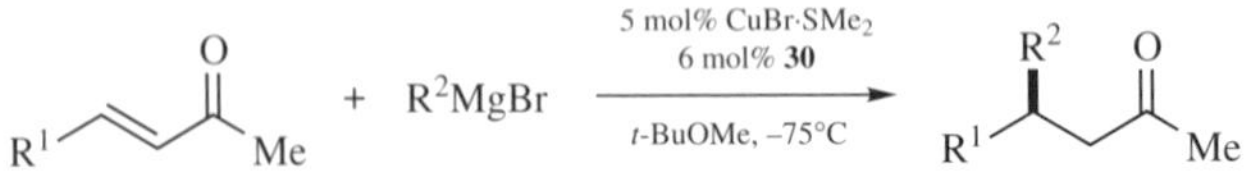

$$R^1\diagdown\diagup\overset{O}{\underset{}{\diagdown}}Me + R^2MgBr \xrightarrow[\text{t-BuOMe, –75°C}]{\begin{array}{c}\text{5 mol\% CuBr·SMe}_2\\ \text{6 mol\% 30}\end{array}} R^1\overset{R^2\ \ O}{\diagdown\diagup\diagdown}Me$$

entry	R^1	R^2	yield (%)	ee (%)
1	Me	n-Bu	78	93
2	n-Pr	n-Bu	91	95
3	n-Pr	Ph	75	76
4	n-Bu	Me	86	98
5	n-Bu	n-Pr	88	91
6	i-Pr	Me	52	94
7	Ph	Me	73	97

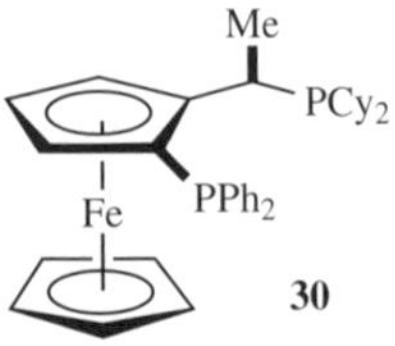

30

Figure 3.8. Scope of Cu/**30**-catalyzed asymmetric 1,4-addition of Grignard reagents to acyclic enones.

Other α,β-Unsaturated Substrates Copper-catalyzed asymmetric 1,4-addition reactions have been applied to other α,β-unsaturated compounds as well. Five- and six-membered lactones can be ethylated in high ee by the use of Cu/**31** (Figure 3.9).[27] This catalyst system is also effective for cyclic enones as shown in entries 3–5. For α,β-unsaturated lactam substrates, ligand ent-**1** can induce high enantios- electivity in the addition of diorganozinc reagents [Eq. (3.8)].[28] Furthermore, by a parallel screening of libraries of ligands, Hoveyda found that ligand **32** is highly effective for the 1,4-addition of diorganozinc reagents to *N*-alkenoyloxazolidinone substrates (Figure 3.10).[29]

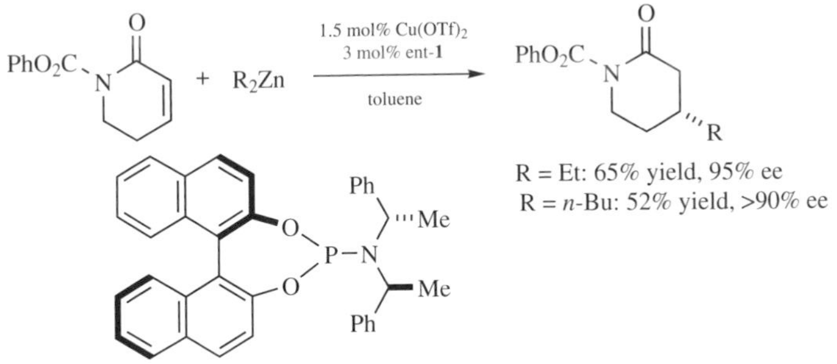

R = Et: 65% yield, 95% ee
R = *n*-Bu: 52% yield, >90% ee   (3.8)

ent-**1**

entry	substrate		temp (°C)	conv. (%)	ee (%)
1	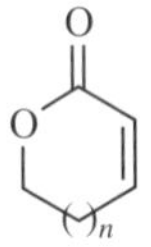	n = 0	−30	70	87
2		n = 1	−50	95	94
3	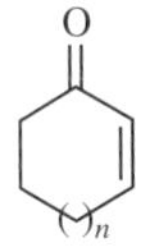	n = 0	−30	98	93
4		n = 1	0	>99	95
5		n = 2	−30	>99	88

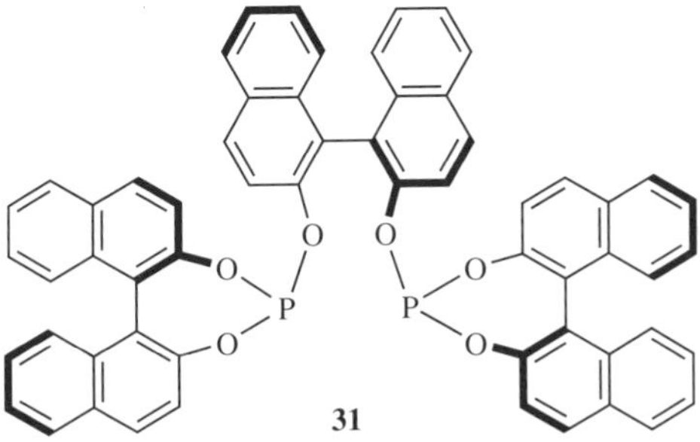

31

Figure 3.9. Scope of Cu/**31**-catalyzed asymmetric 1,4-addition of dietheylzinc.

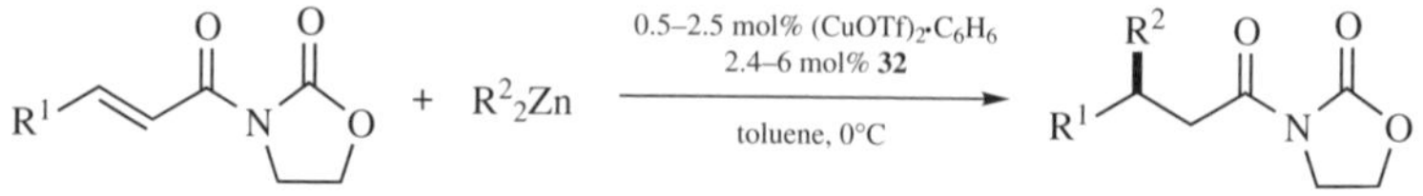

entry	R^1	R^2	yield (%)	ee (%)
1	Me	Et	95	95
2	Me	*i*-Pr	95	76
3	*n*-Pr	Et	86	94
4	$(CH_2)_3OTBS$	Et	95	>98
5	*i*-Pr	Et	88	92

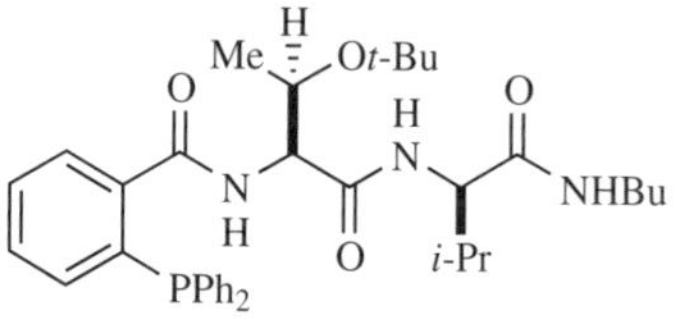

32

Figure 3.10. Scope of Cu/**32**-catalyzed asymmetric 1,4-addition of diorganozinc reagents to *N*-alkenoyloxazolidinones.

Nitroalkenes have also been employed in the copper-catalyzed asymmetric 1,4-addition reactions, and phosphoramidite ligand **1** exhibits high selectivity in several substrates of this class with diethylzinc as the nucleophile (Figure 3.11).[30–32] In the 1,4-addition to nitrocyclohexene, the use of ligand **33** leads to cis-enriched 1,4-adducts in high ee (Figure 3.12).[33] As shown in Eq. (3.9), on changing the workup conditions, a ketone product can be obtained through the Nef reaction.

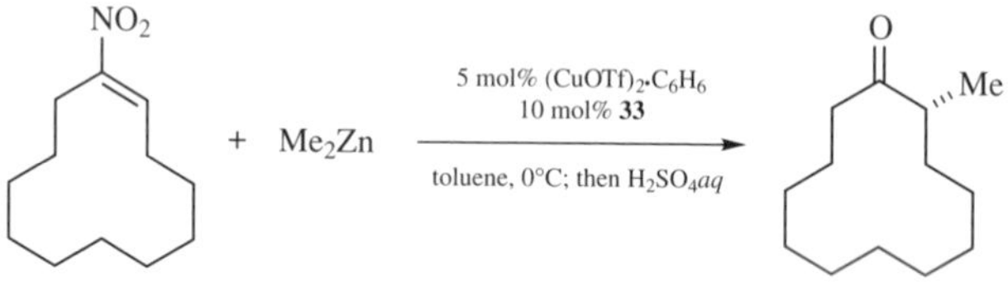

86% yield
96% ee

(3.9)

Ligand **4**, which is highly effective for the 1,4-addition of diethylzinc to 2-cyclohexen-1-one, shows relatively wide scope with respect to the β-substituent in the copper-catalyzed asymmetric 1,4-addition of diethylzinc to acyclic nitroalkenes

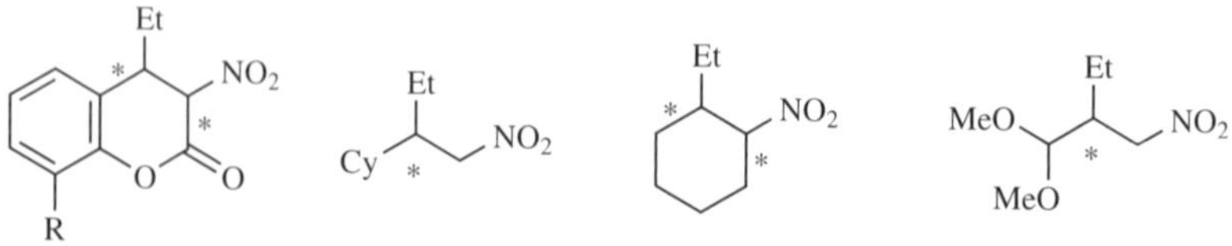

(dr = 10 : 1) [30]
R = H: 92% ee
R = OMe: 90% ee

94% ee [31]

95% ee [8]
90% ee [32]
(cis:trans = 83 : 17)

90% ee [32]

Figure 3.11. Products obtained in ≥90% ee in the copper-catalyzed asymmetric 1,4-addition of diethylzinc to nitroalkenes by the use of ligand **1**.

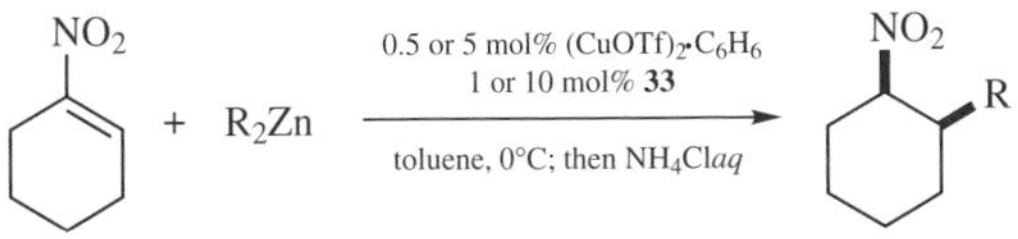

Figure 3.12. Scope of Cu/**33**-catalyzed asymmetric 1,4-addition of diorganozinc reagents to nitrocyclohexene.

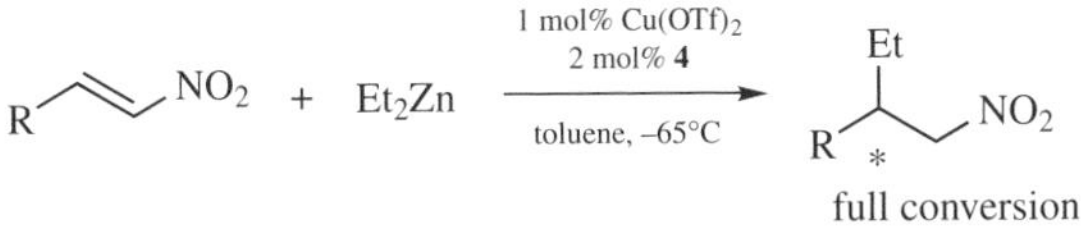

entry	R	yield (%)	syn:anti	ee (%)
1	Me	72	83 : 17	95
2	(CH$_2$)$_3$CHMe$_2$	89	83 : 17	93
3	(CH$_2$)$_4$OAc	76	81 : 19	95

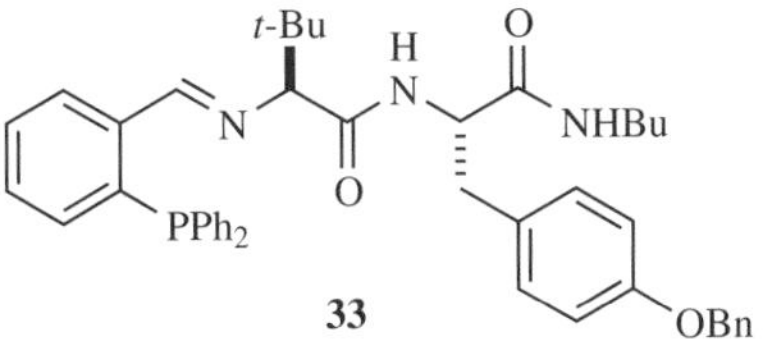

Figure 3.13. Scope of Cu/**4**-catalyzed asymmetric 1,4-addition of diethylzinc to acyclic nitroalkenes.

entry	R	ee (%)
1	Ph	94
2	4-MeOC$_6$H$_4$	99
3	4-FC$_6$H$_4$	91
4	2-CF$_3$C$_6$H$_4$	88
5	thienyl	96
6	CH(OMe)$_2$	96

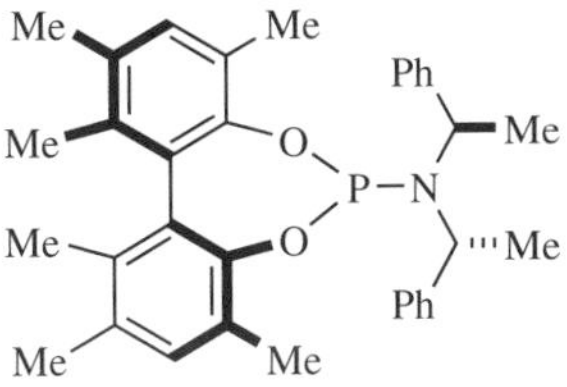

(Figure 3.13).[34] In addition, ligand **34** can achieve high enantioselectivity to a wide range of nitroalkene/diorganozinc combination (Figure 3.14).[35]

3.2.2   Rhodium Catalysis

Addition of Organometallic Reagents to Enones in Aqueous Media   Rhodium-catalyzed 1,4-addition of organometallic reagents to α,β-unsaturated compounds was first developed by Miyaura in 1997.[36] Thus, Rh(acac)(CO)$_2$/dppb was found to catalyze the 1,4-addition of aryl- and alkenylboronic acids to several α,β-unsaturated ketones in water-containing solvents at 50°C. The reaction conditions were successfully modified for the development of an asymmetric variant of this process by Hayashi and Miyaura in 1998.[37] The important points of modification are (1) the use of Rh(acac)(C$_2$H$_4$)$_2$/(S)-binap as a catalyst and

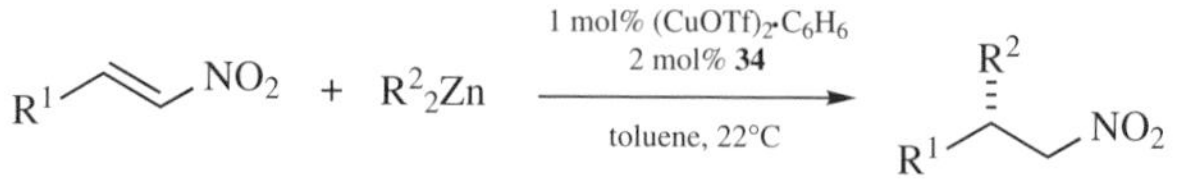

entry	R^1	R^2	yield (%)	ee (%)
1	Ph	Et	79	95
2	Ph	Me	78	92
3	Ph	$(CH_2)_4OAc$	60	89
4	4-MeOC_6H_4	Et	72	95
5	4-ClC_6H_4	Et	84	93
6	2-MeC_6H_4	Et	65	94
7	2-furyl	Et	78	95
8	n-Hept	Et	72	93
9	n-Hept	Me	68	85
10	Cy	Et	52	95

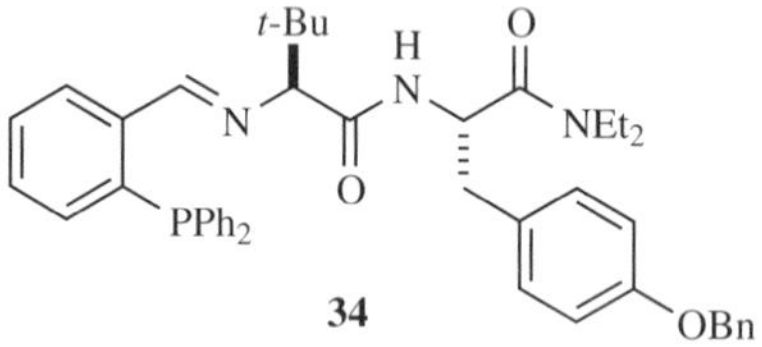

Figure 3.14. Scope of Cu/**34**-catalyzed asymmetric 1,4-addition of diorganozinc reagents to acyclic nitroalkenes.

(2) the use of 1,4-dioxane/H_2O (10/1) as the reaction medium at 100°C. Under these conditions, the scope is rather broad with respect to both the enone and the organoboronic acid, furnishing uniformly high enantioselectivity (Figure 3.15).

The mechanistic investigations of this process revealed that the catalytic cycle involves (1) transmetallation of an aryl or alkenyl group from boron to rhodium; (2) insertion of an enone into the carbon–rhodium bond, generating an oxa-π-allyl-rhodium intermediate; and (3) hydrolysis of this intermediate by water to release the 1,4-adduct along with a hydroxorhodium species (Figure 3.16).[38] During the course of this study, it was found that each step (transmetallation, insertion, and hydrolysis) occurs smoothly at 25°C, but the catalytic reactions with Rh(acac)(C_2H_4)$_2$/(S)-binap require much higher temperatures ($\geq$60°C, typically 100°C). This was ascribed to the presence of acetylacetonato (acac) ligand, which decelerates the step of transmetallation. The catalytic activity was, therefore, improved by the use of acac-free [Rh(OH)((S)-binap)]$_2$ as a catalyst under milder conditions, leading to higher enantioselectivity (Figure 3.17).

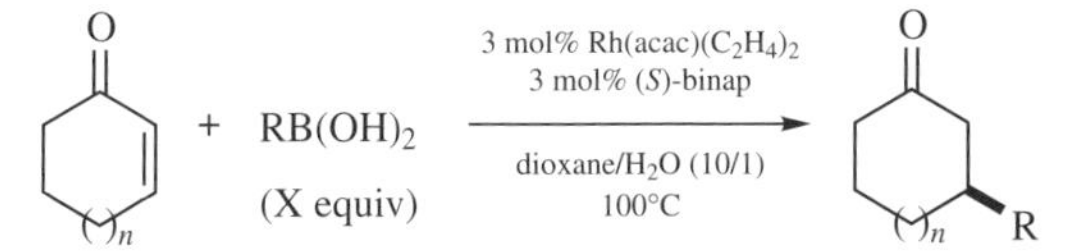

entry	n	R (X)	yield (%)	ee (%)
1	1	Ph (2.5)	93	97
2	1	4-MeC$_6$H$_4$ (5.0)	>99	97
3[a]	1	4-CF$_3$C$_6$H$_4$ (2.5)	70	99
4	1	3-ClC$_6$H$_4$ (5.0)	94	96
5	1	*E*-1-heptenyl (2.5)	88	94
6	0	Ph (1.4)	93	97
7	2	Ph (1.4)	51	93

[a] Reaction in 1-propanol/H$_2$O (10/1).

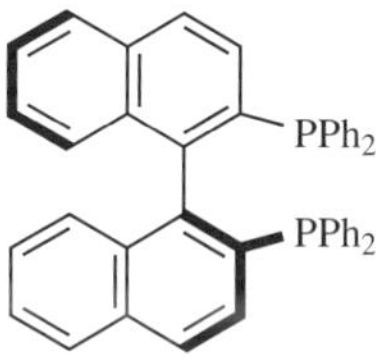

(*S*)-binap

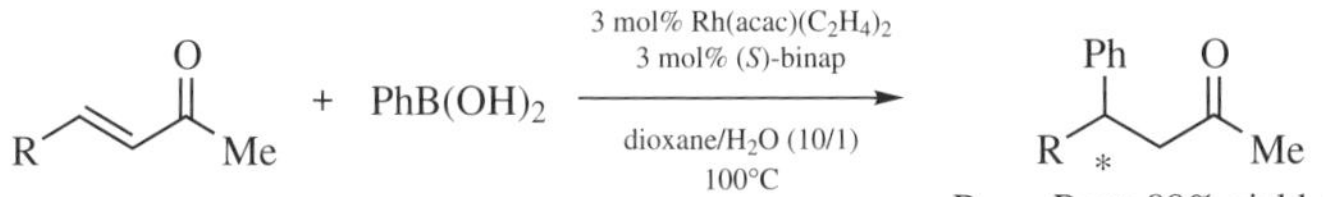

R = *n*-Pent: 88% yield, 92% ee
R = *i*-Pr: 82% yield, 97% ee

Figure 3.15. Scope of Rh/(*S*)-binap-catalyzed asymmetric 1,4-addition of organoboronic acids to α,β-enones.

With regard to the origin of stereoselectivity in these reactions catalyzed by Rh/(*S*)-binap, the stereo-determining step is the insertion of an enone to the carbon–rhodium species coordinated with (*S*)-binap. As shown in Eq. (3.10), for example, 2-cyclohexen-1-one should approach from its 2*si* face to avoid the steric hindrance of the phenyl groups of the ligand, leading to the 1,4-adduct in (*S*) configuration, which is consistent with the observed stereochemical outcome:[37]

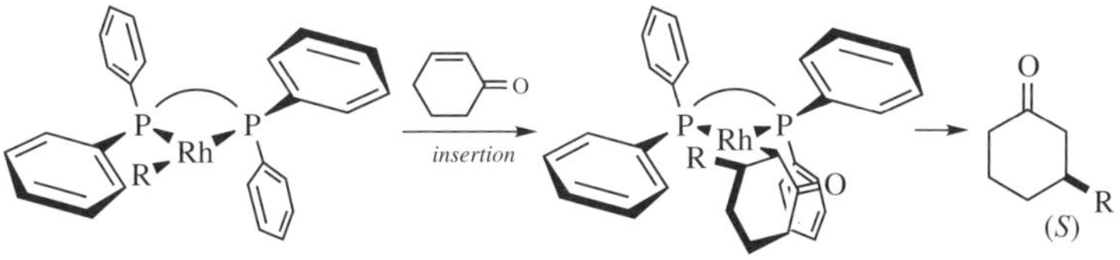

(3.10)

Following the initial work by Hayashi and Miyaura using (*S*)-binap, several other chiral ligands were reported to achieve high enantioselectivity in the rhodium-catalyzed asymmetric 1,4-addition of arylboronic acids to α,β-enones (Figure 3.18).[39–45]

A polymer-supported (*S*)-binap analog **42** was also synthesized and it was successfully utilized in the rhodium-catalyzed asymmetric 1,4-addition reactions in water (Figure 3.19).[46] The stereoselectivities observed in this system are comparable to those obtained in the unsupported Rh/(*S*)-binap system.[37] It was also

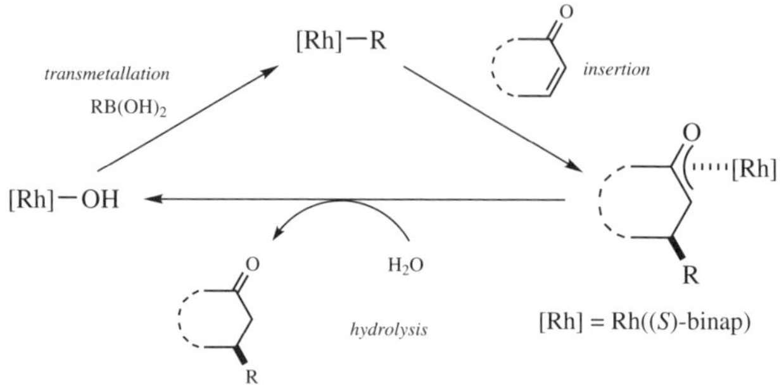

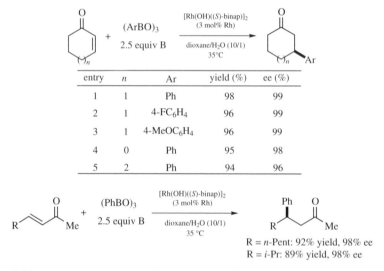

Figure 3.16. Catalytic cycle of Rh/(S)-binap-catalyzed asymmetric 1,4-addition of organoboronic acids to α,β-enones.

entry	n	Ar	yield (%)	ee (%)
1	1	Ph	98	99
2	1	4-FC$_6$H$_4$	96	99
3	1	4-MeOC$_6$H$_4$	96	99
4	0	Ph	95	98
5	2	Ph	94	96

R = n-Pent: 92% yield, 98% ee
R = i-Pr: 89% yield, 98% ee

Figure 3.17. Scope of [Rh(OH)((S)-binap)]$_2$-catalyzed asymmetric 1,4-addition of arylboroxines to α,β-enones.

demonstrated that the catalyst can be recycled several times without any loss of reactivity and enantioselectivity.

By tethering α,β-enones with another carbonyl-containing moiety, Krische developed rhodium-catalyzed asymmetric 1,4-addition/aldol cyclization reactions.[47] These reactions proceed with high diastereo- and enantioselectivity, furnishing structurally complex cyclic compounds in a single step [Eqs. (3.11–3.13)].

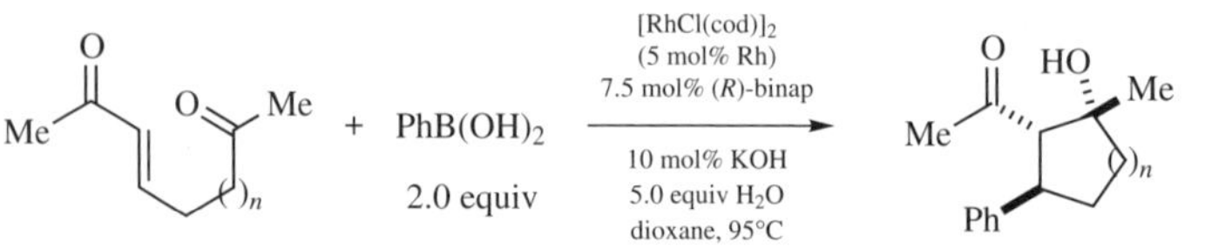

$$(3.11)$$

n = 1 : 88% yield, 94% ee
n = 2 : 69% yield, 95% ee

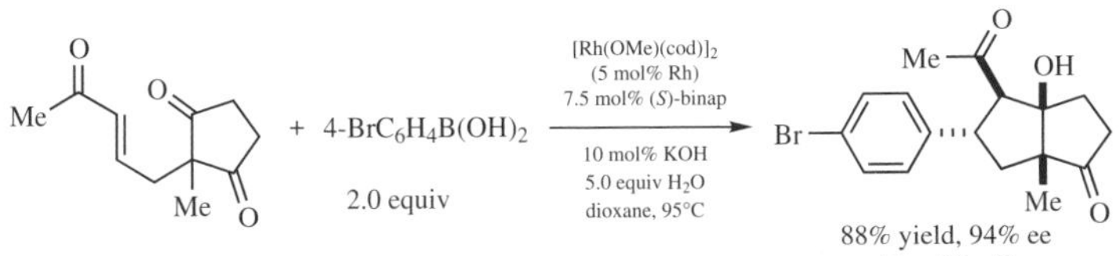

$$ (3.12) $$

88% yield, 94% ee
(dr > 99 : 1)

$$ (3.13) $$

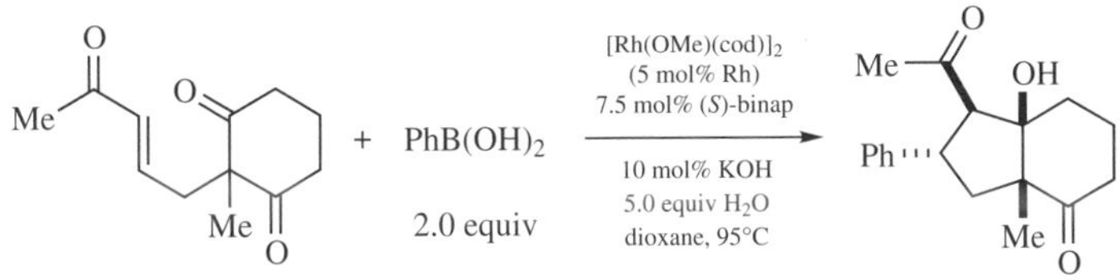

87% yield, 91% ee
(dr > 99 : 1)

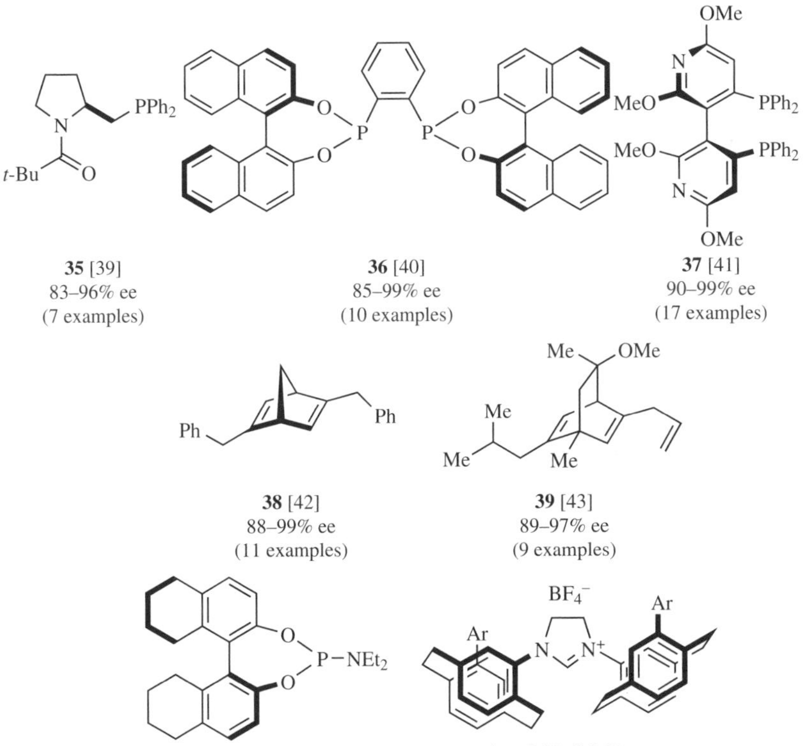

35 [39]
83–96% ee
(7 examples)

36 [40]
85–99% ee
(10 examples)

37 [41]
90–99% ee
(17 examples)

38 [42]
88–99% ee
(11 examples)

39 [43]
89–97% ee
(9 examples)

40 [44]
85–>98% ee
(7 examples)

41 [45]
78–98% ee
(14 examples)

Ar = 2-MeOC$_6$H$_4$

Figure 3.18. Chiral ligands that exhibit high enantioselectivity in the rhodium-catalyzed 1,4-addition of arylboronic acids to α,β-enones.

$$\text{substrate} + \text{PhB(OH)}_2 \xrightarrow[\text{H}_2\text{O, 100°C}]{\substack{\text{3 mol\% Rh(acac)(C}_2\text{H}_4)_2 \\ \text{4.5 mol\% } \mathbf{42}}} \text{product}$$

entry	substrate		yield (%)	ee (%)
1	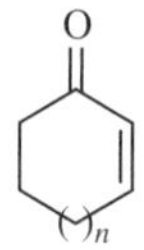	n = 0	95	94
2		n = 1	83	97
3		n = 2	86	95
4	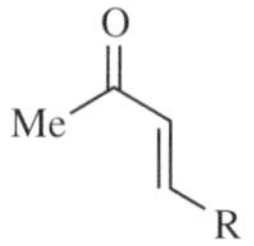	R = n-Pent	80	91
5		R = i-Pr	71	96

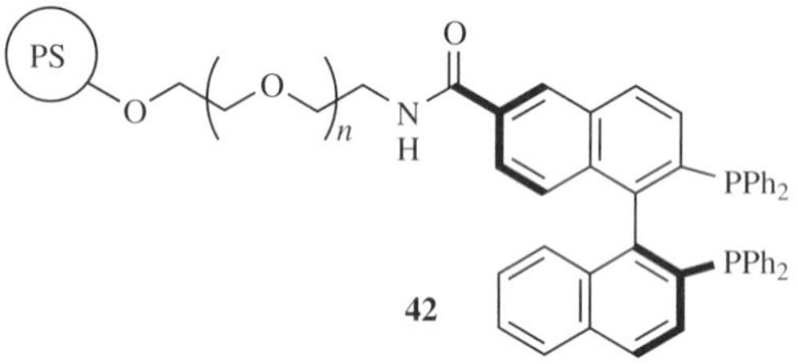

Figure 3.19. Scope of Rh/**42**-catalyzed asymmetric 1,4-addition of phenylboronic acid to α,β-enones.

In addition to organoboronic acids and organoboroxines, other organoboron nucleophiles have been successfully used in the rhodium-catalyzed asymmetric 1,4-addition reactions to α,β-enones. For example, an asymmetric alkenylation of α,β-enones is efficiently achieved by the use of alkenylboronic acid catechol esters, which can be easily prepared by a hydroboration of alkynes with catecholborane (Figure 3.20).[48] The reaction conditions are similar to those for the addition of organoboronic acids,[37] except that the addition of Et$_3$N to the reaction mixture is essential to gain the reactivity with these nucleophiles. Furthermore, LiArB(OMe)$_3$, generated in situ from ArBr + n-BuLi and B(OMe)$_3$, can also be employed as the nucleophile, leading to the 1,4-adducts in high yield and ee even at 0.1 mol% catalyst loading (Figure 3.21).[49] As another type of organoboron nucleophiles, Genet introduced the use of potassium organotrifluoroborates, which are more stable and easier to prepare (from the corresponding organolithium reagents) than organoboronic acids (Figure 3.22).[50] With these organoboron reagents, the 1,4-adducts can be obtained in comparably high yield and ee. It is worth noting that unsubstituted vinyl group can also be installed under these conditions (entry 6), unlike the use of vinylboronic acid, which is known to be unstable in the rhodium-catalyzed 1,4-addition reaction. Potassium aryltrifluoroborates have been employed under Rh/**41** catalysis as well, achieving high enantioselectivity.[45]

Not only organoboron reagents but also organosilicon reagents are suitable nucleophiles in the rhodium-catalyzed asymmetric 1,4-additions to α,β-enones in

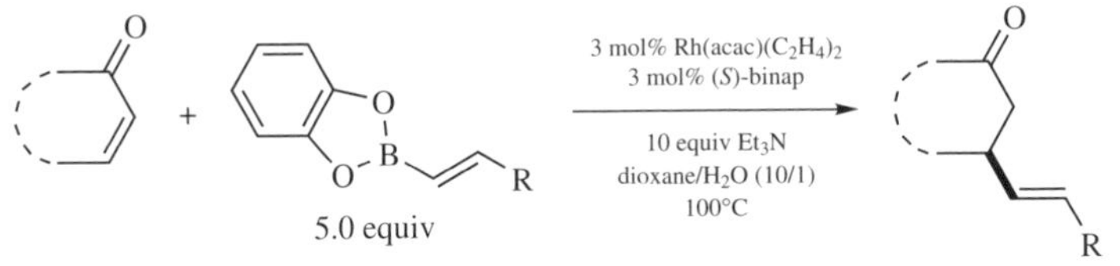

entry	substrate	R	yield (%)	ee (%)
1	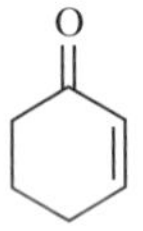	n-Pent	92	96
2		t-Bu	77	91
3		Ph	76	92
4	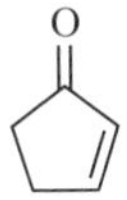	n-Pent	83	98
5	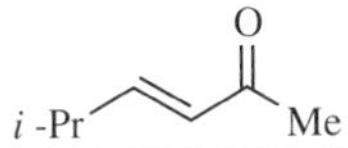	n-Pent	68	81

Figure 3.20. Scope of Rh/(S)-binap-catalyzed asymmetric 1,4-addition of alkenylboronic acid catechol esters to α,β-enones.

water-containing reaction media. Thus, the Rh/(S)-binap system can effectively catalyze the addition of organosiloxanes in high enantioselectivity as well (Figure 3.23).[51] Subsequently, it was also demonstrated that Rh/(S)-binap catalyzes both E-selective hydrosilylation of terminal alkynes and asymmetric 1,4-addition of

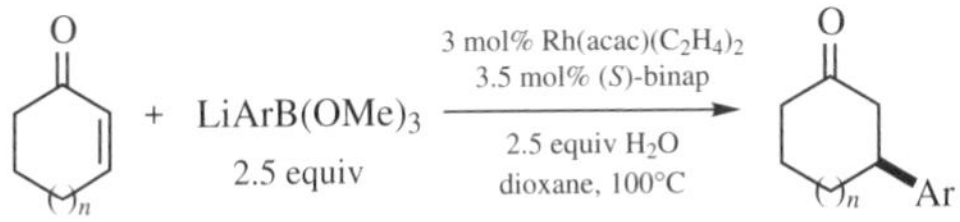

entry	n	Ar	yield (%)^a	ee (%)^a
1	1	Ph	>99 (71)	99 (99)
2	1	4-MeOC$_6$H$_4$	80	98
3	1	4-CF$_3$C$_6$H$_4$	>99 (73)	99 (98)
4	1	2-naphthyl	>99 (96)	99 (99)
5	0	Ph	>99	97
6	0	2-naphthyl	>99	94

^{a} Numbers in parentheses are at 0.1 mol% catalyst loading.

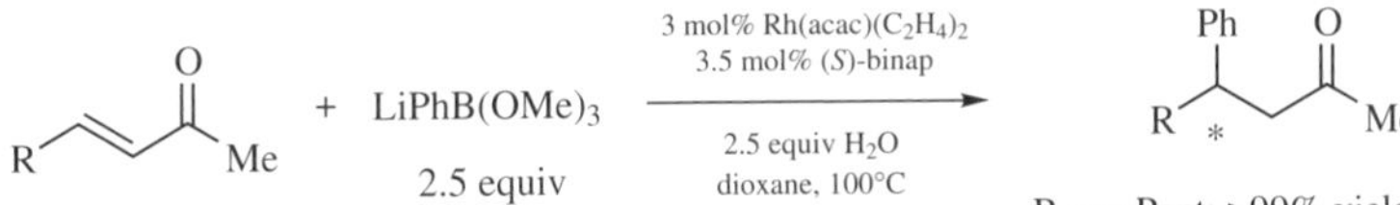

R = n-Pent: >99% yield, 91% ee
R = i-Pr: 75% yield, 97% ee

Figure 3.21. Scope of Rh/(S)-binap-catalyzed asymmetric 1,4-addition of LiArB(OMe)$_3$ to α,β-enones.

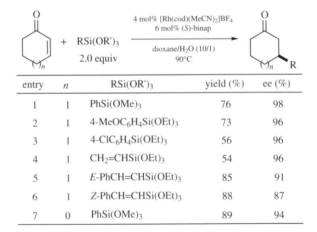

entry	n	R (X)	yield (%)	ee (%)
1	1	Ph (2.0)	99	98
2	1	4-MeOC$_6$H$_4$ (3.0)	97	90
3	1	4-FC$_6$H$_4$ (5.0)	95	98
4	1	3-ClC$_6$H$_4$ (3.0)	96	97
5	1	E-CH=CH(4-MeC$_6$H$_4$) (3.0)	93	95
6	1	vinyl (4.0)	99	92
7	0	Ph (2.0)	99	95
8	2	Ph (2.0)	98	95

Figure 3.22. Scope of Rh/(R)-binap-catalyzed asymmetric 1,4-addition of RBF$_3$K to α,β-enones.

entry	n	RSi(OR')$_3$	yield (%)	ee (%)
1	1	PhSi(OMe)$_3$	76	98
2	1	4-MeOC$_6$H$_4$Si(OEt)$_3$	73	96
3	1	4-ClC$_6$H$_4$Si(OEt)$_3$	56	96
4	1	CH$_2$=CHSi(OEt)$_3$	54	96
5	1	E-PhCH=CHSi(OEt)$_3$	85	91
6	1	Z-PhCH=CHSi(OEt)$_3$	88	87
7	0	PhSi(OMe)$_3$	89	94

R = n-Pr: 87% yield, 91% ee
R = i-Pr: 62% yield, 98% ee

Figure 3.23. Scope of Rh/(S)-binap-catalyzed asymmetric 1,4-addition of organosiloxanes to α,β-enones.

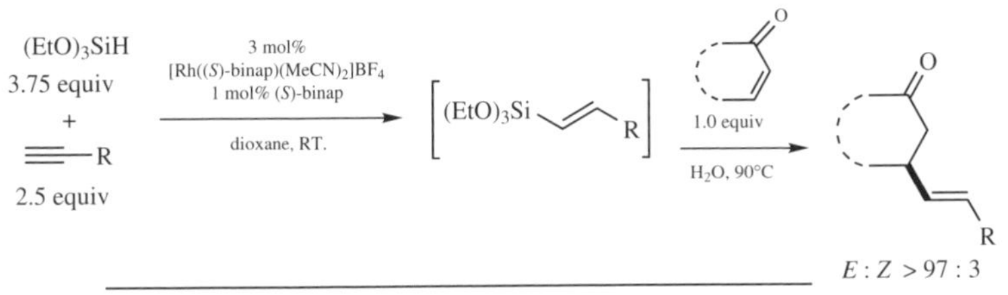

entry	substrate	R	yield (%)	ee (%)
1	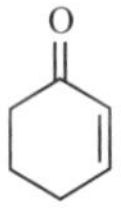	Ph	89	93
2		4-MeOC$_6$H$_4$	66	92
3		1-naphthyl	67	90
4		n-Hex	73	98^a
5		SiEt$_3$	85	98^a
6	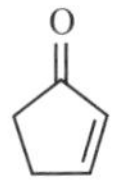	Ph	74	90
7	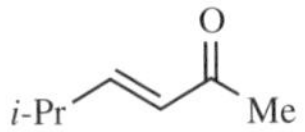	Ph	76	78

^{a} Determined by ^{13}C NMR of diastereomeric cyclic ketals
with (2R,3R)-butanediol.

Figure 3.24. Scope of Rh/(S)-binap-catalyzed tandem hydrosilylation/asymmetric 1,4-addition reactions.

the resulting alkenylsilicon compounds to α,β-enones, and these sequential reactions can be conducted in one pot without isolation of the alkenylsiloxane intermediates (Figure 3.24).[52]

Addition of Organometallic Reagents to Enones in Aprotic Media As represented by the catalytic cycle shown in Figure 3.16, the reactions described so far in aqueous media provide the 1,4-adducts as a hydrolyzed form due to the hydrolysis of oxa-π-allylrhodium intermediates by water. Although these conditions efficiently afford the β-chiral ketones, reaction conditions that give chiral enolates as the products would be more useful. In this context, the use of Ar-9-BBN as the nucleophile in aprotic solvents leads to chiral boron enolates as the 1,4-addition products in high enantioselectivity (Figure 3.25).[53] The boron enolates thus obtained can be further functionalized by the reaction with electrophiles as shown below. On the basis of the mechanistic study, it is proposed that the catalytic cycle of this process involves (1) insertion of an enone into the aryl–rhodium bond, generating an oxa-π-allylrhodium intermediate, and (2) direct transmetallation of an aryl group from boron to this oxa-π-allylrhodium intermediate, releasing the 1,4-adduct as a boron enolate along with an arylrhodium species (Figure 3.26).

A rhodium-catalyzed asymmetric synthesis of chiral titanium enolates by 1,4-addition of aryltitanium reagents to α,β-enones under aprotic conditions has

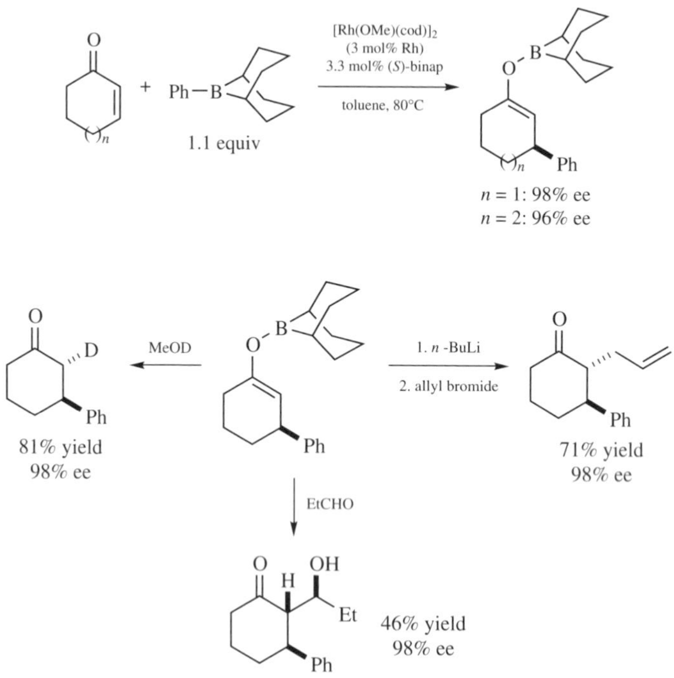

Figure 3.25. Rh/(*S*)-binap-catalyzed asymmetric 1,4-addition of Ar-9-BBN to α,β-enones and subsequent reactions of the chiral boron enolates with electrophiles.

also been developed (Figure 3.27).[54] Treatment of the 1,4-addition products with LiO*i*-Pr, followed by Me₃SiCl, converts the titanium enolates to the corresponding silyl enol ethers, which are isolable and useful synthetic intermediates. Highly enantio-enriched tiatinium enolates obtained by these 1,4-addition reactions can react with some other electrophiles by way of lithium titanates to further

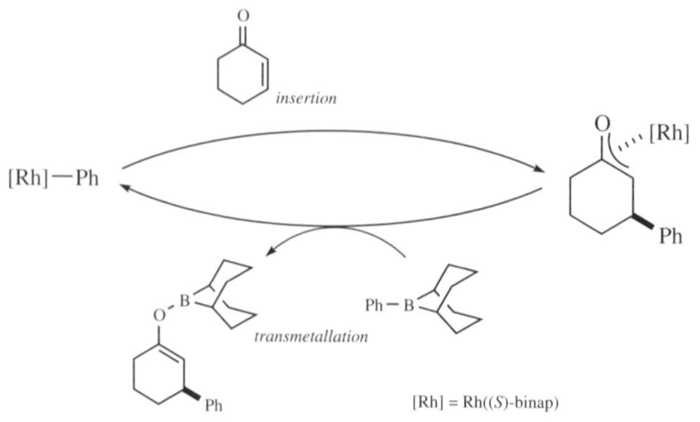

Figure 3.26. Catalytic cycle of Rh/(*S*)-binap-catalyzed asymmetric 1,4-addition of Ar-9-BBN to α,β-enones.

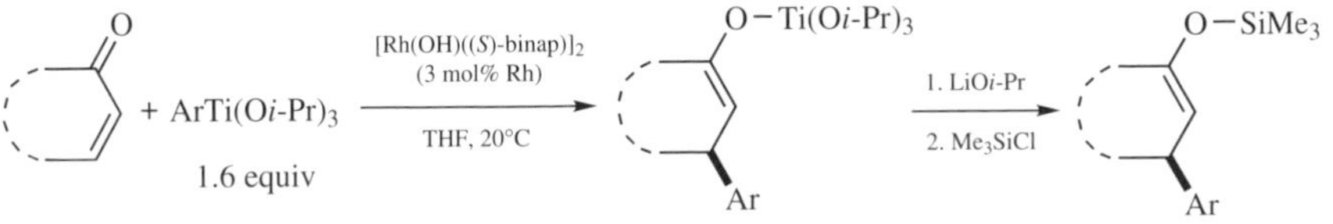

entry	substrate	Ar	yield (%)	ee (%)
1	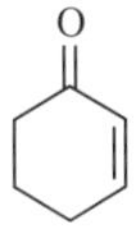	Ph	84	99.5
2		4-FC$_6$H$_4$	68	99
3		4-MeOC$_6$H$_4$	84	99.8
4	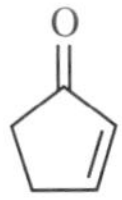	Ph	62	99.8
5	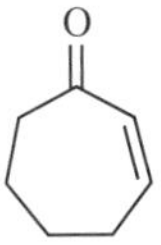	Ph	89	98
6	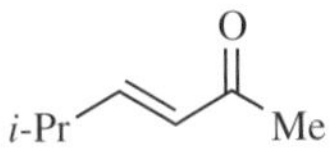	Ph	77^a	99.8

^{a} Product is obtained as a mixture of E / Z (10/9).

Figure 3.27. Scope of Rh/(S)-binap-catalyzed asymmetric 1,4-addition of aryltitanium reagents to α,β-enones.

functionalize the 1,4-adducts [Eq. (3.14)]. A catalytic cycle similar to that of the addition of Ar-9-BBN is proposed on the basis of the mechanistic study.

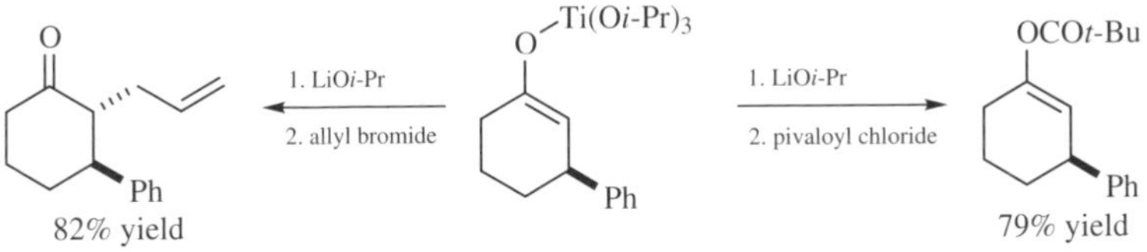

$$(3.14)$$

A similar, but mechanistically different, process has been developed using ArTi(O*i*-Pr)$_4$Li as the nucleophilic component in apotic media under Rh/(S)-binap catalysis.[55] The reaction proceeds smoothly in the presence of Me$_3$SiCl to furnish highly enantio-enriched β-chiral silyl enol ethers (Figure 3.28). Interestingly, the reaction does not occur without Me$_3$SiCl, and the mechanistic study using ^{31}P NMR revealed that the catalytic cycle of this process involves (1) transmetallation of an aryl group from titanium to rhodium; (2) insertion of an enone into the carbon–rhodium bond, generating an oxa-π-allylrhodium intermediate; and (3) silylation of this intermediate by Me$_3$SiCl to release the 1,4-adduct as a silyl enol ether along with a chlororhodium species (Figure 3.29).

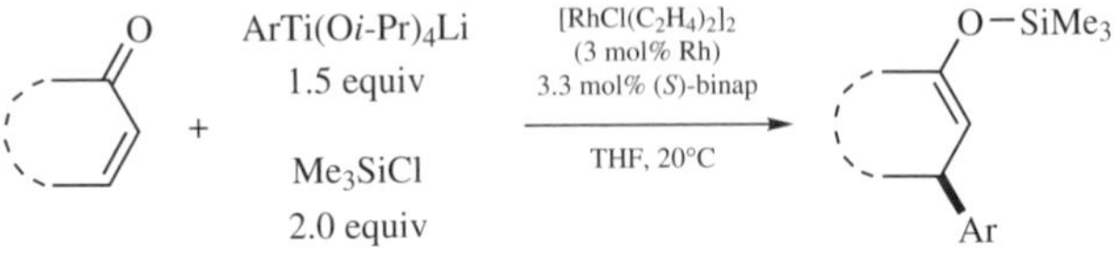

entry	substrate	Ar	yield (%)	ee (%)
1	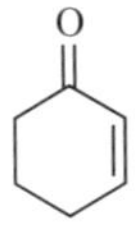 (cyclohexenone)	Ph	90	96
2	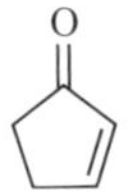 (cyclopentenone)	Ph	91	99
3		4-CF$_3$C$_6$H$_4$	94	97
4		4-MeOC$_6$H$_4$	83	96
5	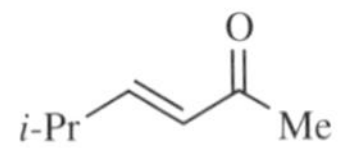 i-Pr...Me	Ph	67^a	90

^{a} Product is obtained as a mixture of E / Z (3/10).

Figure 3.28. Scope of Rh/(S)-binap-catalyzed asymmetric 1,4-addition of ArTi(Oi-Pr)$_4$Li to α,β-enones in the presence of Me$_3$SiCl.

For the asymmetric 1,4-addition of alkenyl groups in aprotic media, alkenyl zirconium reagents can be used, which are generated by hydrozirconation of terminal alkynes (Figure 3.30).[56] Under these conditions, alkenyl groups derived from alkylacetylenes are efficiently installed, but those from arylacetylenes are not as effective (entry 3).

Arylzinc reagents have also been used in the context of rhodium-catalyzed 1,4-addition reactions. Thus, the addition of these reagents to 2,3-dihydro-4-pyridones was investigated for the synthesis of 2-aryl-4-piperidones, which are biologically and synthetically useful compounds (Figure 3.31).[57] For this particular class of substrates, other organometallic reagents, such as arylboronic acids and aryltitanium reagents, are not as effective, and only the use of arylzinc reagents provides the

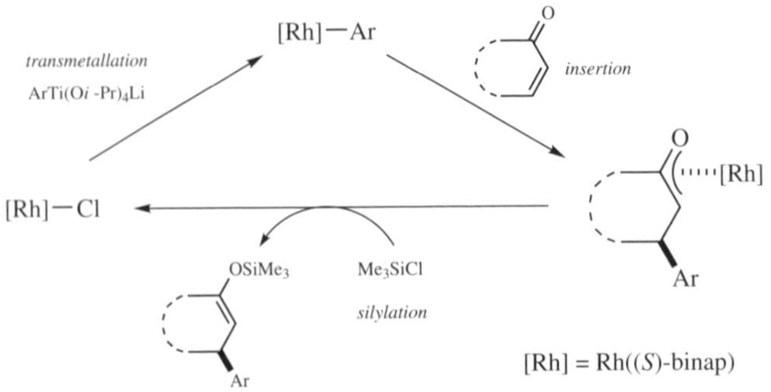

Figure 3.29. Catalytic cycle of Rh/(S)-binap-catalyzed asymmetric 1,4-addition of ArTi(Oi-Pr)$_4$Li to α,β-enones in the presence of Me$_3$SiCl.

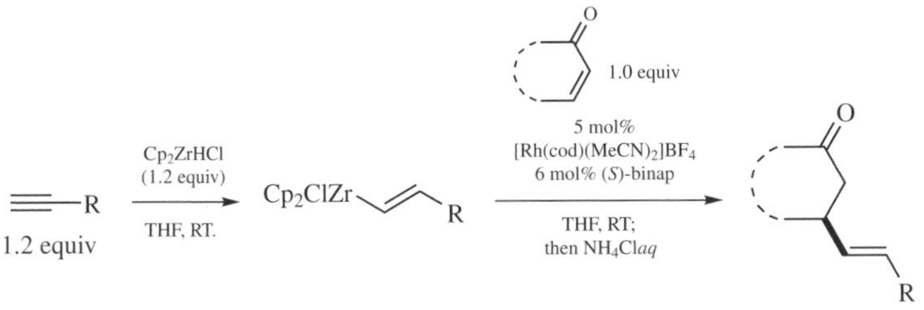

Figure 3.30. Scope of Rh/(S)-binap-catalyzed asymmetric 1,4-addition of alkenylzirconium reagents to α,β-enones.

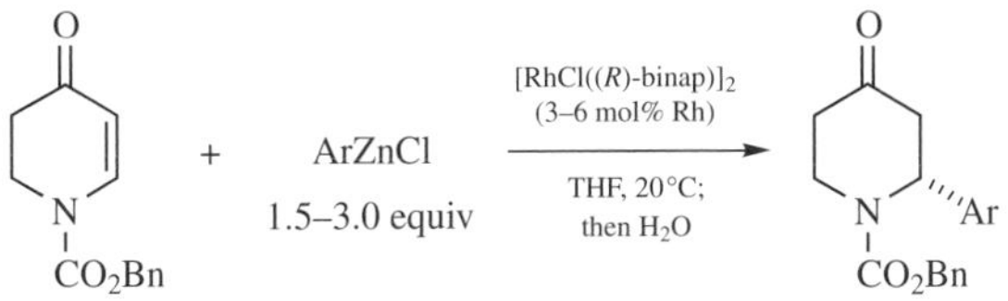

entry	substrate	R	yield (%)	ee (%)
1	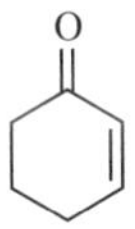	n-Bu	83	96
2		n-Pent	96	99
3		Ph	20	95
4	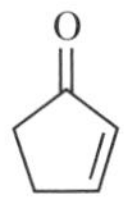	n-Bu	93	86
5	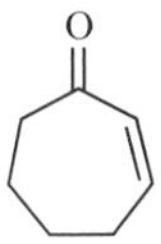	n-Bu	63	95
6	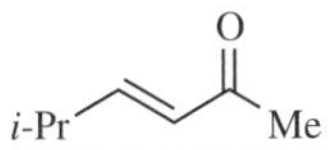	n-Bu	82	71

entry	Ar	yield (%)	ee (%)
1	Ph	95	>99.5
2	$4\text{-MeOC}_6\text{H}_4$	90	99
3	$4\text{-FC}_6\text{H}_4$	91	>99.5
4	$3,5\text{-Me}_2\text{C}_6\text{H}_3$	87	99
5	$2\text{-MeC}_6\text{H}_4$	100	99

Figure 3.31. Scope of Rh/(R)-binap-catalyzed asymmetric 1,4-addition of ArZnCl to 2,3-dihydro-4-pyridones.

desired 1,4-adducts in high yield and ee (typically $\geq 99\%$ ee). Furthermore, arylzinc reagents were found to be effective nucleophiles for simple α,β-enones as well [Eq. (3.15)].

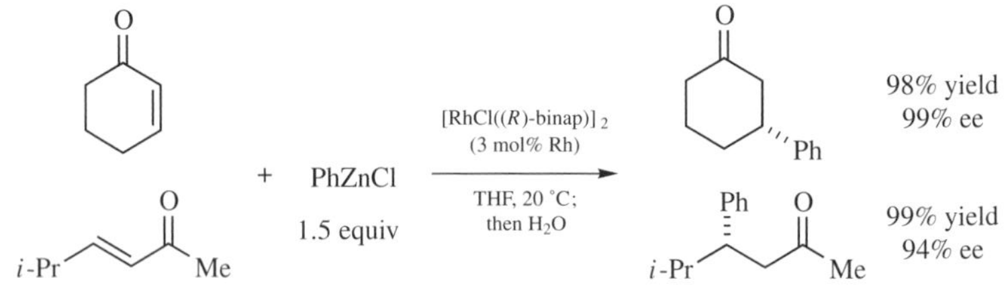

$$\text{(3.15)}$$

1,4-Addition to Other α,β-Unsaturated Compounds In addition to simple α,β-enones, various other α,β-unsaturated compounds have been employed in the rhodium-catalyzed asymmetric 1,4-addition reactions. There are several reports on the usage of α,β-unsaturated esters and amides as substrates in combination with arylboron reagents under aqueous conditions. Hayashi described that Rh/ (S)-binap efficiently catalyzes the addition of LiArB(OMe)$_3$ to acyclic α,β-enoates (Figure 3.32), and higher ee is achieved by increasing the size of the ester group with the maintenance of high reactivity.[58] In contrast, the use of arylboronic acids, instead of LiArB(OMe)$_3$, is less efficient for substrates with a bulky ester group, although similarly high ee is observed (entry 1 vs. entry 2). For cyclic enoate substrates, however, the use of LiArB(OMe)$_3$ leads to low yield of the 1,4-adducts and the addition of arylboronic acids provides a much more efficient process (Figure 3.33). Miyaura independently reported a similar system for the 1,4-addition of arylboronic acids to α,β-enoates in the presence of Rh/(S)-binap.[59]

A cyclic α,β-unsaturated amide can also be employed in the rhodium-catalyzed 1,4-addition reactions (Figure 3.34).[60] The best yield and ee are achieved when ligand **43** is used instead of (R)-binap (entry 1 vs. entry 2). The product thus

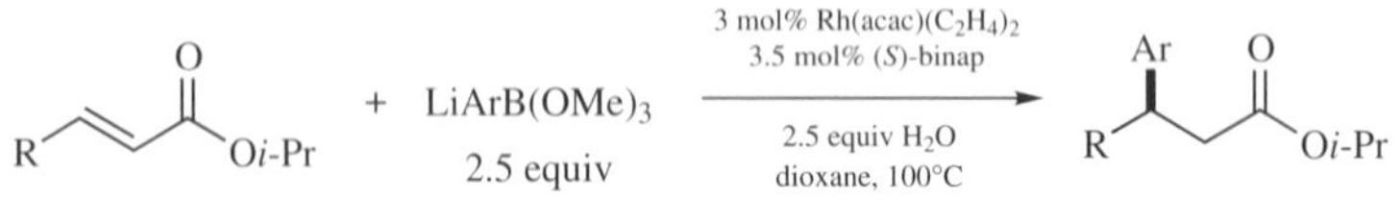

entry	R	Ar	yield (%)	ee (%)
1	n-Pr	Ph	96	95
2[a]	n-Pr	Ph	42	94
3	n-Pr	4-ClC$_6$H$_4$	95	97
4	n-Pr	4-MeC$_6$H$_4$	88	97
5	i-Pr	Ph	64	98

[a] PhB(OH)$_2$ (5.0 equiv) was used instead of LiPhB(OMe)$_3$.

Figure 3.32. Scope of Rh/(S)-binap-catalyzed asymmetric 1,4-addition of LiArB(OMe)$_3$ to acyclic α,β-enoates.

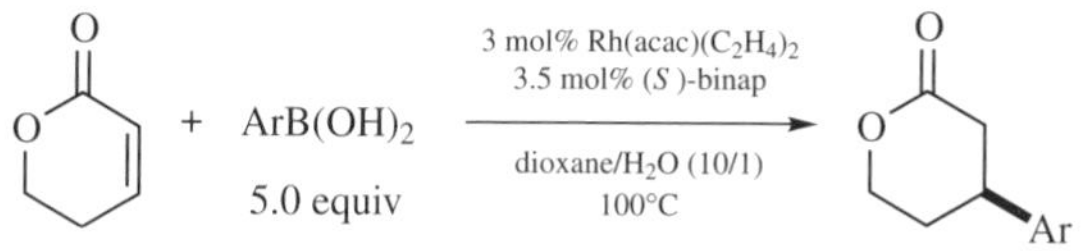

entry	Ar	yield (%)	ee (%)
1[a]	Ph	38	98
2	Ph	94	98
3	4-ClC₆H₄	95	97
4	4-MeC₆H₄	91	97
5	2-naphthyl	93	98

[a] LiPhB(OMe)₃ (2.5 equiv) was used instead of PhB(OH)₂.

Figure 3.33. Scope of Rh/(S)-binap-catalyzed asymmetric 1,4-addition of arylboronic acids to a cyclic α,β-enoate.

obtained with 4-fluorophenyl group constitutes a key intermediate for the synthesis of (−)-paroxetine. Acyclic α,β-unsaturated amides have been used in the presence of a Rh/(S)-binap catalyst as well.[61] After examining various bases, it was found that the addition of K₂CO₃ (0.5 equiv) to the reaction system leads to high yields of the 1,4-adducts (Figure 3.35). Carreira also reported a rhodium-catalyzed asymmetric 1,4-addition of phenylboronic acid to α,β-unsaturated esters and amides by

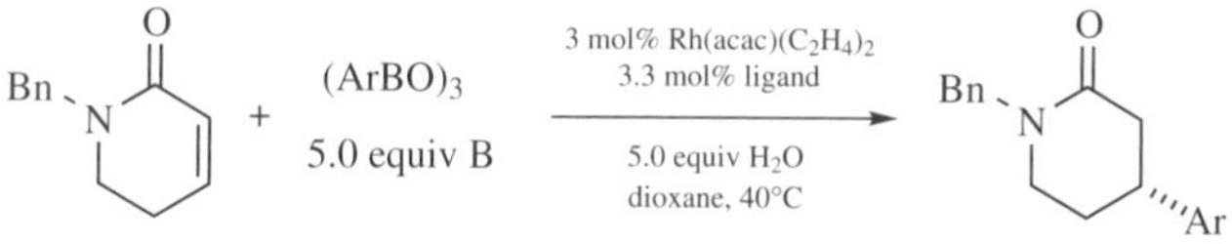

entry	Ar	ligand	yield (%)	ee (%)
1	4-FC₆H₄	(R)-binap	73	98
2	4-FC₆H₄	**43**	82	98
3[a]	4-ClC₆H₄	**43**	88	98

[a] The reaction was carried out at 60°C.

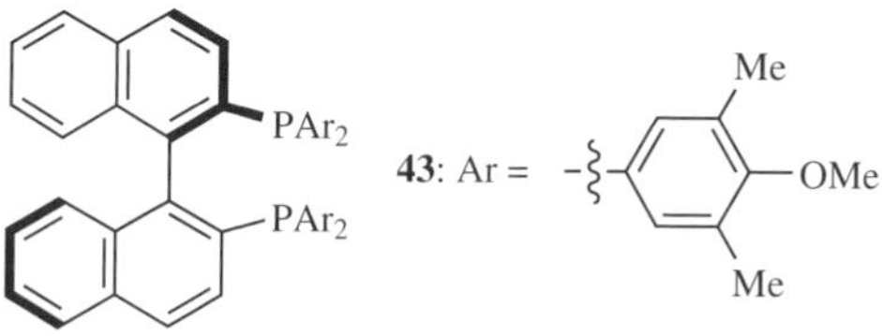

Figure 3.34. Scope of Rh/**43**-catalyzed asymmetric 1,4-addition of arylboroxines to a cyclic α,β-enamide.

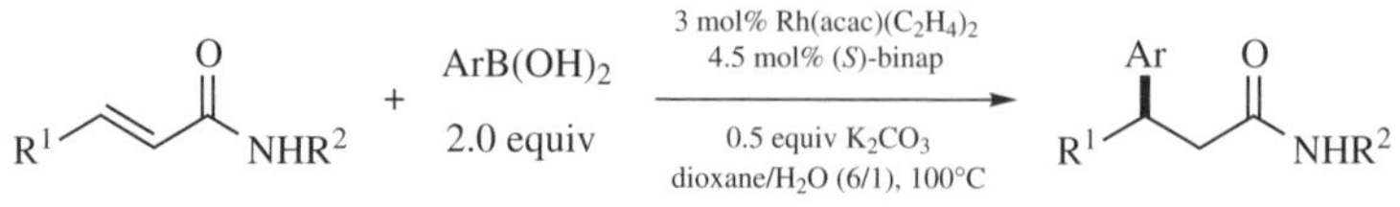

Figure 3.35. Scope of Rh/(S)-binap-catalyzed asymmetric 1,4-addition of arylboronic acids to acyclic α,β-enamides.

entry	R^1	R^2	Ar	yield (%)	ee (%)
1	Me	H	Ph	62	89
2	Me	Ph	Ph	88	90
3	Me	Cy	Ph	80	93
4	Me	Bn	Ph	85	93
5	Me	Bn	4-MeC$_6$H$_4$	74	87
6	Me	Bn	4-CF$_3$C$_6$H$_4$	82	92
7	n-Pent	Bn	Ph	89	91

using chiral diene ligand **39** developed in his group, achieving high enantioselectivity (88–98% ee).[43]

The use of potassium aryltrifluoroborates as nucleophiles in the addition to α,β-unsaturated amides has also been described (Figure 3.36).[62] Various aryl groups can be installed with high ee, although the β-substituent of enamides is limited to methyl group. For the use of other nucleophiles than organoboron species, Rh/(S)-binap can catalyze the addition of phenyltrimethoxysilane to α,β-unsaturated esters and amides in relatively high enantioselectivity as well.[49]

Rhodium-catalyzed asymmetric 1,4-addition reactions have been applied to various other α,β-unsaturated compounds as well. Thus, Rh/(S)-binap can catalyze the asymmetric 1,4-addition of arylboroxines to α,β-unsaturated phosphonates in high yield and ee (Figure 3.37).[63] The key to the success in achieving high yield is the

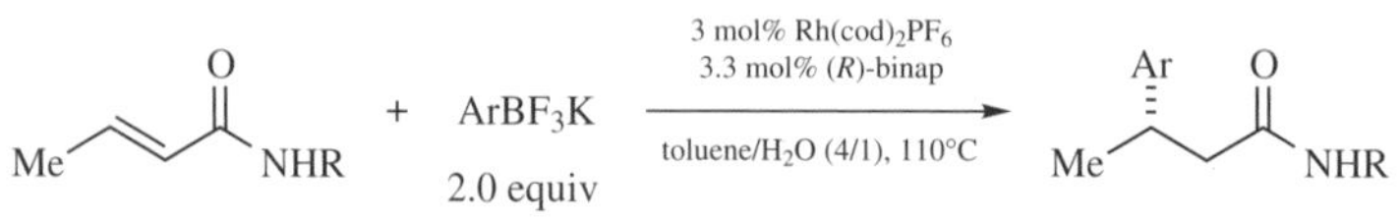

entry	R	Ar	yield (%)	ee (%)
1	Bn	Ph	86	93
2	Bn	4-MeOC$_6$H$_4$	94	95
3	Bn	4-ClC$_6$H$_4$	91	94
4	Bn	3-ClC$_6$H$_4$	81	93
5	i-Pr	2-naphthyl	100	95

Figure 3.36. Scope of Rh/(R)-binap-catalyzed asymmetric 1,4-addition of potassium aryltrifluoroborates to acyclic α,β-enamides.

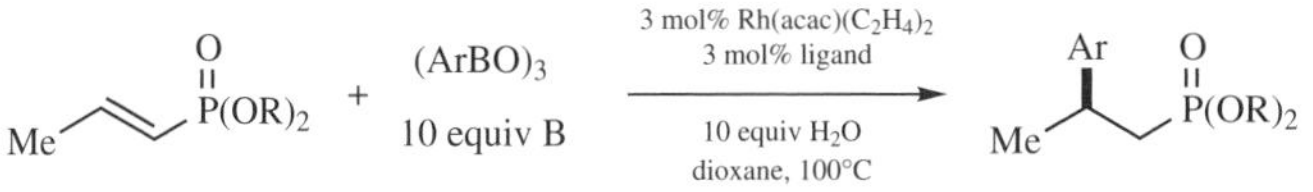

entry	R	Ar	ligand	yield (%)	ee (%)
1	Et	Ph	(*S*)-binap	94	96
2	Et	3-$MeOC_6H_4$	(*S*)-binap	81	95
3	Et	4-$CF_3C_6H_4$	(*S*)-binap	64	96
4	Et	4-MeC_6H_4	(*S*)-binap	84	95
5	Et	4-MeC_6H_4	**44**	88	96
6	Ph	Ph	(*S*)-binap	95	91
7	Ph	Ph	**44**	99	94

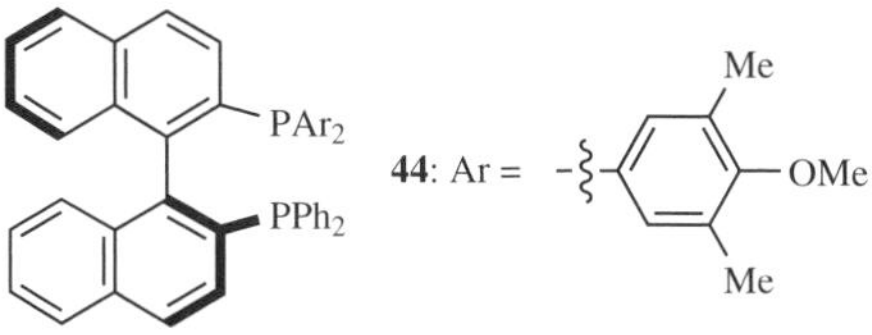

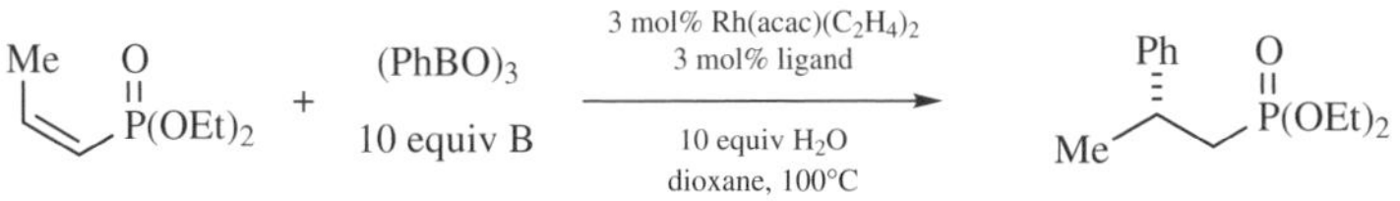

96% yield, 89% ee ((*S*)-binap)
98% yield, 92% ee (ligand **44**)

Figure 3.37. Scope of Rh/(*S*)-binap- or Rh/**44**-catalyzed asymmetric 1,4-addition of arylboroxines to α,β-unsaturated phosphonates.

use of arylboroxines with 1 equiv (to boron) of water in dioxane, and the use of corresponding arylboronic acids in dioxane/water (10/1) results in low yield of the 1,4-adducts (~40% yield). Furthermore, both yield and ee can be slightly improved by the use of ligand **44** instead of (*S*)-binap (entries 4 and 6 vs. entries 5 and 7). It is worth noting that when a (*Z*)-substrate is used, the 1,4-adduct is obtained in the opposite enantiomer enriched. In addition, nitroalkenes have also been employed as substrates in the Rh/(*S*)-binap-catalyzed 1,4-addition reactions.[64] Thus, 1-nitrocyclohexene can be arylated by the use of 5–10 equiv of arylboronic acids, preferentially generating cis isomers (Figure 3.38). These cis enriched products can be isomerized to the thermodynamically more stable trans isomers by treatment with $NaHCO_3$ in refluxing EtOH (cis/trans = 4/96 ~ 2/98).

Fumaric and maleic compounds are also suitable substrates in the rhodium-catalyzed asymmetric 1,4-addition reactions.[65] While phosphorus-based chiral ligands, such as (*R*)-binap, provide low enantioselectivity for these substrates (≤51% ee), chiral diene ligands **38** and **45** are particularly effective for achieving high ee (Figures 3.39 and 3.40).

entry	Ar (X)	yield (%)	*cis/trans*	ee (%)
1	Ph (10)	89	87/13	99
2	4-MeC_6H_4 (10)	89	88/12	98
3	4-$CF_3C_6H_4$ (5)	88	85/15	99
4	3-ClC_6H_4 (5)	89	85/15	99
5	2-naphthyl (10)	84	85/15	98

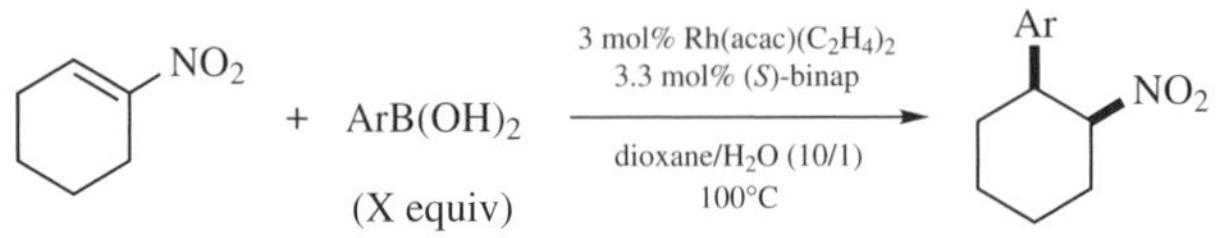

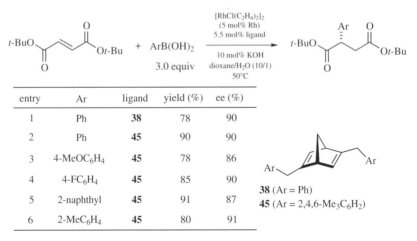

Figure 3.38. Scope of Rh/(*S*)-binap-catalyzed asymmetric 1,4-addition of arylboronic acids to 1-nitrocyclohexene.

As another way of constructing nitrogen-containing heterocycles enantioselectively, the use of 3-alkoxycarbonyl-3-pyrrolines as substrates has also been described in the rhodium-catalyzed asymmetric 1,4-addition reactions (Figure 3.41).[66] Among the conditions examined, it was found that $[Rh(OH)(cod)]_2$/(*S*)-binap catalyst is uniquely effective for achieving high yield of the 1,4-adducts.

By employing *N*-acetamidoacrylates, rhodium-catalyzed 1,4-addition reactions can be applied to the enantioselective synthesis of α-amino acids. Unlike other typical 1,4-addition reactions, which install a stereogenic center at the β-position of the 1,4-adducts, this process deals with the formation of a stereocenter at the α-position, and thus the protonation (hydrolysis) step becomes an important step

entry	Ar	ligand	yield (%)	ee (%)
1	Ph	**38**	78	90
2	Ph	**45**	90	90
3	4-$MeOC_6H_4$	**45**	78	86
4	4-FC_6H_4	**45**	85	90
5	2-naphthyl	**45**	91	87
6	2-MeC_6H_4	**45**	80	91

38 (Ar = Ph)
45 (Ar = 2,4,6-$Me_3C_6H_2$)

Figure 3.39. Scope of Rh/chiral diene-catalyzed asymmetric 1,4-addition of arylboronic acids to di-*tert*-butyl fumarate.

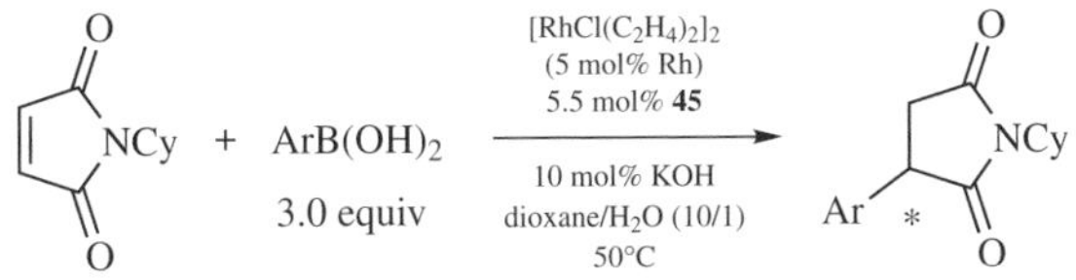

Figure 3.40. Scope of Rh/**45**-catalyzed asymmetric 1,4-addition of arylboronic acids to N-cyclohexylmaleimide.

for controlling the stereochemical outcome [see Figure 3.16 and Eq. (3.10)]. Reetz described one example of this type in 87% ee using ligand **46** in the presence of NaF, although the detailed reaction conditions or the chemical yield were not reported.[40] Subsequently, Darses and Genet reported that relatively high ee can be achieved through optimization of the proton source in the Rh/(R)-binap-catalyzed 1,4-addition of potassium organotrifluoroborates to methyl N-acetamidoacrylate (Figure 3.42).[67] Thus, guaiacol (2-methoxyphenol) was found to be uniquely effective for achieving high ee among those tested.

When α,β-unsaturated sulfones are employed as substrates for the rhodium-catalyzed 1,4-addition of aryltitanium reagents, they provide *cine*-substitution products instead of regular 1,4-addition products [Eq. (3.16)].[68] This new process

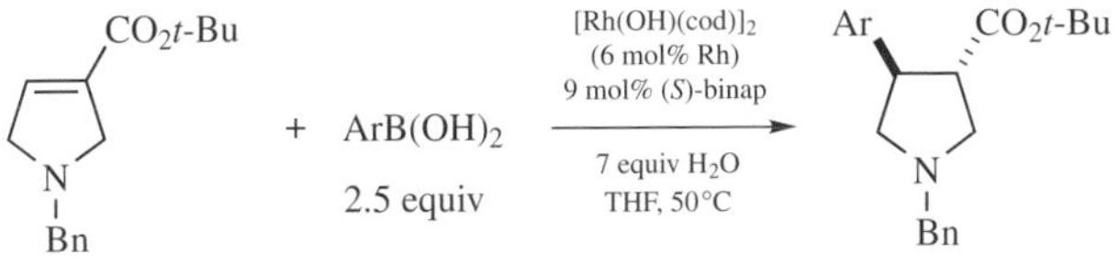

Figure 3.41. Scope of Rh/(S)-binap-catalyzed asymmetric 1,4-addition of arylboronic acids to N-benzyl-3-*tert*-butoxycarbonyl-3-pyrroline.

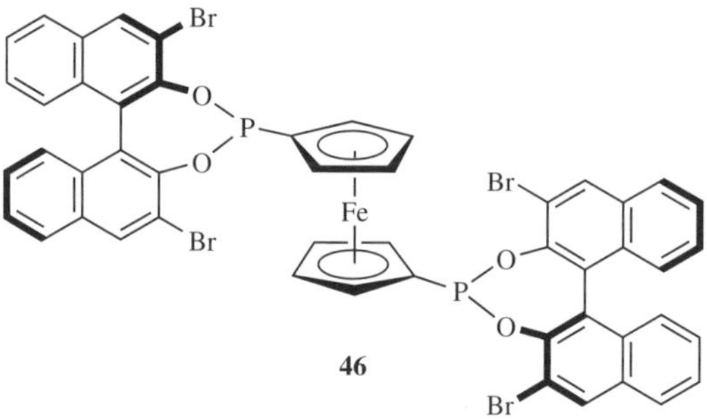

46

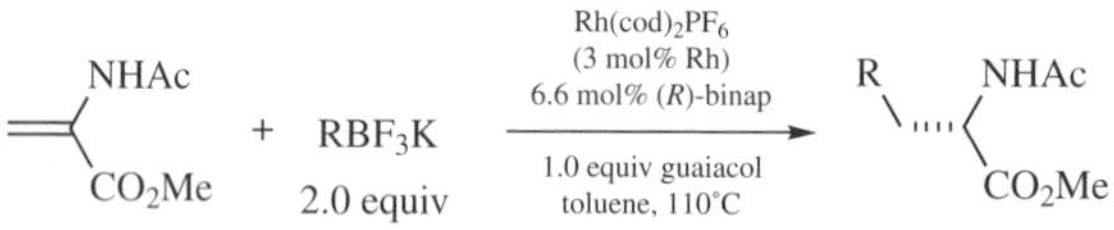

entry	R	yield (%)	ee (%)
1	Ph	89	90
2	4-FC_6H_4	82	83
3	3-$MeOC_6H_4$	89	90
4	2-naphthyl	88	87
5	*E*-CH=CH(4-MeC_6H_4)	96	88

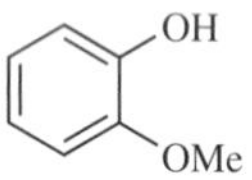

guaiacol

Figure 3.42. Scope of Rh/(*R*)-binap-catalyzed asymmetric 1,4-addition of potassium organotrifluoroborates to methyl *N*-acetamidoacrylate in the presence of guaiacol.

was successfully applied to the enantioselective synthesis of allyl arenes by the use of a cyclic alkenyl sulfone, achieving >99% ee [eq. (3.17)]. In contrast, normal 1,4-addition reactions proceed by the use of α,β-unsaturated 2-pyridyl sulfones in combination with arylboronic acids.[69] It was found that (*S*,*S*)-chiraphos is the most effective ligand in achieving high enantioselectivity (Figure 3.43).

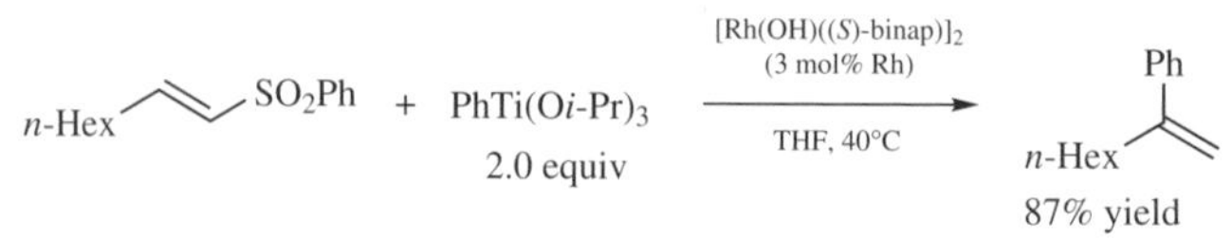

(3.16)

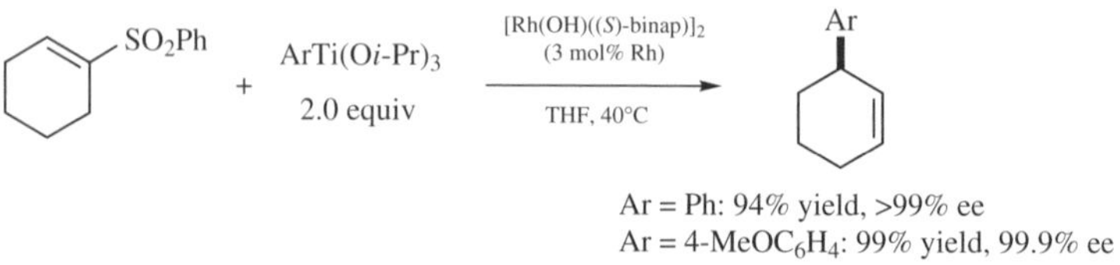

(3.17)

Ar = Ph: 94% yield, >99% ee
Ar = 4-$MeOC_6H_4$: 99% yield, 99.9% ee

Other Related Reactions In addition to simple 1,4-addition reactions described so far, other related rhodium-catalyzed asymmetric carbon–carbon bond-forming processes have also been developed. Lautens reported an asymmetric ring-opening of oxabicyclic alkenes with organoboronic acids.[70] High enantioselectivity can be

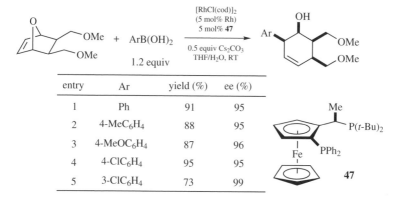

entry	R	Ar	yield (%)	ee (%)
1	Me	Ph	97	81
2	n-Pent	Ph	98	84
3	n-Pent	4-FC$_6$H$_4$	94	87
4	n-Pent	4-MeOC$_6$H$_4$	92	81
5	i-Pr	4-FC$_6$H$_4$	93	85
6	2-naphthyl	4-FC$_6$H$_4$	97	92

Figure 3.43. Scope of Rh/(S,S)-chiraphos-catalyzed asymmetric 1,4-addition of aryl-boronic acids to α,β-unsaturated 2-pyridyl sulfones.

achieved by the use of **47** as a ligand, establishing multiple stereocenters in a single step (Figure 3.44). Hayashi developed a 1,6-addition of aryltitanium reagents to conjugated enynones in the presence of Rh/(R)-segphos to produce axially chiral allenes in good enantioselectivity (Figure 3.45).[71] Hayashi also described an asymmetric arylative cyclization of alkynals with arylboronic acids.[72] High yield and ee are realized by the use of a rhodium/**48** catalyst, while Rh/(S)-binap shows low reactivity with moderate ee (Figure 3.46).

3.3   CROSS-COUPLING REACTIONS

Transition-metal-catalyzed cross-coupling reactions also represent a powerful approach for the construction of carbon–carbon bonds. As a result, these processes have been widely studied in the past few decades. Among the transition metals

entry	Ar	yield (%)	ee (%)
1	Ph	91	95
2	4-MeC$_6$H$_4$	88	95
3	4-MeOC$_6$H$_4$	87	96
4	4-ClC$_6$H$_4$	95	95
5	3-ClC$_6$H$_4$	73	99

Figure 3.44. Scope of Rh/**47**-catalyzed asymmetric ring-opening of oxabicyclic alkenes with organoboronic acids.

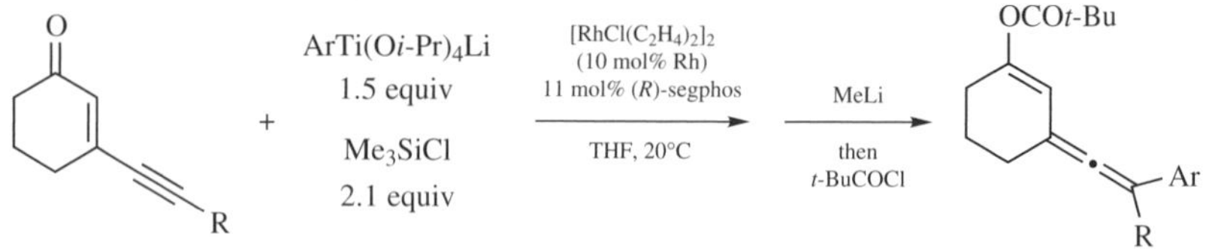

entry	R	Ar	yield (%)	ee (%)
1	*n*-Bu	Ph	85	92
2	*n*-Bu	4-FC_6H_4	85	91
3	*n*-Bu	4-$MeOC_6H_4$	83	93
4	Cy	4-$MeOC_6H_4$	80	80
5	4-$MeOC_6H_4$	Ph	56	75

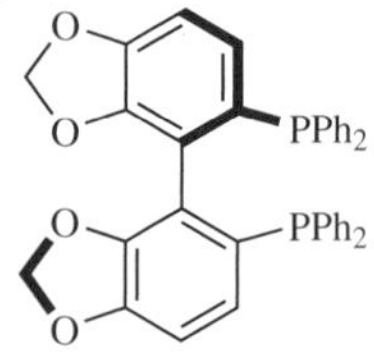

(*R*)-segphos

Figure 3.45. Scope of Rh/(*R*)-segphos-catalyzed asymmetric 1,6-addition of ArTi $(Oi\text{-}Pr)_4Li$ to conjugated enynones in the presence of Me_3SiCl.

employed, the majority of the investigations have focused on the nickel- and palladium-catalyzed cross-couplings of aryl and alkenyl halides with various organometallic reagents.[73] In addition to the simple, nonasymmetric cross-coupling reactions, enantioselective variants of them have also been a topic of interest in organic and organometallic chemistry.[74]

3.3.1 Kumada-Type Cross-Couplings

In the late 1970s and early 1980s, Kumada described nickel-catalyzed asymmetric cross-coupling reactions of 1-arylethylmagnesium chlorides with vinyl bromide in

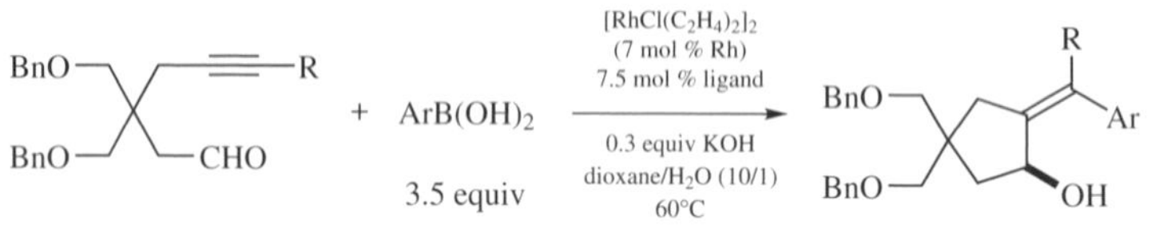

entry	R	Ar	ligand	yield (%)	ee (%)
1	Me	Ph	(*S*)-binap	24	76
2	Me	Ph	**48**	78	95
3	Et	Ph	**48**	89	94
4	Me	4-$MeOC_6H_4$	**48**	71	93
5	Me	4-FC_6H_4	**48**	77	93
6	Me	3-ClC_6H_4	**48**	71	96

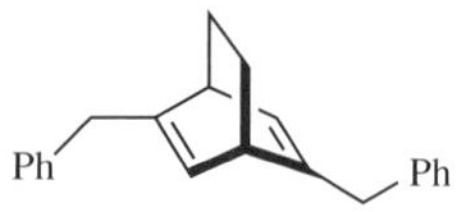

48

Figure 3.46. Scope of Rh/**48**-catalyzed asymmetric arylative cyclization of alkynals with arylboronic acids.

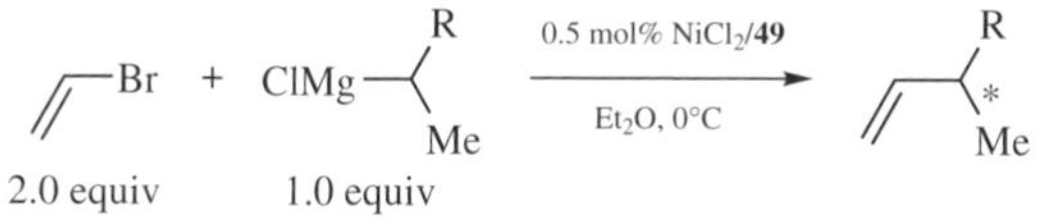

entry	R	yield (%)	ee (%)^a
1	Ph	>95^b	81
2	4-MeC$_6$H$_4$	94	83
3	2-naphthyl	88	72
4	n-hex	45^b	6

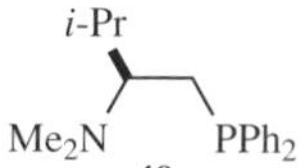

^{a} Calculated from optical rotations.

^{b} Determined by gas–liquid chromatography (GLC)

Figure 3.47. Scope of Ni/**49**-catalyzed asymmetric cross-coupling of vinyl bromide with secondary alkyl Grignard reagents.

the presence of chiral phosphine ligands.[75] It was determined that NiCl$_2$/**49** catalyzes the coupling reactions efficiently, affording the products in good enantiomeric excess (Figure 3.47). The prerequisite of these reactions using racemic secondary alkyl Grignard reagents as a coupling partner is that the equilibration of two enantiomeric forms of the Grignard reagents has to be faster than the actual coupling reaction with vinyl bromide. From the proposed catalytic cycle depicted in Figure 3.48, the origin of high enantioselectivity under Ni/**49** conditions is speculated to be derived from the dimethylamino group of the ligand **49**, which directs the approach of the Grignard reagents to the nickel center in a diastereoselective fashion (formation of **D** via **C**), following the preceeding oxidative addition of vinyl bromide (formation of **B** from **A**). The reductive elimination of diorganonickel species provides the coupling product and regenerates the nickel catalyst (formation of **A** from **D**).

Following the pioneering work by Kumada, there appeared other successful ligands, such as **50–53**, in the nickel-catalyzed asymmetric Grignard cross-coupling reaction of vinyl bromide (or chloride) with 1-phenylethylmagnesium chloride to afford the coupling product in relatively high enantioselectivity (Figure 3.49).[76–79]

Chiral palladium catalysts can also catalyze similar reactions, although secondary organozinc reagents are sometimes used instead of the corresponding Grignard reagents [Eqs. (3.18)-(3.20)].[80–82] In addition, by using Pd/**54** complex as a chiral catalyst, α-(trimethylsilyl)benzyl Grignard reagent can be employed as a nucleophile in the asymmetric cross-coupling with vinyl bromides, giving chiral allylsilanes in high enantiomeric excess (Figure 3.50).[83] The use of E-vinyl bromides is necessary to achieve high stereoselection and the use of corresponding Z-isomer leads to a significant erosion of ee (entry 3 vs. entry 4).

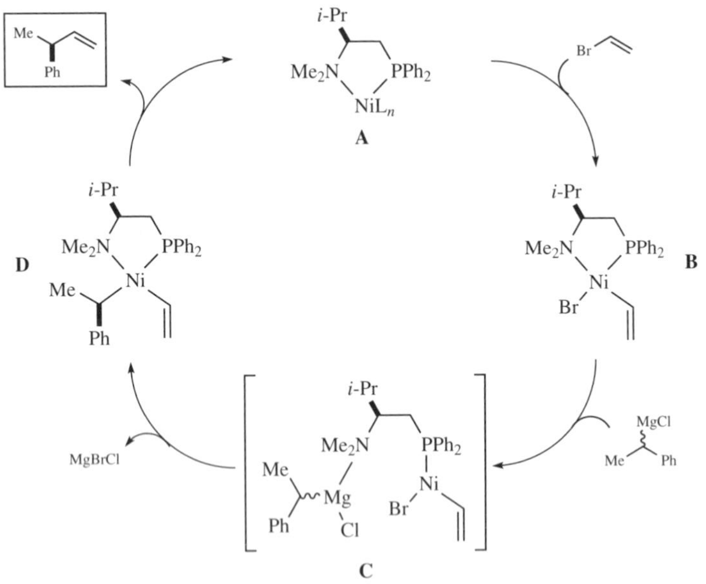

Figure 3.48. Catalytic cycle of Ni/**49**-catalyzed asymmetric cross-coupling of vinyl bromide with secondary alkyl Grignard reagents.

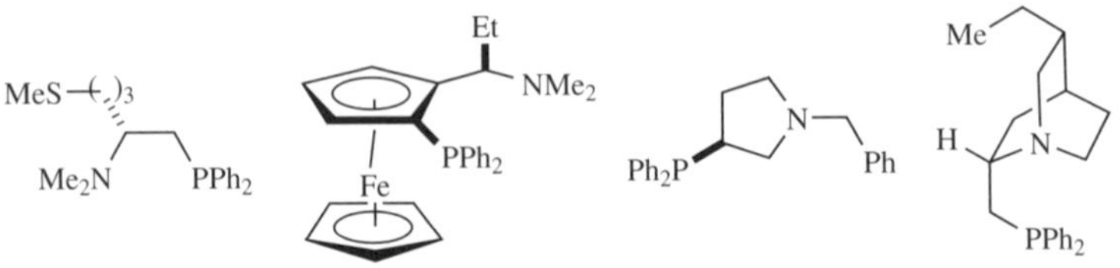

50 [76]	**51** [77]	**52** [78]	**53** [79]
88% ee	82% ee	89% ee	85% ee

Figure 3.49. Chiral ligands that exhibit high enantioselectivity in the nickel-catalyzed Grignard cross-coupling reaction of vinyl bromide (or chloride) with 1-phenylethylmagnesium chloride.

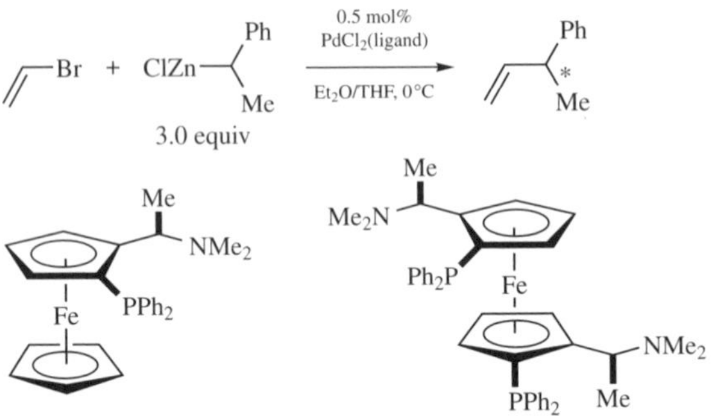

(3.18)

54	**55**
>95% GLC yield	100% yield
85% ee (by optical rotation)	93% ee (by optical rotation)

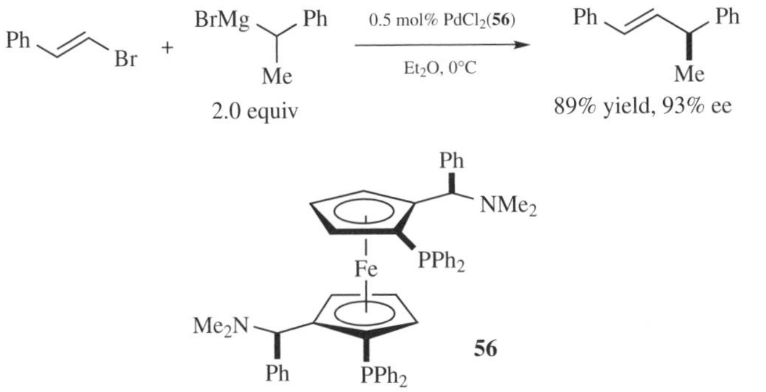

(3.19)

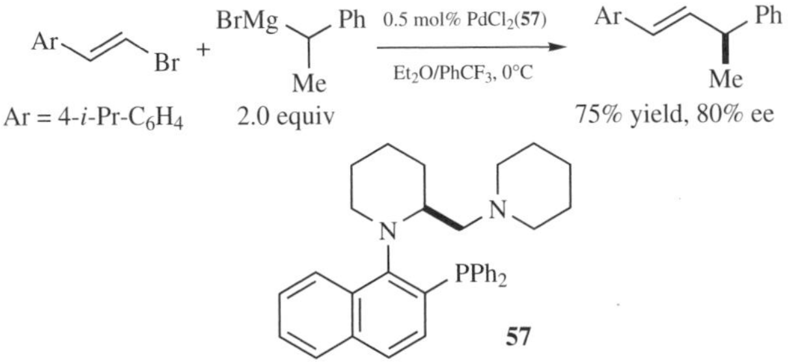

(3.20)

Asymmetric Grignard cross-coupling reactions can also be used to construct axially chiral binaphthyl derivatives, and there have been some reports in the literature on nickel or palladium catalysis to induce such axial chirality. Hayashi and Ito developed asymmetric couplings between 2-substituted-1-bromonaphthalenes and 2-substituted-1-naphthylmagnesium bromides in the presence of Ni/**58**, furnishing enantio-enriched substituted binaphthyls (Figure 3.51).[84] It is worth noting that a coupling between 2-methyl-1-naphthylmagnesium bromide

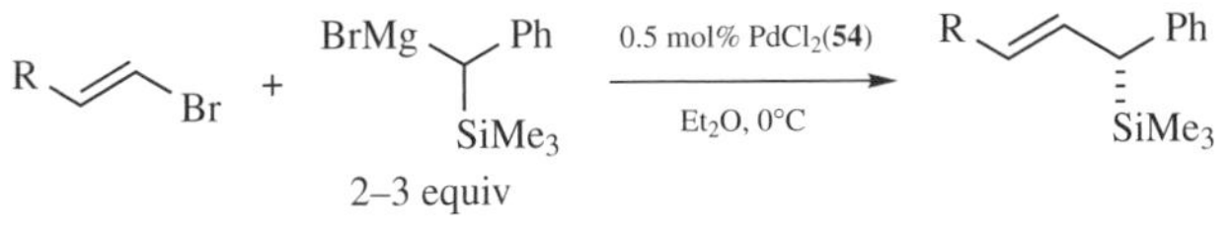

entry	R	yield (%)	ee (%)
1	H	42	95
2	Ph	93	95
3	Me	77	85
4[a]	Me	38	24

[a] Z-1-bromo-1-propene was used.

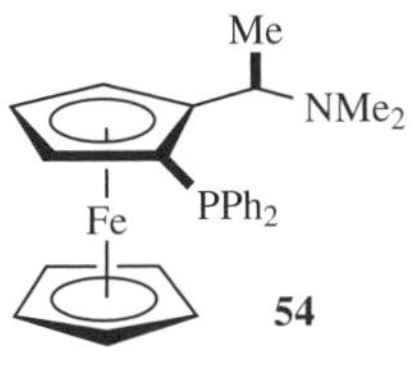

Figure 3.50. Scope of Pd/**54**-catalyzed asymmetric cross-coupling of vinyl bromides with α-(trimethylsilyl)benzylmagnesium bromide.

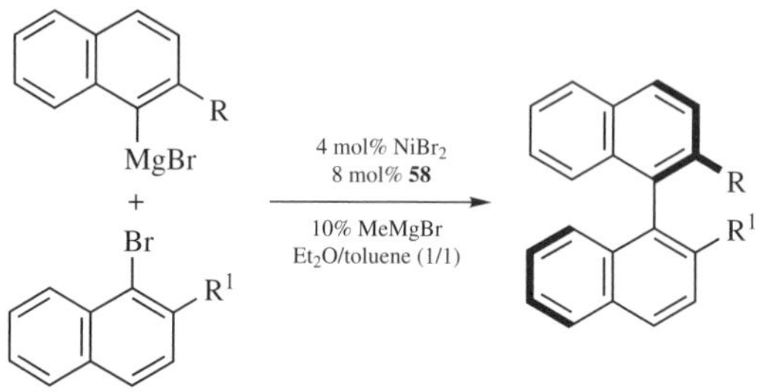

entry	R	R^1	temp (°C)	yield (%)	ee (%)
1	Me	Me	−15	69	95
2	Me	H	−30	92	83
3	H	Me	−10	25	16
4	Et	H	−20	85	77

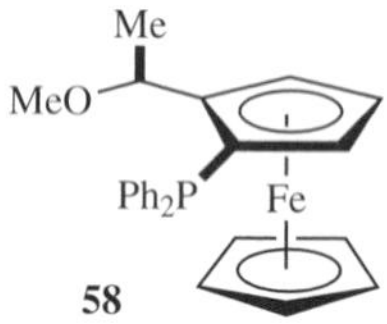

Figure 3.51. Scope of Ni/**58**-catalyzed asymmetric cross-coupling of 2-substituted-1-bromonaphthalenes and 2-substituted-1-naphthylmagnesium bromides.

and 1-bromonaphthalene provides significantly higher ee than does the reverse combination (1-naphthylmagnesium bromide and 2-methyl-1-bromonaphthalene) for the synthesis of 2-methyl-1,1′-binaphthyl (entry 2 vs. entry 3). This observation strongly suggests that the bis(naphthyl)-nickel species, which is generated after oxidative addition and transmetallation, undergoes reductive elimination without epimerization. Highly enantio-enriched 2,2′-dimethyl-1,1′-binaphthyl obtained by this method can be further functionalized as shown in Eq. (3.21), furnishing cyclic amide **59** with no erosion of enantiomeric excess.

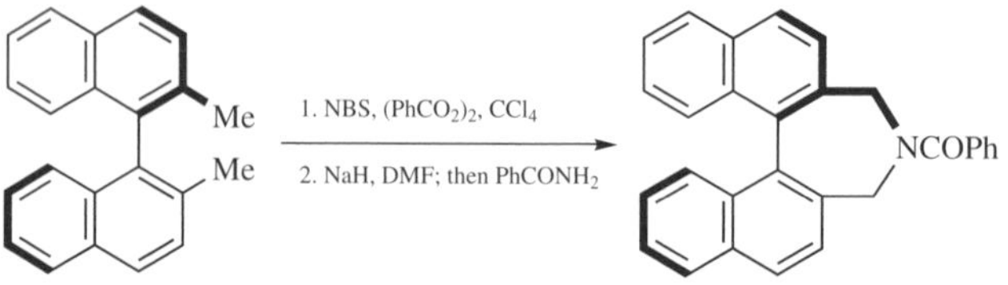

(3.21)

Another way of constructing axially chiral binaphthyls under nickel catalysis involves an enantioselective desymmetrization strategy.[85] Thus, dinaphthothiophene

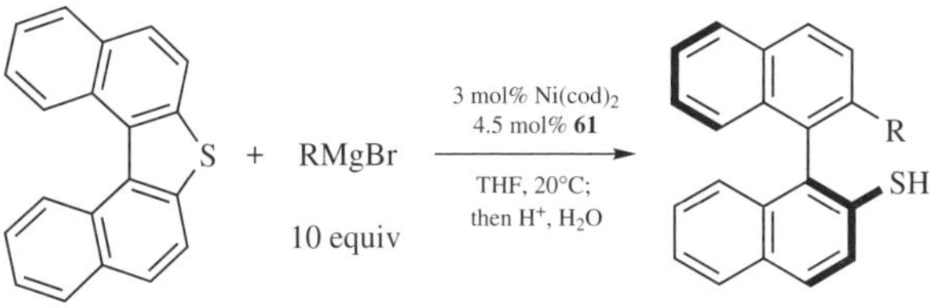

entry	R	yield (%)	ee (%)
1	Ph	92	95
2	4-MeC$_6$H$_4$	97	95
3	4-MeOC$_6$H$_4$	96	93
4	Me^a	54	54

^{a} MeMgI was used at 0 °C.

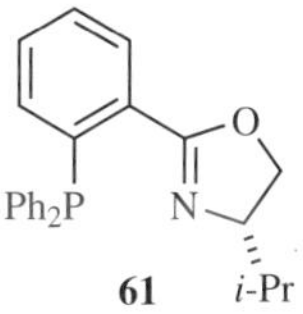

Figure 3.52. Scope of Ni/**61**-catalyzed asymmetric ring-opening of dinaphthothiophene **60** with Grignard reagents.

60 can be enantioselectively ring-opened by Grignard reagents in the presence of Ni/**61** (Figure 3.52). Based on the catalytic cycle described in Figure 3.53, it was proposed that the stereochemical outcome of this coupling reaction is determined at or after the transmetallation step, since the ee of the product depends on the nature of the Grignard reagent (e.g., entry 1 vs. entry 4).

The enantioselective desymmetrization strategy can also be used for the palladium-catalyzed cross-coupling reactions with Grignard reagents. Thus, in the

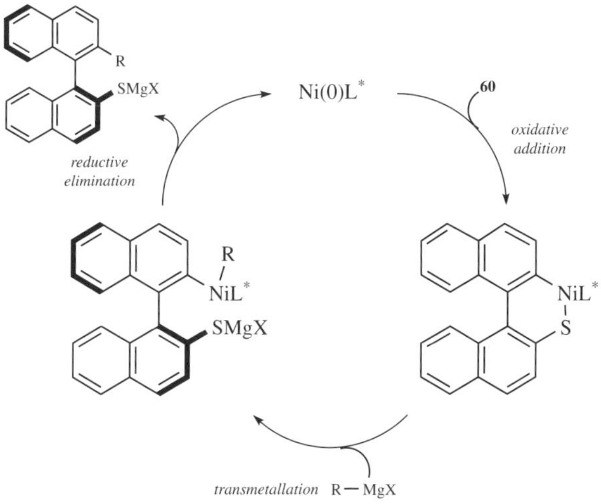

Figure 3.53. Catalytic cycle of Ni/**61**-catalyzed asymmetric ring-opening of **60** with Grignard reagents.

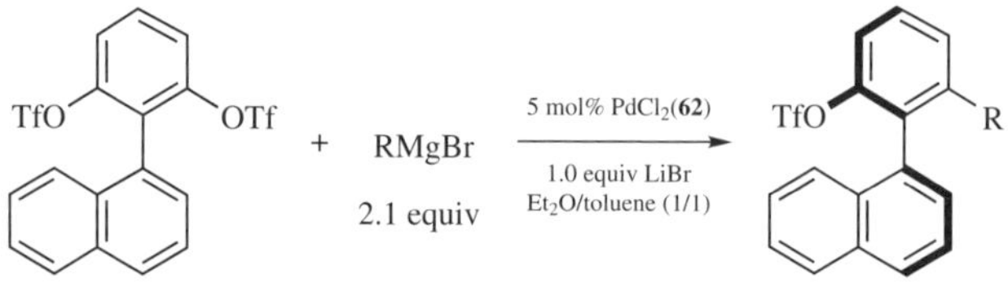

Figure 3.54. Scope of Pd/**62**-catalyzed asymmetric desymmetrization of a biaryl ditriflate with Grignard reagents.

entry	R	temp (°C)	time (h)	yield (%)	ee (%)
1	Ph	−30	48	87	93
2	3-MeC$_6$H$_4$	−20	48	83	90
3	≡—SiPh$_3$	20	6	88	92
4	≡—Ph	20	20	84	86

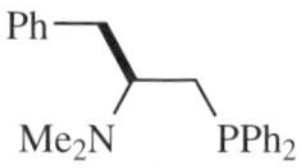

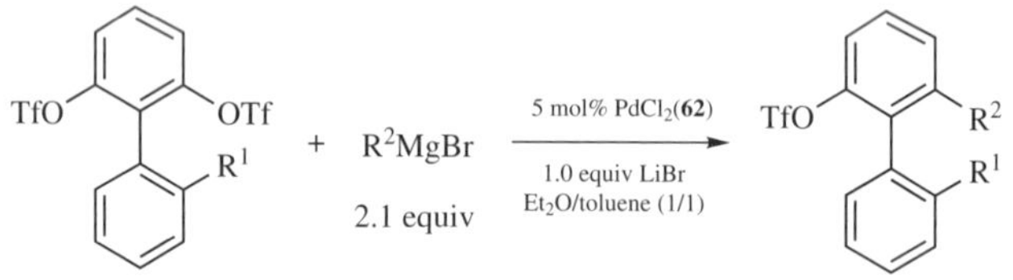

entry	R^1	R^2	temp (°C)	time (h)	yield (%)	ee (%)
1	Me	Ph	−20	48	77	84
2	Me	≡—SiPh$_3$	20	10	87	85
3	Ph	≡—SiPh$_3$	20	48	88	99

Figure 3.55. Scope of Pd/**62**-catalyzed asymmetric desymmetrization of biaryl ditriflates with Grignard reagents.

presence of Pd/**62** catalyst, prochiral biaryl ditriflates are desymmetrized effectively, generating axially chiral biaryls in high enantioselectivity (Figures 3.54 and 3.55).[86]

3.3.2 Suzuki-Type Cross-Couplings

Not only Grignard reagents but also organoboron nucleophiles have been used in the context of asymmetric cross-coupling reactions. Thus, Cammidge reported a palladium-catalyzed asymmetric Suzuki coupling for the construction of axially chiral binaphthyls by the use of several chiral phosphine ligands, and found that the use of ligand ent-**54** leads to the coupling product in as high as 85% ee [Eq. (3.22)].[87] Buchwald also developed a similar transformation by using ligand **63** (Figure 3.56).[88] The reaction proceeds with several ortho-substituted arylboronic acids to provide axially chiral biaryls in high yield and ee.

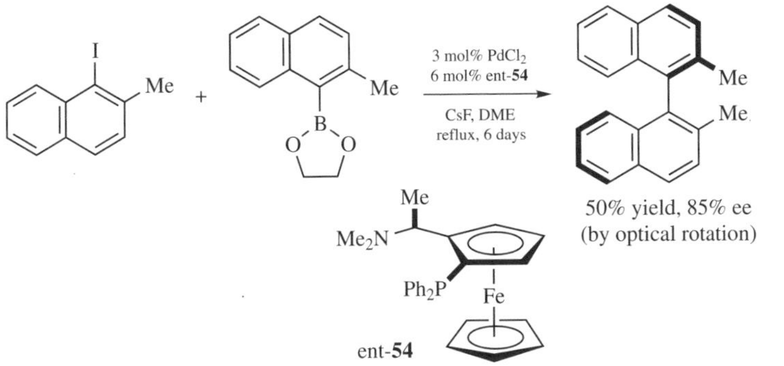

$$(3.22)$$

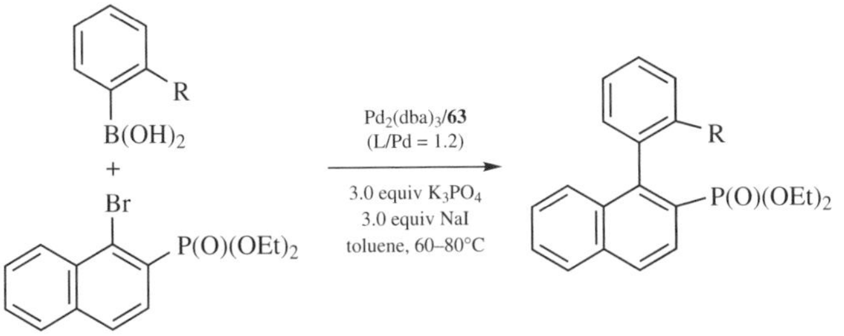

entry	R	mol% Pd	Yield (%)	ee (%)
1	Me	1	93	87
2	Et	2	96	92
3	*i*-Pr	2	89	85
4[a]	Ph	3	74	74

[a] 2.0 equiv of K_3PO_4 was used without NaI.

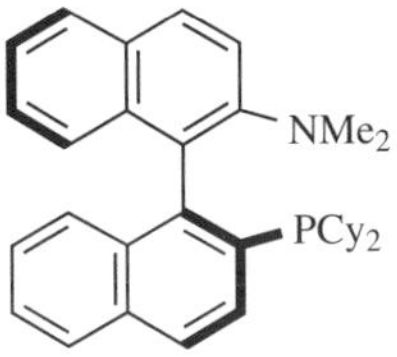

Figure 3.56. Scope of Pd/**63**-catalyzed asymmetric Suzuki cross-coupling of naphthyl bromides with arylboronic acids.

REFERENCES

1. Perlmutter, P. *Conjugate Addition Reactions in Organic Synthesis*, Tetrahedron Organic Chemistry Series Vol. 9, Pergamon, Tarrytown, NY, **1992**.

2. For an overview, see: (a) Tomioka, K.; Nagaoka, Y. In *Comprehensive Asymmetric Catalysis*; Jacobsen, E. N.; Pfaltz, A.; Yamamoto, H. (Eds.), Springer-Verlag, New York, 1999; Chap. 31.1; (b) Yamaguchi, M. In *Comprehensive Asymmetric Catalysis*; Jacobsen, E. N.; Pfaltz, A.; Yamamoto, H. (Eds.), Springer-Verlag; New York, **1999**; Chap. 31.2.

3. For recent reviews, see: (a) Feringa, B. L. *Acc. Chem. Res.* **2000**, *33*, 346–353; (b) Krause, N.; Hoffmann-Roder, A. *Synthesis* **2001**, 171–196; (c) Alexakis, A.; Benhaim, C. *Eur. J. Org. Chem.* **2002**, 3221–3236.

4. For recent reviews, see: (a) Hayashi, T.; Yamasaki, K. *Chem. Rev.* **2003**, *103*, 2829–2844; (b) Fagnou, K.; Lautens, M. *Chem. Rev.* **2003**, *103*, 169–196.

5. Kharasch, M. S.; Tawney, P. O. *J. Am. Chem. Soc.* **1941**, *63*, 2308–2315.

6. Villacorta, G. M.; Rao, C. P.; Lippard, S. J. *J. Am. Chem. Soc.* **1988**, *110*, 3175–3182.

7. Feringa, B. L.; Pineschi, M.; Arnold, L. A.; Imbos, R.; de Vries, A. H. M. *Angew. Chem., Int. Ed. Engl.* **1997**, *36*, 2620–2623.

8. Alexakis, A.; Benhaim, C.; Rosset, S.; Humam, M. *J. Am. Chem. Soc.* **2002**, *124*, 5262–5263.

9. Hua, Z.; Vassar, V. C.; Choi, H.; Ojima, I. *Proc. Natl. Acad. Sci. USA* **2004**, *101*, 5411–5416.

10. Alekakis, A.; Polet, D.; Benhaim, C.; Rosset, S. *Tetrahedron Asym.* **2004**, *15*, 2199–2203.

11. Degrado, S. J.; Mizutani, H.; Hoveyda, A. H. *J. Am. Chem. Soc.* **2001**, *123*, 755–756.

12. Breit, B.; Laungani, A. C. *Tetrahedron Asym.* **2003**, *14*, 3823–3826.

13. Reetz, M. T.; Gosberg, A.; Moulin, D. *Tetrahedron Lett.* **2002**, *43*, 1189–1191.

14. Krauss, I. J.; Leighton, J. L. *Org. Lett.* **2003**, *5*, 3201–3203.

15. Kanai, M.; Tomioka, K. *Tetrahedron Lett.* **1995**, *36*, 4275–4278.

16. Feringa, B. L.; Badorrey, R.; Peña, D.; Harutyunyan, S. R.; Minnaard, A. J. *Proc. Natl. Acad. Sci. USA* **2004**, *101*, 5834–5838.

17. Liang, L.; Chan, A. S. C. *Tetrahedron Asym.* **2002**, *13*, 1393–1396.

18. Peña, D.; López, F.; Harutyunyan, S. R.; Minnaard, A. J.; Feringa, B. L. *Chem. Commun.* **2004**, 1836–1837.

19. Escher, I. H.; Pfaltz, A. *Tetrahedron* **2000**, *56*, 2879–2888.

20. Hu, X.; Chen, H.; Zhang, X. *Angew. Chem. Int. Ed.* **1999**, *38*, 3518–3521.

21. Wan, H.; Hu, Y.; Liang, Y.; Gao, S.; Wang, J.; Zheng, Z.; Hu, X. *J. Org. Chem.* **2003**, *68*, 8277–8280.

22. Morimoto, T.; Mochizuki, N.; Suzuki, M. *Tetrahedron Lett.* **2004**, *45*, 5717–5722.

23. Shintani, R.; Fu, G. C. *Org. Lett.* **2002**, *4*, 3699–3702.

24. Mizutani, H.; Degrado, S. J.; Hoveyda, A. H. *J. Am. Chem. Soc.* **2002**, *124*, 779–781.

25. Degrado, S. J.; Mizutani, H.; Hoveyda, A. H. *J. Am. Chem. Soc.* **2002**, *124*, 13362–13363.

26. López, F.; Harutyunyan, S. R.; Minnaard, A. J.; Feringa, B. L. *J. Am. Chem. Soc.* **2004**, *126*, 12784–12785.

27. Liang, L.; Yan, M.; Li, Y.-M.; Chan, A. S. C. *Tetrahedron Asym.* **2004**, *15*, 2575–2578.

28. Pineschi, M.; Moro, F. D.; Gini, F.; Minnaard, A. J.; Feringa, B. L. *Chem. Commun.* **2004**, 1244–1245.

29. Hird, A. W.; Hoveyda, A. H. *Angew. Chem. Int. Ed.* **2003**, *42*, 1276–1279.

30. Versleijen, J. P. G.; van Lausen, A. M.; Feringa, B. L. *Tetrahedron Lett.* **1999**, *40*, 5803–5806.

31. Alexakis, A.; Benhaim, C. *Org. Lett.* **2000**, *2*, 2579–2581.

32. Duursma, A.; Minnaard, A. J.; Feringa, B. L. *Tetrahedron* **2002**, *58*, 5773–5778.

33. Luchaco-Cullis, C. A.; Hoveyda, A. H. *J. Am. Chem. Soc.* **2002**, *124*, 8192–8193.

34. Choi, H.; Hua, Z.; Ojima, I. *Org. Lett.* **2004**, *6*, 2689–2691.

35. Mampreian, D. M.; Hoveyda, A. H. *Org. Lett.* **2004**, *6*, 2829–2832.

36. Sakai, M.; Hayashi, H.; Miyaura, N. *Organometallics* **1997**, *16*, 4229–4231.

37. Takaya, Y.; Ogasawara, M.; Hayashi, T.; Sakai, M.; Miyaura, N. *J. Am. Chem. Soc.* **1998**, *120*, 5579–5580.

38. Hayashi, T.; Takahashi, M.; Takaya, Y.; Ogasawara, M. *J. Am. Chem. Soc.* **2002**, *124*, 5052–5058.

39. (a) Kuriyama, M.; Tomioka, K. *Tetrahedron Lett.* **2001**, *42*, 921–923; (b) Kuriyama, M.; Nagai, K.; Yamada, K.-i.; Miwa, Y.; Taga, T.; Tomioka, K. *J. Am. Chem. Soc.* **2002**, *124*, 8932–8939.

40. Reetz, M. T.; Moulin, D.; Gosberg, A. *Org. Lett.* **2001**, *3*, 4083–4085.

41. Shi, Q.; Xu, L.; Li, X.; Jia, X.; Wang, R.; Au-Yeung, T. T.-L.; Chan, A. S. C.; Hayashi, T.; Cao, R.; Hong, M. *Tetrahedron Lett.* **2003**, *44*, 6505–6508.

42. Hayashi, T.; Ueyama, K.; Tokunaga, N.; Yoshida, K. *J. Am. Chem. Soc.* **2003**, *125*, 11508–11509.

43. Defieber, C.; Paquin, J.-F.; Serna, S.; Carreira, E. M. *Org. Lett.* **2004**, *6*, 3873–3876.

44. Boiteau, J.-G.; Minnaard, A. J.; Feringa, B. L. *J. Org. Chem.* **2003**, *68*, 9481–9484.

45. Ma, Y.; Song, C.; Ma, C.; Sun, Z.; Chai, Q.; Andrus, M. B. *Angew. Chem. Int. Ed.* **2003**, *42*, 5871–5874.

46. Otomaru, Y.; Senda, T.; Hayashi, T. *Org. Lett.* **2004**, *6*, 3357–3359.

47. (a) Cauble, D. F.; Gipson, J. D.; Krische, M. J. *J. Am. Chem. Soc.* **2003**, *125*, 1110–1111; (b) Bocknack, B. M.; Wang, L.-C.; Krische, M. J. *Proc. Natl. Acad. Sci. USA* **2004**, *101*, 5421–5424.

48. Takaya, Y.; Ogasawara, M.; Hayashi, T. *Tetrahedron Lett.* **1998**, *39*, 8479–8482.

49. Takaya, Y.; Ogasawara, M.; Hayashi, T. *Tetrahedron Lett.* **1999**, *40*, 6957–6961.

50. (a) Pucheault, M.; Darses, S.; Genet, J.-P. *Tetrahedron Lett.* **2002**, *43*, 6155–6157; (b) Pucheault, M.; Darses, S.; Genêt, J.-P. *Eur. J. Org. Chem.* **2002**, 3552–3557.

51. Oi, S.; Taira, A.; Honma, Y.; Inoue, Y. *Org. Lett.* **2003**, *5*, 97–99.

52. Otomaru, Y.; Hayashi, T. *Tetrahedron Asym.* **2004**, *15*, 2647–2651.

53. Yoshida, K.; Ogasawara, M.; Hayashi, T. *J. Org. Chem.* **2003**, *68*, 1901–1905.

54. Hayashi, T.; Tokunaga, N.; Yoshida, K.; Han, J. W. *J. Am. Chem. Soc.* **2002**, *124*, 12102–12103.

55. Tokunaga, N.; Yoshida, K.; Hayashi, T. *Proc. Natl. Acad. Sci. USA* **2004**, *101*, 5445–5449.

56. Oi, S.; Sato, T.; Inoue, Y. *Tetrahedron Lett.* **2004**, *45*, 5051–5055.

57. Shintani, R.; Tokunaga, N.; Doi, H.; Hayashi, T. *J. Am. Chem. Soc.* **2004**, *126*, 6240–6241.

58. Takaya, Y.; Senda, T.; Kurushima, H.; Ogasawara, M.; Hayashi, T. *Tetrahedron Asym.* **1999**, *10*, 4047–4056.

59. Sakuma, S.; Sakai, M.; Itooka, R.; Miyaura, N. *J. Org. Chem.* **2000**, *65*, 5951–5955.

60. Senda, T.; Ogasawara, M.; Hayashi, T. *J. Org. Chem.* **2001**, *66*, 6852–6856.

61. Sakuma, S.; Miyaura, N. *J. Org. Chem.* **2001**, *66*, 8944–8946.

62. Pucheault, M.; Michaut, V.; Darses, S.; Genet, J.-P. *Tetrahedron Lett.* **2004**, *45*, 4729–4732.

63. Hayashi, T.; Senda, T.; Takaya, Y.; Ogasawara, M. *J. Am. Chem. Soc.* **1999**, *121*, 11591–11592.

64. Hayashi, T.; Senda, T.; Ogasawara, M. *J. Am. Chem. Soc.* **2000**, *122*, 10716–10717.

65. Shintani, R.; Ueyama, K.; Yamada, I.; Hayashi, T. *Org. Lett.* **2004**, *6*, 3425–3427.

66. Belyk, K. M.; Beguin, C. D.; Palucki, M.; Grinberg, N.; DaSilva, J.; Askin, D.; Yasuda, N. *Tetrahedron Lett.* **2004**, *45*, 3265–3268.

67. Navarre, L.; Darses, S.; Genet, J.-P. *Angew. Chem. Int. Ed.* **2004**, *43*, 719–723.

68. Yoshida, K.; Hayashi, T. *J. Am. Chem. Soc.* **2003**, *125*, 2872–2873.

69. Mauleon, P.; Carretero, J. C. *Org. Lett.* **2004**, *6*, 3195–3198.

70. Lautens, M.; Dockendorff, C.; Fagnou, K.; Malicki, A. *Org. Lett.* **2002**, *4*, 1311–1314.

71. Hayashi, T.; Tokunaga, N.; Inoue, K. *Org. Lett.* **2004**, *6*, 305–307.

72. Shintani, R.; Okamoto, K.; Otomaru, Y.; Ueyama, K.; Hayashi, T. *J. Am. Chem. Soc.* **2005**, *127*, 54–55.

73. For recent reviews on cross-coupling reactions, see: (a) *Handbook of Organopalladium Chemistry for Organic Synthesis*; Negishi, E. (Ed.), Wiley-Interscience, New York, **2002**, Vols. 1–2; (b) *J. Organomet. Chem.* **2002**, *653* (Special Issue: 30 Years of the Cross-Coupling Reaction); (c) *Topics Curr. Chem.* **2002**, *219* (Cross-Coupling Reactions); (d) Littke, A. F.; Fu, G. C. *Angew. Chem. Int. Ed.* **2002**, *41*, 4176–4211; (e) for recent advances in alkyl–alkyl cross-couplings, see: Cárdenas, D. J. *Angew. Chem. Int. Ed.* **2003**, *42*, 384–387 and references cited therein.

74. For recent reviews on asymmetric cross-coupling reactions, see: (a) Hayashi, T. In *Handbook of Organopalladium Chemistry for Organic Synthesis*; Negishi, E. (Ed.), Wiley-Interscience, New York, **2002**, Vol. 1, pp. 791–806; (b) Hayashi, T. *J. Organomet. Chem.* **2002**, *653*, 41–45; (c) Ogasawara, M.; Hayashi, T. In *Catalytic Asymmetric Synthesis,* 2nd ed., Ojima, I. (Ed.), Wiley-VCH, New York, **2000**, pp. 651–674; see also: (d) Tietze, L. F.; Ila, H.; Bell, H. P. *Chem. Rev.* **2004**, *104*, 3453–3516.

75. (a) Hayashi, T.; Tajika, M.; Tamao, K.; Kumada, M. *J. Am. Chem. Soc.* **1976**, *98*, 3718–3719; (b) Hayashi, T.; Fukushima, M.; Konishi, M.; Kumada, M. *Tetrahedon Lett.* **1980**, *21*, 79–82; (c) Hayashi, T.; Konishi, M.; Fukushima, M.; Kanehira, K.; Hioki, T.; Kumada, M. *J. Org. Chem.* **1983**, *48*, 2195–2202.

76. Vriesema, B. K.; Kellogg, R. M. *Tetrahedron Lett.* **1986**, *27*, 2049–2052.

77. Ohno, A.; Yamane, M.; Hayashi, T.; Oguni, N.; Hayashi, M. *Tetrahedron Asym.* **1995**, *6*, 2495–2502.

78. Nagel, U.; Nedden, H. G. *Chem. Ber.* **1997**, *130*, 535–542.

79. Pellet-Rostaing, S.; Saluzzo, C.; Ter Halle, R.; Breuzard, J.; Vial, L.; Le Guyader, F.; Lamaire, M. *Tetrahedron Asym.* **2001**, *12*, 1983–1985.

80. (a) Hayashi, T.; Hagihara, T.; Katsuro, Y.; Kumada, M. *Bull. Chem. Soc. Jpn.* **1983**, *56*, 363–364; (b) Hayashi, T.; Yamamoto, A.; Hojo, M.; Ito, Y. *J. Chem. Soc., Chem. Commun.* **1989**, 495–496.

81. Schwink, L.; Knochel, P. *Chem. Eur. J.* **1998**, *4*, 950–968.

82. Horibe, H.; Fukuda, Y.; Kondo, K.; Okuno, H.; Murakami, Y.; Aoyama, T. *Tetrahedron* **2004**, *60*, 10701–10709.

83. (a) Hayashi, T.; Konishi, M.; Ito, H.; Kumada, M. *J. Am. Chem. Soc.* **1982**, *104*, 4962–4963; (b) Hayashi, T.; Konishi, M.; Okamoto, Y.; Kabeta, K.; Kumada, M. *J. Org. Chem.* **1986**, *51*, 3772–3781.

84. Hayashi, T.; Hayashizaki, K.; Kiyoi, T.; Ito, Y. *J. Am. Chem. Soc.* **1988**, *110*, 8153–8156.

85. Shimada, T.; Cho, Y.-H.; Hayashi, T. *J. Am. Chem. Soc.* **2002**, *124*, 13396–13397.

86. (a) Hayashi, T.; Niizuma, S.; Kamikawa, T.; Suzuki, N.; Uozumi, Y. *J. Am. Chem. Soc.* **1995**, *117*, 9101–9102; (b) Kamikawa, T.; Hayashi, T. *Tetrahedron* **1999**, *55*, 3455–3466; (c) Kamikawa, T.; Uozumi, Y.; Hayashi, T. *Tetrahedron Lett.* **1996**, *37*, 3161–3164.

87. Cammidge, A. N.; Crépy, K. V. L. *Chem. Commun.* **2000**, 1723–1724.

88. Yin, J.; Buchwald, S. L. *J. Am. Chem. Soc.* **2000**, *122*, 12051–12052.

4

ACTIVATION OF SMALL MOLECULES (C=O, HCN, RN=C, AND CO₂)

KYOKO NOZAKI

Department of Chemistry and Biotechnology, The University of Tokyo, Bunkyo-ku, Tokyo, Japan

4.1 INTRODUCTION

Several catalytic reactions are known for the addition of small molecules, such as carbon monoxide, carbon dioxide, hydrogen cyanide, and their isoelectronic analogs, to carbon–carbon or carbon–heteroatom multiple bonds. Asymmetric catalysis is of particular interest synthetically for such an addition reaction because it is possible to introduce a functional group and a stereogenic center in one step. Asymmetric addition of hydrogen cyanide to an aldehyde is one of the oldest targets for asymmetric catalysis, which has been studied since 1912.[1] Although some of the reactions discussed in this chapter have rich historical background, the author pays more attention to the state of the art of catalyst design, rather than surveying the historical developments.

The asymmetric reactions discussed in this chapter may be divided into three different types of reaction, as (1) hydrometallation of olefins followed by the C–C bond formation, (2) two C–C bond formations on a formally divalent carbon atom, and (3) nucleophilic addition of cyanide or isocyanide anion to a carbonyl or its analogs (Scheme 4.1). For reaction type 1, here described are hydrocarbonylation represented by hydroformylation and hydrocyanation. As for type 2, Pauson–Khand reaction and olefin/CO copolymerization are mentioned. Several nucleophilic additions to aldehydes and imines (or iminiums) are described as type 3.

New Frontiers in Asymmetric Catalysis, Edited by Koichi Mikami and Mark Lautens
Copyright © 2007 John Wiley & Sons, Inc.

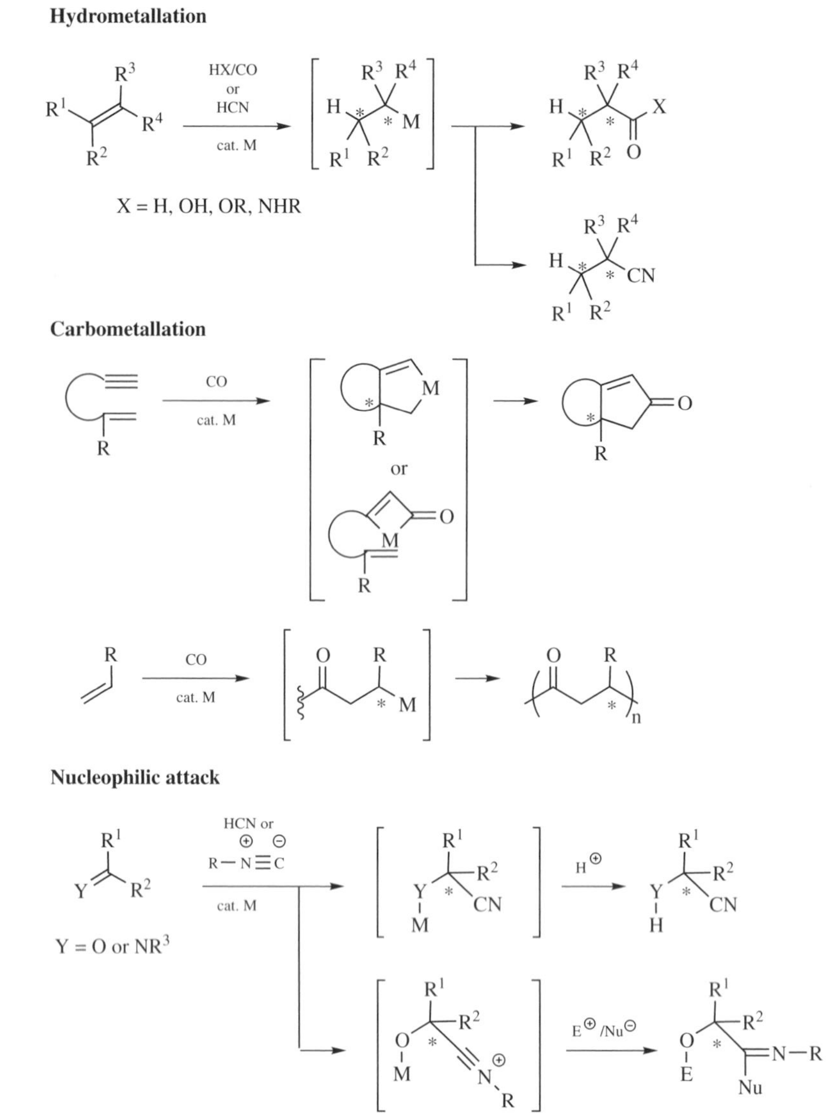

Hydrometallation

X = H, OH, OR, NHR

Carbometallation

Nucleophilic attack

Y = O or NR^3

Scheme 4.1. Addition of CO, HCN, and RNC to carbon–carbon and carbon–heteroatom double bonds.

4.2 ASYMMETRIC HYDROFORMYLATION OF OLEFINS

Transition-metal-catalyzed carbonylations are of great importance in organic synthesis as a powerful tool to prepare a variety of carbony compounds. Among them, hydroformylation has been most extensively studied not only in the laboratory but also in industry. Industrial production of alkanals from 1-alkenes is one of the biggest commercial processes of homogeneous catalysis.[2] The catalysts mostly employed are $HCo(CO)_4$, $HRh(CO)_4$, or their phosphine or phosphite-modified complexes. In contrast, Rh-chiral phosphine or phosphite ligands are of the first choice for the recent asymmetric hydroformylation. Modified Pt complexes are also sometimes employed as catalysts.

4.2.1 The Mechanism of Hydroformylation

The reaction mechanism described in Scheme 4.2 is now accepted for hydroformylation of olefins catalyzed by rhodium complexes of chelating diphosphines and diphosphites (L—L).[3–5] For platinum catalysts, a comprehensive review by Mortreux may be consulted.[6] The cycle on the right of Scheme 4.2 gives branched aldehydes, and the cycle on the left provides linear aldehydes. In this mechanism (1) $HRh(CO)_2L_2$ (18-electron species) is formed as a key precursor; (2) dissociation of CO from this complex generates a coordinatively unsaturated 16-electron species $HRh(CO)L_2$ that is the active catalyst; (3) coordination of an olefin, followed by olefin insertion into the Rh—H bond, takes place to form alkyl-$Rh(CO)L_2$ complex; (4) coordination of carbon monoxide is followed by migratory insertion of the alkyl group to one of the coordinated carbon monoxides; and (5) oxidative addition of molecular hydrogen, followed by the reductive elimination to give the aldehyde and regeneration of the active catalyst $HRh(CO)L_2$, which completes the catalytic cycle. It has been shown that the hydrometalation of olefin (step 3) proceeds through complete cis addition, and the subsequent migratory insertion of carbon monoxide (step 4) takes place with retention of configuration.[3] The rate-determining step varies depending on the

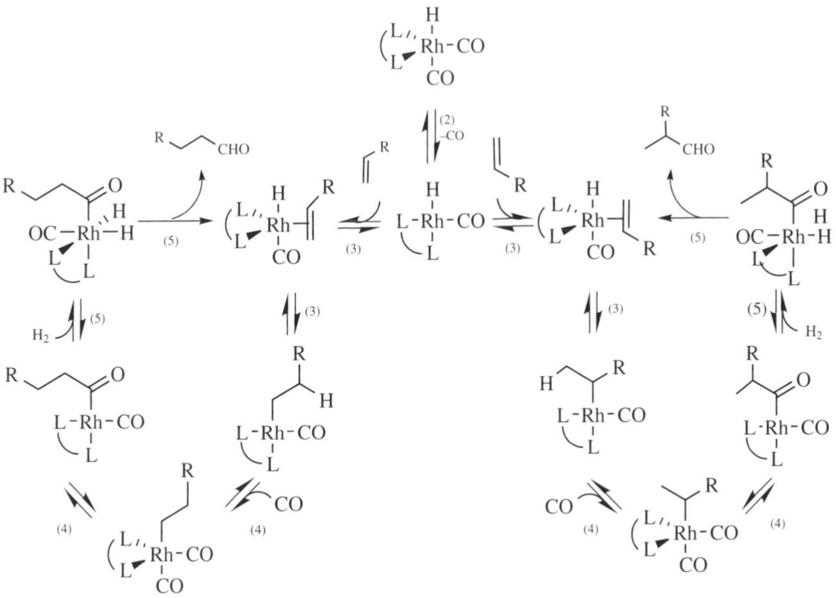

Scheme 4.2

catalysts and reaction conditions, such as ligand, catalyst concentration, and partial pressures of H$_2$ and CO. As far as asymmetric catalysis is concerned, the reaction rate is reported to vary with the olefin concentration, suggesting that the olefin coordination/insertion is the key step for the enantiofacial selection.[3,7]

4.2.2 Scope and Limitation of Asymmetric Hydroformylation

Asymmetric Hydroformylation of Vinylarenes α-Arylpropanals, the products of asymmetric hydroformylation of vinylarenes, serve as useful intermediates for pharmaceutical drugs.[8] For example, (S)-2-arylpropanals can be oxidized to the corresponding (S)-2-arylpropanoic acids, such as (S)-ibuprofen (Ar = 4-isobutylphenyl), (S)-naproxen (Ar = 6-methoxynaphthalen-2-yl), and (S)-suprofen (Ar = 4-(2-thienylcarbonyl)phenyl) (see later in chapter, Scheme 4.4). Styrene is thus one of the most popular substrates used to test new catalyst systems. Representative ligands and their use as Pt or Rh complexes in the asymmetric hydroformylation are summarized in Figure 4.1 and Table 4.1. (See also Scheme 4.3.)

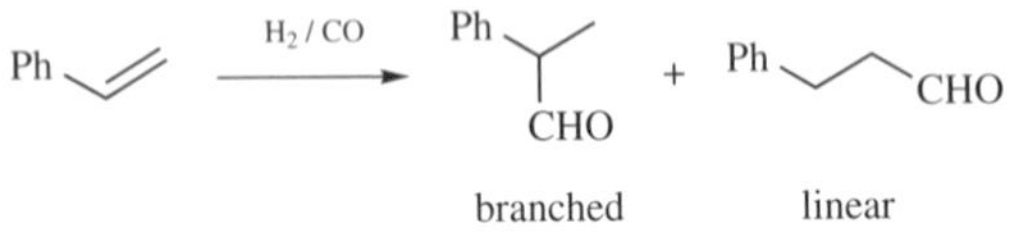

Scheme 4.3

Good regioselectivity (branched/linear = 80/20) and enantioselectivity (85% ee) for the formation of (S)-1-phenylpropanal are achieved using PtCl$_2$/SnCl$_2$/(R,R)-BCO-DBP (**1**), but the selectivity for aldehydes is rather low (67%) and a substantial amount of hydrogenated compound, ethylbenzene, was formed as a byproduct (entry 1).[9] The reaction catalyzed by PtCl$_2$/SnCl$_2$/(S,S)-BPPM (**2**) gives the branced aldehyde with lower regioselectivity (branched/linear = 31/69) and enantiopurity (77% ee) (entry 2).[10,11] Racemization of the branced product was observed because of the acidic conditions in this system. In order to avoid this racemization, CH(OEt)$_3$ was used as solvent so that the branched aldehyde was converted in situ to the corresponding diethyl acetal. This protocol has improved the enantioselectivity dramatically (>96% ee), but the regioselectivity remains low (branched/linear = 33/67) and the reaction rate is very low (entry 3).[11] In contrast to the Pt-catalyzed reactions, the use of Rh-catalysts result in the exclusive formation of aldehydes. Chiral diphosphite **3** is successfully used in the Rh(I) complex-catalyzed asymmetric hydroformylation of styrene, achieving excellent regioselectivity (branched/linear = 98/2) as well as enantioselectivity (90% ee) at 25°C (entry 4).[12] Rhodium catalyst with chiral diphosphite **4** has also attained 87% ee and excellent regioselectivity at 25°C, although the conversion is low (26%) (entry 5).[13] Rhodium complex–catalyzed reaction of styrene with chiral phosphine–phosphite ligands, (R,S)-BINAPHOS (**5a**) and its derivative **5b** give excellent enantioselectivities (94 and 98% ee) with high branched aldehyde selectivity (entries 6,8).[14–16] For the Rh-BINAPHOS (**5a**)–catalyzed reaction, a turnover frequency of 100 (mol

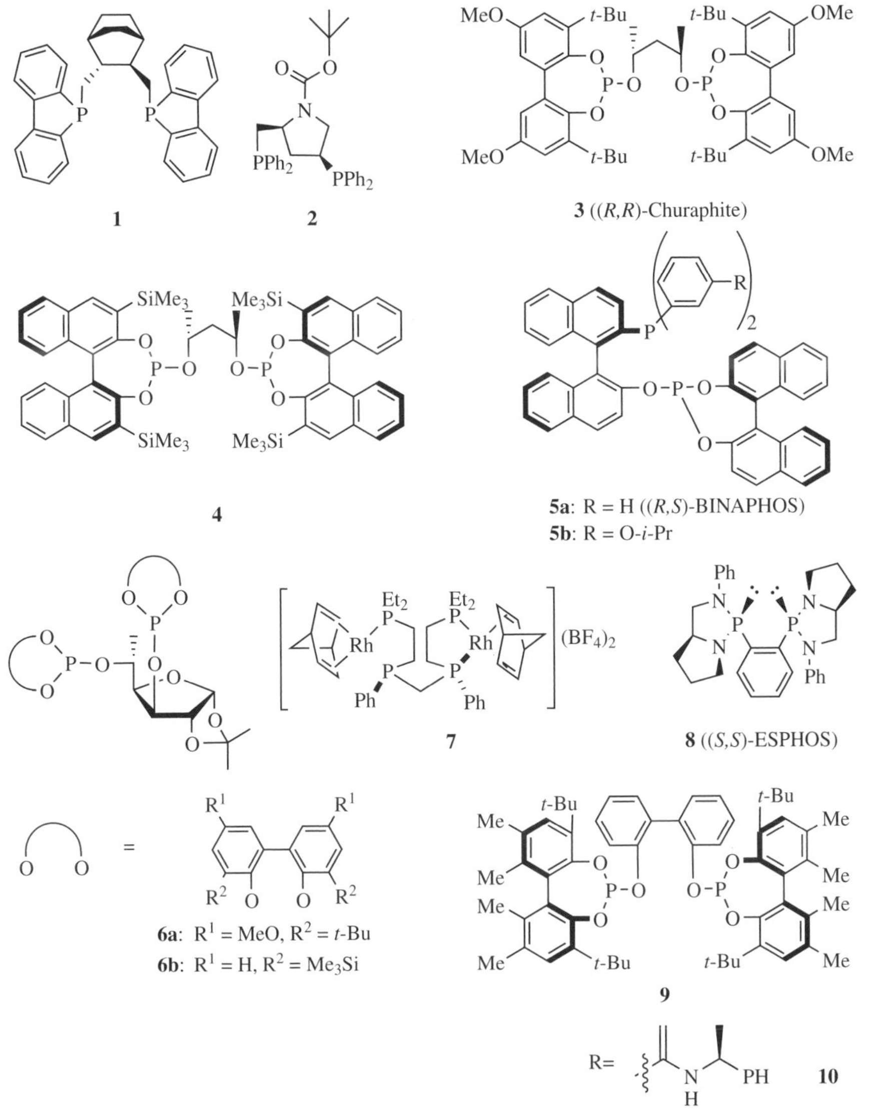

Figure 4.1. Chiral ligands used in asymmetric hydroformylation.

aldehydes)•(mol Rh)$^{-1}$(h)$^{-1}$ is achieved at 40°C under a total pressure of 10 atm (H$_2$/CO = 1/1) (entry 7).[7a] Recently developed diphosphites **6a**[17] and **6b**[18] bearing sugar frameworks provide the branched aldehydes in >98/2 branch selectivity with 90 and 93% ee, respectively.[19]

(S)-2-(4-Isobutylphenyl)propanal with 92% ee is obtained from p-isobutylstyrene using the Rh–BINAPHOS catalyst, which is the precursor of anti-inflammatory drug (S)-ibuprofen.[14] In a similar manner, the precursor of (S)-naproxen is obtained

TABLE 4.1. Hydroformylation of Styrene by Chiral Pt(II) or Rh(I) Complexes

Entry	Metal/(Additive)/ Chiral Ligand	Pressure (atm/atm)	Temperature (°C)	TOF^a × 10^{-3}	Branched/ Linear Ratio	% ee of Branched (Configuration)	Reference
1	PtCl$_2$/SnCl$_2$/**1**^{b}	70/140	60	41	80/20	85 (S)	9
2	PtCl$_2$/SnCl$_2$/**2**	81/81	60	4.0	31/69	77 (S)	11
3	PtCl$_2$/SnCl$_2$/**2**^{c}	81/81	60	0.27	33/67	≥96 (S)	10,11
4	Rh(acac)(CO)$_2$/**3**	19/19	25	^{e}	98/2	90 (S)	12
5	Rh(acac)(CO)$_2$/**4**	10/10	15	11	94/6	86 (S)	13
6	Rh(acac)(CO)$_2$/**5a**	50/50	60	>4.7	88/12	94 (R)	14
7	Rh(acac)(CO)$_2$/**5a**	5/5	40	100	90/10	94 (R)	7a
8	Rh(acac)(CO)$_2$/**5b**	10/10	30	1.2	94/6	98 (R)	16
9	Rh(acac)(CO)$_2$/**6a**	10/10	20	18	98.6/1.4	90	17
10	Rh(acac)(CO)$_2$/**6b**	10/10	20	11	98.8/1.2	93	18

^{a}Turnover Frequency (TOF): (mol product) · (mol metal)$^{-1}$ h^{-1}.
^{b}Selectivity to aldehyde was 67%. Ethylbenzene (33%) was formed as byproduct.
^{c}Triethyl orthoformate is used as a solvent. Product aldehydes are obtained as diethylacetals.
^{d}0.11 g·mol L^{-1} h^{-1} at [Rh] = 250 ppm.
^{e}Not reported.

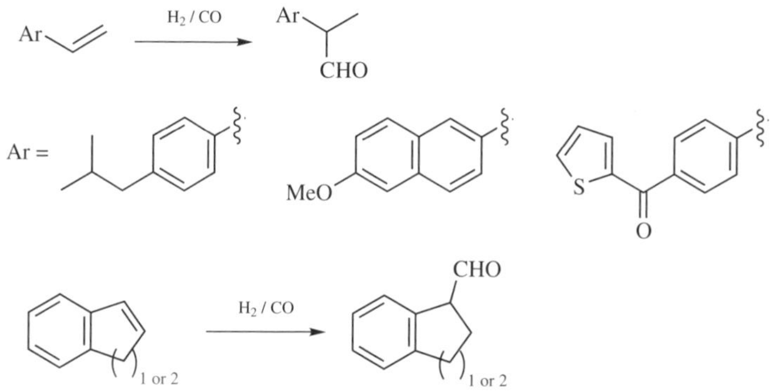

Scheme 4.4

with 85% ee and excellent regioselectivity in the reaction of 6-methoxy-2-vinylnaphthalene catalyzed by Rh-(diphosphite **3**) complex.[12] Using PtCl$_2$/SnCl$_2$/**1** in (EtO)$_3$CH, the precursor of (S)-suprofen (4-(2-thienylcarbonyl)benzoic acid) was obtained in high ee (>96%) but the catalytic activity was extremely low (15% conversion in 143 h).[11] Pentafluorostyrene, 1-propenylbenzenes, indene and 1,2-dihydronaphthalene were all converted into the corresponding α-arylaldehydes, with high regio- and enantioselectivities.[15] (See Scheme 4.4.)

Asymmetric Hydroformylation of Other Olefins   In contrast to vinylarenes, the asymmetric hydroformylation of aliphatic alkenes, especially 1-alkenes, is still very challenging, although the Rh–BINAPHOS (**5a**) catalyst has made significant improvement in enantioselectivity as compared to that achieved by previous catalysts. Representative results are summarized in Scheme 4.5. The reaction of

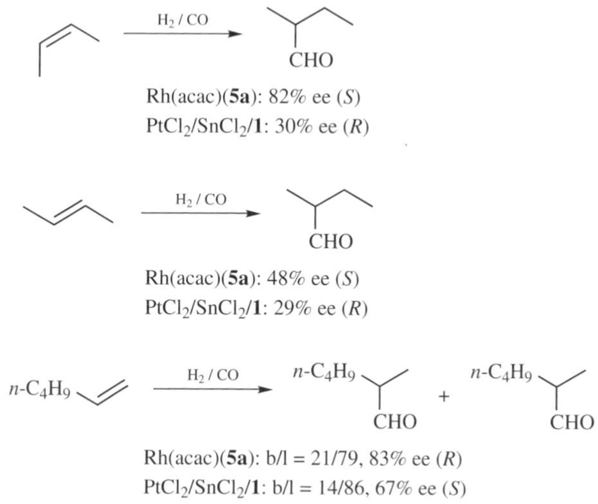

Rh(acac)(**5a**): 82% ee (S)
PtCl$_2$/SnCl$_2$/**1**: 30% ee (R)

Rh(acac)(**5a**): 48% ee (S)
PtCl$_2$/SnCl$_2$/**1**: 29% ee (R)

Rh(acac)(**5a**): b/l = 21/79, 83% ee (R)
PtCl$_2$/SnCl$_2$/**1**: b/l = 14/86, 67% ee (S)

Scheme 4.5

(Z)-2-butene catalyzed by Rh(acac)(CO)$_2$/**5a** gives (S)-2-methylbutanal as single product with 82% ee.[14] The absence of pentanal formation in this reaction unambiguously excludes the possibility of isomerization of 2-butene to 1-butene under the reaction conditions. In contrast, the formation of 1-pentanal is observed (2-methylbutanal (30% ee)/1-pentanal = 87/13) in the reactions catalyzed by PtCl$_2$/SnCl$_2$/BCO-DBP (**1**). As compared to the asymmetric hydroformylation of (Z)-2-butene, the reaction of its (E)-isomer proceeds more slowly, resulting in lower enantioselectivity, 48% ee with Rh/**5a** and 29% ee with Pt/**1**. The asymmetric hydroformylation of 1-alkenes suffers from low regioselectivity for the formation of branched aldehydes (branch/linear = 21/79) although the enantioselectivity has reached 83% ee.[20]

Asymmetric hydroformylation of conjugated dienes affords unsaturated chiral aldehydes. Promising results were reported using Rh(acac)(CO)$_2$/**5a** as the catalyst; the corresponding β,γ-unsaturated aldehydes were obtained with high enantiopurity ($\leq$ 96% ee) and regioselectivity (78–95% branched).[21] Asymmetric hydroformylation of functionalized alkenes can serve as a useful method for the syntheses of polyfunctionalized intermediates to biologically active compounds and materials. Representative results are listed in Scheme 4.6. Reaction of vinyl acetate catalyzed by Rh(acac)(CO)$_2$/**5a** gives (S)-2-acetoxypropanal with 92% ee accompanied by a small amount of 3-acetoxypropanal (14%) in quantitative yield.[14] 2-Acetoxypropanal can be readily converted to lactic acid derivatives that acts a monomer of biodegradable or bioabsorbable polymers. Threonine and its derivatives can also be synthesized from these aldehydes. Chiral bimetallic catalyst Rh$_2$(allyl)$_2$(et,ph-P4)/

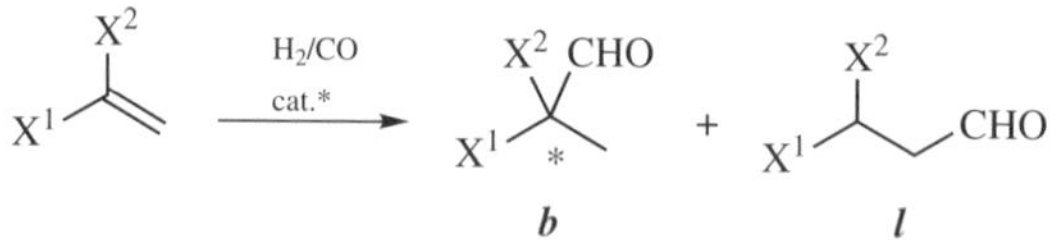

X^1	X^2	cat	b/l	% ee of b
cyclohexen-1-yl	H	Rh(acac)(CO)$_2$/**5a**	87/13	96 (S)
CH$_3$COO	H	Rh(acac)(CO)$_2$/**5a**	86/14	92 (S)
CH$_3$COO	H	**7**	80/20	85 (nr)
CH$_3$COO	H	Rh(acac)(CO)$_2$/**8**	94/6	90 (S)
CH$_3$COO	H	Rh(acac)(CO)$_2$/**10**	97/3	96 (S)
C(O)(C$_6$H$_4$)C(O)N	H	Rh(acac)(CO)$_2$/**5a**	89/11	85 (S)
CF$_3$	H	Rh(acac)(CO)$_2$/**5a**	95/5	93 (S)
COOMe	NHCOMe	RhH(CO)(PPh$_3$)$_3$/(R,R)-DIOP	100/0	59 (R)
COOMe	CH$_2$COOMe	PtCl$_2$/SnCl$_2$/(R,R)-DIOP	0/100	82 (R)
CH$_2$CN	H	Rh(acac)(CO)$_2$/**5a**	73/27	77 (S)
CH$_2$CN	H	Rh(acac)(CO)$_2$/**9**	95/5	80 (S)

Scheme 4.6

HBF$_4$ (**7**) exhibits excellent catalytic activity (TOF $= 125$ h^{-1}) under low pressure (6 atm) to give the branched aldehyde with 85% ee.[22] Rhodium complex of bis(diazapospholidine) ligand ESPHOS (**8**) is a highly selective hydroformylation catalyst particular for vinyl acetate providing the branched aldehyde with 90% ee.[23] Very recently, even higher selectivity of 96% ee was reported by Landis and Klosin using bis-3,4-diazaphospholane ligand **10**.[24] Reaction of N-vinylphthalimide catalyzed by Rh-**5a** complex gives N-(1-formylethyl)phthalimide that can be transformed to alanine and its derivatives. Reaction of trifluoropropene with the same catalyst gives 2-trifluoromethylpropanal with 93% ee,[14] which is known to be a   versatile intermediate to CF$_3$-containing amino acids and peptides of biological interest. Therefore, the asymmetric hydroformylation of trifluoropropene is a convenient way to synthesize such CF$_3$–amino acids and peptides.[25] N-Acethyldehydroalanine methyl ester is converted to the corresponding tertiary aldehyde with 100% selectivity and 59% ee using RhH(CO)(PPh$_3$)$_3$/(R,R)-DIOP.[26] The reaction of dimethyl itaconate catalyzed by PtCl$_2$/SnCl$_2$/(R,R)-DIOP gives the linear aldehyde with 82% ee, but the selectivity to aldehyde is only 35%.[27] Aldehyde obtained by asymmetric hydroformylation of allyl cyanide,[28] (R)- and (S)-3-cyano-2-methylpropanals, are both important intermediates to a novel tachykinin NK1 receptor antagonist developed by Takeda[29] and a nonpeptide gonadotropin-releasing hormone antagonist reported by Merck,[30] respectively (Scheme 4.7).

Reaction of cinnamyl alcohol catalyzed by Rh(acac)(CO)$_2$/**5a** gives the product as lactol (1:1 mixture of diastereomers at the anomeric carbon) with high enantioselectivity (88% ee).[31] The enantiopurity of the lactol is determined by oxidizing the lactol to the corresponding lactone. In the same manner, homoallyl alcohol is converted to the corresponding α-methyl-γ-butyrolactone with 73% ee via a lactol).[31] However, the regioselectivity of the reaction is not favorable to the formation of α-methyl-γ-butyrolactone, forming achiral δ-lactol as the major product. (See Scheme 4.8.)

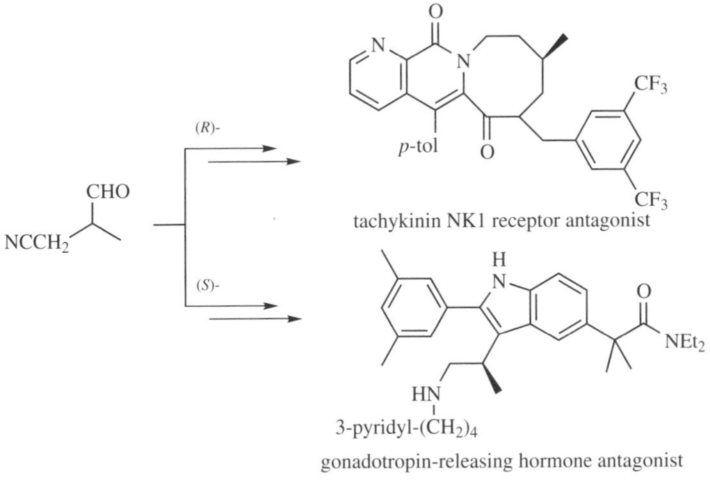

tachykinin NK1 receptor antagonist

gonadotropin-releasing hormone antagonist

Scheme 4.7

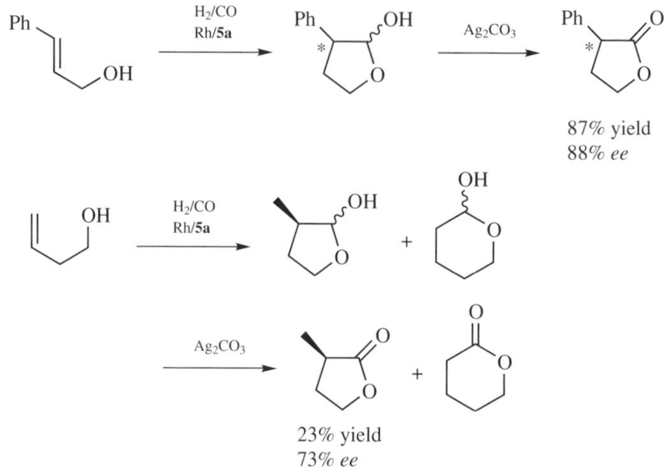

Scheme 4.8

Application to the Synthesis of Complex Molecules   One goal of homogeneous asymmetric catalysis is its applications to the production of fine chemicals such as pharmaceuticals, agrochemicals, flavors, and fragrances. As mentioned above, asymmetric hydroformylation of vinylarenes and allyl cyanide find their applications as the intermediates to drugs. In addition, asymmetric hydroformylation is now applicable to substrates with the more complex structures. For example, the hydroformylation of (3S,4R)-3-{(S)-1-(*tert*-butyldimethylsilyloxy)ethyl}-4-vinyl-β-lactam with Rh/**5a** catalyst affords the desired product β-product, its epimer, the α-product, and their linear isomer in a ratio of 51:4:45 (Scheme 4.9). Using chiral phosphine–phosphinites, an even better result (92% de; β:α:linear = 71:3:26) was achieved.[32] The aldehydes thus obtained can be oxidized to the corresponding carboxylic acids without epimerization. Asymmetric synthesis of 1β-methylcarbapenem is of much interest because of its high potential as antibacterial antibiotic drug, and the β-product is an attractive intermediate for this family of compounds. Asymmetric hydroformylation is also used in total synthesis of a natural product. As shown in Scheme 4.10, asymmetric hydroformylation of conjugated diene is employed as a key step for the

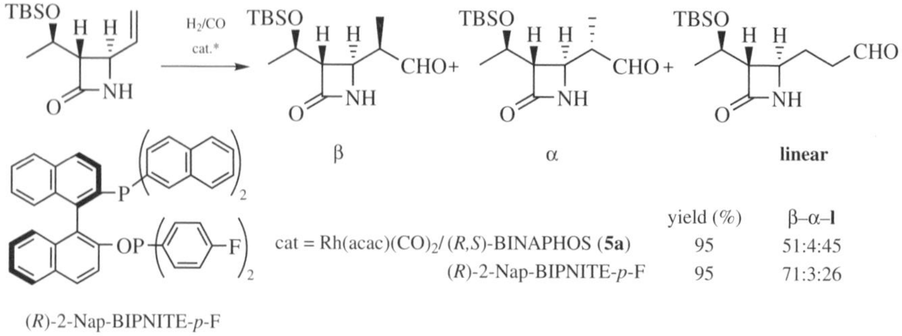

	yield (%)	β–α–l
cat = Rh(acac)(CO)$_2$/ (R,S)-BINAPHOS (**5a**)	95	51:4:45
(R)-2-Nap-BIPNITE-*p*-F	95	71:3:26

(R)-2-Nap-BIPNITE-*p*-F

Scheme 4.9

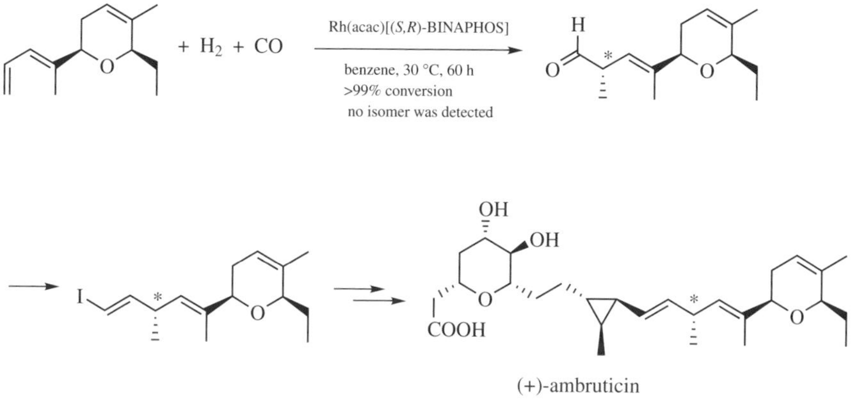

Scheme 4.10

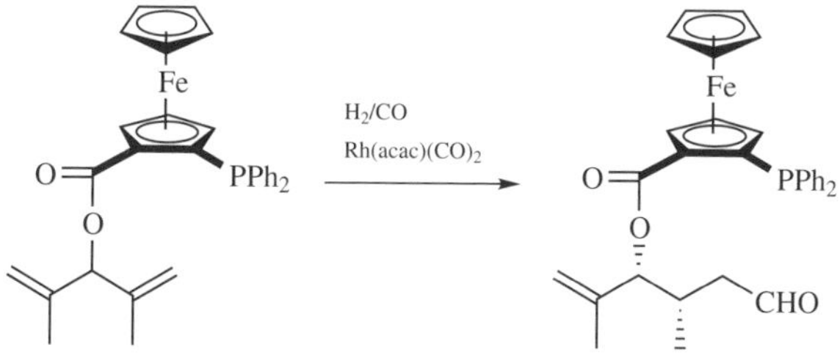

Scheme 4.11

total synthesis of (+)-ambruticin, an antifugal agent isolated from fermentation extracts of the myxobacterium *Polyangium cellulosum*.[33] Diastereoselective hydroformylation is an important issue for the synthesis of complex molecules and is summarized in a review article.[34] As a unique example of the diastereoselective hydroformylation, desymmetrization of achiral carbinol is cited as Scheme 4.11.[35]

4.2.3 "Greener" Catalysts in Asymmetric Hydroformylation

Catalyst separation is often one of the major problems in homogeneous catalysis. Extensive research has been devoted to developing new methods that combine the ease of catalyst recovery associated with heterogeneous systems with the more desirable activity and selectivity obtained with homogeneous catalysts. The BPPM-Pt catalyst immobilized on a polymer support, PtCl$_2$/SnCl$_2$/polymer–BPPM, is reported to exhibit virtually the same regio- as well as enantioselectivity as that achieved by its homogeneous counterpart).[36,37] Immobilization of (R,S)-BINAPHOS (5a) at the 6′-position of the phosphine unit to highly crosslinked polystyrene (55% crosslinking) enabled its use under the continuous flow of substrates.[38] When the substrate is less volatile, supercritical carbon dioxide is used as a cosolvent for the flow system. Asymmetric hydroformylation in supercritical carbon dioxide is also reported using BINAPHOS derivative bearing perfluoroalkyl groups.[39,5] Biphasic reaction systems using water-soluble Rh complexes have been reported but the selectivities are rather low so far.

4.3   ASYMMETRIC HYDROCARBOHYDROXYLATION AND RELATED REACTIONS

4.3.1   Asymmetric Hydrocarbalkoxylation of Alkenes

Asymmetric hydrocarbalkoxylation of alkenes has been studied since early 1970s, but the number of papers published on this subject is much less than that of asymmetric hydroformylation. Representative results on the asymmetric hydrocarbalkoxylation and hydrocarbohydroxylation are summarized in Scheme 4.12. The reaction of 2-phenylpropene with CO and *tert*-butyl alcohol catalyzed by PdCl$_2$/DBP-DIOP (**11a**) (238 atm of CO at 100°C) provides *tert*-butyl 3-phenylbutanoate with 69% ee at 8% conversion).[40] A closely related catalyst system, PdCl$_2$/DIOP (**11b**), catalyzes the reactions of 2-phenylpropene and methyl methacrylate at 100–120°C and 380–400 atm of CO to give the corresponding linear esters in 59% ee and 49% ee, respectively).[41] Using monodentate ligands asymmetric hydrocarbalkoxylation of styrene is accomplished under much milder conditions to realize ~50% ee. For example, a combination of Pd$_2$(dba)$_3$/neomenthyldiphenyl-phosphine (**12**)/trifuloroacetic acid,[42] or an acid-free system PdCl$_2$/2-(dicyclopentylphosphino)-2′-methoxy-1,1′-binaphthyl (**13**)[43] can be used at 50°C (CO 1 atm) or at 40°C (CO 30 atm), respectively. Higher ee's of 86% and 99% have been reported using the catalyst systems Pd(OAc)$_2$/BPPFA (**14**)/*p*-toluenesulfonic acid[44] and PdCl$_2$/CuCl$_2$/**15**.[45] A catalyst system PdCl$_2$/CuCl$_2$/ (*R*)-1,1′-binaphthyl-2,2′-diyl hydrogen phosphate, (*R*)-(−)-BNPPA (**16**) was introduced in 1990.[46] The reactions of 4-isobutylstyrene and 6-methoxy-2-naphthylethene give the corresponding methyl esters of (*S*)-ibuprofen with 83–84% ee and (*S*)-naproxen with 91% ee, respectively.[47]

Several related reations are summarized in Scheme 4.13. When a nucleophilic —OH or —NH$_2$ group exists at the proper position of the olefinic substrate, cyclohydrocarbonylation provides a lactone or a lactam. Allyl alcohols[48,49] and homoallyl amines[50] are converted into γ-lactones and δ-lactams in high enantiomeric excesses. Thiol is employable as a nucleophile for hydrocarbonylation of a C—C double bond. A successful catalytic carbonylation is reported for hydrothiocarbonylation.[51]

4.3.2   Asymmetric Oxidative Hydrocarbalkoxylation of Alkenes

Under oxidation conditions, a C—C double bond can be functionalized by either two alkoxycarbonyl groups or one alkoxycarbonyl group and one heteroatom. As shown in Scheme 4.14, two ester groups are successfully introduced to styrene in an enantioselective manner, producing a phenylsuccinic ester using a Pd/MeO–BIPHEP complex.[52] *meso*-Diols are converted into cyclic ethers in an asymmetric manner when catalyzed by Pd/chiral bisoxazoline.[53] Intramolecular aminopalladation followed by carbomethoxylation gives an cyclic amino ester in moderate ee when catalyzed by a Pd/bis(isoxazoline) complex.[54]

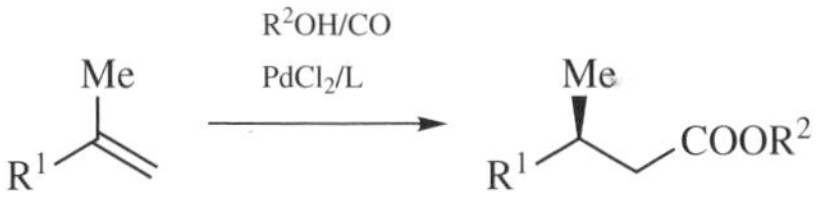

R² = OH/CO
PdCl₂/L

Me
$\underset{R^1}{\diagdown}$

$\longrightarrow$

Me
R^1 COOR²

R¹ = Ph, R² = t-Bu, L = **11a**:   69% ee
R¹ = Ph, R² = t-Bu, L = **11b**:   59% ee
R¹ = COOMe, R² = Me, L **11b**:   49% ee

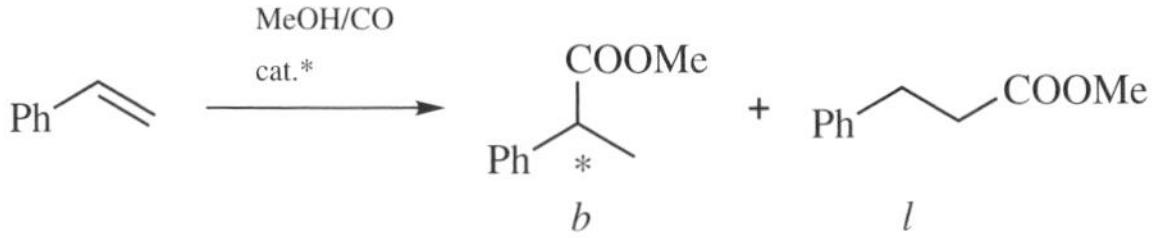

MeOH/CO
cat.*

Ph $\diagup\diagdown$

$\longrightarrow$

COOMe
Ph $*$
b

+   Ph COOMe
l

Pd(dba)₂/TfOH/**12**   94%, b/l = 94/6, 52% ee (nr)
PdCl₂/**13**   72%, b/l = 100/0, 46% ee (nr)
Pd(OAc)₂/TsOH/**14**   17%, b/l = 44/56, 86% ee (S)
PdCl₂/CuCl₂/**15**   97%, b/l = 96/4, 99% ee (S)

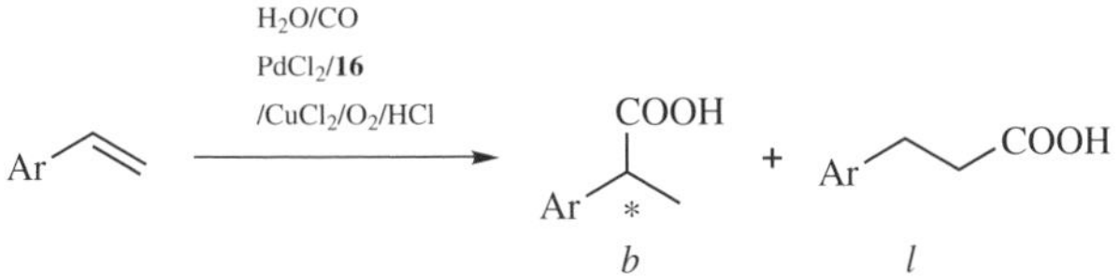

H₂O/CO
PdCl₂/**16**
/CuCl₂/O₂/HCl

Ar $\diagup\diagdown$

$\longrightarrow$

COOH
Ar $*$
b

+   Ar COOH
l

Ar = 4-i-Bu-C₆H₄-   89%, b.l = 100/0, 83% ee (S)
Ar = 6-MeO-2-naphthyl   71%, b.l = 100/0, 85% ee (S)

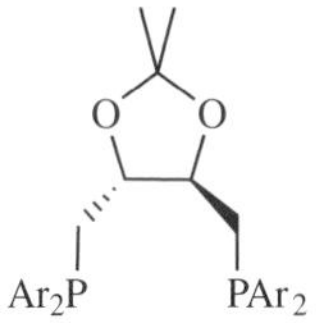

11a: Ar₂ = o-biphenylene
11b: Ar = Ph

Ar₂P   PAr₂

11a: Ar₂ = o-biphenylene
11b: Ar = Ph

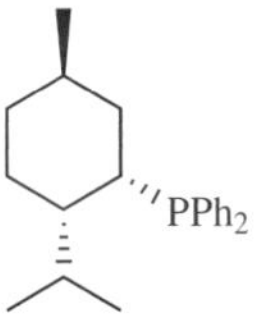

PPh₂

12

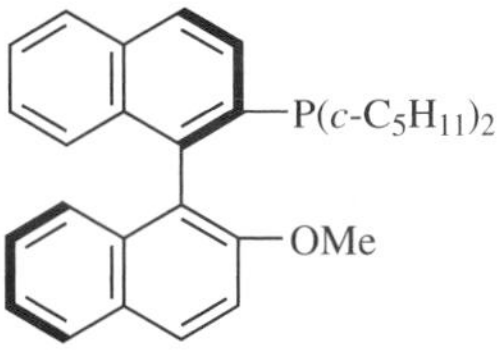

P(c-C₅H₁₁)₂
OMe

13

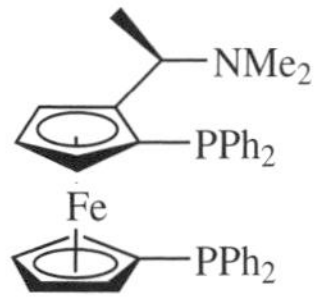

NMe₂
PPh₂
Fe
PPh₂

14

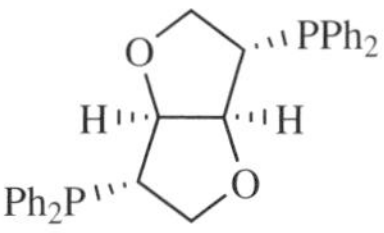

O   PPh₂
H   H
Ph₂P   O

15

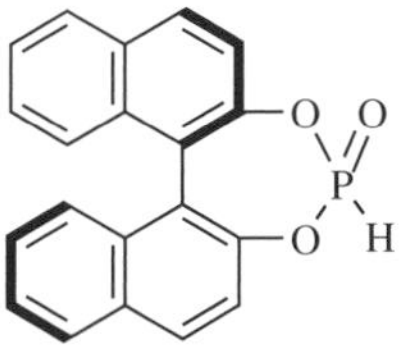

O   O
P
O   H

16

Scheme 4.12

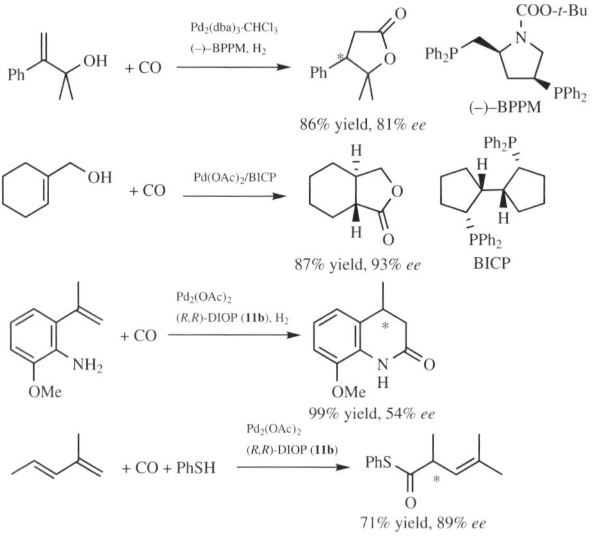

Scheme 4.13

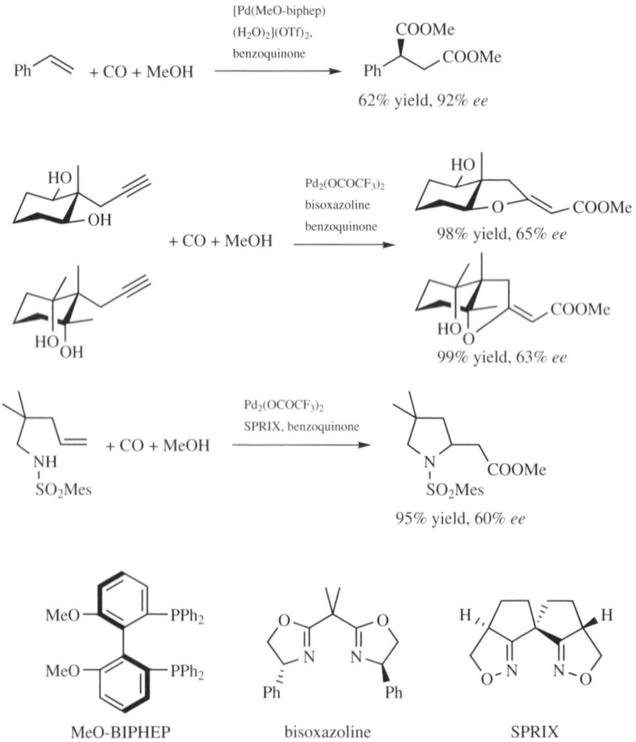

Scheme 4.14

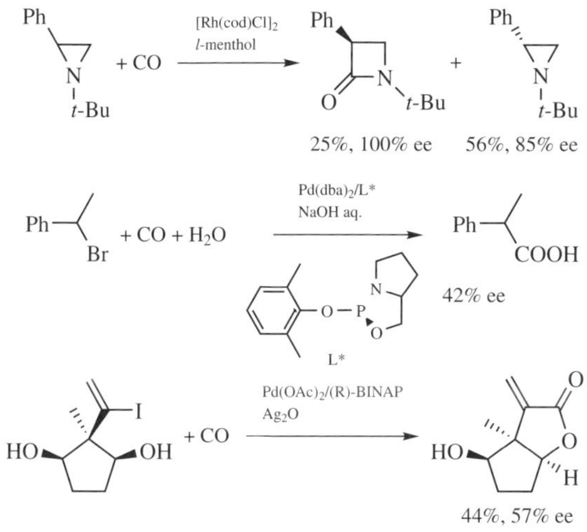

Scheme 4.15

4.3.3 Asymmetric Carbonylation of Carbon–Heteroatom Bonds

A carbon–oxygen bond and a carbon–nitrogen bond in an epoxide and an aziridine can be cleaved by metal-catalyzed carbonylation giving a lactone and a lactam, respectively. Highly efficient kinetic resolution has been achieved in the reaction of racemic aziridines catalyzed by [Rh(COD)Cl]$_2$ in the presence of *l*-menthol (3 equivalents to the aziridine).[55] A carbon–halogen bond can be converted into the corresponding carbon–carbonyl bond via Pd-catalyzed carbonylation. A benzylic halide, (1-bromoethyl)benzene, is converted into the corresponding α-arylpropanoic acid[56] Asymmetric desymmetrization, namely, enantiotopic group discrimination, has been applied to cyclocarbonylations for the syntheses of lactones using chiral Pd–diphosphine catalysts. For example, the reactions of cyclopentane-1,3-diol bearing vinyl halide gives the corresponding bicyclic α-methylenelactones with moderate enantiopurity.[57] (See Scheme 4.15.)

4.4 ASYMMETRIC KETONE FORMATION FROM CARBON–CARBON MULTIPLE BONDS AND CO

4.4.1 Asymmetric Pauson–Khand Reaction

Transition-metal-promoted cycloaddition is of much interest as a powerful tool for synthesis of carbocyclic structure in a single step. Utilization of carbon monoxide as a component of the cycloaddition reaction is now widely known as the *Pauson–Khand* reaction, which results in cyclopentenone formation starting from an alkyne, an alkene, and carbon monoxide mediated by cobalt catalyst.[58] Although mechanistic understanding is limited, a commonly accepted mechanism is shown in Scheme 4.16. Formation of dicobalt-alkyne complex followed by alkene

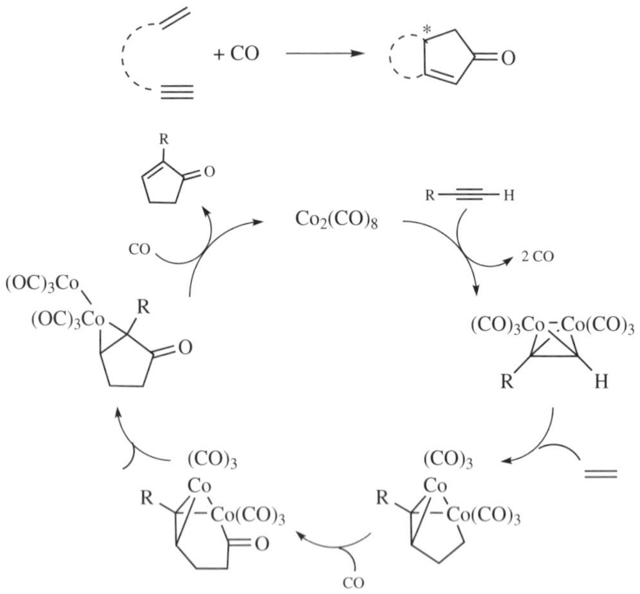

Scheme 4.16

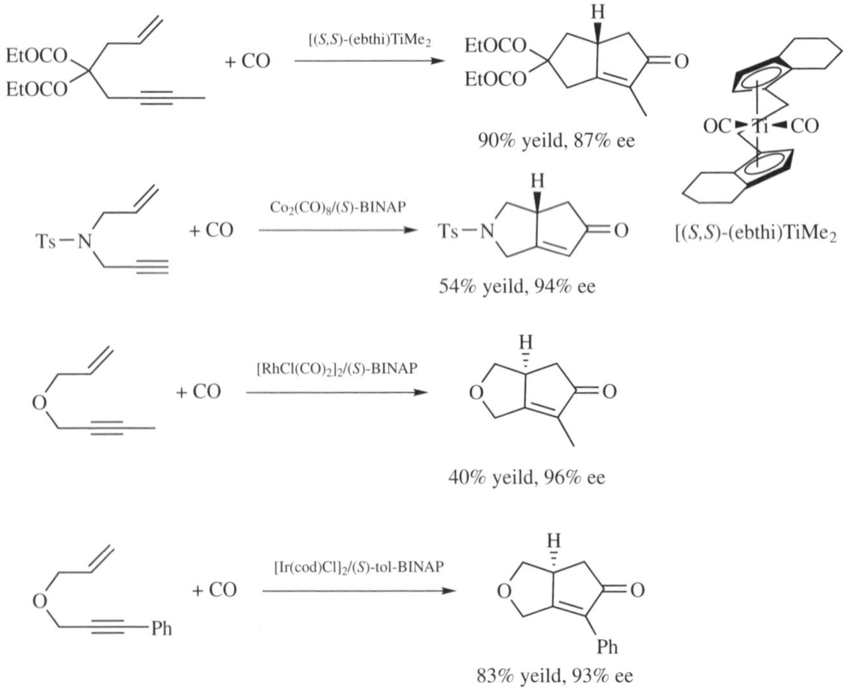

Scheme 4.17

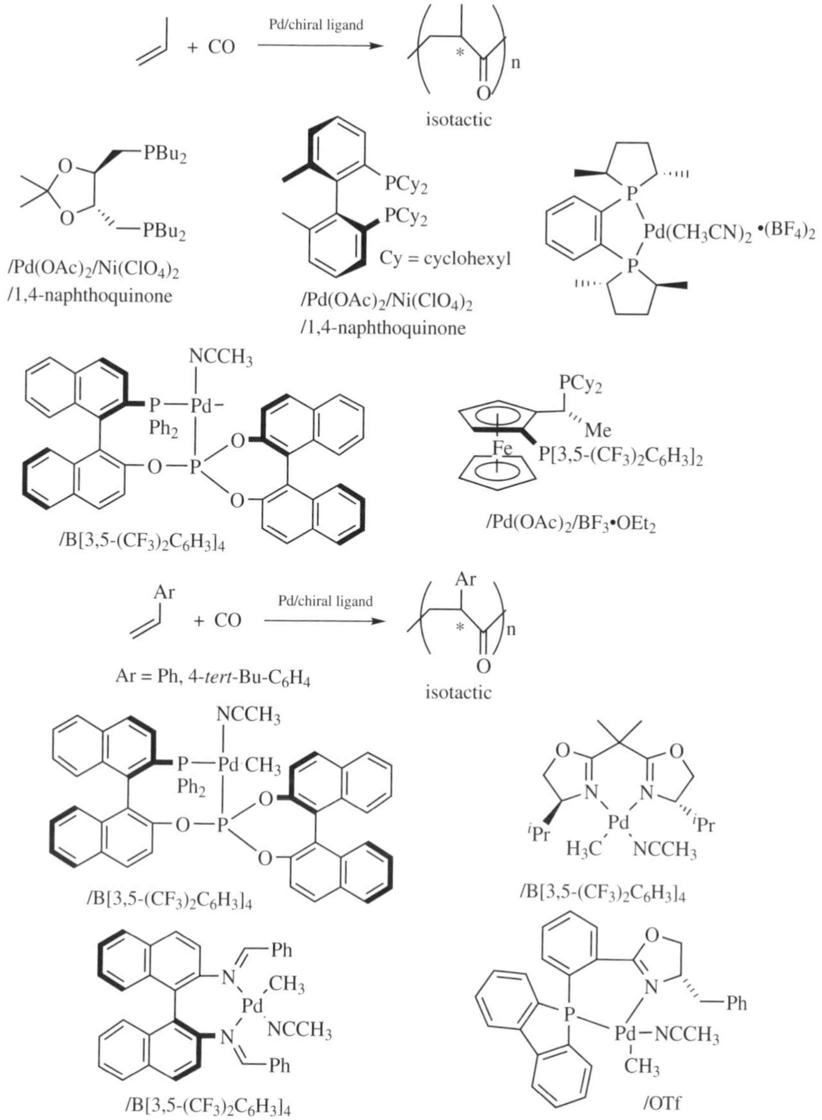

Pd/chiral ligand

+ CO

isotactic

/Pd(OAc)$_2$/Ni(ClO$_4$)$_2$
/1,4-naphthoquinone

PCy$_2$
PCy$_2$
Cy = cyclohexyl

/Pd(OAc)$_2$/Ni(ClO$_4$)$_2$
/1,4-naphthoquinone

Pd(CH$_3$CN)$_2$ •(BF$_4$)$_2$

/B[3,5-(CF$_3$)$_2$C$_6$H$_3$]$_4$

PCy$_2$
Me
Fe P[3,5-(CF$_3$)$_2$C$_6$H$_3$]$_2$

/Pd(OAc)$_2$/BF$_3$•OEt$_2$

Ar

+ CO

Pd/chiral ligand

Ar

isotactic

Ar = Ph, 4-tert-Bu-C$_6$H$_4$

/B[3,5-(CF$_3$)$_2$C$_6$H$_3$]$_4$

^{i}Pr H$_3$C NCCH$_3$ ^{i}Pr

/B[3,5-(CF$_3$)$_2$C$_6$H$_3$]$_4$

/B[3,5-(CF$_3$)$_2$C$_6$H$_3$]$_4$

/OTf

Scheme 4.18

insertion at the less hindered site gives a cobaltacycle. Migratory insertion of CO followed by reductive elimination liberates a cyclopentenone.

In the 1990s, successful catalytic asymmetric intermolecular Pauson–Khand reactions (PKR) were reported.[59,60] The representative catalysts are summarized in Scheme 4.17. The first catalytic asymmetric PKR was reported in 1996 using a chiral titanium complex.[61] Later, cobalt catalyst modified with chiral diphosphines[62] and diphosphites[63] have been reported. More recently, examples using rhodium[64] and iridium[65] complexes of diphosphines have been reported.

4.4.2 Asymmetric Alternating Copolymerization of Olefins with CO

Unlike the formation of cyclopentenone from alkyne, alkene, and CO, the reaction of alkenes with CO gives an alternating copolymer, which is a 1,4-polyketone. Stereoregular alternating copolymerization of a 1-alkene (e.g., propene and styrene) with CO provides a polyketone which contains asymmetric centers in the main-chain. Thus, it is possible to synthesize optically active polyketone starting from prochiral olefins if a chiral catalyst is employed.[66,67] Asymmetric alternating propylene/CO copolymerization is realized using chiral phosphine-based ligands, while asymmetric alternating styrene/CO copolymerization takes place using sp^2-nitrogen based ligands as summarized in Scheme 4.18. The enantiomeric excess of the asymmetric centers is estimated to be at least 95% for most of the copolymers based on model studies on the shorter oligomers.

4.4.3 Asymmetric Polymerization of Isocyanide

Unlike CO, it is possible to polymerize isocyanides (R—N=C), isoelectronic analogs to CO. When R is a bulky group, such as *tert*-Bu, the polymer forms a stable helical structure. Asymmetric catalytic polymerization has been reported for *t*-Bu—NC using [Ni(η3-allyl)(N-trifluoroacetyl-proline)]$_2$ providing (M)-helical polymer with 69% ee.[68] The more stable helical polymer was prepared from 1,2-diisocyanobenzene derivative initiated by a chiral Pd complex.[69] (See Scheme 4.19.)

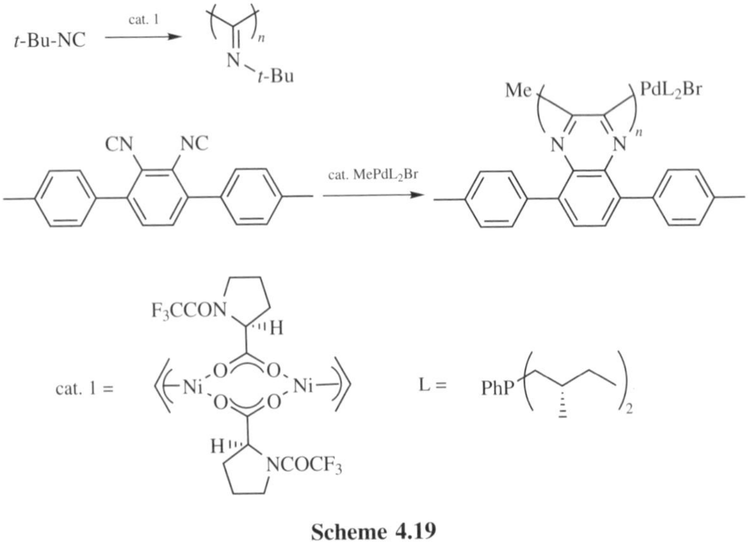

Scheme 4.19

4.5   ASYMMETRIC HYDROCYANATION OF OLEFINS

Over a billion pounds of adiponitrile is produced each year by hydrocyanation of butadiene. Asymmetric addition of hydrogen cyanide has been studied for activated olefins such as norbornene and vinylarenes using nickel complexes of phosphite-based ligands.[70] Optically active α-arylpropionitriles, the products of asymmetric hydrocyanation of vinylarenes, can be transformed into α-arylpropanoic acids, known as *antiinflammatory drugs*. The highest enantioselectivities have been obtained for the asymmetric hydrocyanation of vinylarenes using carbohydrate-derived phosphinite–Ni catalysts. Thus, hydrocyanation of 2-methoxy-6-vinylnaphthalene gives the corresponding nitrile in up to 91% ee.[71,72] A proposed catalytic cycle is shown in Scheme 4.20. The reductive elimination of alkyl cyanide form (L)(CN)Ni-CH(Me)(Ar), which forms a π-benzylic complex, is the irreversible step. Prompted by this success, several investigators applied a few bisphosphine ligands have to the Ni-catalyzed asymmetric hydrocyanation reactions.[73,74]

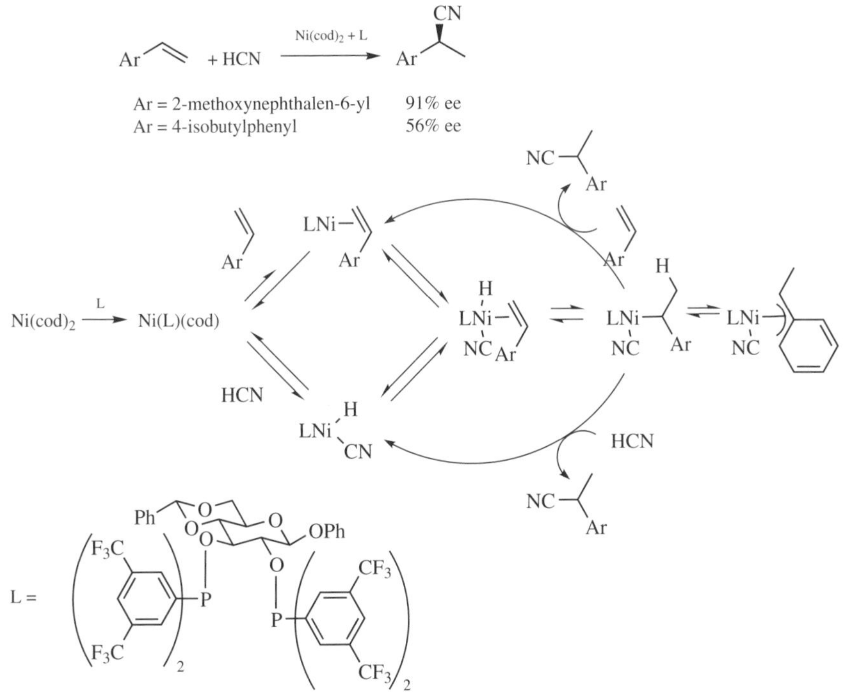

Scheme 4.20

4.6 ASYMMETRIC ADDITION OF CYANIDE AND ISOCYANIDE TO ALDEHYDES OR IMINES

The first attempt for the asymmetric hydrocyanation of aldehyde was reported in 1912 using an optically active natural alkaloid as a base catalyst.[75] Among several organocatalysts examined by now, a cyclic dipeptide composed of (S)-phenylalanine and (S)-histidine is the most effective one for the reaction of benzaldehyde, with hydrogen cyanide giving the corresponding (R)-cyanohydrin in 97% ee.[76] Numbers of chiral Lewis acids have been employed for the cyanation of aldehydes, especially using trimethylsilyl cyanide as the cyanide source. A titanium complex of a Schiff base is also useful for the reaction.[77] Some other Lewis acid complexes have been successfully employed for the TMS-CN addition to aldehydes; examples are, tin, aluminum, magnesium, and rhenium complexes.[78] (See Scheme 4.21.)

Optically active α-amino acids are prepared by a cyanide addition to imines, known as the *Strecker reaction*. Several organobase catalysts and metal complex catalysts have been successfully applied to the asymmetric catalytic Strecker amino

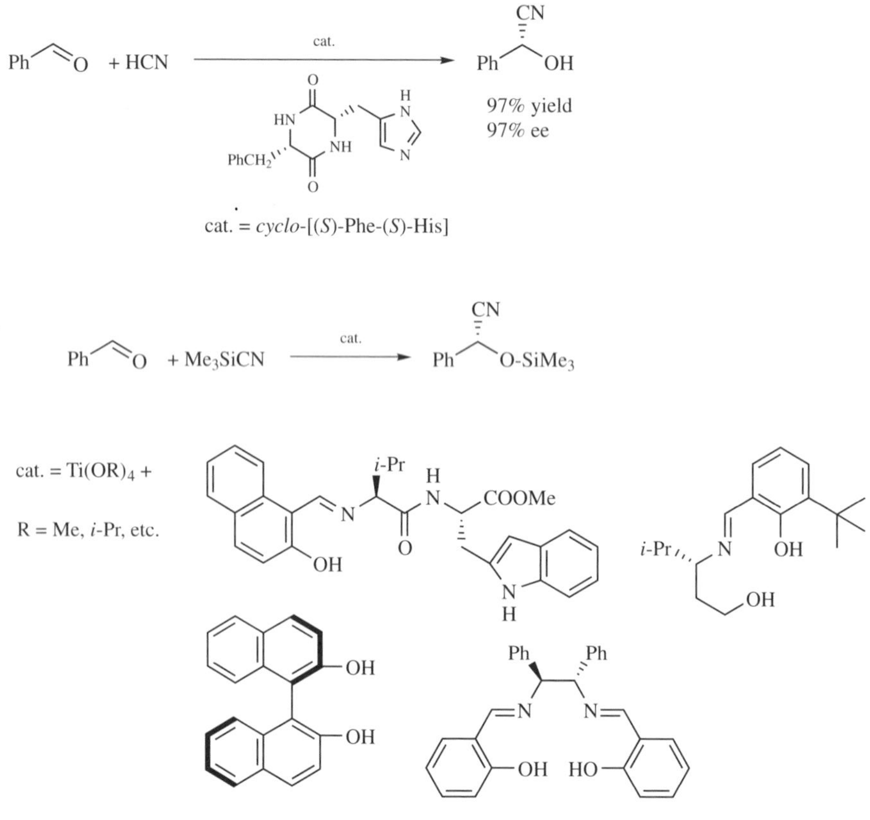

Scheme 4.21

nitrile synthesis. The first example was reported in 1996 using a cyclic dipeptide bearing a guanidine unit.[79] Imines derived from arene carboaldehydes are mostly converted into the corresponding amino nitriles in high ee percentages. Chiral guanidine is applicable to aliphatic sybstrates, and a precursor for the pharmaceutically important L-*tert*-leucine has been obtained in ~95% yield with 84% ee.[80] Chiral urea catalysts were developed via high-throughput screening of resin-bound derivatives.[81] (See Scheme 4.22.)

Representative metal complexes employed for the catalytic asymmetric Strecker reaction are summarized in Figure 4.2.[82] Aluminum-, titanium-, lanthanoid-, and zirconium-based catalysts are highly efficient. Direct one-pot synthesis starting from aldehydes, and amines is reported using the Zr complex described in Figure 4.2.[83]

Not only cyanide but also an isocyanide behaves as a nucleophile to attack a carbonyl compound or an imine that is prepared in situ from an carbonyl compound.[84,85] In these reactions, an isocyanide is a synthetic equivalent to an aminocarbonyl anion. Asymmetric version of this reaction appeared in 2003.[86,87] Using a combination of Lewis acid $SiCl_4$ and a Lewis base chiral bisphosphoramide, the corresponding α-hydroxyamide is obtained in 96% yield with >98% ee (Scheme 4.23).

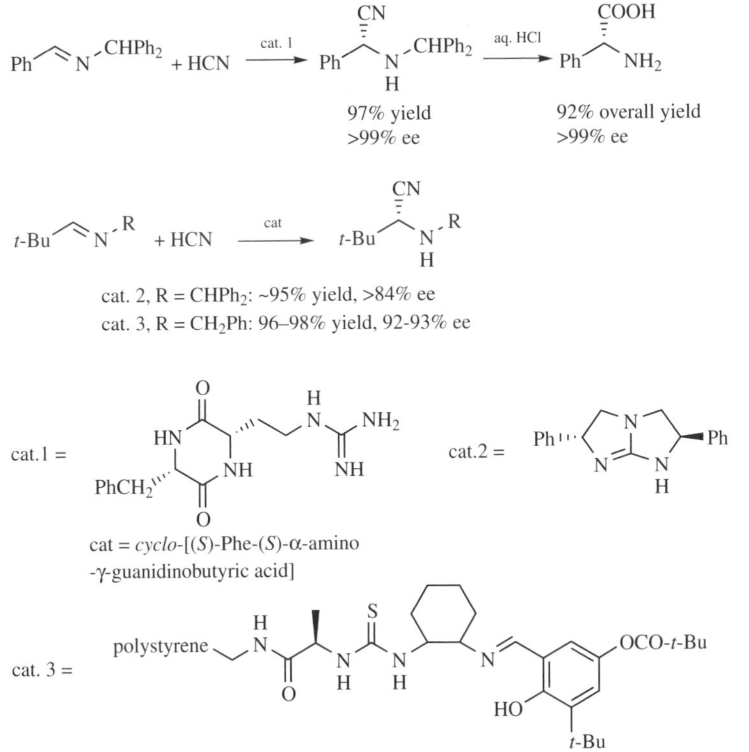

Scheme 4.22

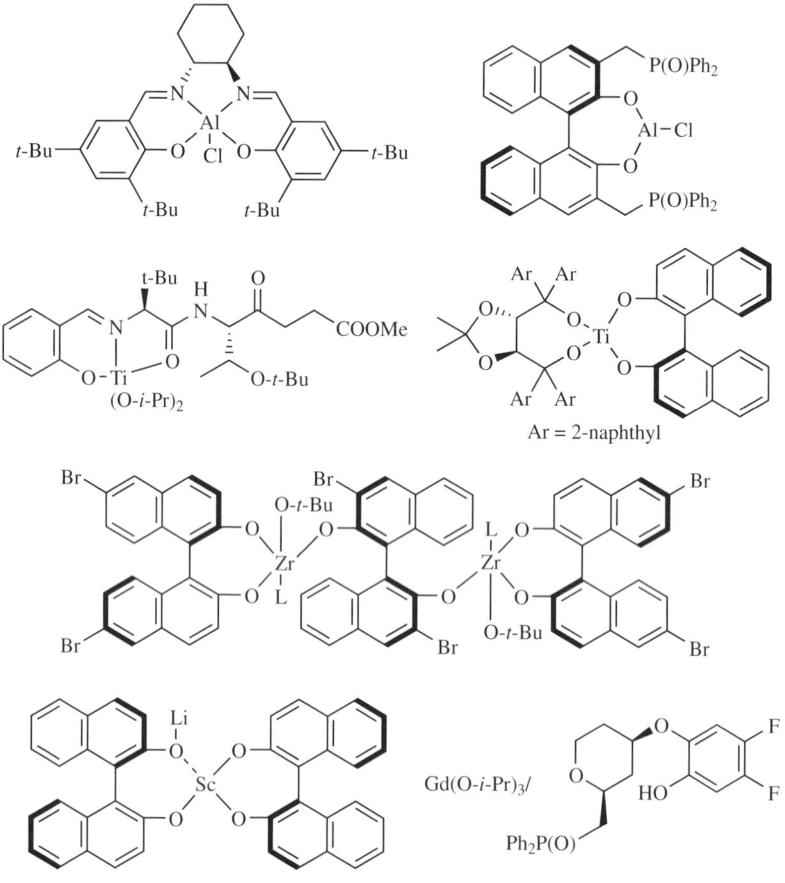

Figure 4.2. Representative metal complex catalysts for asymmetric Strecker reaction.

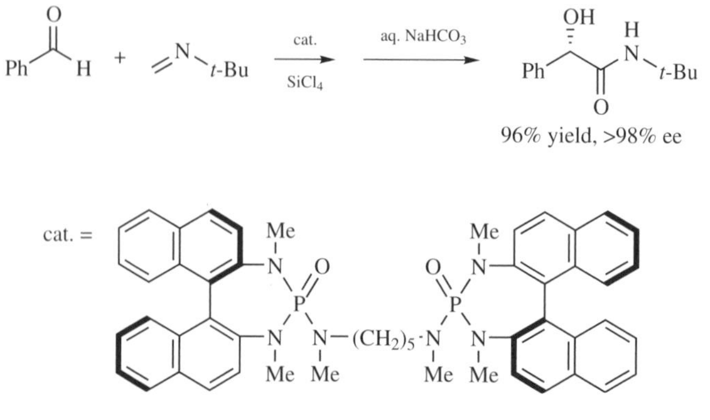

96% yield, >98% ee

Scheme 4.23

4.7 ASYMMETRIC ADDITION OF CARBON DIOXIDE

Carbon dioxide is one of the most abundant carbon resources on earth. It reacts with an epoxide to give either a cyclic carbonate or a polycarbonate depending on the substrates and reaction conditions. Kinetic resolution of racemic propylene oxide is reported in the formation of both cyclic carbonate and polycarbonate. The k_{rel} value defined as $\ln[1\text{-(conversion)}(1+\%ee)]/\ln[1\text{-(conversion)}(1\%\ ee)]$ reached 6.4 or 5.6 by using a Co(OTs)–salen complex with tetrabutylammonium chloride under neat propylene oxide[88] or using a combination of a Co–salen complex and a chiral DMAP derivative in dichloromethane,[89] respectively.

Kinetic resolution of propylene oxide in its alternating copolymerization with CO_2 is performed using similar Co–salen complexes. Reaction conditions,

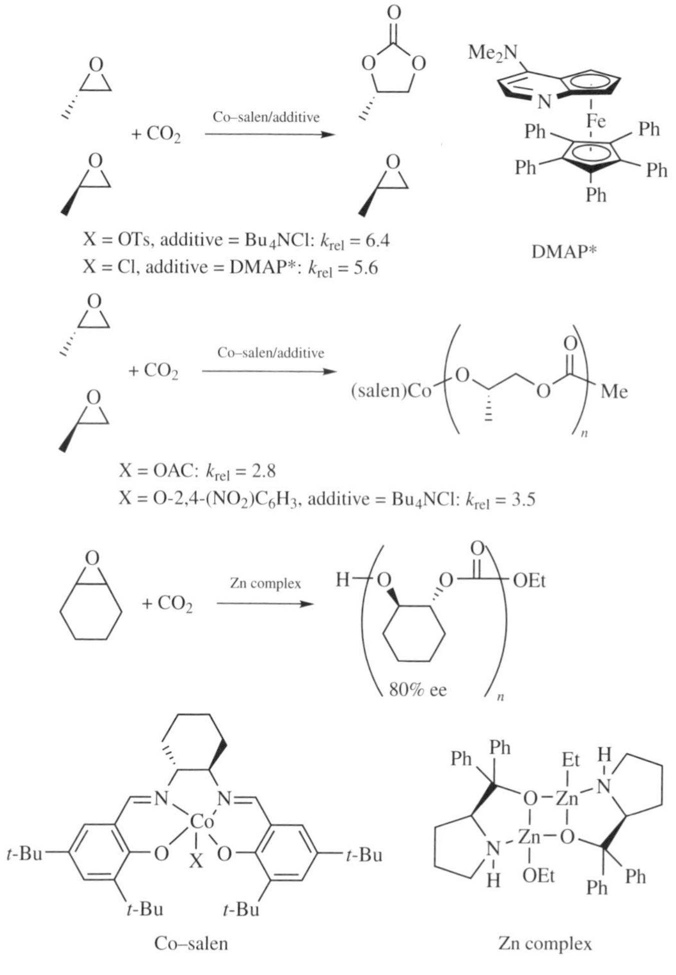

X = OTs, additive = Bu$_4$NCl: $k_{rel} = 6.4$
X = Cl, additive = DMAP*: $k_{rel} = 5.6$

DMAP*

X = OAC: $k_{rel} = 2.8$
X = O-2,4-(NO$_2$)C$_6$H$_3$, additive = Bu$_4$NCl: $k_{rel} = 3.5$

80% ee

Co–salen

Zn complex

Scheme 4.24

Co(OAr)–salen complex [Ar = 2,4-$(NO_2)_2C_6H_3$] with tetrabutylammonium chloride under neat propylene oxide, quite similar to the conditions for the cyclic carbonate synthesis, give polycarbonate with k_{rel} of 3.5.[90] Without any additives, the use of Co(OAc)-salen provides the polycaronate with k_{rel} of 2.8.[91]

It is also possible to desymmetrize a meso epoxide in the alternating copolymerization. Thus, asymmetric alternating copolymerization of cyclohexene oxide with CO_2 catalyzed by a dimeric zinc complex provides a polycarbonate in which the diol unit is optically active with 80% ee.[92,93] (See Scheme 4.24.)

4.8 CONCLUSION AND OUTLOOK

Since the discovery and development of highly efficient Rh catalysts with chiral diphosphites and phosphine–phosphites in the 1990s, the enantioselectivity of asymmetric hydroformylation has reached the equivalent level to that of asymmetric hydrogenation for several substrates. Nevertheless, there still exist substrates that require even further development of more efficient chiral ligands, catalyst systems, and reaction conditions. Diastereoselective hydroformylation is expected to find many applications in the total synthesis of complex natural products as well as the syntheses of biologically active compounds of medicinal and agrochemical interests in the near future. Advances in asymmetric hydrocarboxylation has been much slower than that of asymmetric hydroformylation in spite of its high potential in the syntheses of fine chemicals.

The asymmetric catalytic Pauson–Khand reaction met success in the late 1990s. Not only the conventional Co catalyst but also other metal complexes, such as Ti, Rh, and Ir, are applicable to the reaction. Asymmetric hydrocyanation of vinylarenes is accomplished using Ni complex of chiral diphosphite. Further studies on the scope and limitation are expected.

Asymmetric addition of small-molecule nucleophiles to carbonyl groups and their derivatives are catalyzed by either Lewis acids or Lewis bases. Carbon dioxide is now a promising building block for asymmetric organic synthesis.

Several asymmetric polymerization of small molecules have been reported applying the asymmetric catalysis methodology to polymer synthesis. Unique physical properties due to the mainchain stereoregularity are envisioned.

REFERENCES

1. Pfaltz, A. In *Comprehensive Asymmetric Catalysis*, Vol. 1, Chap. 1; Jacobsen, E.; Phaltz, A.; Yamamoto, H. (Eds.), Springer-Verlag, Berlin, **1999**.

2. Cornils, B.; Herrmann, W. A. (Eds.), *Applied Homogeneous Catalysis with Organometallic Compounds*, Vol. 1, VCH, Weinheim, **1996**.

3. van Leeuwen, P. W. N. M.; Claver, C. (Eds.) *Rhodium Catalyzed Hyfroformylation*, Kluwer Academic Publishers, Dordrecht, **2001**.

4. Ojima, I.; Tsai, C.-Y.; Tzamarioudaki, M.; Bonafoux, D. In *Organic Reactions*, Vol. 56, Overmann, L. E. (Ed.), **2000**.

5. Nozaki, K.; Ojima, I. In *Catalytic Asymmetric Synthesis*, 2nd ed., Chap. 7, Ojima, I. (Ed.), Wiley-VCH, New York, **2000**.

6. Agbossou, F.; Carpentier, J.-F.; Mortreux, A. *Chem. Rev.* **1995**, *95*, 2485.

7. (a) Horiuchi, T.; Shirakawa, E.; Nozaki, K.; Takaya, H. *Organometallics*, **1997**, *16*, 2981; (b) Nozaki, K.; Matsuo, T.; Shibahara, F.; Hiyama, T. *Organometallics* **2003**, *22*, 594.

8. Botteghi, C.; Paganelli, S.; Schionato, A.; Marchetti, M. *Chirality* **1991**, 355.

9. Consiglio, G.; Nefkens, S. C. A.; Borer, A. *Organometallics* **1991**, *10*, 2046.

10. Parrinello, G.; Stille, J. K. *J. Am. Chem. Soc.* **1987**, 109, 7122.

11. Stille, J. K.; Su, H.; Brechot, P.; Parrinello, G.; Hegedus, L. S. *Organometallics* **1991**, *10*, 1183.

12. Babin, J. E.; Whiteker, G. T. U.S. Patent **1994**, 5,360,938; *Chem Abstr.* **1994**, *122*, 186609.

13. Buisman, G. J. H.; van der Veen, L. A.; Klootwijk, A.; de Lang, W. G. J.; Kamer, P. C. J.; van Leeuwen, P. W. N. M.; Vogt, D. *Organometallics* **1997**, *16*, 2929.

14. Nozaki, K.; Sakai, N.; Nanno, T.; Higashijima, T.; Mano, S.; Horiuchi, T.; Takaya, H. *J. Am. Chem. Soc.* **1997**, *119*, 4413.

15. Nozaki, K.; Takaya, H.; Hiyama, T. *Top. Catal.* **1998**, *4*, 175–185.

16. Nozaki, K.; Matsuo, T.; Shibahara, F.; Hiyama, T. *Adv. Synth. Catal.* **2001**, *343*, 61.

17. Diéguez, M.; Pàmies, O.; Ruiz, A.; Castillon, S.; Claver, C. *Chem. Eur. J.* **2001**, *7*, 3086.

18. Diéguez, M.; Pàmies, O.; Ruiz, A.; Claver, C. *New J. Chem.* **2002**, *7*, 827.

19. Diéguez, M.; Pàmies, O.; Claver, C. *Tetrahedron Asym.* **2004**, *15*, 2113.

20. Nozaki, K.; Nanno, T.; Takaya, H. *J. Organomet. Chem.* **1997**, *527*, 103.

21. Horiuchi, T.; Ohta, T.; Shirakawa, E.; Nozaki, K.; Takaya, H. *Tetrahedron* **1997**, *53*, 7795.

22. Stanley, G. G. In *Catalysis of Organic Reactions*; Scaros, M. G.; Prunier, M. L. (Eds.), Marcel Dekker, New York, **1995**, p. 363.

23. Breeden, S.; Cole-Hamilton, D. J.; Foster, D. F.; Schwarz, G. J.; Wills, M. *Angew. Chem. Int. Ed.* **2000**, *39*, 4106.

24. Clark, T. P.; Landis, C. R.; Freed; S. L.; Klosin, J.; Abboud, K. *J. Am. Chem. Soc.* **2005**, *127*, 5040.

25. Ojima, I. *Chem. Rev.* **1988**, *88*, 1011.

26. (a) Gladiali, S.; Pinna, L. *Tetrahedron Asym.* **1990**, *1*, 693; (b) Gladiali, S.; Pinna, L. *Tetrahedron Asym.* **1991**, *2*, 623.

27. Kollár, L.; Consiglio, G.; Pino, P. *Chimia* **1986**, *40*, 428.

28. Lambers-Verstappen, M. M. H.; de Vries, J. G. *Adv. Synth. Catal.* **2003**, *345*, 478.

29. (a) Ikeua, Y.; Ishimaru, T.; Doi, T.; Kawada, M.; Fujishima, A.; Natsugari, H. *Chem. Commun.* **1998**, 2141; (b) Natsugari, H.; Ikeura, Y.; Kamo, I.; Ishimaru, T.; Ishichi, Y.; Fujishima, A.; Tanaka, T.; Kasahara, F.; Kawada, M.; Doi, T. *J. Med. Chem.* **1999**, *42*, 3982.

30. Simeone, J. P.; Bugianesi, R. L.; Ponpiom, M. M.; Goulet, M. T.; Levorse, M. S.; Desai, R. C. *Tetrahedron Lett.* **2001**, *42*, 6459.

31. Nozaki, K.; Li, W.; Horiuchi, T.; Takaya, H. *Tetrahedron Lett.* **1997**, *38*, 4611.

32. Nozaki, K.; Li, W.; Horiuchi, T.; Takaya, H.; Saiyo, T.; Yoshida, A.; Matsumura, K.; Kato, Y.; Imai, T.; Miura, T.; Kumobayashi, H. *J. Org. Chem.* **1996**, *61*. 7658.

33. Liu, P.; Jacobsen, E. N. *J. Am. Chem. Soc.* **2003**, *123*, 10772.

34. Breit, B. *Acc. Chem. Res.* **2003**, *36*, 264

35. Breit, B.; Breuninger, D. *J. Am. Chem. Soc.* **2004**, *126*, 10244

36. Parrinello, G.; Stille, J. K. *Polym. Prepr.* **1986**, *27*, 9.

37. Stille, J. K.; Parrinello, G. J. *Mol. Catal.* **1983**, *21*, 203.

38. Shibahara, F.; Nozaki, K.; Hiyama, T. *J. Am. Chem. Soc.* **2003**, *125*, 8555.

39. Kainz, S.; Leitner, W. *Catal. Lett.* **1998**, *55*, 223.

40. Hayashi, T.; Tanaka, M.; Ogata, I. *Tetrahedron Lett.* **1978**, 3925

41. Consiglio, G. *Adv. Chem. Ser.* **1982**, *196*, 371-388; Consiglio, G.; Pino, P. *Chimia* **1976**, *30*, 193.

42. Commetti, G.; Chiusoli, G. P. *J. Organometal. Chem.* **1982**, *236*, C31.

43. Kawashima, Y.; Okano, K.; Nozaki, K.; Hiyama, T. *Bull. Chem. Soc. Jpn.* **2004**, *77*, 347.

44. Oi, S.; Nomura, M.; Aiko, T.; Inoue, Y. *J. Mol. Catal.* **1997**, *115*, 289.

45. Zhou, H.; Hou, J.; Chen, J.; Lu, S.; Fu, H.; Wang, H. *J. Organometal. Chem.* **1997**, *543*, 227

46. Alper, H.; Hamel, N. J. *J. Am. Chem. Soc.* **1990**, *112*, 2803.

47. Alper, H.; Hamel, N. *J. Chem. Soc. Chem. Commun.* **1990**, 135.

48. El Ali, B.; Alper, H. *Synlett* **2000**, 161.

49. Cao, P.; Zhang, X. *J. Am. Chem. Soc.* **1999**, *121*, 7708.

50. Okuro, K.; Kai, H.; Alper, H. *Tetrahedron Asym.* **1997**, *8*, 2307.

51. Xio, W.-J. Alper, H. *J. Org. Chem.* **2001**, *66*, 6229.

52. Sperrle, M. Consiglio, G. *J. Mol. Cat. A Chem.* **1999**, *143*, 263.

53. Kato, K.; Tanaka, M.; Yamamoto, Y.; Akita, H. *Tetrahedron Lett.* **2002**, *43*, 1511.

54. Shinohara, T.; Arai, M. A.; Wakita, K.; Arai, T.; Sasai, H. *Tetrahedron Lett.* **2003**, *44*, 711.

55. Calet, S.; Urso, F.; Alper, H. *J. Am. Chem. Soc.* **1989**, *111*, 931

56. Azoumanian, H.; Buono, G.; Choukrad, M.; Petrigani, J.-F. *Organometallics* **1988**, *7*, 59.

57. Suzuki, T.; Uozumi, Y.; Shibasaki, M. *J. Chem. Soc., Chem. Commun.* **1991**, 1593.

58. Khand, I. U.; Knox, G. R.; Pauson, P. L.; Watts, W. E. *J. Chem. Soc. Perkin Trans 1*, **1973**, 975

59. Gibson, S. E.; Stevenazzi, A. *Angew. Chem. Int. Ed.* **2003**, *42*, 1800.

60. Blanco-Urgoiti, J. Anorbe, L.; Pérez-Serrano, L.; DomÀnguez, G.; Pérez-Castells, *J. Chem. Soc. Rev.* **2004**, *33*, 32.

61. (a) Hicks, F. A.; Buchwald, S. L. *J. Am. Chem. Soc.* **1996**, *118*, 11688; (b) Hicks, F. A.; Buchwald, S. L. *J. Am. Chem. Soc.* **1999**, *121*, 7026.

62. Hiroi, K.; Watanabe, T.; Kawagishi, R.; Abe, I. *Tetrahedron Lett.* **2000**, *41*, 891; idem, *Tetrahedon Asym.* **2000**, *11*, 797.

63. (a) Sturla, S. J.; Buchwald, S. L. *J. Org. Chem.* **2002**, *67*, 3398; (b) Jeong, N.; Sung, B. K.; Choi, Y. K. *J. Am. Chem. Soc.* **2000**, *122*, 6771.

64. Schmid, T. M.; Consiglio, G. *Chem. Commun.* **2004**, 2318.

65. Shibata, T.; Takagi, K. *J. Am. Chem. Soc.* **2000**, *122*, 9852.

66. Nozaki, K. In *Catalytic Synthesis of Alkene-Carbon Monoxide Copolymerrs and Cooligomers*, Chap. 7, Sen, A. (Ed.), Kluwer Academic Publishers, Dordrecht, **2003**.

67. Nozaki, K.; Hiyama, T. J. *Organomet. Chem.* **1999**, *576*, 248.

68. Kamer, P. C. J.; Nolte, R. J. M.; Drenth, W. J. *Am. Chem. Soc.* **1988**, *110*, 6818.

69. Ito, Y.; Ihara, E.; Murakami, M. *Angew. Chem. Int. Ed.* **1992**, *31*, 1509.

70. Rajanbabu, T. V.; Casalnuovo, A. L. In *Comprehensive Asymmetric Catalysis*, Vol. I, Chap. 10, Jacobsen, E. N.; Pfalz, A.; Yamamoto, H. (Eds.), Springer-Verlag; Berlin, 1999.

71. Rajanbabu, T. V.; Casalnuovo, A. L. J. *Am. Chem. Soc.* **1992**, *114*, 6265.

72. Casalnuovo, A. L.; Rajanbabu, T. V.; Ayers, T. A.; Warren, T. H. J. *Am. Chem. Soc.* **1994**, *116*, 9869.

73. Yan, M.; Xu, Q.-Y.; Chan, A. S. C. *Tetrahedron Asym.* **2000**, *11*, 845.

74. Goestz, W.; Kamer, P. C. J.; van Leeuwen, P. W. N. M.; Vogt, D. *Chem. Eur. J.* **2001**, *7*, 1614.

75. Bredig, G.; Fiske, P. S. *Biochem Z* **1912**, *46*, 7.

76. Tanaka, K.; Mori, A.; Inoue, S. J. *Org. Chem.* **1990**, *55*, 181.

77. Mori, A.; Inoue, S. In *Comprehensive Asymmetric Catalysis*, Vol. II, Chap. 28, Jacobsen, E. N.; Pfalz, A.; Yamamoto, H. (Eds.), Springer-Verlag, Berlin, **1999**.

78. Gregory, R. J. H. *Chem. Rev.* **1999**, *99*, 3649.

79. Iyer, M. S.; Gigstad, K. M.; Namdev, N. D.; Lipton, M. J. *Am. Chem. Soc.* **1996**, *118*, 4910.

80. Corey, E. J.; Grogan, M. *Org. Lett.* **1999**, *1*, 157.

81. Sigman, M. S.; Vachal, P.; Jacobsen, E. N. *Angew. Chem. Int. Ed. Engl.* **2000**, *39*, 1279.

82. Gröger, H. *Chem. Rev.* **2003**, *103*, 2795.

83. Ishitani, H.; Komiyama, S.; Kobayashi, S. *Angew. Chem. Int. Ed. Engl.* **1998**, *37*, 3186.

84. Passerini, M. *Gazz. Chim. Ital.* **1921**, *51*, 126.

85. Ugi, I.; Meyr, R.; Fetzer, U.; Steinbrückner, C. *Angew. Chem.* **1959**, *71*, 386.

86. Denmark, S. E.; Fan, Y. J. *Am. Chem. Soc.* **2003**, *125*, 7825.

87. Kusebauch, U.; Beck, B.; Messer, K.; Hertweck, E.; Dömling, A. *Org. Lett.* **2003**, *5*, 4021.

88. Lu, X.-B.; Liang, B.; Zhang, Y.-J.; Tian, Y.-J.; Wang, Y.-M.; Bai, C.-X.; Wang, H.; Zhang, R. J. *Am. Chem. Soc.* **2004**, *126*, 3732.

89. Paddock, R.; Nguyen, S. B. *Chem. Commun.* **2004**, 1622.

90. Lu, X.-B.; Wang, Y. *Angew. Chem. Int. Ed.* **2004**, *43*, 3574.

91. Qin, Z.; Thomas, C. M.; Lee, S.; Coates, G. W. *Angew. Chem. Int. Ed.* **2003**, *42*, 5484.

92. (a) Nakano, K.; Nozaki, K.; Hiyama, T. J. *Am. Chem. Soc.* **2003**, *125*, 5501; (b) Nozaki, K.; Nakano, K.; Hiyama, T. J. *Am. Chem. Soc.* **1999**, *121*, 11008.

93. Cheng, M.; Darlimg, N. A.; Lobkovsky, E. B.; Coates, G. W. *Chem. Commun.* **2000**, 2007.

5

ASYMMETRIC SYNTHESIS BASED ON CATALYTIC ACTIVATION OF C—H BONDS AND C—C BONDS

ZHIPING LI AND CHAO-JUN LI

Department of Chemistry, McGill University, Montreal, Quebec, Canada

5.1 INTRODUCTION

In nature, most chiral compounds exist in enantiomerically pure forms. The absolute configurations of the stereogenic centers are also critical to their bioactivities. Therefore, the development of synthetic methods for optically active compounds is an everlasting aim in organic synthesis. Through the efforts of numerous chemists, a turning point in catalytic asymmetric synthetic chemistry appeared in the 1980s. The most notable advances are the asymmetric epoxidation of alkenes and asymmetric hydrogenation reactions. As a result, Sharpless, Noyori, and Knowles were honored by the 2001 Noble prize in chemistry. Following decades of development, many compounds could now be obtained in good to excellent enantiomeric excess by asymmetric synthesis.

The activation of C—H bonds[1] and C—C bonds[2] has attracted much attention in both academic and industrial laboratories because of their potential economic and ecological advantages. In the field of asymmetric synthesis, enantioselective catalytic C—X bond formation via the activation of C—H bonds and/or C—C bonds should have a great impact on asymmetric synthesis in both theory and practice. In theory, it is interesting to see how these very unreactive bonds can react preferentially in the presence of more reactive bonds with asymmetric control. In a practical sense, such C—H and C—C bonds are equivalent to the C—M bonds in organometallic reactions and would turn the corresponding stoichiometric amounts of metal into catalytic amounts. Conceptually, there are two fundamental ways to

New Frontiers in Asymmetric Catalysis, Edited by Koichi Mikami and Mark Lautens
Copyright © 2007 John Wiley & Sons, Inc.

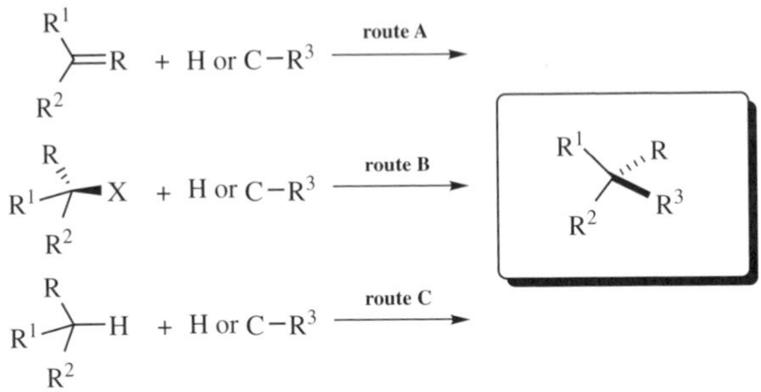

Scheme 5.1. Methods for constructing chiral carbon centers via C–H or C–C bond activation.

construct a chiral carbon center based on C–H/C–C activations: (1) nucleophilic addition to a prochiral carbon center, such as a double bond, by a nucleophile that was generated via activation of C–H or C–C bonds in situ (Scheme 5.1, route A) and (2), through nucleophilic substitutions of a chiral carbon center via activation of C–H bonds or C–C bonds (Scheme 5.1, route B). This is because most asymmetric syntheses are based on the reaction of double bonds (prochiral faces) or chiral carbon centers, as they are converted into chiral carbon centers. Another even more challenging way to build a chiral carbon center is the double activation of C–H or C–C bonds (Scheme 5.1, route C).

We first present some representative results via catalytic asymmetric activation of C–H and C–C bonds in organic synthesis. In the next section, we will summarize the representative strategies, substrates, and chiral ligands that are used in asymmetric activation of C–H and C–C bonds. More detailed discussions and related results, can be found in the references cited.

5.2   ASYMMETRIC SYNTHESIS VIA ACTIVATION OF C–H BONDS

5.2.1   Formation of C–C Bonds

C–C bond formation plays a fundamental role in organic synthesis. In spite of great success in asymmetric epoxidations and hydrogenations, highly efficient asymmetric C–C bond formations are still in great demand. In this section, we will present the results of asymmetric C–C bond formation via the activation of C–H bonds according to the types of C–H bond in the substrates (sp C–H bonds, sp^2 C–H bonds, and sp^3 C–H bonds).

5.2.2.1   Activation of sp C–H Bonds   The activation of sp C–H bonds has been extensively studied for many years. However, asymmetric additions of sp C–H bonds to unsaturated bonds have been studied only since 2003 or so.[3] Since the work of Yamaguchi[4] and later Carreira,[5] the use of stoichiometric (and more recently catalytic) Lewis acids in the presence of an excess amount of base for alkyne–aldehyde additions has also been investigated extensively; these studies

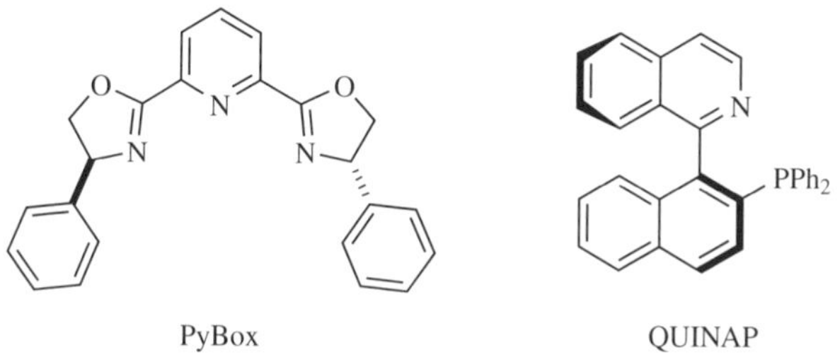

PyBox   QUINAP

Figure 5.1. PyBox and QUINAP.

are less related to C–H activation and will not be included in this chapter. For the direct reaction of terminal alkyne under mild conditions, copper complexes have proved to be effective catalysts in the activation of sp C–H bonds. Various chiral copper catalysts catalyzed various asymmetric alkyne addition reactions. PyBox and QUINAP (Figure 5.1) were widely used as chiral ligands in these reactions. The most recognized process is the enantioselective synthesis of propargyl amines. Among the conditions reported the combinations of CuOTf/PyBox and CuBr/QUINAP have been shown to be excellent chiral catalysts in the asymmetric synthesis of propargyl derivatives.

CuOTf/PyBox System The first direct asymmetric addition of alkynes to imines, generated from aldehydes and amines in situ, was reported by using copper salts in the presence of chiral PyBox ligand (Scheme 5.2).[6] The products were obtained in good yields and excellent enantioselectivities in most cases. When toluene was used as solvent, up to 93% yield and 99% ee were obtained. Up to 99.5% ee was obtained when the reaction was carried out in 1,2-dichloroethane. The reaction can also be performed in water smoothly, and good enantioselectivities (78–91% ee) were obtained.

With the success in asymmetric C–C bond formations based on the addition of sp C–H bonds to double bonds, an even bigger challenge is to achieve enantioselective C–C bond formation based on the double activation of the sp C–H bonds of alkynes and the sp^3 C–H bonds of prochiral CH$_2$ groups (Scheme 5.3), designated as

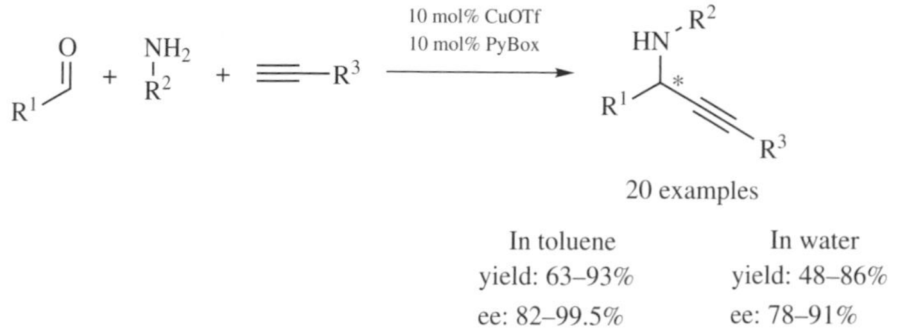

20 examples

In toluene   In water
yield: 63–93%   yield: 48–86%
ee: 82–99.5%   ee: 78–91%

Scheme 5.2. Asymmetric addition of alkynes to imines generated from aldehydes and amines.

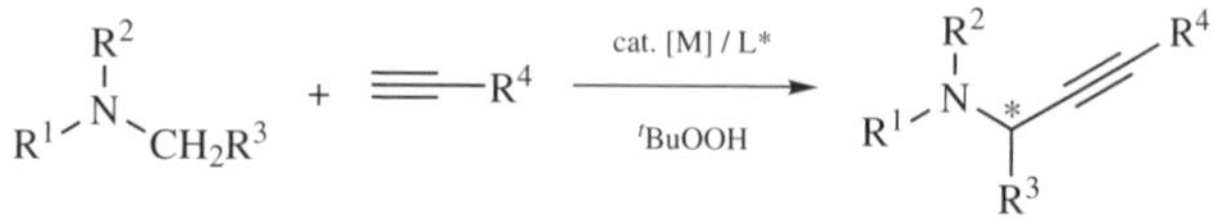

Scheme 5.3. Alkynylation of sp^3 C—H bonds via the CDC reactions.

cross-dehydrogenative coupling (CDC).[7] Such an asymmetric reaction would provide an even simpler and effective catalytic method to construct chiral propargyl amines.

Various copper salts and chiral ligands were examined under different conditions, such as different solvents and reaction temperatures. Again, PyBox together with CuOTf was found to be the best catalyst in this novel type of asymmetric C—C bond formation reaction (Scheme 5.4). The reaction was performed at 50°C, and a variety of substrates were examined. For aromatic substituted alkynes, the reaction provided both good yields and enantiomeric excesses. For aliphatic substituted alkynes, moderate or low enantiomeric excess were obtained. Studies showed that the 4-substituted methoxy group on an aryl ring (R^1) did not influence the enantioselectivity of the reaction. Interestingly, the presence of an ortho methoxy substituent group on an aryl ring (R^1) did improve the enantiomeric excess up to 74%. The enhanced enantioselectivity is due to either the coordination of the oxygen in the ortho methoxy substituent to copper or the steric effect of the ortho substituent on aryl ring.

CuBr/QUINAP System   The CuBr/QUINAP system was initially used in the enantioselective synthesis of proparyl amines via the reaction of alkynes and enamines (Scheme 5.5).[8] It was rationalized that the enamines reacted with protons in terminal alkynes in the presence of copper catalyst to form zwitterionic intermediates in which both the generated iminiums and alkyne anions coordinate to the copper metal center. After an intermolecular transfer of the alkyne moiety to the iminium ion, the desired products were released and the catalyst was regenerated. The combination of CuBr as catalyst and the chiral ligand QUINAP is crucial for the good reactivities and enantioselectivities seen in the reaction. Another potential

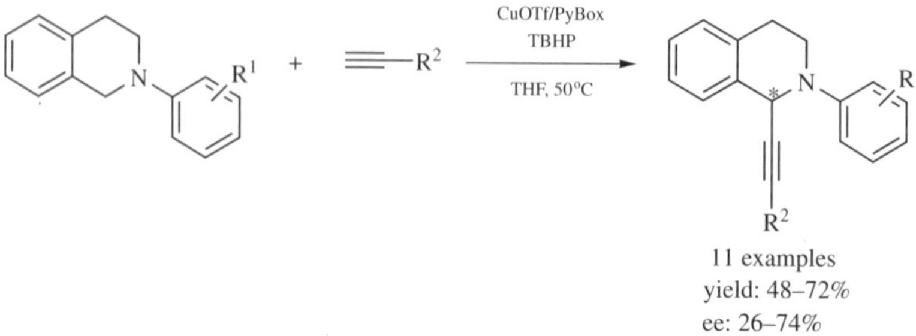

11 examples
yield: 48–72%
ee: 26–74%

Scheme 5.4. Asymmetric synthesis of propargyl amines via the CDC reactions.

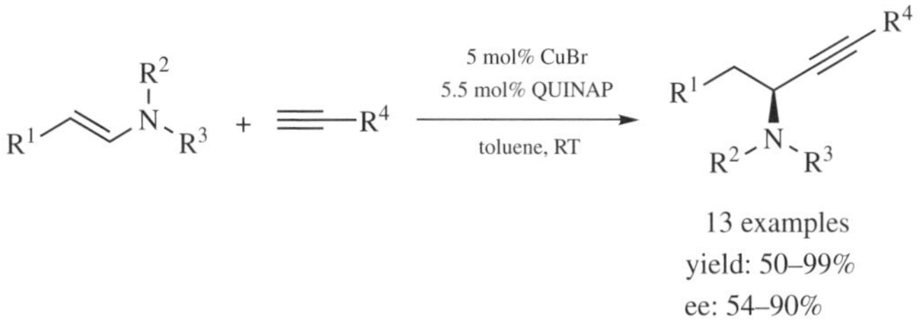

13 examples
yield: 50–99%
ee: 54–90%

Scheme 5.5. Asymmetric alkynylation of enamines.

advantage of the reaction is to use a readily removable protecting group such as an allyl or a benzyl group.

Similar to the CuOTf/PyBox system, the CuBr/QUINAP system also gave high enantioselectivities of the three component reactions to construct propargyl amines from aldehydes, amines, and alkynes (Scheme 5.6).[9] In this system various aldehydes including aromatic aldehydes and aliphatic aldehydes could be used and a wide range of chiral propargyl amines were prepared in good yields and enantioselectivities. Mechanistic studies showed that the dimeric Cu/QUINAP complex is the catalytically active species that differs from the previous reaction.

The asymmetric alkynylation of isoquinoline iminium ion was reported in the presence of CuBr/QUINAP system (Scheme 5.7).[10] Various alkynyl tetrahydroisoquinolines were obtained in excellent yields and enantiomeric excesses. A natural product, homolaudanosine, was synthesized by the reduction of the obtained propargyl tetrahydroisoquinoline.

Although the CuBr/QUINAP system is an effective chiral catalyst in alkynylation of imines or iminiums, one drawback is that enantiopure QUINAP is quite expensive. An analog of QUINAP, PINAP (Figure 5.2), was readily synthesized and found as a very effective chiral ligand in the direct addition of alkynes to iminiums generated from aldehydes and secondary amines in situ (Scheme 5.8).[11]

Other Systems   In contrast to the highly successful alkynylation of imines, copper catalysts failed in the asymmetric alkynylation of aldehydes. On the other hand, the combination of various Lewis acids and chiral amines were studied extensively to

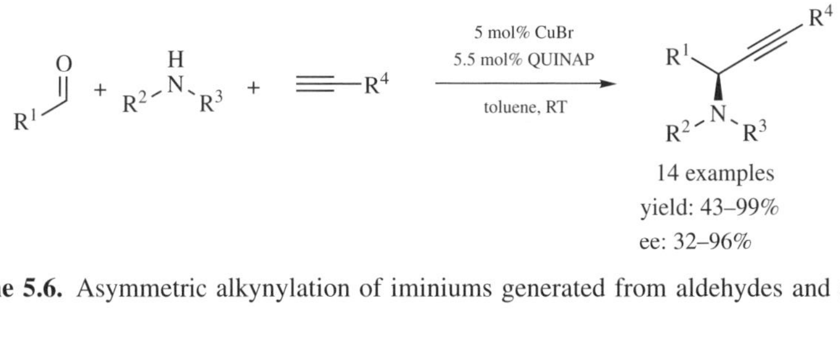

14 examples
yield: 43–99%
ee: 32–96%

Scheme 5.6. Asymmetric alkynylation of iminiums generated from aldehydes and amines.

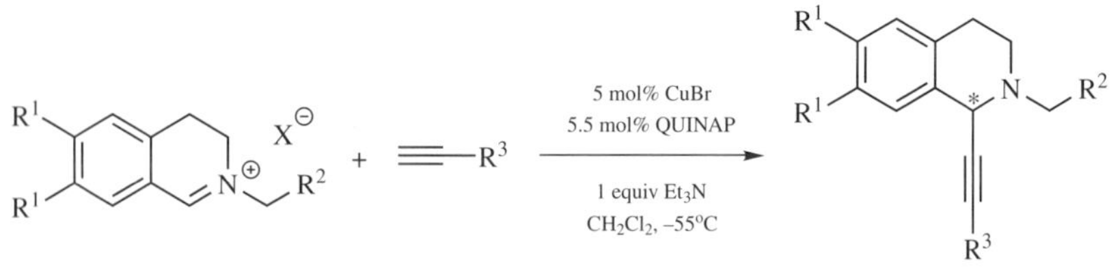

10 examples
yield: 67–95%
ee: 70–99%

Scheme 5.7. Asymmetric alkynylation of isoquinoline iminiums.

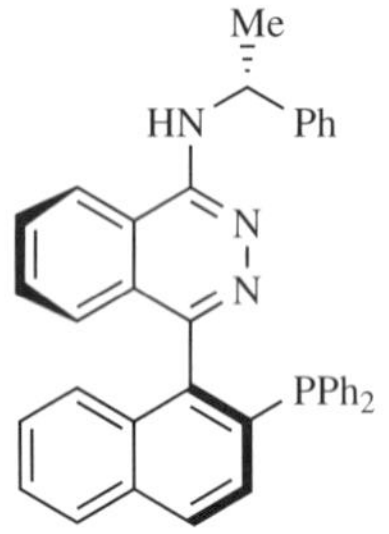

Figure 5.2. PINAP.

obtain enantiomerically pure propargyl alcohols. The $Zn(OTf)_2$/*N*-methyl ephedrine (Figure 5.3) system was successfully used in enantioselective alkylation of aliphatic aldehydes (Scheme 5.9).[12] The desired proparylic alcohols were obtained with good yields and excellent enantioselectivies. These reactions also tolerate moisture and air. An IR study of the reaction of acetylenes and $Zn(OTf)_2$ in the presence of amine bases provided that Zn–acetylide species was generated in situ.[13] This catalytic system was less effective for aromatic substrates. The Lewis acid systems are less related to the topic of this chapter and will not be discussed in detail.

A $InBr_3$/(*R*)-BINOL system was also successfully used in the enantioselective alkynylation of both aliphatic and aromatic aldehydes (Scheme 5.10).[14] This reaction has the following advantages: (1) the broad scope of substrates and (2)

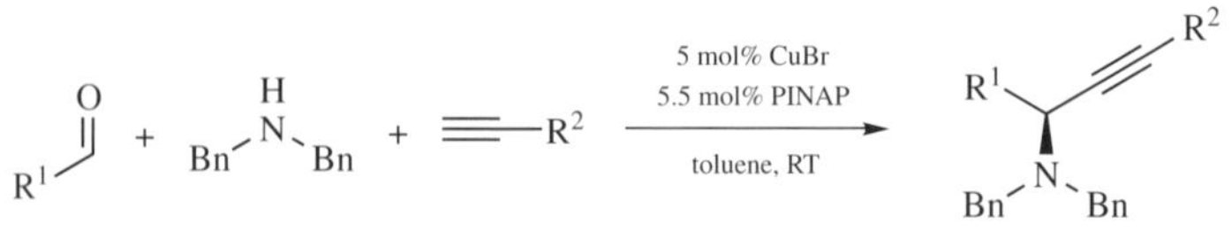

6 examples
yield: 72–88%
ee: 90–99%

Scheme 5.8. CuBr/PINAP-catalyzed asymmetric synthesis of propargyl amines.

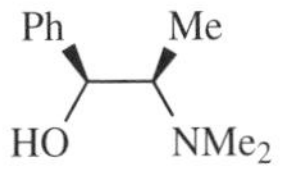

Ph   Me

HO   NMe$_2$

Figure 5.3. *N*-Methylephedrine.

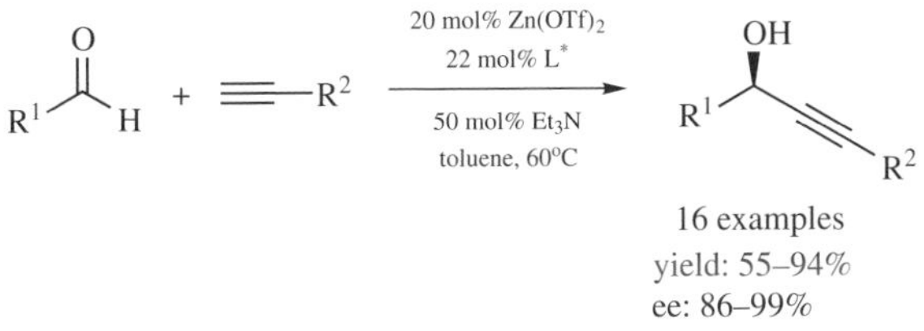

Scheme 5.9. Zn-catalyzed asymmetric synthesis of propargyl alcohols.

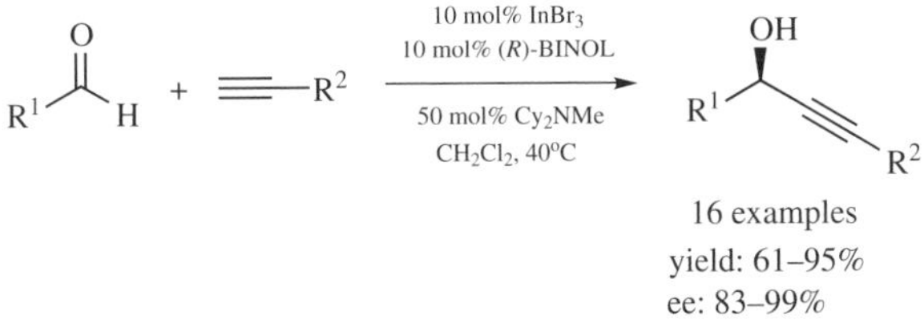

Scheme 5.10. In-catalyzed asymmetric synthesis of propargyl alcohols.

the easily available and relatively cheap chiral ligand, BINOL. The bifunctional character of the indium(III) catalyst, which activates both the soft nucleophiles (alkynes) and the hard electrophiles (aldehydes), was shown by in situ IR and NMR spectroscopic studies.

Among the catalytic asymmetric alkyne additions to the sp^2 carbon center, such as carbonyl, imines, and iminiums, truly metal-catalyzed alkyne addition to alkenes is rare.[15] By using a PINAP derivative (Figure 5.4), Cu-catalyzed Michael addition

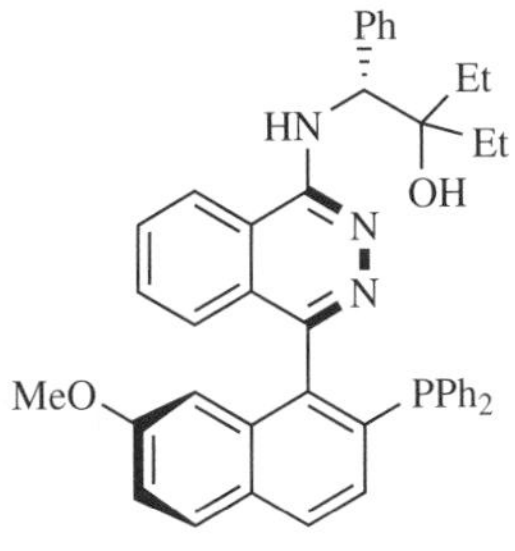

Figure 5.4. PINAP.

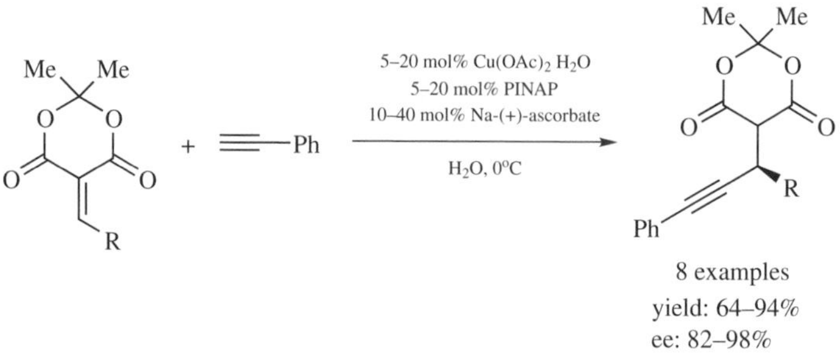

Scheme 5.11. Asymmetric alkyne addition to activated alkenes.

of alkynes to Meldrum's acid derivatives afforded the desired adducts with excellent yields and enantioselectivities (Scheme 5.11).[16] Water was used not only as a solvent but also as an activator to generate the reactive copper species in these reactions. Aliphatic alkynes were not good substrates for this reaction under the current reaction conditions.

5.2.1.2 Activation of sp^2 C–H Bonds

Although there has been significant progress in asymmetric Friedel–Crafts reactions, there are few examples of asymmetric synthesis based on sp^2 C–H bond activation. Enantioselective cyclization of aromatic imines with intramolecular double bonds was achieved (Scheme 5.12).[17] This is the first highly enantioselective C–C bond formation based on catalytic aromatic C–H bond activation. The chiral monodentate phosphorus ligands (Figure 5.5) were crucial for these reactions because chelating phosphines prevented these reactions. The enantioselectivity of the reaction was depended on the chiral ligands when different substrates were used.

Asymmetric intramolecular alkylation via rhodium-catalyzed C–H bond activation was used in the first total synthesis of (+)-lithospermic acid as a key step (Scheme 5.13).[18] Initial studies of asymmetric intramolecular alkylation were unsuccessful by using various chiral ligands. An alternative approach using chiral auxiliaries generated from chiral amines was tested. Various chiral amines were examined and (R)-(−)-aminoindane gave both good yields and diastereoselectivities. After recrystalization, the desired product was obtained in 56% yield and 99% ee. Although the desired absolute stereoconfiguration of C20 in the final product was

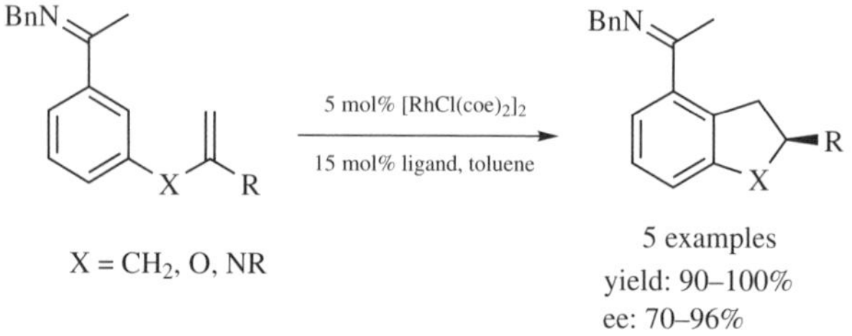

Scheme 5.12. Asymmetric Fridel-Crafts reactions via C–H bond activation.

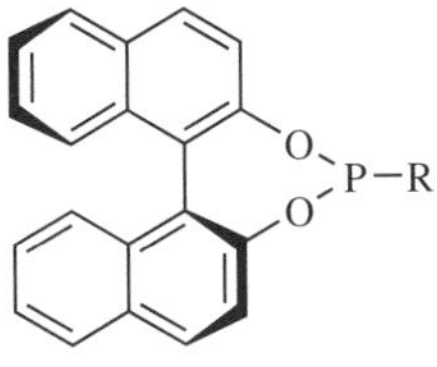
O
P—R
O

(*S*,*R*,*R*)-L, R = $N(CHCH_3Ph)_2$
(*S*,*S*,*S*)-L, R = $N(CHCH_3Ph)_2$

Figure 5.5. Chiral monodentate phosphorus.

opposite to the obtained intermediate, fortunately, the subsequent Knovenagel condensation epimerized the C20 position to give the desired stereochemistry of the final product.

An Ir(I)-catalyzed asymmetric intermolecular hydroarylation of norbornene with benzamide was reported in good to excellent enantiomeric excess, albeit in low yields, via the aryl C—H activation (Scheme 5.14).[19] In some cases, the hydroamination products of norbornene were also formed in high enantioselectivities.

5.2.1.3 Activation of sp^3 C—H Bonds Asymmetric reactions based on the activation of sp^3 C—H bonds were thoroughly studied in the metal-carbennoid-induced

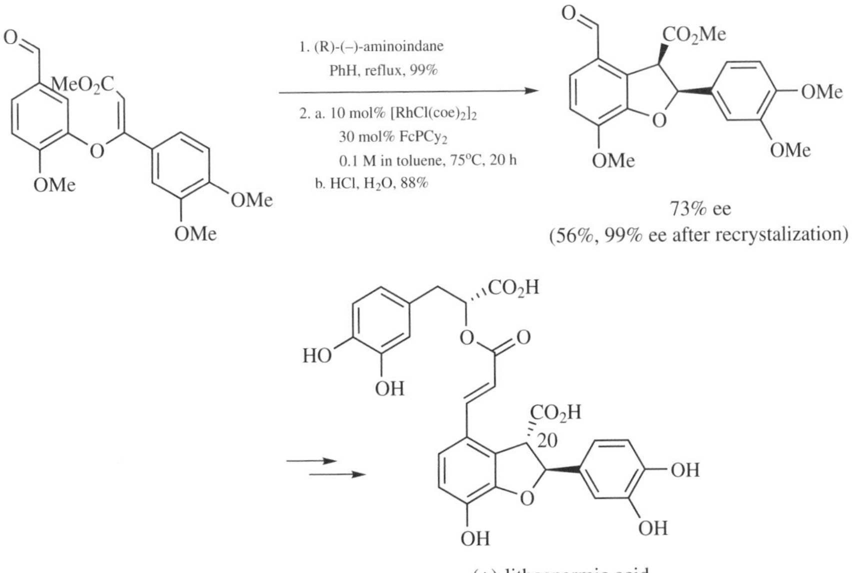
O
MeO2C
O
OMe
OMe
OMe
1. (R)-(–)-aminoindane
PhH, reflux, 99%
2. a. 10 mol% [RhCl(coe)2]2
30 mol% FcPCy2
0.1 M in toluene, 75°C, 20 h
b. HCl, H2O, 88%
O
CO2Me
O
OMe
OMe
OMe
73% ee
(56%, 99% ee after recrystalization)
CO2H
O
O
HO
OH
CO2H
20
O
OH
OH
OH
(+)-lithospermic acid
10 steps and 5.9% overall yield

Scheme 5.13. Total synthesis of (+)-lithospermic acid via C—H bond activation.

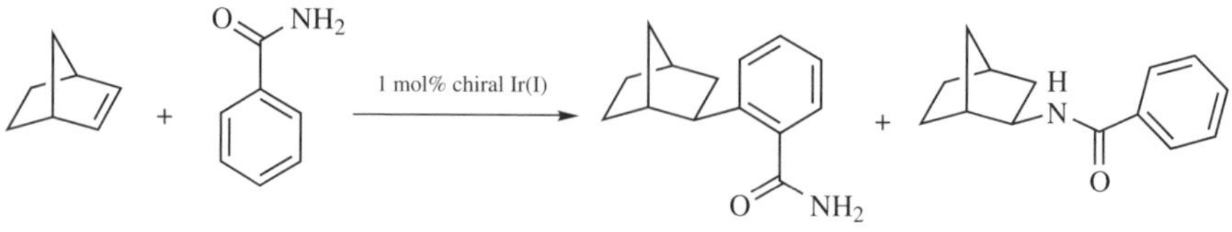

5 examples
yield: 4–35%
ee: 81–94%

3 examples
yield: 5–50%
ee: 65–79%

Scheme 5.14. Ir(I)-catalyzed asymmetric hydroarylation of norbornene.

C—H insertion.[20] Both intramolecular and intermolecular enantioselective carbenoid C—H activation were efficiently applied in various substrates and generally gave quite good yields and enantioselectivities. In intramolecular C—H insertions, four-, five- and six-membered rings are formed, and five-membered rings are predominant in most cases. In intermolecular C—H activations, benzylic, allylic, or α-heteroatom C—H bonds as well as cyclic compounds generally gave high enantioselectivities and high yields. By choosing well-defined chiral Rh catalysts and diazo compounds, products of various classical organic reactions, such as the aldol reactions, the Mannich reactions, the Michael reactions, and the Claisen rearrangement, can be achieved. There have been two excellent reviews on this subject.[18] In this section, we will review some of the latest advances.

Asymmetric activation of the C—H bonds in benzyl silyl ethers was achieved by using Hashimoto's N-phthaloyl-based $Rh_2((S)\text{-PTTL})_4$ catalyst (Figure 5.6) in high diastereoselectivities and enantioselectivities (Scheme 5.15).[21] The well-established dirhodium tetraprolinates such as $Rh_2((S)\text{-DOSP})_4$ and $Rh_2((R)\text{-DOSP})_4$ catalysts, which generally are excellent catalysts for asymmetric C—H bond activation, were not suitable catalysts in these reactions.

In contrast to the asymmetric activation of C—H bonds in benzyl silyl ethers, the dirhodium tetraprolinate, $Rh_2(S\text{-DOSP})_2$ (Figure 5.7), was found to be an efficient catalyst in an enantioselective C—H activation of acetals (Scheme 5.16).[22] Interestingly, when the acetals had a methoxy substituent on the aromatic ring, the Stevens rearrangement was a main competing side reaction of the C—H activation of acetals.

If vinylcarbenoids were used, highly diastereoselective (>98% de) and enantioselective (95.1–99.6% ee) activation of C—H bonds of 1,2-dihydronaphthalenes were

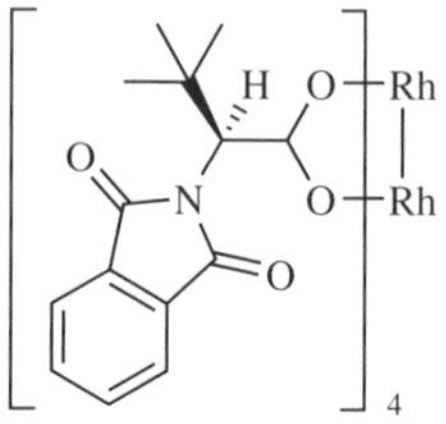

Figure 5.6. $Rh_2(S\text{-PTTL})_4$.

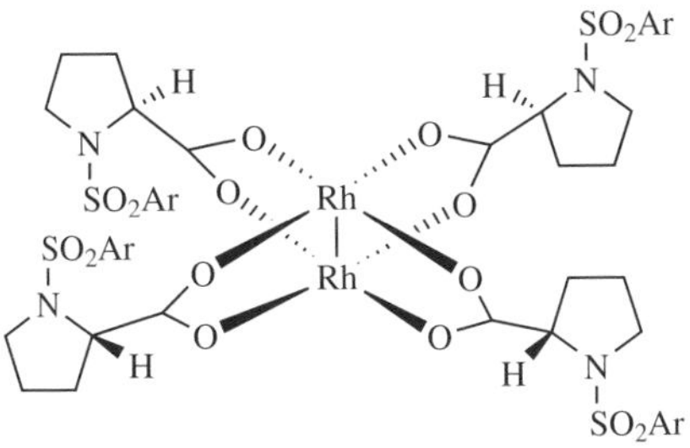

Figure 5.7. Rh$_2$(S-DOSP)$_4$.

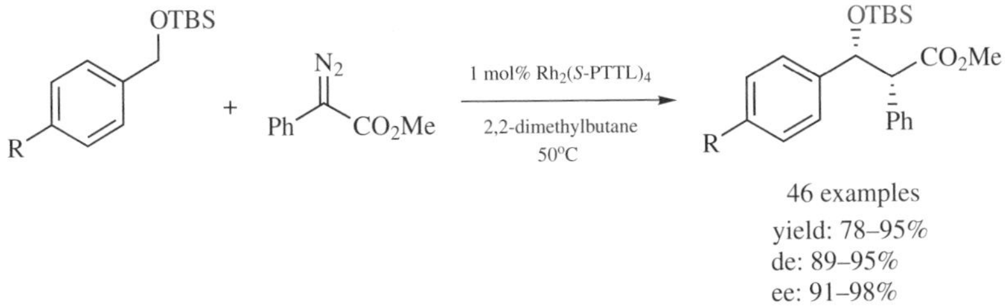

46 examples
yield: 78–95%
de: 89–95%
ee: 91–98%

Scheme 5.15. Asymmetric activation of C–H bonds in benzyl silyl ethers.

observed by using Rh$_2$(S-DOSP)$_2$ catalyst via a combined C–H activation/Cope rearrangement followed by a retro-Cope rearrangement (Scheme 5.17).[23] The excellent stereoselectivity is most likely due to the relatively rigid structures of 1,2-dihydronaphthalenes. This methodology was successfully applied in an enantioselective synthesis of (+)-erogorgiaene (Scheme 5.18).[24] The method allowed us to construct three stereogenic centers in one step in excellent enantioselectivities.

Another approach for the chemoselective and asymmetric iodination of unactivated C–H bonds was reported with a palladium catalyst using a chiral auxiliary (Scheme 5.19).[25] Excellent diastereoselectivities were induced by chelating the auxiliary to the palladium catalyst center followed by an electrophilic C–H activation and iodination. Studies showed that I$_2$ acts as both the reactant and the activator to form the reactive catalyst precursor, Pd$_3$(OAc)$_3$. After the reaction was completed, the formed PdI$_2$ was precipitated from the solution and could be reused several times without losing reactivity and selectivity.

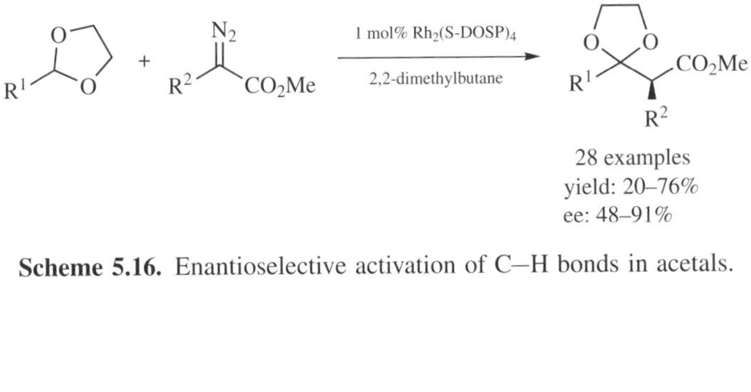

28 examples
yield: 20–76%
ee: 48–91%

Scheme 5.16. Enantioselective activation of C–H bonds in acetals.

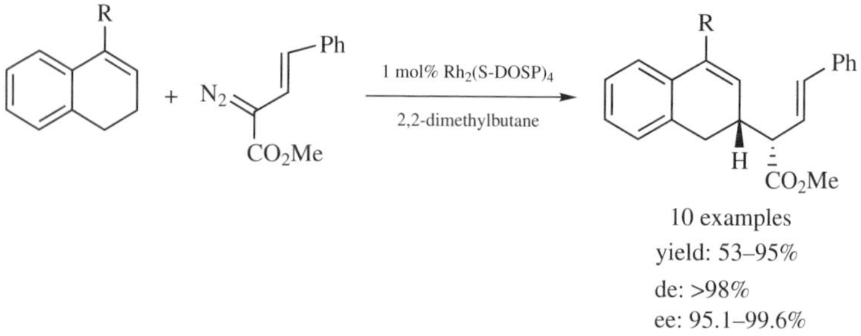

10 examples
yield: 53–95%
de: >98%
ee: 95.1–99.6%

Scheme 5.17. Asymmetric activation of C–H bonds of 1,2-dihydronaphthalenes.

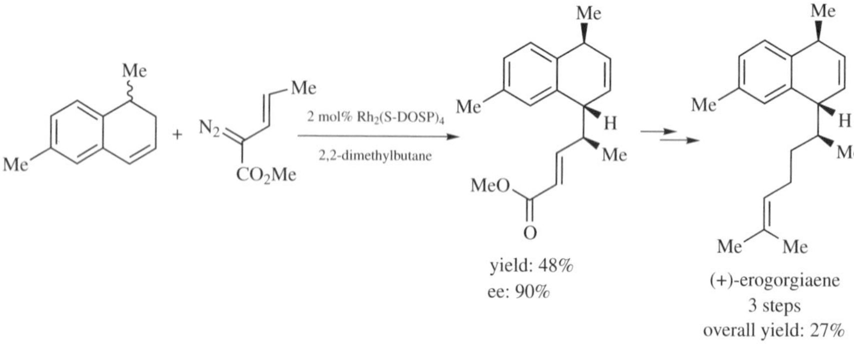

yield: 48%
ee: 90%

(+)-erogorgiaene
3 steps
overall yield: 27%

Scheme 5.18. Enantioselective total synthesis of (+)-erogorgiaene.

1,3-Dicarbonyl compounds are widely used in organic synthesis as activated nucleophiles. Because of the relatively high acidity of the methylenic C–H of 1,3-dicarbonyl compounds, most reactions involving 1,3-dicarbonyl compounds are considered to be nucleophilic additions or substitutions of enolates. However, some experimental evidence showed that 1,3-dicarbonyl compounds could react via C–H activations. Although this concept is still controversial, it opens a novel idea to consider the reactions of activated C–H bonds. The chiral bifunctional Ru catalysts were used in enantioselective C–C bonds formation by Michael addition of 1,3-dicarbonyl compounds with high yields and enantiomeric excesses.[26]

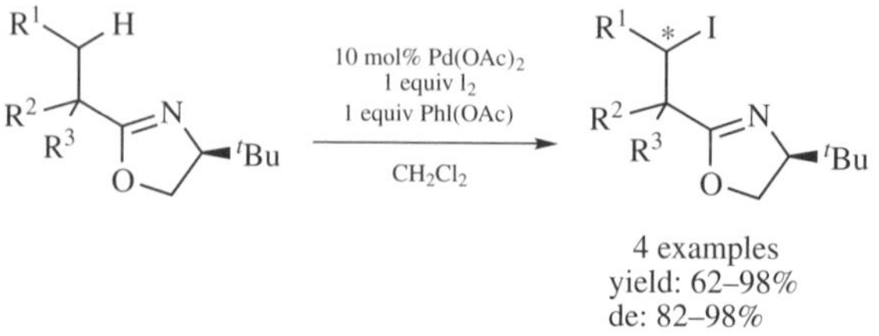

4 examples
yield: 62–98%
de: 82–98%

Scheme 5.19. Asymmetric Iodination of C–H bonds.

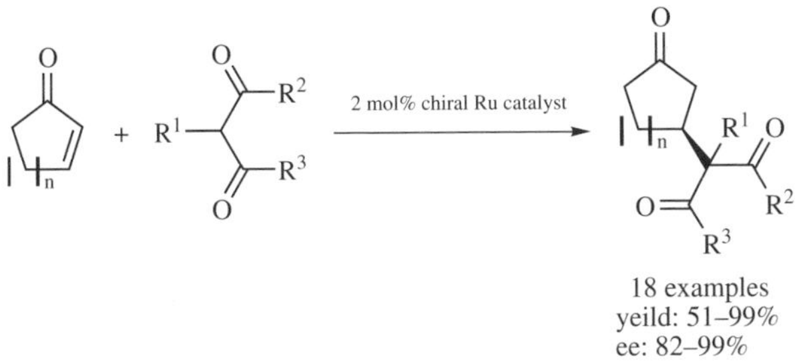

18 examples
yeild: 51–99%
ee: 82–99%

Scheme 5.20. Asymmetric Michael addition of 1,3-dicarbonyl compounds.

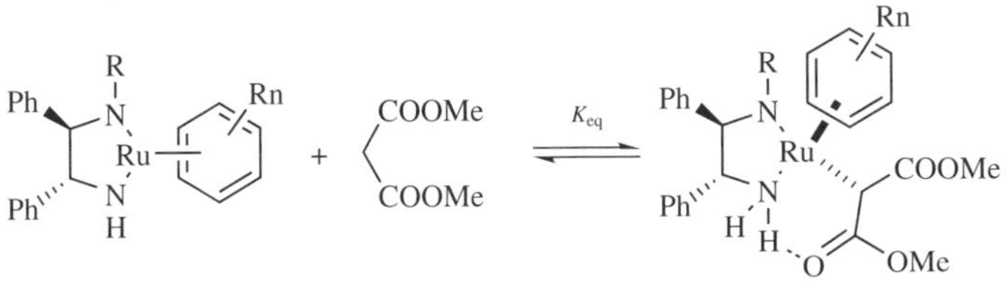

Scheme 5.21. C—H bond activation of 1,3-dicarbonyl compounds.

When cyclic enones were used as Michael acceptors, both malonates and acetoacetates gave impressive yields and enantioselectivities of the desired Michael addition products (Scheme 5.20).[27] ^{1}H NMR spectra and single-crystal X-ray data supported the following ruthenium intermediate (Scheme 5.21) and transition state (Figure 5.8).

When nitroalkenes were used as Michael acceptors, high yields and enantioselectivities of the desired Michael addition products were also obtained (Scheme 5.22).[28] In these reactions, a well-defined chiral Ru amido complex (Figure 5.9) was an efficient catalyst. The mild reaction conditions and high reactivities and stereoselectivities allowed a large-scale reaction in the presence 1 mol% Ru catalyst. By using a chiral Pd(II) catalyst, an asymmetric allylic arylation was reported by Mikami and coworkers to give the cross-coupling product via the activation of both allylic C—H and aryl C—H bonds in moderate enantioselectivity (Scheme 5.23).[29]

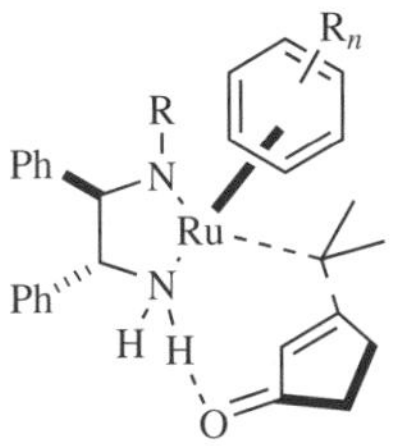

Figure 5.8. Proposed transition state.

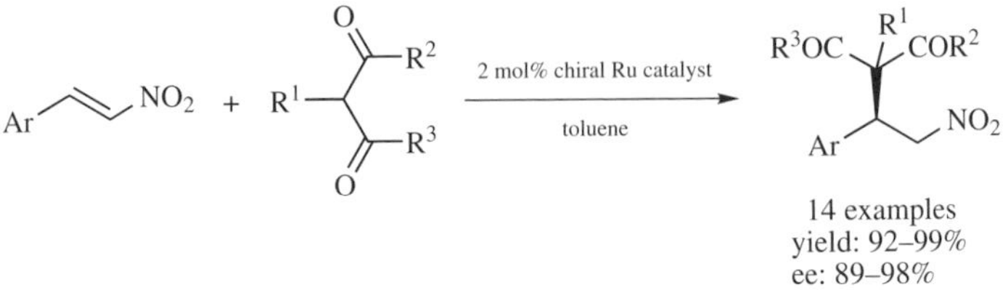

Scheme 5.22. Asymmetric addition to nitroalkenes.

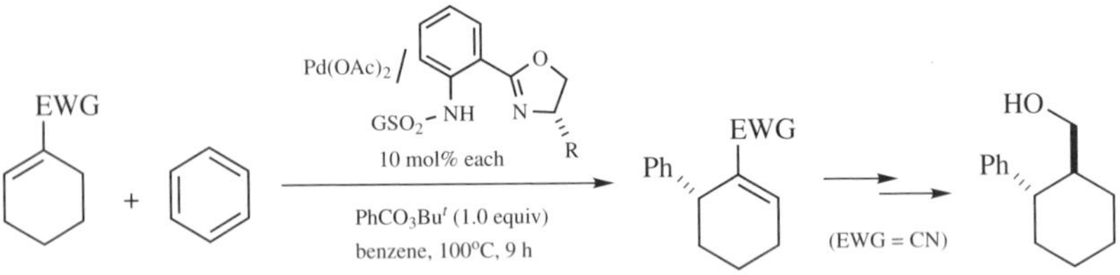

Scheme 5.23. Asymmetric allylic arylation.

5.2.2   Formation of C—O Bonds

Asymmetric allylic oxidation[30] and benzylic oxidation[31] (Kharasch–Sosnovsky reaction)[32] are important synthetic strategies for constructing chiral C—O bonds via C—H bond activation.[33] In the mid-1990s, the asymmetric Kharasch–Sosnovsky reaction was first studied by using chiral C_2-symmetric bis(oxazoline)s.[34] Later various chiral ligands, based mainly on oxazoline derivatives[35] and proline derivatives,[36] were used in such asymmetric oxidation. Although many efforts have been made to improve the enantioselective Kharasch–Sosnovsky oxidation reaction, most cases suffered from low to moderate enantioselectivities or low reactivities.

When *tert*-butyl *p*-nitrobenzoate was used, the desired allylic esters were obtained in high enantiomeric excesses, albeit in low yields, in the presence of copper/bisoxazoline (Scheme 5.24).[37] The reason for high enantioselectivities was attributed to the use of *tert*-butyl *p*-nitrobenzoate and *gem*-dialkyl bisoxazo-

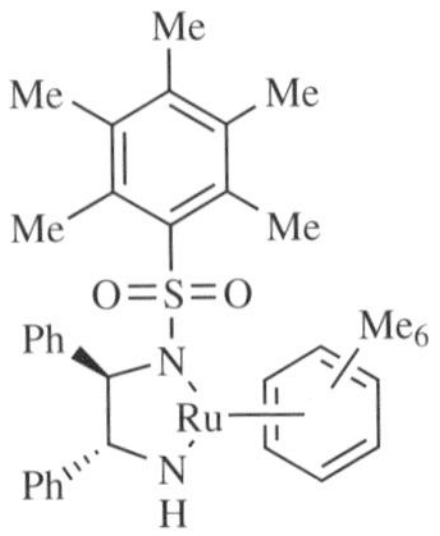

Figure 5.9. Chiral Ru catalysts.

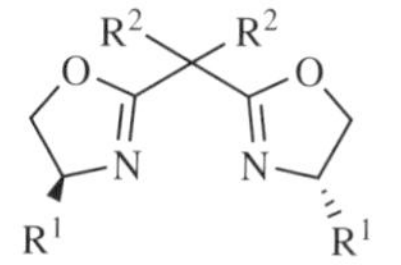

Figure 5.10. Bisoxazolines.

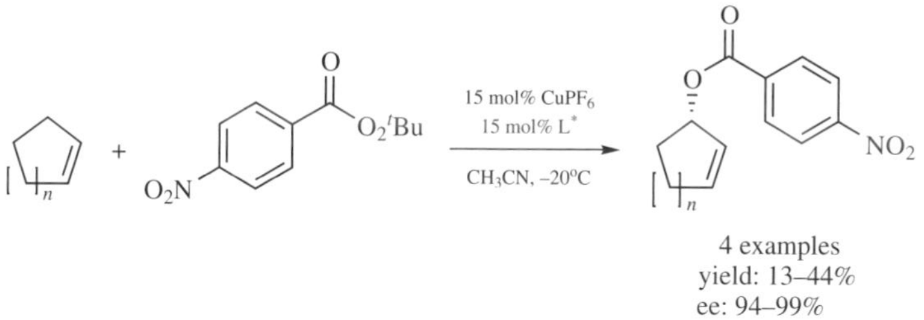

4 examples
yield: 13–44%
ee: 94–99%

Scheme 5.24. Asymmetric allylic oxidation.

lines (Figure 5.10). Mechanistic investigation showed that allylic radicals were initially formed in the presence of the copper catalyst and the peroxide esters which then "rebounded" to the copper metal center. A subsequent intramolecular transfer of the ester group generated the desired products and regenerated the catalyst.

Asymmetric hydroxylation of alkanes to afford chiral alcohols is an attractive method to build C–O bonds. Asymmetric hydroxylation of benzylic C–H bonds was reported by using (salen)manganese(III) complex (Figure 5.11) in good enantioselectivity (Scheme 5.25).[38] The oxidation is proposed via a radical intermediate initiated by Mn(salen) catalyst. The radical intermediate in a solvent cage was expected to delay the radical decay and thus gave good enantioselectivities. Using a similar Mn(salen) as a catalyst, desymmetric hydroxidation of *meso*-tetrahydrofuran derivatives was achieved in good enantioselectivities and moderate yields (Scheme 5.26).[39]

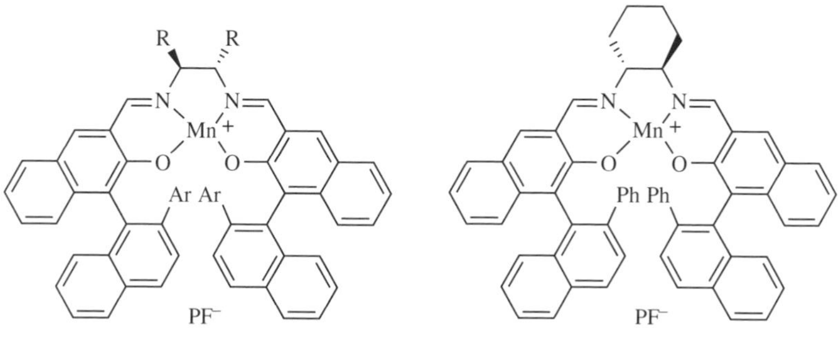

Figure 5.11. (Salen)Mn(III) complexes.

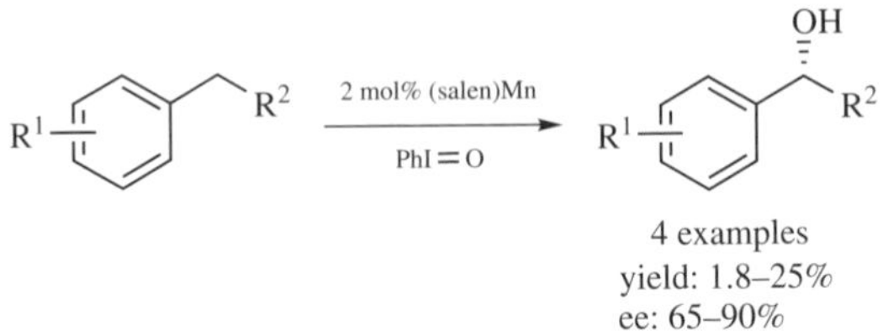

Scheme 5.25. Asymmetric hydroxylation of benzylic C—H bonds.

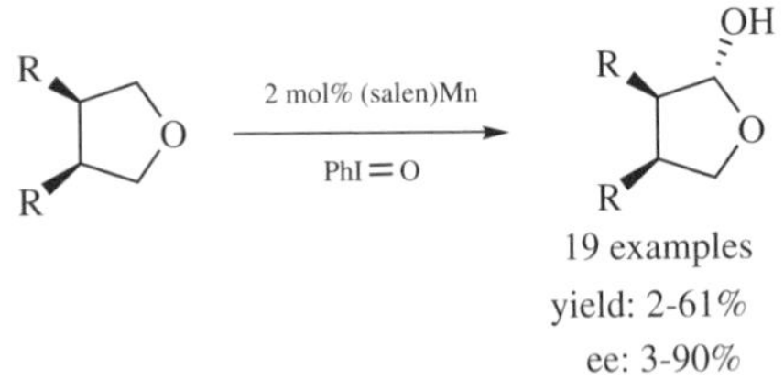

Scheme 5.26. Desymmetric hydroxidation of *meso*-tetrahydrofurans.

5.2.3 Formation of C—N Bonds

In contrast to extensive investigations of C—O bond formations via C—H bond activation, C—N bond formations are much less studied. Asymmetric syntheses of C—N bonds are generally carried out through amination of unsaturated double bonds. This is partly due to the fact that the activation of nitrogen partners is more difficult than oxygen partners. Therefore, the key for C—N bonds formation via C—H bond activation is the activation of the nitrogen partners. The general strategy for the activation of nitrogen partners is to use hypervalent iodine compounds and to form carbenoids and nitrenes, which lead to insertions into C—H bonds.

Asymmetric amidation of sp^3 C—H bonds was reported in good yields and moderate enantioselectivities (Scheme 5.27).[40] When benzylic or allylic C—H bonds were used, similar results were also obtained.[41] In these reactions the prepared nitrenes, PhI=NTs, and/or "PhI(OAc)$_2$+NH$_2$Ts" were used as nitrogen atom transfer sources. The studies showed that Ru=NTs was formed in situ and acted as a possible active intermediate when a ruthenium catalyst was used (Figure 5.12), whereas a radical intermediate might be involved when a manganese catalyst was used.

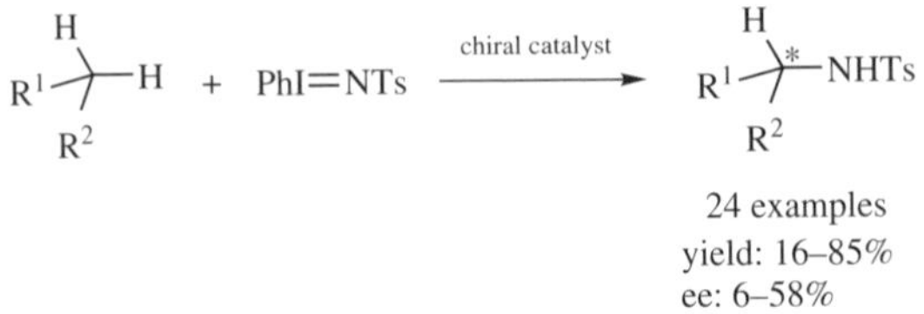

Scheme 5.27. Asymmetric amidation of sp^3 C—H bonds.

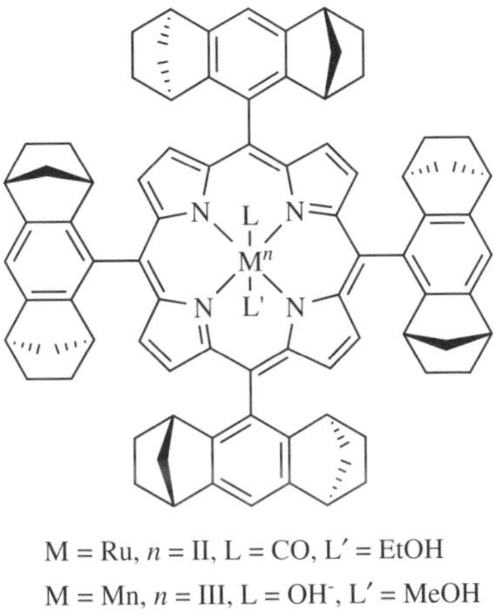

M = Ru, n = II, L = CO, L' = EtOH

M = Mn, n = III, L = OH⁻, L' = MeOH

Figure 5.12. Chiral Ru and Mn complexes.

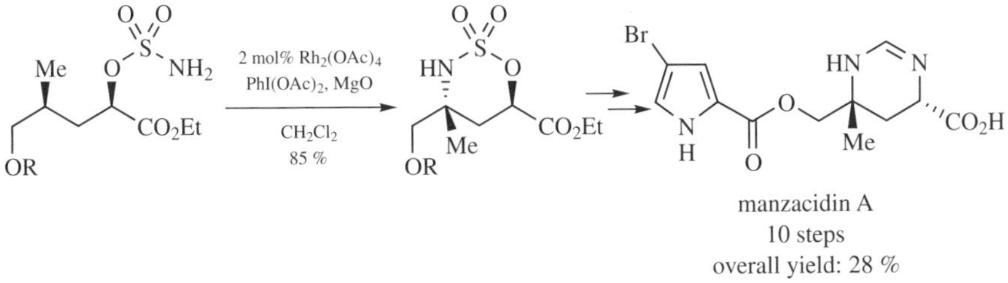

manzacidin A
10 steps
overall yield: 28 %

Scheme 5.28. A synthesis of manzacidin A via stereospecific C–H activation.

Another type of C–N bond formation was achieved via stereospecific C–H activation in the synthesis of bromopyrrole alkaloids, manzacidins A and C (Schemes 5.28 and 5.29).[42]

5.3 ASYMMETRIC SYNTHESIS VIA ACTIVATION OF C–C BONDS

Compared with the activation of C–H bonds, the activation of C–C bonds is rarely used in asymmetric synthesis. Two main factors contributed to this difficulty:

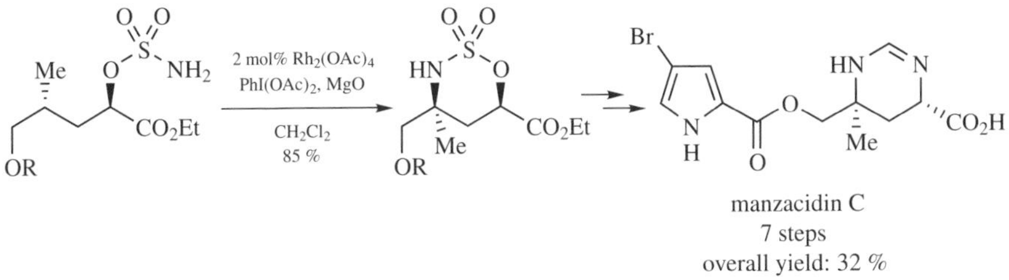

manzacidin C
7 steps
overall yield: 32 %

Scheme 5.29. A synthesis of manzacidin C via stereospecific C–H activation.

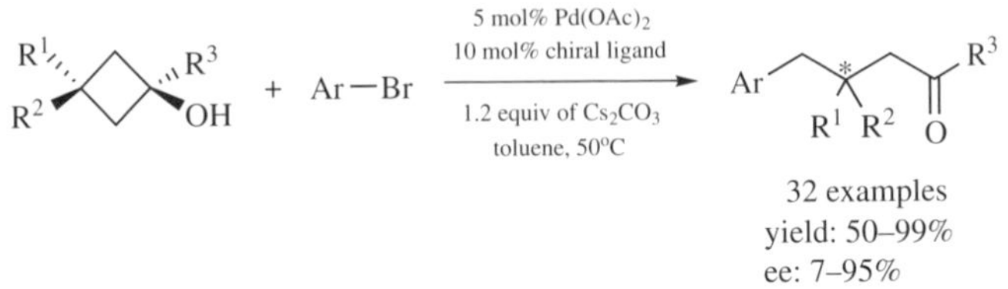

Scheme 5.30. Pd-catalyzed enantioselective C—C bond cleavage.

(1) the inertness of C—C bonds and (2) the more favorable reductive elimination, the reverse reaction. Accordingly, cyclic strained compounds were generally used in C—C bond activations.

5.3.1   Enantioselective C—C Bond Cleavage

By using a Pd-catalyzed enantioselective C—C bond cleavage, γ-arylated ketones were obtained in excellent yields and enantioselectivities (Scheme 5.30).[43] The enantiodetermining step of these reactions occurred at the stage of β-carbon elimination of the Pd(II)-alcoholate intermediates. A variety of chiral ligands were tested and showed that a N,P-bidentate ligand bearing a ferrocene (Figure 5.13) was a more effective chiral ligand than the others. The ferrocene planar chirality played a dominant role in the enantioselective C—C bond cleavage. When vinylating reagents and propargylic acetates were used instead of aryl bromide, the desired vinylation and allenylation products were also obtained in good yields and enantioselectivities.

5.3.2   Formation of C—C Bonds

Cyanation of aldehydes and ketones is an important chemical process for C—C bond formation.[44] Trimethylsilyl cyanide and/or HCN are commonly used as cyanide sources. The intrinsic toxicity and instability of these reagents are problematic in their applications. Acetyl cyanide and cyanoformates were used as cyanide sources in the enantioselective cyanation of aldehydes catalyzed by a chiral Ti complex and Lewis base (Scheme 5.31).[45] The Lewis base was necessary for the good yields and selectivities of these reactions. The desired products were obtained in the presence of 10 mol% triethyl amine and 5 mol% chiral titanium catalyst (Figure 5.14). Various aliphatic and aromatic aldehydes could be used in these reactions.

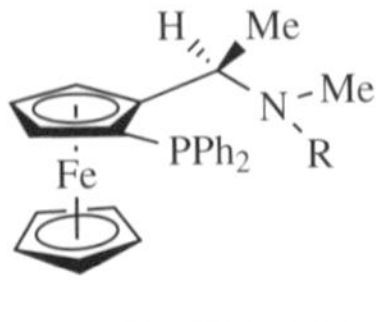

Figure 5.13. Chiral ligands.

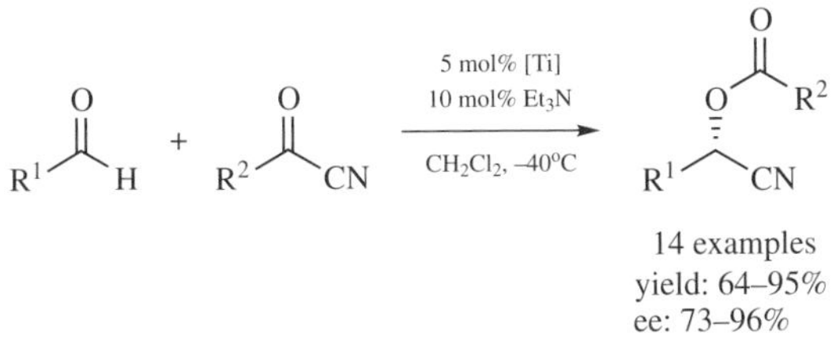

Scheme 5.31. Ti-catalyzed enantioselective cyanation of aldehydes.

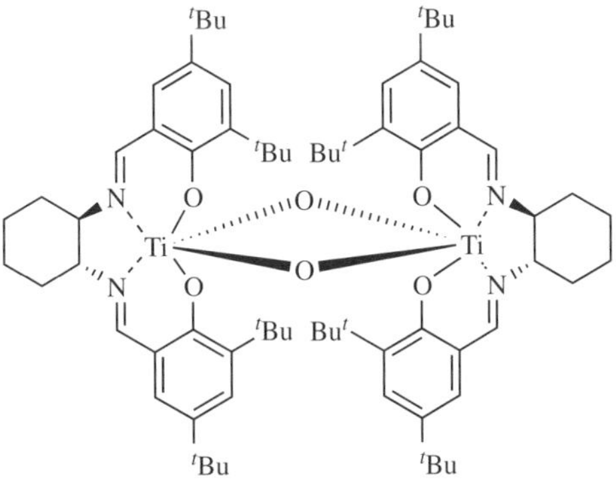

Figure 5.14. Chiral Ti catalyst.

When ethyl cyanoformate was used as the cyanide source and a heterobimetallic YLi$_3$(binaphthoxide) complex (YLB) (Figure 5.15) was used as catalyst, asymmetric cyanoethoxycarbonylations of aldehydes were achieved in high yields and enantioselectivities in the presence of three achiral additives: water, tris(2,6-dimethoxyphenyl)phosphine oxide, and butyl lithium (Scheme 5.32).[46] Mechanistic

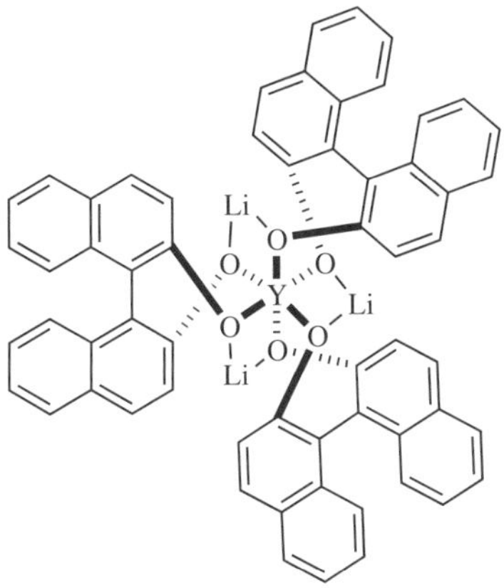

Figure 5.15. YLB.

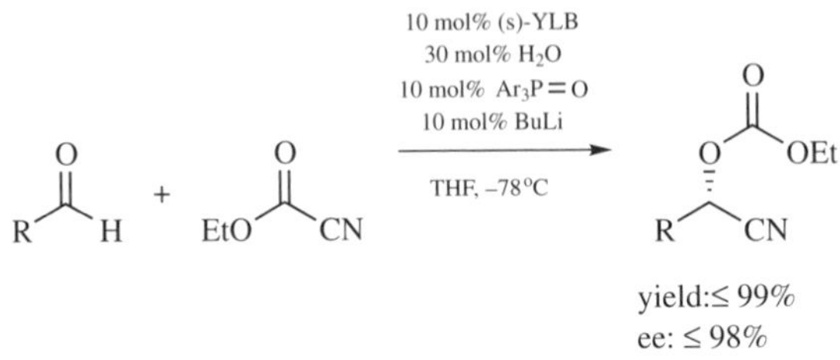

Scheme 5.32. Catalytic heterobimetallic asymmetric cyanation of aldehydes.

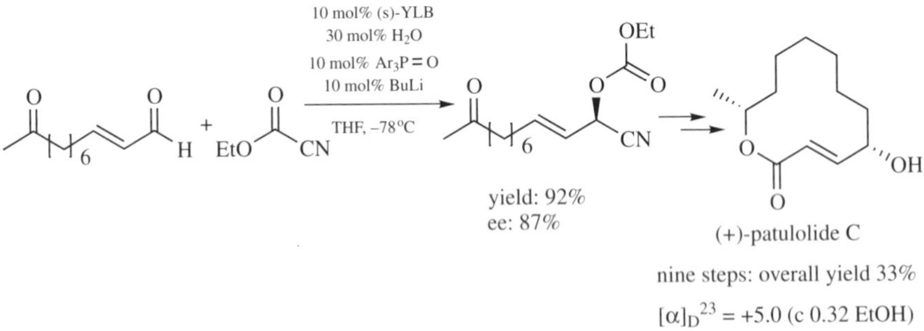

Scheme 5.33. Enantioselective total synthesis of (+)-patulolide C.

investigations found that the three achiral additives help to generate the active nucleophile. The generated active species reacted with aldehydes and then quenched by ethyl cyanoformate to give the desired products and regenerated the active catalyst.

This method was successfully used in a concise catalytic enantioselective total synthesis of (+)-patulolide C by serving as a key asymmetric C—O bond formation step (Scheme 5.33).[47]

In addition to metal catalysts, organocatalysts could also be used in asymmetric cyanation reactions. Chiral Lewis bases, modified cinchona alkaloids, catalyzed asymmetric cyanation of ketones by using ethyl cyanoformate as the cyanide source (Scheme 5.34).[48] Similar to metal-catalyzed reactions, ethyl cyanoformate was first activated by chiral Lewis bases to form active nucleophiles. Various acyclic and cyclic dialkyl ketones were transformed into the desired products. Because of using

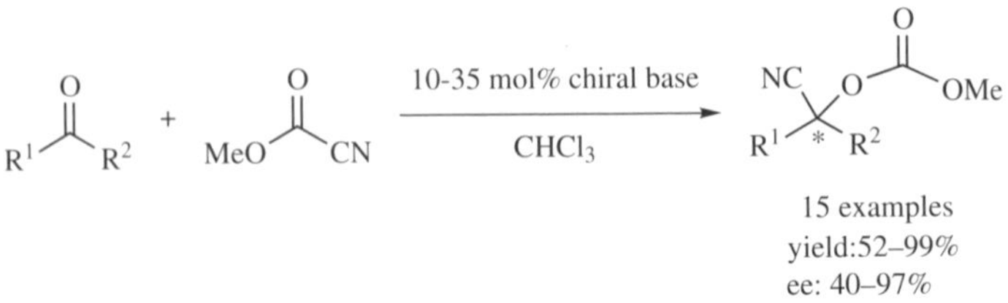

Scheme 5.34. Chiral base-catalyzed asymmetric cyanation of ketones.

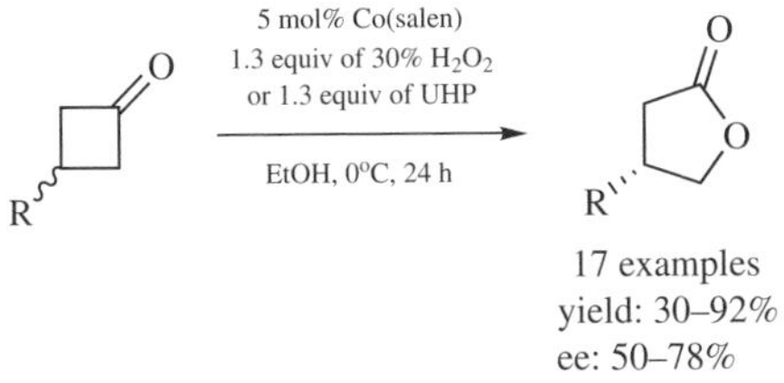

Scheme 5.35. Co-catalyzed asymmetric Baeyer–Villiger reaction of ketones.

base catalysts, these reactions could tolerate acid-sensitive function groups, such as acetals or ketals.

5.3.3   Formation of C—O Bonds

The Baeyer–Villiger (BV) reaction of ketones with peroxides is an important C—O bond formation method (Scheme 5.35).[49] H_2O_2 and O_2 are promising green oxidants in such an oxidation.[50] A variety of metal catalysts, organocatalysts and biocatalysts have been used in the BV reaction. Enantioselective BV reactions were reported since the mid-1990s.[51] Low reactivities and/or enantioselectivities are still hampering the applications of asymmetric BV reactions. The cobalt–salen catalyst (Figure 5.16) is one of the most effective catalysts in metal-catalyzed asymmetric BV oxidation of prochiral cyclobutanones thus far.[52]

5.4   CONCLUSIONS AND OUTLOOK

This chapter reviewed the recent developments on the activations of C—H and C—C bonds for asymmetric synthesis. Although the scope of this subject is still quite limited, the excellent research results obtained thus far are a promising start in this fascinating area. Asymmetric synthesis of propargyl amines and propargyl alcohols are the most successful examples in this area; however, asymmetric cross dehydrogenative coupling reactions still awaits improvement. Although more and more

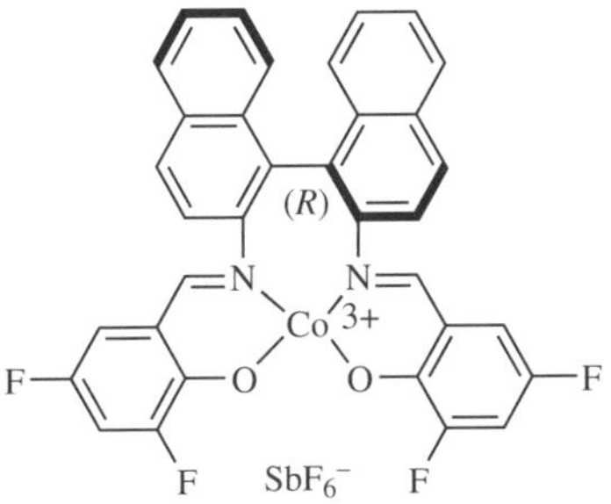

Figure 5.16. Co(salen).

examples of sp^2 C—H activation are being reported, enantioselective examples of such reactions are still quite rare. For asymmetric synthesis based on the activation of sp^3 C—H bonds, most examples are still limited to the metal–carbennoid-induced C—H insertion, and moderate enantioselectivities have been obtained by CDC reactions. Given that C—O and C—N bonds widely exist in nature, asymmetric C—H bond oxidation and amination are in high demand for asymmetric synthesis and are still challenging. Asymmetric synthesis based on C—H and C—C bond activation is promising for efficiently building chiral carbon centers in the near future.

ACKNOWLEDGMENTS

We are grateful to the Canada Research Chair (Tier I) foundation (to CJL), the CFI, NSERC, Merck Frosst and CIC(AstraZeneca/Boehringer Ingelheim/Merck Frosst) for support of our research.

REFERENCES

1. For representative reviews, see: (a) Ritleng, V.; Sirlin, C.; Pfeffer, M. *Chem. Rev.* **2002**, *102*, 1731–1770; (b) Jia, C.; Kitamura, T.; Fujiwara, Y. *Acc. Chem. Res.* **2001**, *34*, 633–639; (c) Dyker, G. *Angew. Chem. Int. Ed.* **1999**, *38*, 1698–1712; (d) Shilov, A. E.; Shul'pin, G. B. *Chem. Rev.* **1997**, *97*, 2879–2932; (e) Arndtsen, B. A; Bergman, R. G.; Mobley, T. A.; Peterson, T. H. *Acc. Chem. Res.* **1995**, *28*, 154–162.

2. For a representative review, see: Jun, C.-H. *Chem. Soc. Rev.* **2004**, *33*, 610–618.

3. For asymmteric addition of alkynyl C—H bonds, see: (a) Wei, C.; Li, Z.; Li, C.-J. *Synlett* **2004**, 1472–1483; (b) Lu, G.; Li, Y.-M.; Li, X.-S.; Chan, A. S. C. *Coord. Chem. Rev.* **2005**, *249*, 1736–1745.

4. Yamaguchi, M.; Hayashi, A.; Minami, T. *J. Org. Chem.* **1991**, *56*, 4091–4092.

5. Frantz, D. E.; Fassler, R.; Carreira, E. M. *J. Am. Chem. Soc.* **1999**, *121*, 11245–11246.

6. (a) Wei, C.; Mague, J. T.; Li, C.-J. *Proc. Natl. Acad. Sci. USA* **2004**, *101*, 5749–5754; (b) Wei, C.; Li, C.-J. *J. Am. Chem. Soc.* **2002**, *124*, 5638–5639.

7. Li, Z.; Li, C.-J. *Org. Lett.* **2004**, *6*, 4997–4999.

8. Koradin, C.; Polborn, K.; Knochel, P. *Angew. Chem., Int. Ed.* **2002**, *41*, 2535–2538.

9. Gommermann, N.; Koradin, C.; Polborn, K.; Knochel, P. *Angew. Chem., Int. Ed.* **2003**, *42*, 5763–5766.

10. Taylor, A. M.; Schreiber, S. L. *Org. Lett.* **2006**, *8*, 143–146.

11. Knopfel, T.; Aschwanden, P.; Ichikawa, T.; Watanabe, T.; Carreira, E. M. *Angew. Chem., Int. Ed.* **2004**, *43*, 5971–5973.

12. Anand, N. K.; Carreira, E. M. *J. Am. Chem. Soc.* **2001**, *123*, 9687–9688.

13. Fassler, R.; Tomooka, C. S.; Frants, D. E.; Carreira, E. M. *Proc. Natl. Acad. Sci. USA* **2004**, *101*, 5843–5845.

14. Takita, R.; Yakura, K.; Ohshima, T.; Shibasaki, M. *J. Am. Chem. Soc.* **2005**, *127*, 13760–13761.

15. Knai, M.; Shibasaki, M. In *Catalytic Asymmetric Synthesis*, 2nd ed., Ojima, I. (Ed.), Wiley-VCH, New York, **2000**, pp. 569–592.

16. Knopfel, T. F.; Zarotti, P.; Ichikawa, T.; Carreira, E. M. *J. Am. Chem. Soc.* **2005**, *127*, 9682–9683.

17. Thalji, R. K.; Ellman, J. A.; Bergman, R. G. *J. Am. Chem. Soc.* **2004**, *126*, 7192–7193.

18. O'Malley, S. J.; Tan, K. L.; Watzke, A.; Bergman, R. G.; Ellman, J. A. *J. Am. Chem. Soc.* **2005**, *127*, 13496–13497.

19. Aufdenblatten, R.; Diezi, S.; Togni, A. *Monatsh. Chem.* **2000**, *131*, 1345–1350.

20. For excellent reviews, see: (a) Davies, H. M. L.; Nikolai, J. *Org. Biomol. Chem.* **2005**, *3*, 4176–4187; (b) Davies, H. M. L.; Beckwith, R. E. J. *Chem. Rev.* **2003**, *103*, 2861–2903.

21. Davies, H. M. L.; Hedley, S. J.; Bohall, B. R. *J. Org. Chem.* **2005**, *70*, 10737–10742.

22. Davies, H. M. L.; Yang, J.; Nikolai, J. *J. Organmet. Chem.* **2005**, *690*, 6111–6124.

23. Davies, H. M. L.; Jin, Q. *J. Am. Chem. Soc.* **2004**, *126*, 10862–10863.

24. Davies, H. M. L.; Walji, A. M. *Angew. Chem. Int. Ed.* **2005**, *44*, 1733–1735.

25. Giri, R.; Chen, X.; Yu, J.-Q. *Angew. Chem. Int. Ed.* **2005**, *44*, 2112–2115.

26. Ikariya, T.; Murata, K.; Noyori, R. *Org. Biomol. Chem.* **2006**, *4*, 393–406.

27. Watanabe, M.; Murata, K.; Ikariya, T. *J. Am. Chem. Soc.* **2003**, *125*, 7508–7509.

28. Watanabe, M.; Ikagawa, A.; Wang, H.; Murata, K.; Ikariya, T. *J. Am. Chem. Soc.* **2004**, *126*, 11148–11149.

29. Mikami, K.; Hatabu, M.; Terada, M. *Chem. Lett.* **1999**, 55–56.

30. Eames, J.; Watkinson, M. *Angew. Chem. Int. Ed.* **2001**, *40*, 3567–3571.

31. Katsuki, T. *Synlett* **2003**, 281–297.

32. (a) Khrasch, M. S.; Sosnovsky, G. *J. Am. Chem. Soc.* **1958**, *80*, 756; (b) Khrasch, M. S.; Sosnovsky, G. Yang, N. C. *J. Am. Chem. Soc.* **1959**, *81*, 5819–5824.

33. Andrus, M. B.; Lashley, J. C. *Tetrahedron* **2002**, *58*, 845–866.

34. (a) Gokhale, A. S.; Minidis, A. B. E.; Pfaltz, A. *Tetrahedron Lett.* **1995**, *36*, 1831–1834; (b) Kawasaki, K.; Tsumura, S.; Katsuki, T. *Synlett* **1995**, 1245–1246; (c) Andrus, M. B.; Argade, A. B.; Chen, X.; Pamment, M. G. *Tetrahedron Lett.* **1995**, *36*, 2945–2948.

35. Kawasaki, K.; Tsumura, S.; Katsuki, T. *Synlett* **1995**, 1245–1246.

36. Zondervan, C.; Feringa, B. L. *Tetrahedron Asym.* **1996**, *7*, 1895–1898.

37. Andrus, M. B.; Zhou, Z. *J. Am. Chem. Soc.* **2002**, *124*, 8806–8807.

38. Hamada, T.; Irie, R.; Mihara, J.; Hamachi, K.; Katsuki, T. *Tetrahedron* **1998**, *54*, 10017–10028.

39. (a) Miyafuji, A.; Ito, K.; Katsuki, T. *Heterocycles* **2000**, *52*, 261–272; (b) Miyafuji, A.; Katsuki, T. *Tetrahedron* **1998**, *54*, 10339–10348; (c) Miyafuji, A.; Katsuki, T. *Synlett* **1997**, 836–838.

40. Zhou, X.-G.; Yu, X.-Q.; Huang, J.-S.; Che. C.-M. *Chem. Commun.* **1999**, 2377–2378.

41. Liang, J.-L.; Huang, J.-S.; Yu, X.-Q.; Zhu, N.; Che, C.-M. *Chem. Eur. J.* **2002**, *8*, 1563–1572.

42. Wehn, P. M.; Du Bois, J. *J. Am. Chem. Soc.* **2002**, *124*, 12950–12951.

43. Mutsumura, S.; Maeda, Y.; Nishimura, T.; Uemura, S. *J. Am. Chem. Soc.* **2003**, *125*, 8862–8869.

44. (a) Jacobsen, E. N.; Pfaltz, A.; Yamamoto, H. *Comprehensive Asymmetric Catalysis*; Springer, Berlin, **1999**; (b) Gregory, R. J. H. *Chem. Rev.* **1999**, *99*, 3649–3682.

45. Lundgren, S.; Wingstrand, E.; Penhoat, M.; Moberg, C. *J. Am. Chem. Soc.* **2005**, *127*, 11592–11593.

46. (a) Yamagiwa, N.; Tian, J.; Matsunaga, S.; Shibasaki, M. *J. Am. Chem. Soc.* **2005**, *127*, 3413–3422; (b) Tian, J.; Yamagiwa, N.; Matsunaga, S.; Shibasaki, M. *Angew. Chem. Int. Ed.* **2002**, *41*, 3636–3638.

47. Tian, J.; Yamagiwa, N.; Matsunaga, S.; Shibasaki, M. *Org. Lett.* **2003**, *5*, 3021–3024.

48. Tian, S.-K.; Deng, L. *J. Am. Chem. Soc.* **2001**, *123*, 6195–6196.

49. (a) Krow, G. R. *Org. React.* **1993**, *43*, 251–798; (b) Strukul, G. *Angew. Chem., Int. Ed.* **1998**, *37*, 1198–1209; (c) Renz, M.; Meunier, B. *Eur. J. Org. Chem.* **1999**, 737–750.

50. Ten Brink, G.-J.; Arends, I. W. C. E.; Sheldon, R. A. *Chem. Rev.* **2004**, *104*, 4105–4123.

51. (a) Gusso, A.; Baccin, C.; Pinna, F.; Strukul, G. *Organometallics* **1994**, *13*, 3442–3451; (b) Bolm, C.; Schlingloff, G.; Weickhardt, K. *Angew. Chem. Int. Ed.* **1994**, *33*, 1848–1849.

52. Uchida, T.; Katsuki, T. *Tetrahedron Lett.* **2001**, *42*, 6911–6914.

6

RECENT PROGRESS IN THE METATHESIS REACTION

MIWAKO MORI

Health Sciences University of Hokkaido, Ishikari-Toubetsu, Hokkaido, Japan

6.1 INTRODUCTION

Metathesis has been the most useful reaction in synthetic organic chemistry since the mid-1990s.[1] In this reaction, bond fission of a double bond occurs and a new double bond is simultaneously formed:

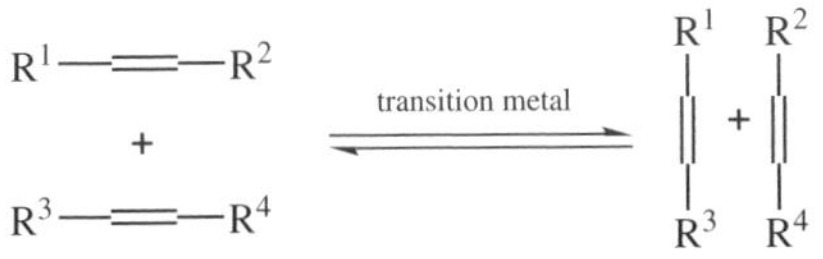

Since the alkene formed in this reaction can further react with other alkenes, many products should be formed in the cross-metathesis (CM). Therefore, in the early days, only ring-closing metathesis (RCM) of diene was investigated. It is known that the reaction is catalyzed by a transition metal. Pioneering work on olefin metathesis was undertaken by Villemin[2a] and Tsuji,[2b] who reported the synthesis of lactones using alkene metathesis:

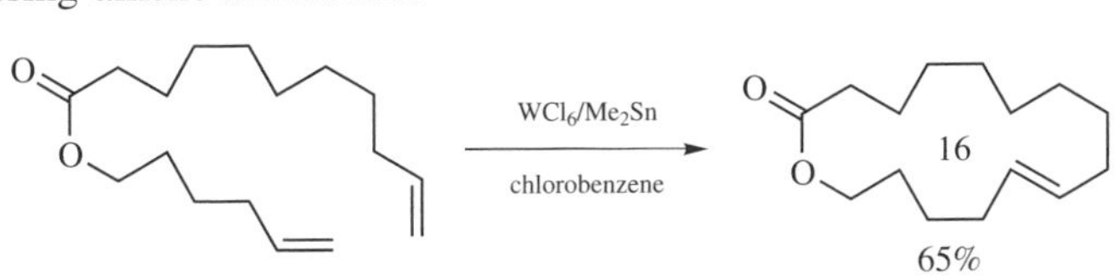

(6.1)

New Frontiers in Asymmetric Catalysis, Edited by Koichi Mikami and Mark Lautens
Copyright © 2007 John Wiley & Sons, Inc.

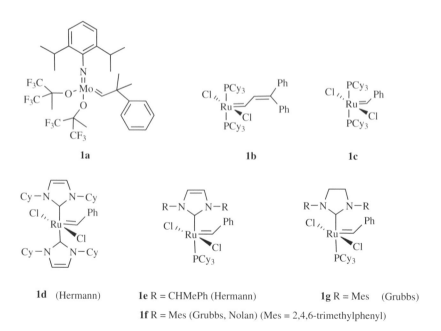

$$ \xrightarrow[\text{benzene}]{\text{WCl}_6/\text{Cp}_2\text{TiMe}_2} \qquad (6.2) $$

R = C$_8$H$_{17}$ 18%

Since Schrock[3a] and Grubbs[4a] discovered molybdenum and ruthenium carbene complexes **1a** and **1b** in 1990 and 1992, metathesis reactions have been a useful tool in synthetic organic chemistry. The ruthenium carbene complex **1c**, reported by Grubbs in 1995,[4b] is commercially available, and various cyclic compounds have been synthesized using olefin metathesis. In 1999, Hermann,[5] Nolan,[6] and Grubbs[7] developed novel ruthenium carbene complexes **1d–1g** that have a heterocyclic carbene as a ligand. These catalysts are very reactive toward olefin metathesis compared with the first-generation ruthenium catalysts **1b** and **1c**. Using these catalysts, cross-metathesis (CM) of alkene is possible.

Metathesis reactions are now widely used in natural product synthesis. Novel retrosynthetic analyses were developed because a carbon–carbon single bond can be formed after hydrogenation of a double bond constructed by metathesis. Although many types of metathesis are now known, the reaction is classified by olefin, enyne, and alkyne metatheses in this chapter.

1a   **1b**   **1c**

1d (Hermann)   **1e** R = CHMePh (Hermann)   **1g** R = Mes   (Grubbs)

1f R = Mes (Grubbs, Nolan) (Mes = 2,4,6-trimethylphenyl)

Figure 6.1

6.2   OLEFIN METATHESIS

6.2.1   Ring-Closing Olefin Metathesis

In 1992, Grubbs[8] reported the synthesis of carbo- and heterocycles from a tethered diene using molybdenum carbene complex **1a** discovered by Schrock.[3a] This is the first example of transition-metal-catalyzed olefin metathesis by a metal carbene complex [Eq. (6.3)]. The reaction proceeds via [2+2] cycloaddition of diene **2** and molybdenum methylidene carbene complex generated from **1a** to form metalacyclobutane, which gives a molybdenum carbene complex by ring opening. Reaction with an olefin intramolecularly produces metalacyclobutane, and ring opening of this complex gives cyclic compound **3** and the molybdenum methylidene carbene complex is regenerated.

$$(6.3)$$

Later Grubbs discovered ruthenium carbene complex **1b**[4a] and used it for a metathesis reaction to synthesize cyclic compounds **5a–d** [Eqs. (6.4) and (6.5)].[9] In 1995, Grubbs found that ruthenium benzylidene carbene complex **1c**,[4b] which is now commercially available, has the same reactivity as that of **1b**. Many researchers have therefore used this complex for olefin metathesis, and this reaction has been useful for the synthesis of carbo- and heterocyclic compounds and fused bicyclic compounds [Eq. [6.6)]:[10]

$$(6.4)$$

$$(6.5)$$

$$(6.6)$$

$$n = 0, m = 2; 87\%$$
$$n = 1, m = 1; 95\%$$

Crimmins reported the synthesis of enantiomerically pure cyclopentenol derivative **8** from diene **6** using ring-closing olefin metathesis followed by treatment with
LiBH$_4$ [Eq. (6.7)]:[11]

$$(6.7)$$

A medium-sized ring compound, whose synthesis is difficult by traditional methods, can be easily synthesized by ring-closing metathesis (RCM). However a substrate having no substituents on the chain does not afford the cyclic compound.
Cyclohexane derivative **9a** having trans substituents gives fused six- and eight-
membered ring compound **10a** in good yield, while **9b** having cis substituents
affords bicyclic compound **10b** in moderate yield [Eqs. (6.8) and (6.9)].[12] Using
this procedure, Nolan synthesized functionalized eight-membered ring compound
10c from 6-deoxy-6-iodoglycoside **11** [Eq. (6.10)]:[13]

$$(6.8)$$

$$(6.9)$$

$$(6.10)$$

Fürstner succeeded in the synthesis of dactylol using ring-closing metathesis of **9d**.[14] Crimmins synthesized enantioselectively diene **9e**, and RCM of **9e** gave eight-membered ring compound **10e**, which is an intermediate for the synthesis of laurencin [Eq. (6.12)]:[15]

$$(6.11)$$

$$(6.12)$$

For the synthesis of C_2-symmetric 1,4-diol, Evans planned temporary silicon-tethered ring-closing metathesis, and asymmetric synthesis of D-altritol was achieved [Eq. (6.13)]:[16]

(6.13)

Metathesis of a diene containing silicon atoms gives an interesting cyclic compound, whose silicon moiety can be converted into the hydroxyl group by Tamao oxidation [Eqs. (6.14) and (6.15)]:[17]

(6.14)

(6.15)

However, the reactivities of ruthenium carbene complexes **1b** and **1c** are not high, and RCM of the substituted diene does not give the desired cyclized compound (Table 6.1, entry 3). Therefore, more reactive ruthenium carbene complexes are synthesized. In 1999, Hermann,[5] Nolan[6] and Grubbs[7] found second-generation ruthenium carbene complexes **1d–g**. They have an *N*-heterocyclic carbene as a ligand and are effective for RCM of disubstituted alkenes (Table 6.1, entries 3 and 4).[18]

Grubbs investigated the reason for the higher reactivity of second-generation ruthenium carbene complex **1g** in comparison to that for the first-generation ruthenium catalyst **1c**.[19] The initial step of this reaction is dissociation of a phosphine ligand from a ruthenium metal of **I**, and then an olefin coordinates to the ruthenium metal to give complex **III** [Eq. (6.16)]. [2+2] Cycloaddition occurs to give ruthenacyclobutane **IV**, whose ring opening gives an alternative olefin, and methylidene ruthenium carbene complex **V** is regenerated. Surprisingly, the dissociation of the phosphine ligand from the ruthenium metal in **1c** having two phosphine ligands is faster than that in **1g** having a phosphine and *N*-heterocyclic carbene. However, coordination of the olefin to the resulting complex **II** having an *N*-heterocyclic

TABLE 6.1. RCM at 45 °C Utlizing 5 mol% Catalysts 1c and 1f

Entry	Substrate	Product	1c	1f
1			quant	quant
2			82	quant
3			0	95
4	$E=CO_2Et$		0	40

carbene ligand is overwhelmingly faster than that to **II** possessing a phosphine ligand. Thus, the overall reaction rate using **1g** is faster than that using **1c**. A higher reaction temperature is required when the second-generation ruthenium carbene complex is used to promote dissociation of the phosphine ligand from **1g**:

$$(6.16)$$

Enol ether or silyl enol ether as an alkene in RCM can be used by the second-generation ruthenium carbene complex **1g**. It affords the cyclic enol ether or silyl enol ether, which gives a cyclic ketone [Eqs. (6.17) and (6.18)]:[20]

$$(6.17)$$

19a R = Me   95%
     R = TBS   95%

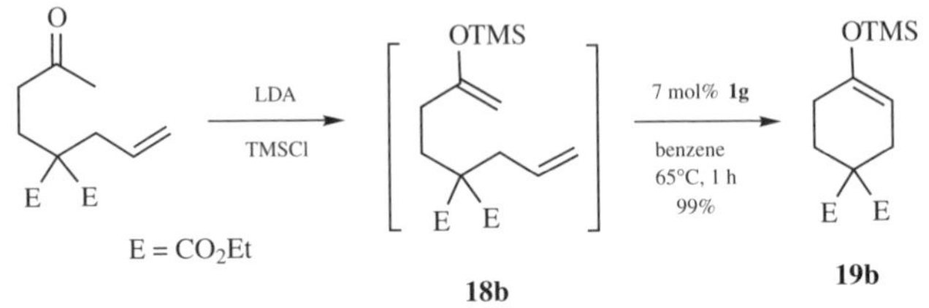

$$\text{(6.18)}$$

E = CO$_2$Et

18b   **19b**

A concise stereoselective synthesis of a myoinositol derivative has been achieved by ring-closing metathesis of diene **21** prepared from a readily available bis-Weinreb amide **20** of D-tartrate [Eq. (6.19)].[21]

A variety of cycloheptenols **24** can be synthesized in enantiomerically pure form with high chemical yields by ring-closing metathesis of acyclic, chiral polyoxygenated 1,8-nonadiene precursors **23** derived from carbohydrates [Eq. (6.20)].[22]

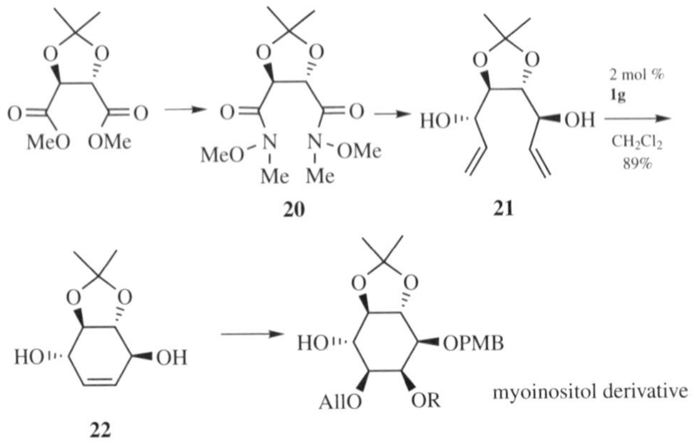

20   **21**

22

myoinositol derivative

$$\text{(6.19)}$$

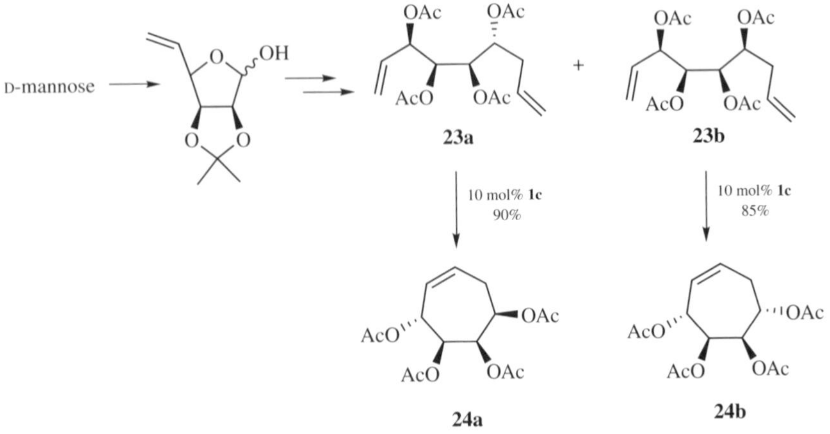

D-mannose

23a   **23b**

10 mol% **1c**
90%

10 mol% **1c**
85%

24a   **24b**

$$\text{(6.20)}$$

The reaction of **25** with **1c** gives lactone **26**, but the yield is only 40%. It was thought that the carbonyl oxygen should coordinate to the ruthenium metal to give

27. When the reaction is carried out using **1c** in the presence of Ti(O^{i}Pr)$_4$, the yield is improved to 72% [Eq. (6.21)]:[23]

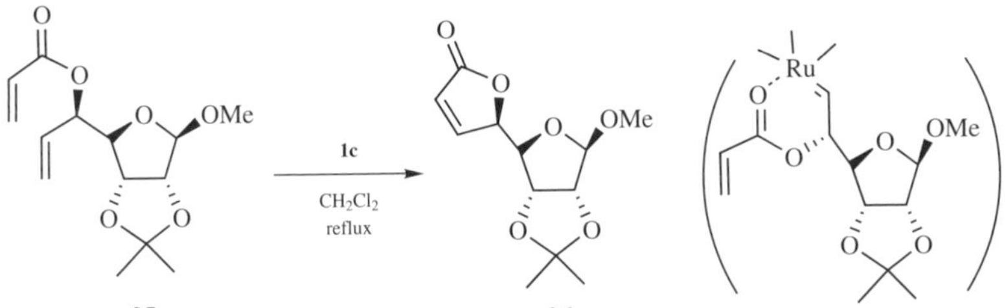

25 **26** **27** (6.21)

no additive; 15 h, 40%
Ti(O^{i}Pr)$_4$ (30 mol%); 15 h, 72%

In a similar manner, a new strategy to access the fumagilol skeleton was reported. RCM of diene **29**, which was synthesized by the Evans aldol reaction, was carried out using **1c** in the presence of Ti(O^{i}Pr)$_4$ to give a key cyclohexanone intermediate **30** [Eq. (6.22)]. This compound was readily converted to fumagilol:[24]

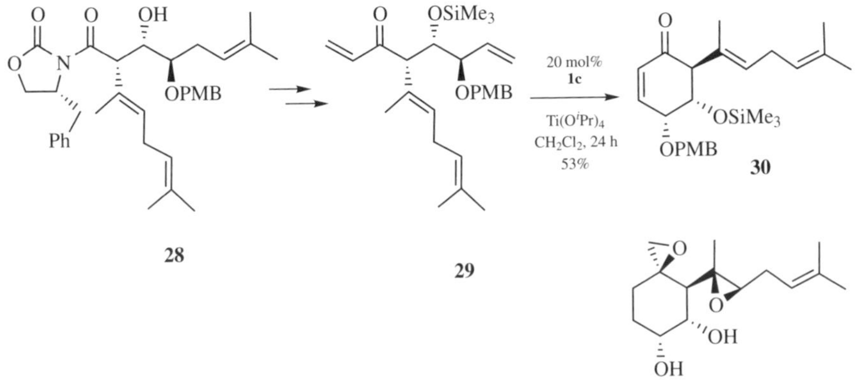

28 **29** **30**

fumagilol

(6.22)

Nicolaou reported a novel method for the synthesis of polycyclic compounds in which the six-membered ring of polycyclic compound **32** was constructed by RCM [Eq. (6.23)]:[25]

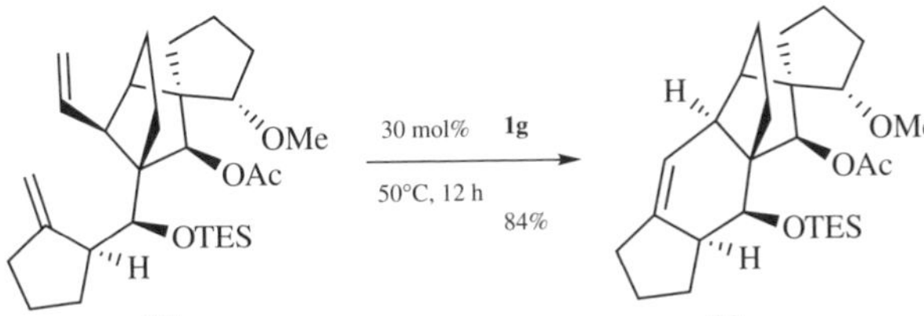

31 **32** (6.23)

Hirama et al. synthesized polyether **34** by RCM of diene **33**,[26a] and they succeeded in the total synthesis of ciguatoxin using this procedure:[26b]

$$(6.24)$$

Although RCM gives brilliant results for the synthesis of medium-ring carbocycles, it is also effective for the synthesis of macrocyclic lactone as shown in Eqs. (6.25)–(6.27).[27] Prior to RCM, macrolactonization was the most common method for the synthesis of macrocyclic lactone. However, we can now obtain the desired macrocyclic lactone from diene having an ester moiety in a chain by RCM followed by hydrogenation:

$$(6.25)$$

$$(6.26)$$

$$(6.27)$$

Approaches to the total synthesis of complex natural products using RCM have been investigated. Martin et al. succeeded in the construction of the core 13-membered ring from **37** and 8-membered ring from diene **39** using RCM in manzamine synthesis, and they then achieved the total synthesis of manzamine using RCM as the key steps:[28]

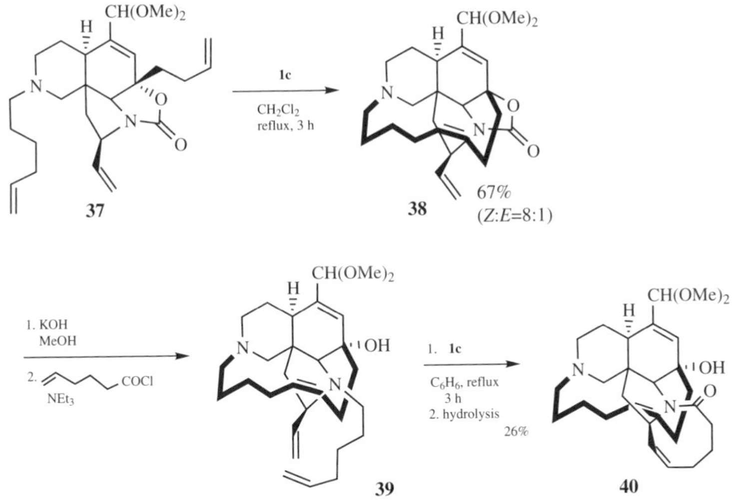

The construction of a complete carbocyclic skeleton of ingenol using ring-opening metathesis (ROM) and ring-closing metathesis (RCM) has been reported.[29] ROM of the cyclopentene ring of 41 smoothly proceeds under ethylene gas to give diene 42 in high yield, and a seven-membered ring is constructed by RCM of diene 43:

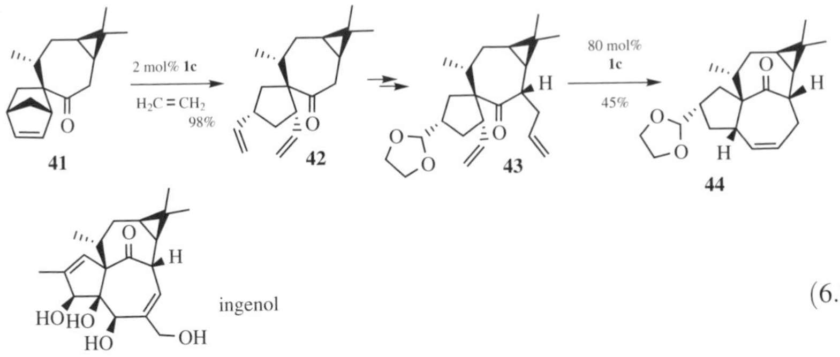

An interesting technique for the synthesis of crown ethers has been developed using template-directed RCM or ROM by Grubbs.[30] Various crown ethers 46 have been prepared from linear polyether 45 in the presence of an appropriate metal [Eq. (6.30)]. For the synthesis of 14-membered crown ether, the Li cation gives good results. Surprisingly, the polyether 45 can be synthesized by ROM of the crown ether 46a using 1c at room temperature in high yield, and the crown ether 46a is reproduced from the polyether 45 using RCM in the presence of the Li cation

[Eq. (6.31)]:

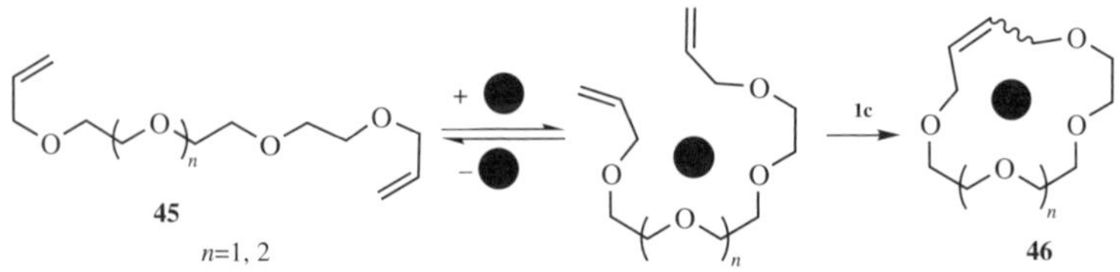

(6.30)

45
$n=1, 2$

Substrate	Template	Yield (%)	cis:trans
45a $n=1$	—	39	38:62
	LiClO$_4$	95	100:0
	NaClO$_4$	42	62:38
	KClO$_4$	36	36:64
45b $n=2$	—	57	26:74
	LiClO$_4$	89	61:39
	NaClO$_4$	90	68:32
	KClO$_4$	64	25:75

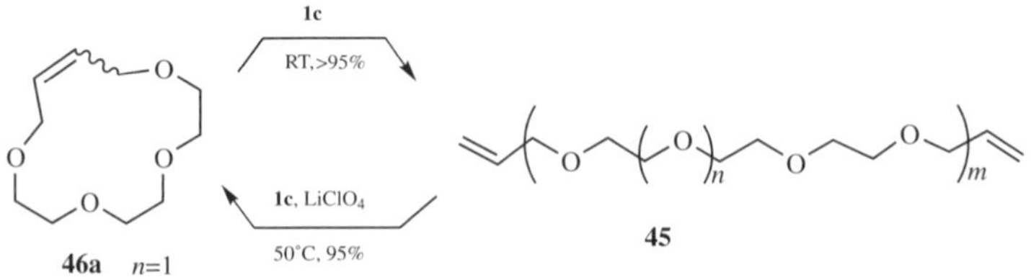

(6.31)

46a $n=1$

A novel synthetic method of a catenane by RCM was reported.[31] RCM of the diene **47** having the ligand part in a chain in the presence of the metal gives cate-nane **48** containing the metal. Dissociation of the metal gives catenane **49**. When diene **47a** is treated with **1c** in the presence of Cu$^+$, the ligand part in the chain coordinates to the metal by crossing of two dienes **47a**, and RCM of the diene moi-ety gives catenane **48a** containing Cu$^+$ in high yield [Eq. (6.32)]:

SCHEMATIC DRAWING FOR SYNTHESIS OF CATENANES

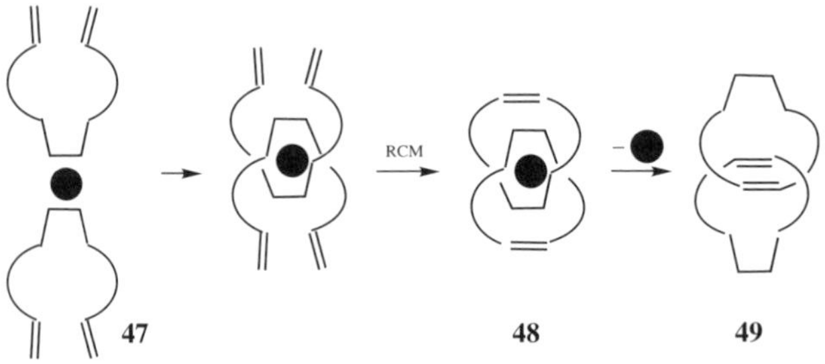

$$(6.32)$$

90%

47a   ● = Cu+   **48a**

6.2.2   Cross-Metathesis (CM) of Diene

Since the product alkenes further react with the starting alkenes in the cross-metathesis, a mixture of various alkenes should be produced. Therefore, only RCM of the olefin is investigated using the first-generation ruthenium catalyst. Following the discovery of the second-generation ruthenium catalysts, cross-metathesis was investigated. If the product is not further reacted with a carbene complex, CM of olefins should be realized. Grubbs showed that CM of a terminal alkene with an alkene **50b** having electron-withdrawing group gives CM product **51b** in high yield [Eq. (6.34)].[32] In this reaction, an intermediary ruthenium carbene complex **1i** is formed from **50b** and benzylidene carbene complex, and reacts with the methylene part on the cyclohexane to give ruthenacyclobutane. Ring opening of this complex affords **51b** and methylidene carbene complex **1h**, which reacts with an alkene part of **50b** to afford **1i**. Thus, the catalytic cycle is established.

$$(6.33)$$

50a   **51a**

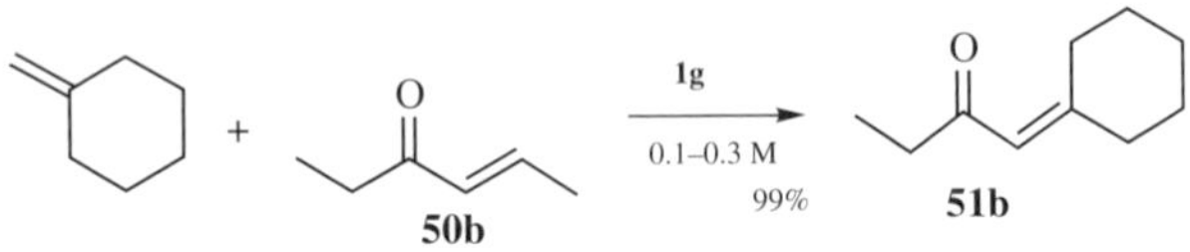

$$ \text{(equation scheme with } \mathbf{50b} \xrightarrow[\substack{0.1-0.3\ M \\ 99\%}]{\mathbf{1g}} \mathbf{51b}) \tag{6.34}$$

[Ru]=CHCOX **1i**

[Ru]=CHPh

X=H, R, OR, OH

1g

1i + ⟶ 51b + Ru= **1h**

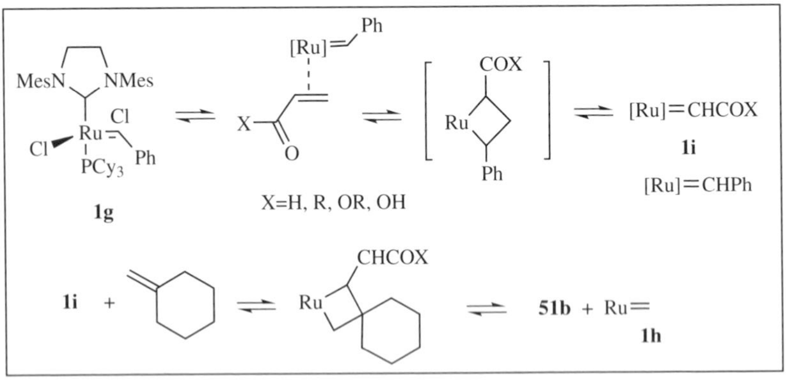

TABLE 6.2. CM with α-Functionalized Olefin Using 1g

Entry	Olefin	α-Functionalized Olefin	Product	Yield (%) (Ratio)
1	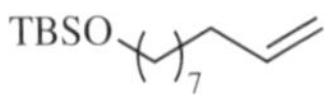 TBSO(·)$_7$⟶	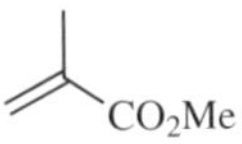 CO$_2$Me	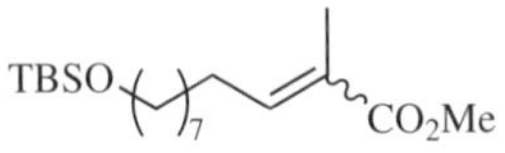 TBSO(·)$_7$ ⟶ CO$_2$Me	62 (20:1)
2	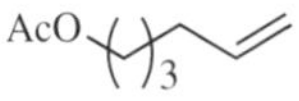 AcO(·)$_3$⟶	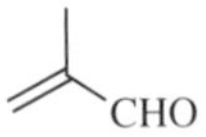 CHO	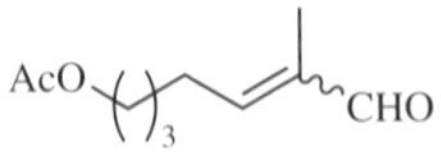 AcO(·)$_3$ ⟶ CHO	92 (20:1)
3	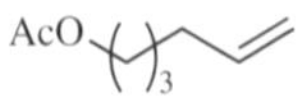 AcO(·)$_3$⟶	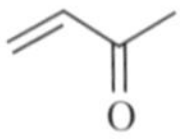 O	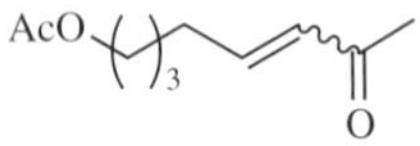 AcO(·)$_3$ ⟶ O	95 (20:1)
4	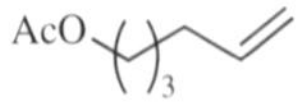 AcO(·)$_3$⟶	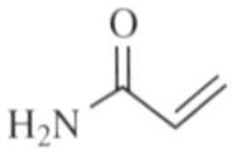 H$_2$N─O	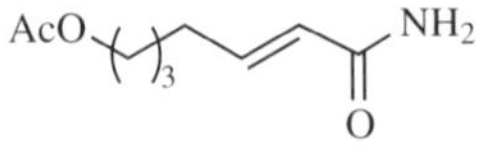 AcO(·)$_3$ ⟶ NH$_2$ O	89
5	 TBSO(·)$_4$⟶	 SO$_2$Ph	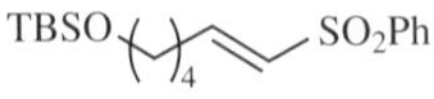 TBSO(·)$_4$ ⟶ SO$_2$Ph	85
6	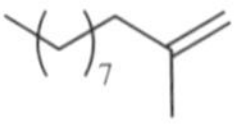 (·)$_7$	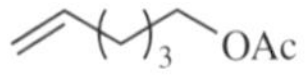 (·)$_3$OAc	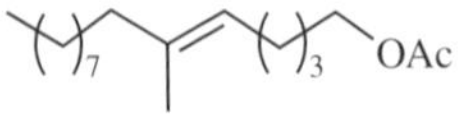 (·)$_7$ ⟶ (·)$_3$OAc	60 (2.3:1)
7	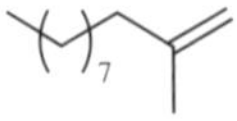 (·)$_7$	 AcO─OAc	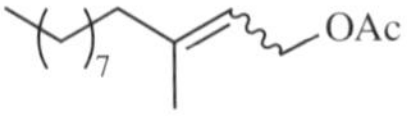 (·)$_7$ ⟶ OAc	53 (2.5:1)

The cross-metathesis of terminal alkenes and functionalized alkenes is shown in Table 6.2. In each case, a CM product is obtained in high yield and an *E*-isomer is formed predominantly.[33]

In traditional synthetic organic chemistry, the Wittig reaction plays an important role in carbon–carbon bond extension from the carbonyl group. CM is an attractive alternative for carbon–carbon extension from a terminal alkene. In fact, a pyrrolidine ring of anthramycin derivative **55** has been constructed by RCM of **52**, and the sidechain has been extended by CM of terminal alkene of **54** with ethyl acrylate.[34] In the CM, ruthenium carbene complex **1j**, reported by Blechert,[35] gives a good result since the ligand of the catalyst easily dissociated from the ruthenium metal at room temperature:

$$(6.35)$$

6.2.3   Ring-Opening Metathesis (ROM)–Ring-Closing Metathesis (RCM) of Alkene

RCM of a substrate having more than three alkenes proceeds via tandem reactions to give various cyclic compounds. It is denoted as ROM-RCM or ROM-CM. Grubbs reported ROM-RCM of a disubstituted cycloalkene having an alkene moiety in each tether [Eqs. (6.36)–(6.39)].[36] In this reaction, two alternative rings are formed. The reaction would proceed via the formation of terminal ruthenium carbene complex **VI**, generated by the reaction of **56** and **1h**. This carbene complex **VI** reacts with a cycloalkene part to form an alternative carbene complex **VII**, which gives another ring by RCM:

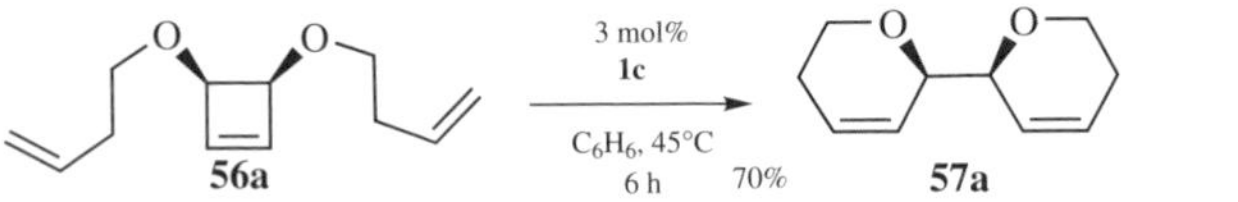

(6.36)

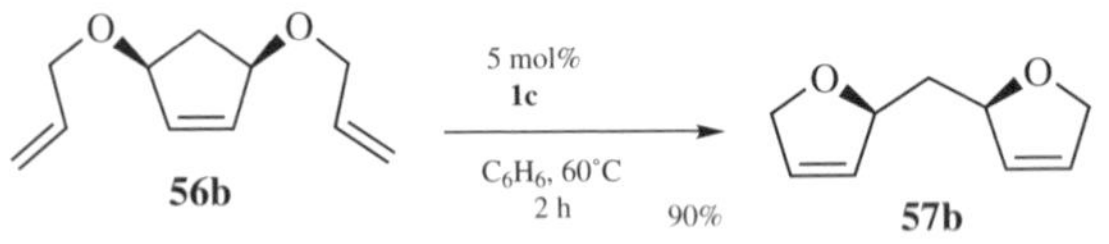

(6.37)

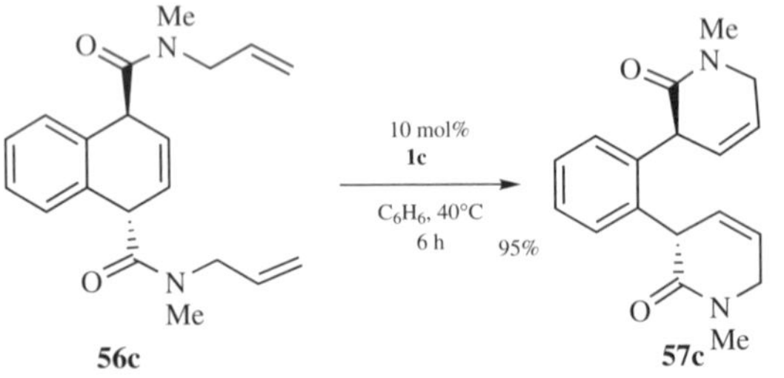

(6.38)

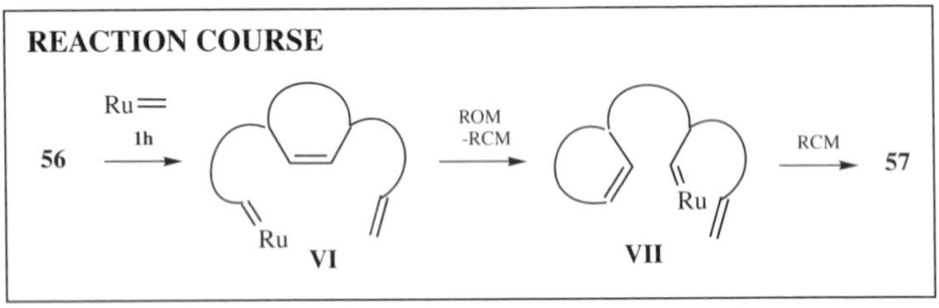

REACTION COURSE

(6.39)

Snapper reported ROM-CM of cyclobutene. ROM-CM of cyclobutene and 1-octene using **1b** gives cyclopentene derivative **59a** [Eq. (6.40)].[37a] On the other hand, cyclobutene rings of fused four- and five-membered ring compounds **58b** and **58c** are opened by an alkene moiety in a tether by ROM to produce cyclopentane derivatives **59b** and **59c** having furan and pyran rings [Eqs. (6.41) and (6.42)].[37b] The ring size of the heterocycle formed in this reaction corresponds to the length of the sidechain on cyclobutene:

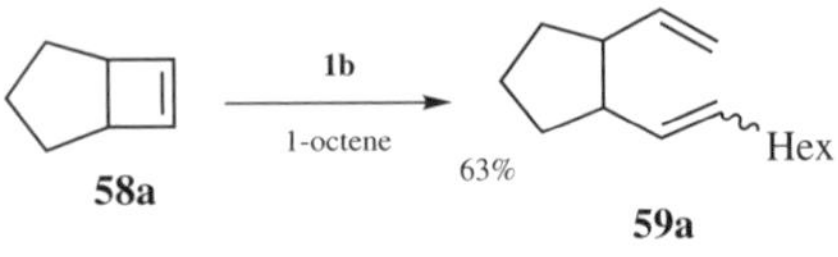

(6.40)

$$\text{(6.41)}$$

$$\text{(6.42)}$$

As an application of this procedure, Snapper succeeded in the formal total synthesis of (+)-astericanolide.[37c] Treatment of iron complex **62**, which is prepared from cyclopentenol derivative **60** and iron complex **61**, with Me$_3$NO gives fused cyclobutene derivative **63**. ROM of cyclobutene of **63** using **1g** under ethylene gas smoothly proceeds to produce an eight-membered ring of **64** via Cope rearrangement. The resulting product **64** is converted into Wender's intermediate for the synthesis of (+)-astericanolide:

$$\text{(6.43)}$$

Since it is known that the cyclopentene ring of norbornene can be easily opened by methylidene carbene complex **1h**, bicyclic compound **66** has been synthesized from norbornene derivative **65** having an alkene part in a sidechain in the presence of alkene and a catalytic amount of **1c**.[38] The size of the ring constructed in this reaction corresponds to the length of the sidechain on norbornene **65**. As the reaction course, two possible pathways should be considered: (1) route A, where the reaction starts by cycloaddition of terminal alkene in the sidechain and ruthenium methylidene carbene complex **1h** to form metal carbene **VIII**, which reacts with the olefin of the cyclopentene to give bicyclic compound **X** bearing metal carbene (this reacts with ethylene to give compound **66**); and (2) route B, where the reaction proceeds via ring opening of the five-membered ring by methylidenene carbene **1h** to give metal carbene **IX** and **IX′**. RCM of **IX** gives **66**. Furthermore, each metal carbene **IX** or **IX′** reacts with ethylene to produce triene **67** and RCM of **67** gives bicyclic compound **66**:

$$(6.44)$$

1a	R = CH$_2$TMS	83%
1c	R = CH$_2$TMS	71%
1c	R = H	88%

$$(6.45)$$

ROM-RCM of cyclopentenyl amine derivative **68a** bearing an alkene moiety on the nitrogen gives piperidine derivative **69a** [Eq. (6.46)].[39a] Various heterocycles should be synthesized by change of the carbon chain length on the nitrogen. (−)-Halosaline has been synthesized from **68b** using this procedure [Eq. (6.47)].[39b]

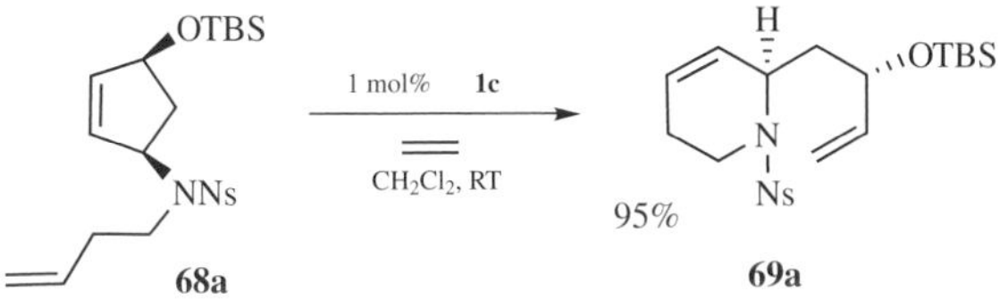

68a **69a**

(6.46)

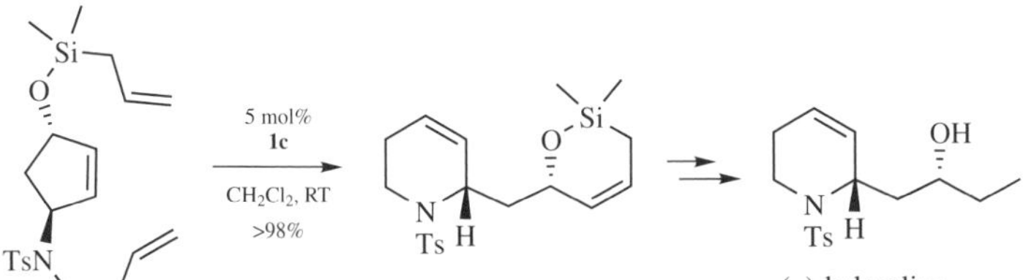

68b **69b** (−)-halosaline

(6.47)

Hoveyda et al. reported a novel method for synthesizing of chromene **71** by ROM-RCM of cycloalkene **70** bearing the phenyl ether at the 3-position [Eq. (6.48)].[40] The yield is improved when the reaction is carried out under ethylene gas. In the case of cyclopentene **70a** ($n = 0$) or cyclohexene **70b** ($n = 1$), the yield is poor because the starting cycloalkene is in a state of equilibrium with the product and a thermodynamic product should be formed under these reaction conditions. They obtained enantiomerically pure cycloheptene derivative (*S*)-**70e** using zirconium–catalyzed kinetic resolution of **70e** developed by their group, and chromene **71c** was synthesized as a chiral form via ROM-RCM using **1b** [Eq. (6.49)]:

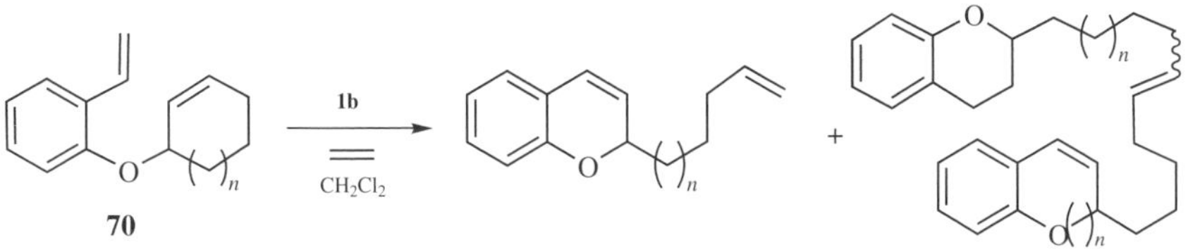

70

		mono	:	dimer
n=0	**71a**	0%	:	0%
n=1	**71b**	35%	:	2% (65% recovered **70b**)
n=2	**71c**	92%	:	2%
n=3	**71d**	90%	:	2%

(6.48)

$$(6.49)$$

Wright succeeded in ROM-RCM of substrate **72a** having a bicyclo[3.2.1] ring sys-tem to form pyran **73a** having a spiro ring in 82% yield [Eq. (6.50)].[41] Reaction of **72a** with **1b** would give ruthenium carbene complex **72a-Ru**, which reacts with an alkene part of furan to give highly strained ruthenacyclobutane. Ring opening of this complex gives ruthenium carbene complex, which reacts with a terminal alkene part of **72a** to give **73a** and ruthenium carbene complex **72a-Ru** is regenerated. In a similar manner, **72b** having bicyclo[2.2.1] gives furan derivative **73b** in high yield [Eq. (6.51)]:

$$(6.50)$$

$$(6.51)$$

In these reactions, it is not clear which alkene—the terminal alkene or the cycloalkene—reacts with methylidene carbene complex **1h** in the initial step, and it is difficult to determine the real reaction course. However, it is interesting that a complex molecule can be synthesized by one step in high yield.

6.2.4 Catalytic Asymmetric Olefin Metathesis

Asymmetric Synthesis Using a Chiral Molybdenum Catalyst   In olefin metathesis, a double bond is cleaved and a double bond is formed. Thus, a chiral carbon center is not constructed in the reaction. To realize the asymmetric induction by ring-closing metathesis, there are two procedures: a kinetic resolution and desymmetrization of symmetric prochiral triene. Various molybdenum complexes are synthesized in order to explore the viability of these approaches (Figure 6.2).

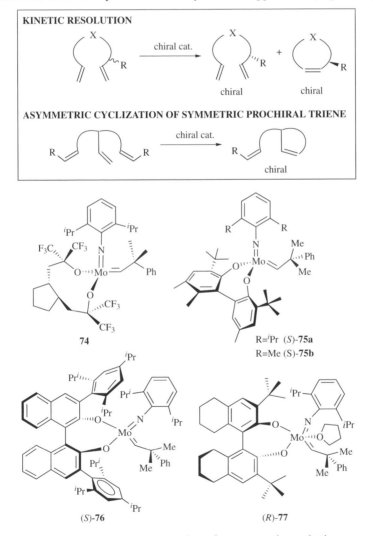

Figure 6.2 Molybdenum catalysts for asymmetric synthesis.

Grubbs synthesized molybdenum catalyst **74** and reported the first example of asymmetric induction by kinetic resolution of diene using RCM with complex **74**.[42] After 90% conversion of the reaction of diene with **74**, an enantiomeric excess (ee) of the recovered staring material **78** showed 84% and an absolute configuration is determined to be *S*:

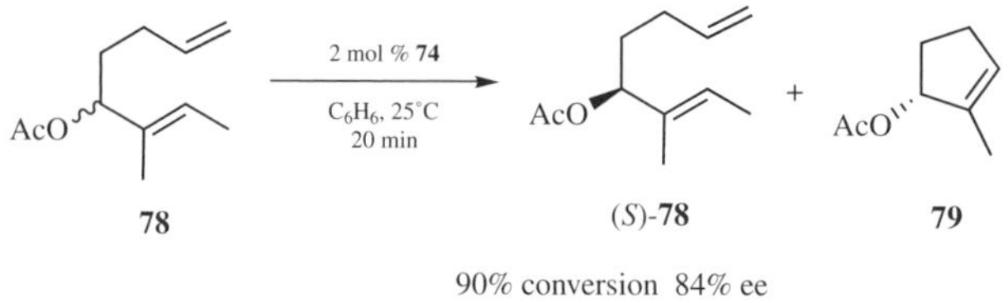

$$\text{(6.52)}$$

90% conversion 84% ee

Hoveyda and Schrock reported a noteworthy result in the field of asymmetric olefin metathesis. They synthesized many chiral molybdenum catalysts and examined the effectiveness of these catalysts in asymmetric induction. In the kinetic resolution of diene using BIPOL-type molybodenum catalyst **75a**, cyclopentene derivative **81a** with 93% ee was obtained in 43% yield after 80% conversion [Eq. (6.53)].[43] In the case of the formation of six-membered ring compound **81b**, BIPOL-type complex **75a** did not give a good result, but BINOL-type complex **76** showed high enantioeselection [Eq. 6.54)].[44] This catalyst **75a** is very effective for desymmetrization of symmetric triene **80c**, and furan derivative **81c** with 99% ee is obtained in 86% yield [Eq. (6.55)]:[45]

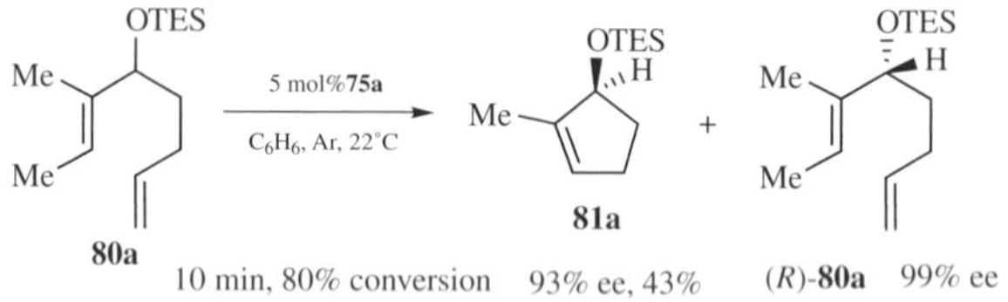

$$\text{(6.53)}$$

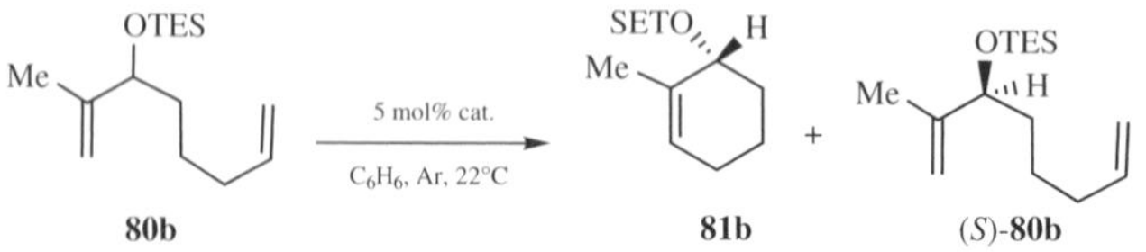

10 min, 80% conversion   93% ee, 43%   (*R*)-**80a**   99% ee

$$\text{(6.54)}$$

cat.		k_{rel}
75a	conv. 58%	4
76	conv. 77%	24

$$\text{(6.55)}$$

86%, 99% ee

Tandem catalytic asymmetric ring-opening metathesis–cross-metathesis (AROM-CM) of norbornene derivative **82a** can be realized using molybdenum catalyst **75a** in the presence of styrene, and cyclopentene derivative **83a** with 98% ee is obtained in 39% yield [Eq. (6.56)].[46] In the case of AROM-CM of cyclobutene derivative **82b**, furan derivative **83b** with 92% ee was obtained in 69% yield [Eq. (6.57)].[47] In this case, [2+2] cycloaddition of cyclobutene **82b** and methylidene carbene complex should occur as the initial step to give **XI**, and then ring-closing metathesis gives five-membered ring compound **83b** not six-membered ring compound **83b'**. 1,1-Disubstituted alkene of **83b** should prevent further ring closing metathesis:

$$(6.56)$$

$$(6.57)$$

Enantioselective desymmetrization of dienes by tandem Mo-catalyzed ring-opening/ring-closing metathesis of **82c** proceeds in the presence of diallylether to give bicyclic compound **83c** with 92% ee in 54% yield [Eq. (6.58)].[47] Addition of diallylether accelerates the formation of a chiral methylidene–molybdenum carbene complex. On the other hand, the reaction of **84** with **75a** gives oligomeric products. Thus, at first, **84** is treated with **1c** under ethylene gas to give cyclopentane derivative **82d**, and then asymmetric ring-closing metathesis of **82d** is carried out using **75a** to give **83d** with high ee [Eq. (6.59)]:

$$(6.58)$$

$$(6.59)$$

Enantioselective cyclic ether is synthesized by molybdenum-catalyzed olefin metathesis. Cyclopentene derivative **85a** is reacted with 5 mol% of chiral molybdenum catalyst **76** to give pyran derivative **86a** in high yield and high ee [Eq. (6.60)].[48] BIPOL-type complex **75** is suitable for asymmetric synthesis of a five-membered ring compound, while BINOL-type catalyst **76** gives a good result for the formation of a six-membered ring. AROM of substrate **85** can lead to the preferential formation of **XII** or **XII'**, and RCM of each complex gives **86** or *ent*-**86**. The process of AROM is reversible; thus, the reaction is more complicated than that of a cyclobutene- or norbornene-containing substrate. Initial AROM may be enantioselective, but the rate of the reverse reaction to a cyclopentene ring of the resulting **XII** or **XII'** would be faster than that of the conversion to **86** or *ent*-**86**. As another route, reaction of a terminal alkene in a tether with methylidene carbene complex gives **XIII**, which then undergoes ARCM to give **86** or *ent*-**86**. To address these issues, RCM of **87a** using **76** is carried out, but the ee of **86a** is lower than that in a previous experiment [Eq. (6.61)]. This means that AROM or ARCM involving the terminal olefin is partially responsible for the levels of asymmetric induction. Unsaturated pyran **86a** is converted into the lactone moiety **90** of the anti-HIV agent tipranavir [Eq. (6.62)]:

$$(6.60)$$

$$(6.61)$$

50°C, 98% conv. 35% ee
80°C, 93%   74% ee

$$(6.62)$$

tipranavir

Enantiomerically enriched medium-ring heterocycles are synthesized using the same protocol. Silicon or nitrogen containing heterocycle **92a**[49] or **92b**[50] is synthesized using **75a** [Eqs. (6.63) and (6.64)]. Eight-membered nitrogen heterocycles

92c can be synthesized from **91c** using modified BIPOL-type complex **75c**:[50]

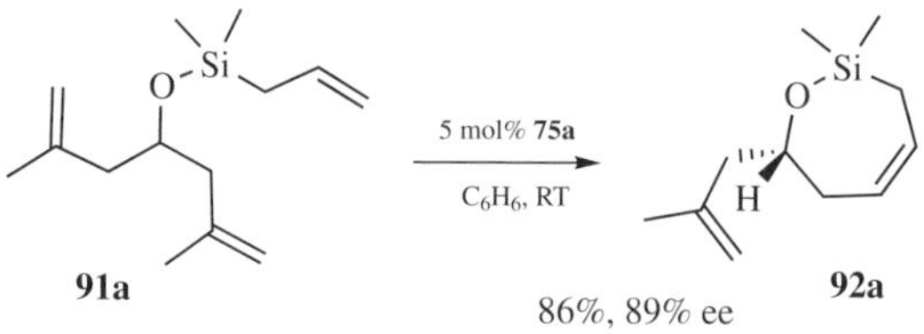

$$(6.63)$$

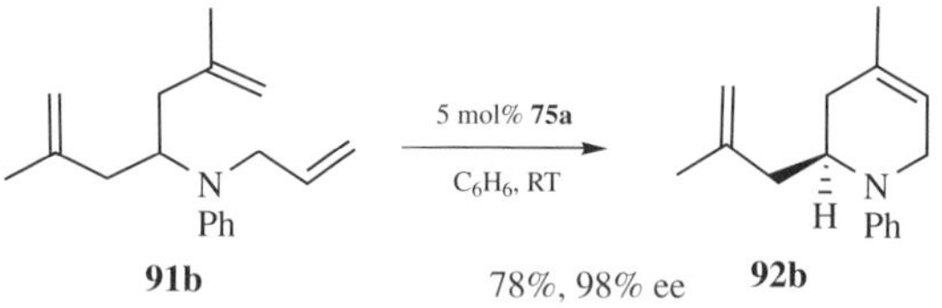

91a

86%, 89% ee

92a

$$(6.64)$$

91b

78%, 98% ee   **92b**

91c

98% ee, 93%   **92c**

75c

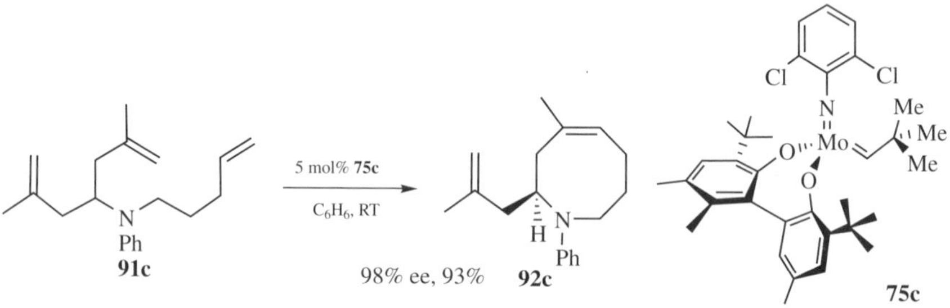

$$(6.65)$$

The catalytic enantioselective ring-closing metathesis of various polyenes affords unsaturated furans, pyrans, and siloxanes efficiently:[51]

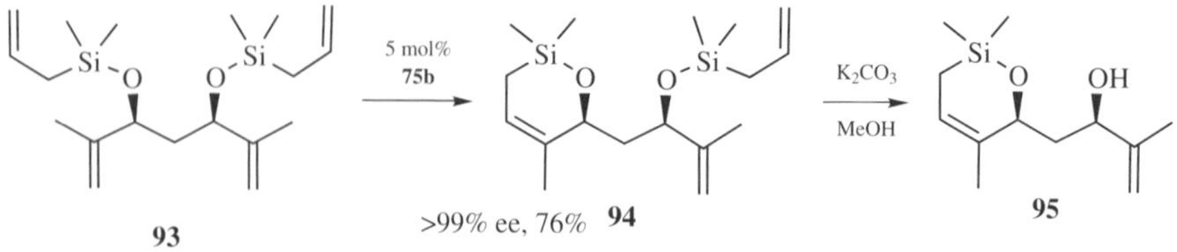

93

>99% ee, 76%   **94**

95

$$(6.66)$$

The 6,8-dioxabicyclo[3.2.1]octane skeleton has been a common structural subunit in natural products. A conceptually new strategy affording these structures is described by Burk et al. [Eqs. (6.67) and (6.68)].[52] For the syntheses of (+)-*exo*-brevicomin, they used desymmetrization of triene **97**, derived from diol **96** with C_2 symmetry, via ring-closing metathesis. Enantiomerically enriched (+)-*endo*-brevicomin is synthesized by employing desymmetrization of trienes **100** derived from diol **99** with meso symmetry using **75a**:

$$(6.67)$$

$$(6.68)$$

The new catalyst is developed from commercially available starting materials and can be used in situ without isolation, to effect enantioselective olefin metathesis [Eq. (6.69)].[53] An active catalyst solution can be accessed by premixing of readily available (*R*)-**102**, and a commercially available Mo-triflate **103** and can be used directly without further purification. This catalyst solution promotes ARCM reactions with equal or higher levels of efficiency and selectivity than does **75a**. Desymmetrization of **80c** with THF solution of **77** gave furan derivative **81c** with 88% ee. Use of **75a** gave **81c** with 93% ee [Eq. (6.70)]; similarly, AROM-CM of norbornene **82e** gave cyclopentane derivative (*R*)-**83e** with 98% ee:

$$(6.69)$$

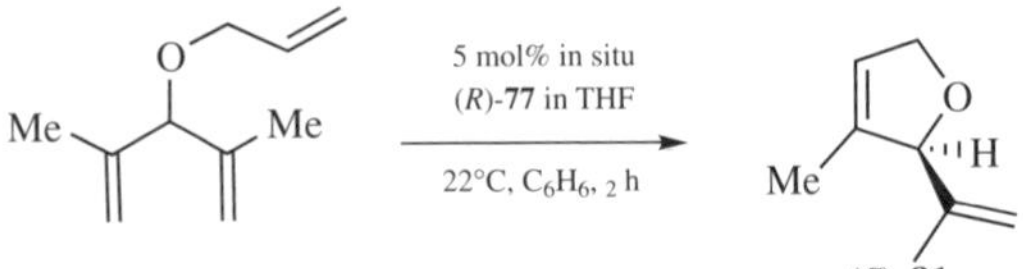

$$(6.70)$$

(S)-**81c**
88% ee, 80% yield

with 5 mol% (S)-**75a**   93% ee, 86% yield

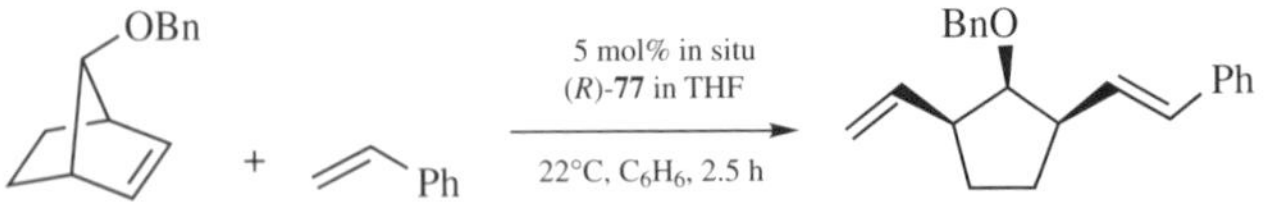

$$(6.71)$$

(R)-**83e**
86%, >98% ee, >98% trans

5 mol% (S)-**75a**   87%, >98% ee, >98% trans

A chiral polymer-bound metathesis catalyst has been developed. The supported chiral complex **75d** shows appreciable levels of reactivity and excellent enantioselectivity.[54] This complex **75d** can be recycled and easily removed from unpurified mixtures. In the first and second cycles of the recycle experiment, almost the same reactivity has been shown. In the third cycle, high enantioselection and conversion are still obtained, but catalyst activity is notably diminished:

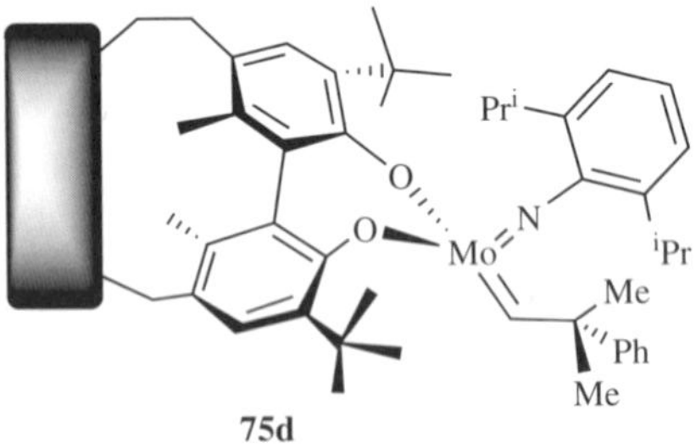

75d

recyclable; pure product obtained by filtration

$$(6.72)$$

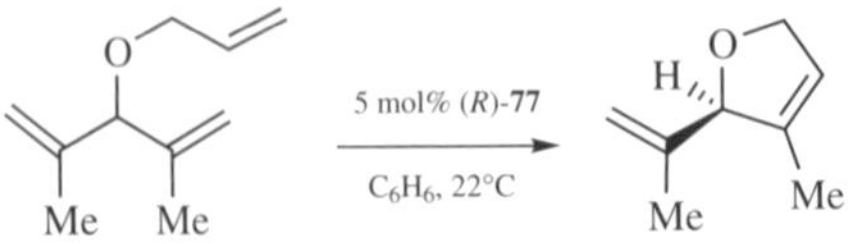

80c   (R)-**81c**

Cycle 1:
97% conv. 2 h, 89% ee

Cycle 2:
69% conv. 2 h, 90% ee

Cycle 3:
78% conv. 24 h, 89% ee

Asymmetric Olefin Metathesis Using Ruthenium Catalysts   Ruthenium catalysts for asymmetric synthesis have also been developed (Figure 6.3).

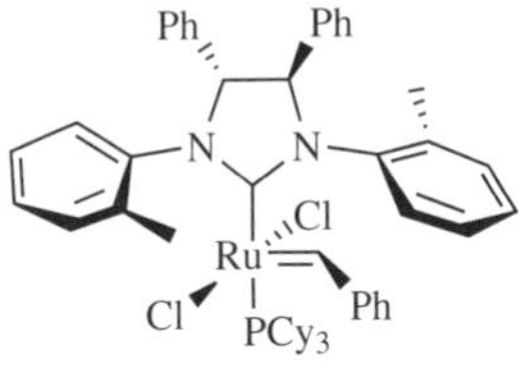

Figure 6.3 Ruthenium catalysts for asymmetric synthesis.

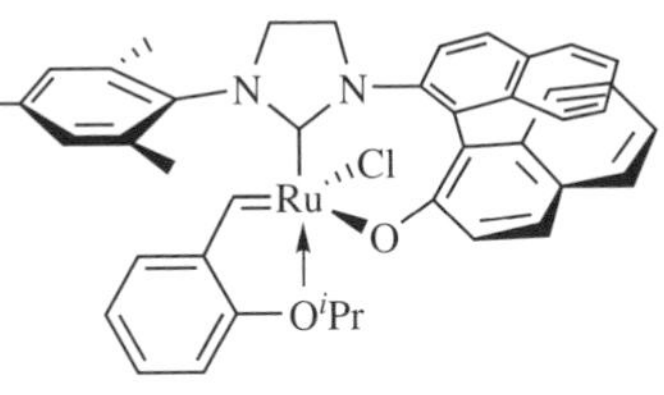

Grubbs synthesized ruthenium carbene complex **104a**, and a high enantiomeric excess (up to 90%) was observed in the desymmetrization of prochiral trienes **80c**:[55]

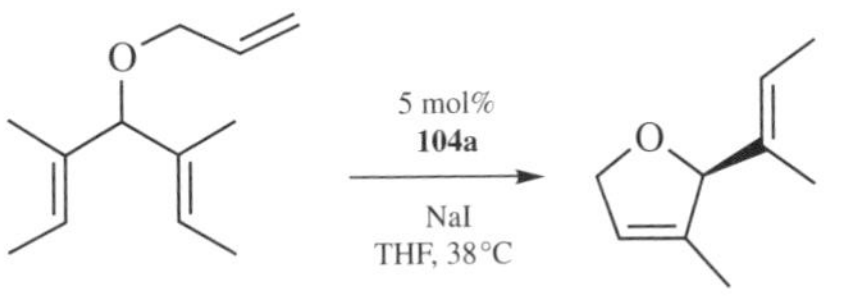

$$(6.73)$$

80c   (S)-**81c**
                                                      82% conv. 90% ee

Hoveyda synthesized recyclable ruthenium catalyst **104b** for enantioselective olefin metathesis. This catalyst is very effective for AROM-CM and can be recovered after chromatography [Eqs. (6.74)–(6.76)].[56] The recovered catalyst can be reused without significant loss of enantioselectivity and with similar reactivity:

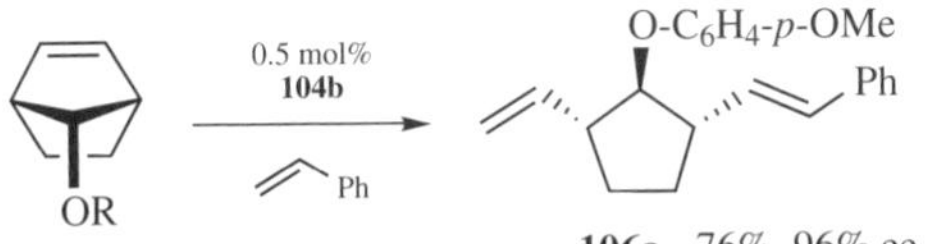

$$(6.74)$$

105a   **106a**   76%, 96% ee

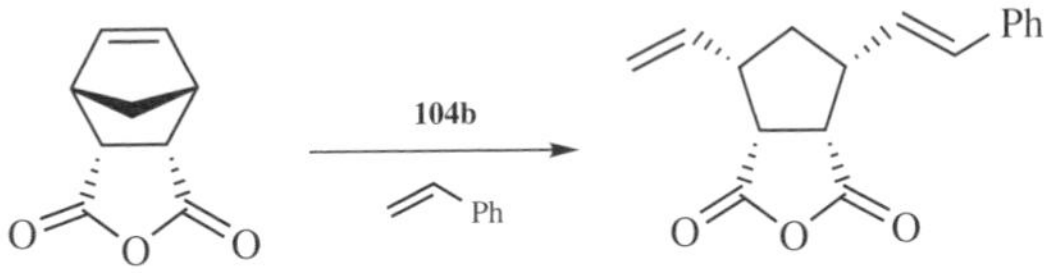

$$(6.75)$$

105b   **106b**   98%, 80% ee

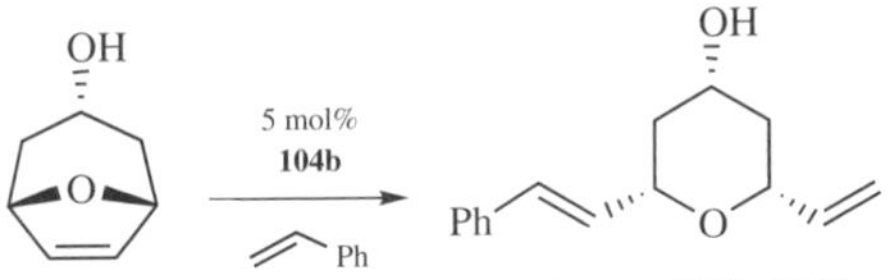

$$(6.76)$$

105c   **106c**   80%, 94% ee

6.3 ENYNE METATHESIS

6.3.1 Ring-Closing Enyne Metathesis

Ring-closing metathesis of an enyne, which has double and triple bonds in the molecule, is a remarkable reaction which is useful in synthetic organic chemistry. In enyne metathesis, the double bond is cleaved and carbon–carbon bond formation occurs between the double and triple bonds. The cleaved alkylidene part is moved to the alkyne carbon. Thus, the cyclized compound formed in this reaction has a diene moiety [Eq. (6.77)].[57] The reaction is also called "skeletal rearrangement" and is induced by Pt, Pd, Ga, and Ru catalysts:

$$ \text{(6.77)} $$

Mori developed this reaction and synthesized five- to nine-membered ring compounds **108a–d** from the corresponding enynes **107a–d**.[58] It is interesting that medium-sized ring compounds **108e–h** can also be synthesized using this method.[58d,e] The possible reaction course is shown in Figure 6.4. The reaction proceeds via [2+2] cycloaddition of an alkyne part of **107** and methylidene carbene complex **1h** to form ruthenacyclobutene **XIV**, whose ring opening gives ruthenium carbene **XV**. Reaction of **XV** with the tethered alkene part gives ruthenacyclobutane **XVI**, and the subsequent ring opening gives cyclized compound **108**, and ruthenium methylidene complex **1h** is regenerated. If this reaction proceeds via [2+2] cycloaddition of an alkene part and **1h**, a similar pathway should be considered.

$$ \text{(6.78)} $$

108a	$n=0$	R = H	90%
108b	$n=1$	R = Me	91%
108c	$n=2$	R = CH$_2$OAc	86%
108d	$n=3$	R = CH$_2$TBS	77%

108e 95%   **108f** 84%   **108g** 97%   **108h** 74% single isomer

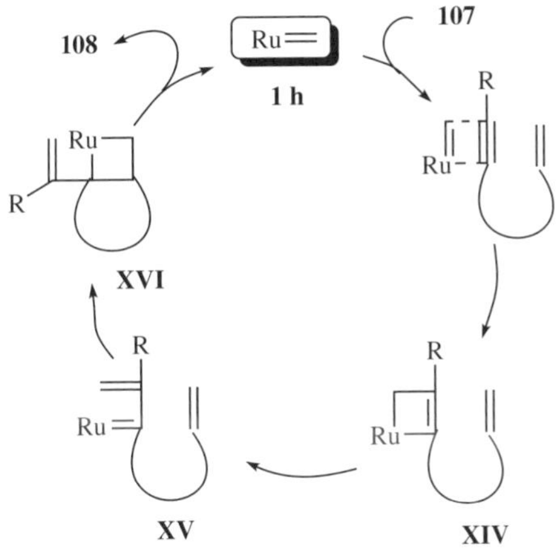

Figure 6.4. Possible reaction course.

A total synthesis of (−)-stemoamide is achieved from (−)-pyroglutamic acid using ruthenium-catalyzed enyne metathesis as a key step in 14 steps in 9% overall yield [Eq. (6.79)]:[59]

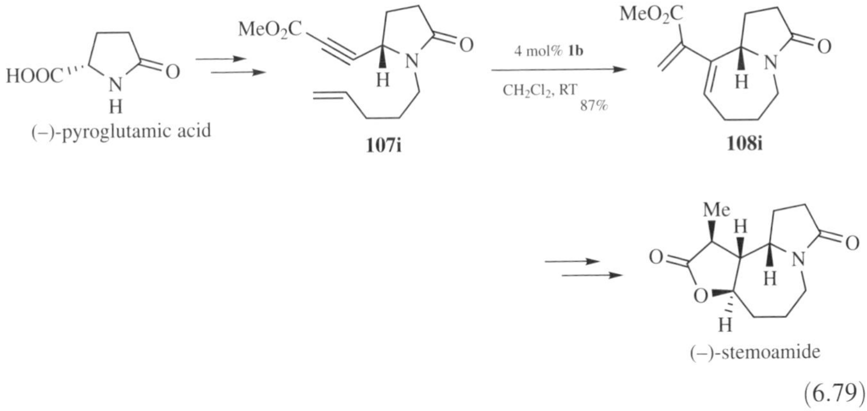

(6.79)

The reaction rate of enyne **107j** having a terminal alkyne is very slow, and the starting material is recovered [Eq. (6.80)].[60] Presumably, the terminal alkene of the product **108j** should further react with ruthenium carbene complex **1h** to form **XVII**, whose ruthenium carbene should be coordinated by the olefin in the pyrrolidine ring. Thus, the catalytic activity of **1h** should be decreased. If complex **XVII** reacts with ethylene, **108j** and methylidene ruthenium carbene complex **1h** should be regenerated. On the basis of this idea, the reaction was carried out under ethylene gas (1 atm) and desired compound **108j** is obtained in high yield:[59]

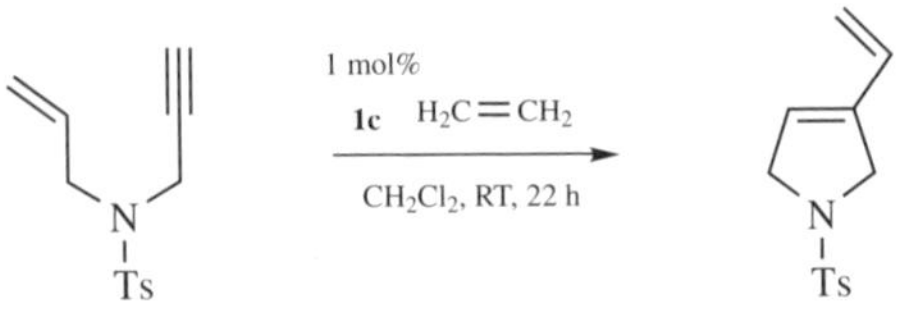

107j **108j** (6.80)

under Ar	21%
under ethylene	90%

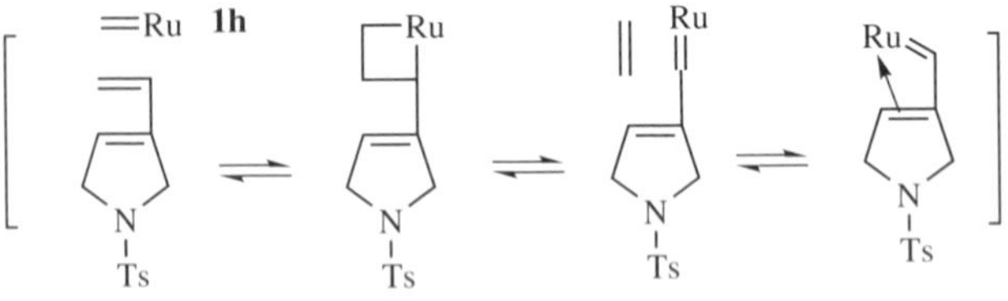

108j **XVII**

When metathesis of 1,6-enyne having 1,1-disubstituted alkene **109a** is reinvestigated using the second-generation ruthenium carbene complex **1f**, two products, **110a** and **111a**, are obtained [Eq. (6.81)].[61] Use of **1g** for this reaction gives the same compounds **110a** and **111a**. Another enyne, **109b**, also gives two products, **110b** and **111b** [Eq. (6.82)]. One is a five-membered ring compound **110**, which was obtained previously using the first-generation ruthenium carbene complex, and the other is a six-membered ring compound **111**. When an alkyne reacts with ruthenium methylidene carbene complex, there are two pathways. Thus, two products should be formed through each pathway. Although it is not clear why two products are formed when the second-generation ruthenium carbene complexes **1f** and **1g** are used, an enyne having a terminal alkyne or alkene gives only five-membered ring compound **110**:[61b]

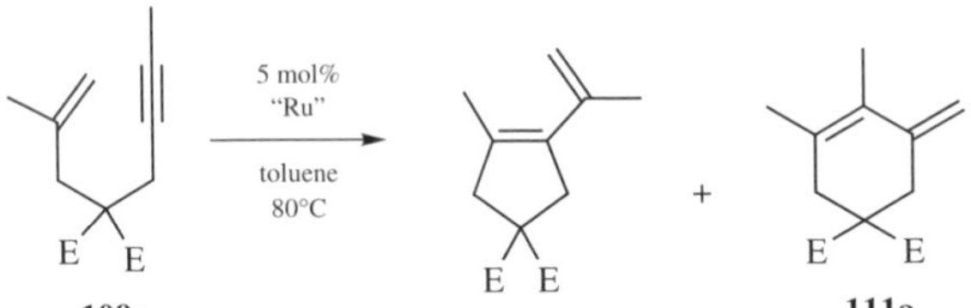

109a **110a** **111a** (6.81)

	"Ru"		
$E = CO_2Et$	**1f**	43%	42%
	1g	42%	41%

$$(6.82)$$

Enyne **112a** having a boron moiety on the alkyne gives vinyl boron **113a** [Eq. (6.83)], whose carbon–boron bond can be converted to a carbon–carbon bond. Enyne metathesis followed by Diels–Alder reaction of the resulting diene with DMAD gives tricyclic compound **114** [Eq. (6.84)].[62]

$$(6.83)$$

$$(6.84)$$

Metathesis of enyne **115** bearing nitrogen[63a] or oxygen[63b] on alkyne smoothly proceeded to give cyclized compound **116** having an enamide or an enol ether moiety in the molecule. These compounds can be converted into various functionalized compounds:

$$(6.85)$$

$$(6.86)$$

$$(6.87)$$

6.3.2 Ring-Opening Metathesis (ROM)–Ring-Closing Metathesis (RCM) of Cycloalkene-Yne

ROM-RCM of cycloalkene-yne **119** having a substituent at the 3-position of the cycloalkene would give a polymer because ruthenium carbene complex **XVIII** generated in this reaction could react with the starting alkyne. If this reaction is carried out under ethylene gas, the cyclized compound **120** should be formed by the reaction of **XVIII** with ethylene [Eq. (6.88)]. On the basis of this idea, ROM-RCM-CM of cycloalkene-yne **119** was carried out under ethylene gas:[64]

$$(6.88)$$

When a CH_2Cl_2 solution of **119a** is stirred in the presence of 10 mol% of **1c** under ethylene gas (1 atm) at room temperature, pyrrolidine derivative **120a** is obtained in high yield. Various cycloalkene-ynes **119b-c** are examined, and pyrrolidine derivative **120b–c** is formed in each case. Formally, the double bonds of cycloalkene and ethylene are cleaved, and each methylidene part of ethylene is combined with the cycloalkene and alkyne carbons, respectively, and bond formation between the double and triple bonds occurs to give pyrrolidine derivative **120**:

	ring size	n	time (h)	yield (%)
119a	6	1	4	78
119b	7	2	1	70
119c	8	3	1	75

$$(6.89)$$

Blechert carried out ROM-RCM-CM of cycloalkene-yne **121** in the presence of alkene instead of ethylene to give **122**:[65]

$$(6.90)$$

If ROM-RCM of cycloalkene-yne **123**, which has a substituent at the 2-position of cycloalkene, is carried out under ethylene gas, what compound is formed? In this reaction, ruthenium carbene **XIX** would be formed via [2+2] cycloaddition of ruthenium methylidene carbene and alkyne as shown in Eq. (6.91). If **XIX** reacts with an olefin intramolecularly or ethylene, bicyclic compound **124** or triene **125** would be formed:[66]

$$(6.91)$$

When a CH_2Cl_2 solution of cyclohexene **123a** is stirred in the presence of **1g** under ethylene gas, the expected bicyclic compound **124a** is obtained in only 14% yield. The other compounds obtained were one-carbon shortened bicyclic compound **124b** and the dimeric compound **126** in 57% and 26% yields, respectively [Eq. (6.92)]. When the dimeric compound **126** was treated with **1g** under similar reaction conditions, **124b**, **124a** and **125a** were obtained in 39%, 21% and 9% yields, respectively, and the starting material **126** was recovered in 10% yield [Eq. (6.93)]. Presumably, the reaction of intermediate ruthenium carbene complex **XIX** with ethylene gives **125a**, and isomerization of the double bond of **125a** occurs and then RCM should give one-carbon shortened bicyclic compound **124b** because **124b** is thermodynamically more stable than **124a** under these reaction conditions. ROM-RCM of cyclopentene-yne **123b** under ethylene gas gives fused bicyclic compound **124b** in a quantitative yield. Formally, in this reaction, carbon–carbon bond formation occurs between each double bond and each triple bond. The initial cyclopentene ring is converted into a two-carbon enlarged cycloheptadiene ring, and the other ring size corresponds to the chain length between the alkyne carbon and alkene carbon of the substituent [Eq. (6.94)]:[66,64b]

$$(6.92)$$

$$(6.93)$$

124b 39%   **124a** 21%   **125a** 9%   10%

$$(6.94)$$

In order to synthesize an isoquinoline derivative using this procedure, the initial cycloalkene should be cyclobutene and the carbon chain length between alkyne and alkene carbons should be 6. Applying this idea, when a CH_2Cl_2 solution of cyclobutene-yne **123c** was refluxed in the presence of **1g** under ethylene gas, isoquinoline derivative **124c** was obtained in 60% yield in one step [Eq. (6.95)]. Similarly, cyclic amino acid **124d** and isoquinolone derivative **124e** were obtained from the corresponding cyclobutene derivatives **123d** and **123e** in high yields [Eqs. (6.96) and (6.97)]:[67]

$$(6.95)$$

$$(6.96)$$

$$(6.97)$$

6.3.3 Dienyne Metathesis

Dienyne metathesis was discovered by Grubbs as one of the early reactions in development of olefin metathesis.[68] The reaction is very interesting, and a bicyclic compound is synthesized from a dienyne in one step [Eq. (6.98)]. Treatment of dienyne **127a** gives bicyclic compound **128a** in high yield. Similarly, dienyne **127b** gives bicyclic compounds **128b** and **129b** in 86% yield (ratio of 1 : 1) [Eq. (6.99)]. Presumably, this reaction starts by [2+2] cycloaddition of terminal alkene and methylidene carbene complex to give **XX**. The resulting ruthenium carbene **XX** reacts with the alkyne intramolecularly to produce ruthenacyclobutene **XXI**, and ring opening of this gives ruthenium carbene **XXII**, which reacts with alkene to give ruthenacyclobutane **XXIII**. From this complex, bicyclic compounds can be produced:

$$(6.98)$$

$$(6.99)$$

POSSIBLE REACTION COURSE

127   **128, 129**

XX   **XXI**   **XXII**   **XXIII**

Boyer synthesized bicyclic compound **128c** from dienyne **127c** in high yield by this approach.[69] Dienyne **127d**, protected by the TES group, gave bicyclic compound **128d** in high yield [Eq. (6.100)]. Mori succeeded in the total synthesis of erythrocarine using dienyne metathesis as a key step.[70a] Reaction of dienyne **127e** with **1c** gives **128e** and **128e'** in quantitative yield (ratio of 1 : 1).

Deacetoxylation of **128e** gave erythrocarine [Eq. (6.101)]. Using a similar procedure, Hatakeyama succeeded in the total synthesis of erythravine:[70b]

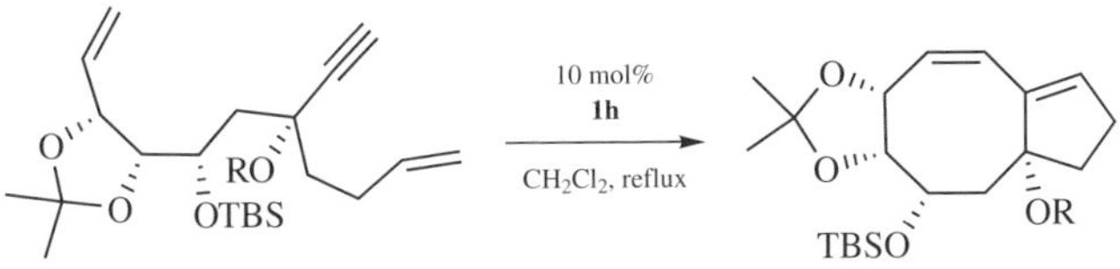

(6.100)

127c R = H
127d R = TES

128c R = H, 23 h, 79%
128d R = TES, 17 h, 96%

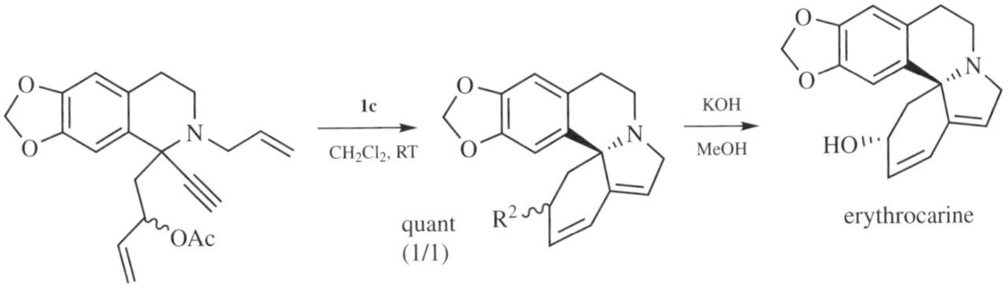

127e

128e $R^2 = \alpha$-OAc
128e′ $R^2 = \beta$-OAc

(6.101)

Group-selective enyne metathesis of dienyne **127f** having a large substituent on the alkyne proceeds in the presence of alkene **130** to give small ring compound **131** as the major product:[71]

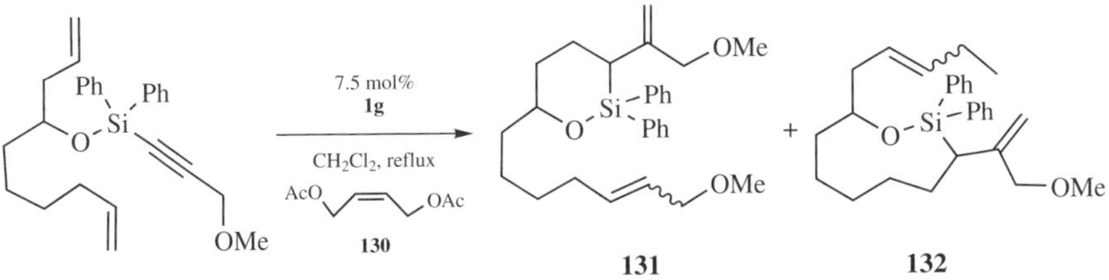

127f **131** **132**

60% (5:1)

(6.102)

Diyne-ene metathesis has also been reported. The reaction of diyne-ene **133a** with **1c** in the presence of terminal alkene gives triene **134a** [Eq. (6.103)].[72] Intramolecular diyne-ene metathesis gives tricyclic compounds **134b**

[Eq. (6.104)]:[73]

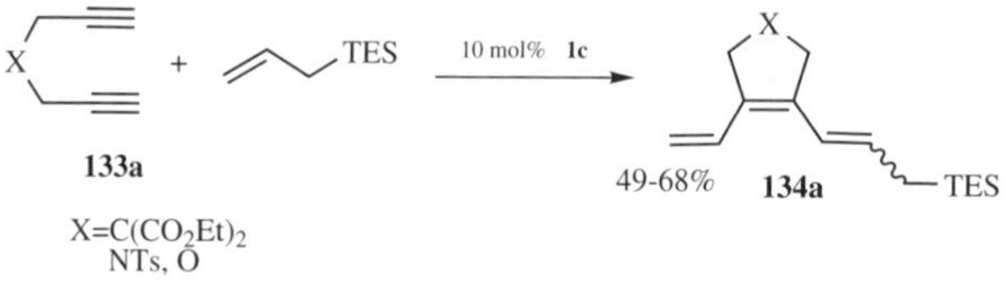

$$(6.103)$$

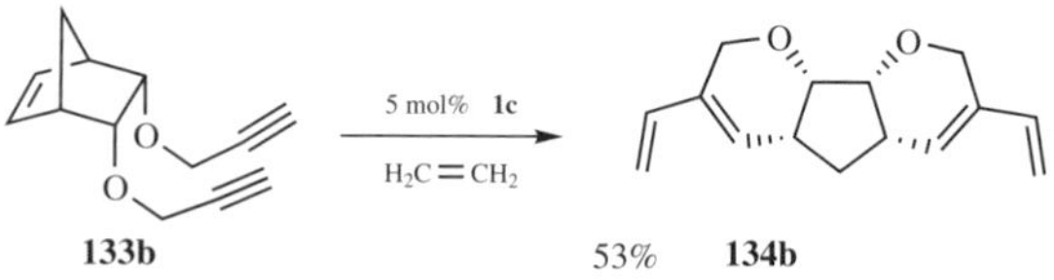

133a

X=C(CO₂Et)₂
NTs, O

$$(6.104)$$

133b   53%   134b

Tandem cyclization of poly (ene-yne) has been achieved by Grubbs. In this reaction, a complex molecule is synthesized by one step from enyne **135** having alkenes and alkynes at appropriate positions [Eqs. (6.105)–(6.106)]:[74]

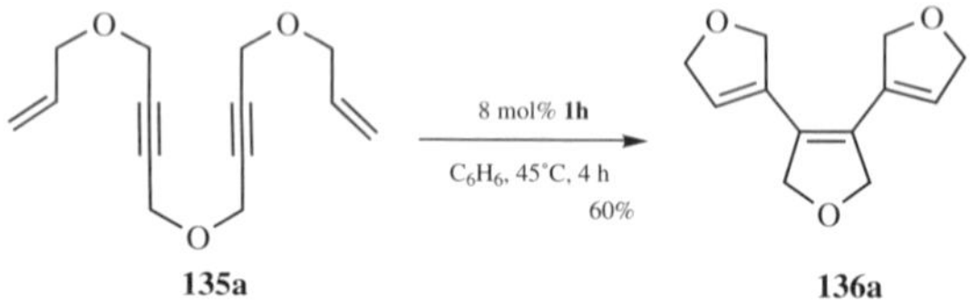

$$(6.105)$$

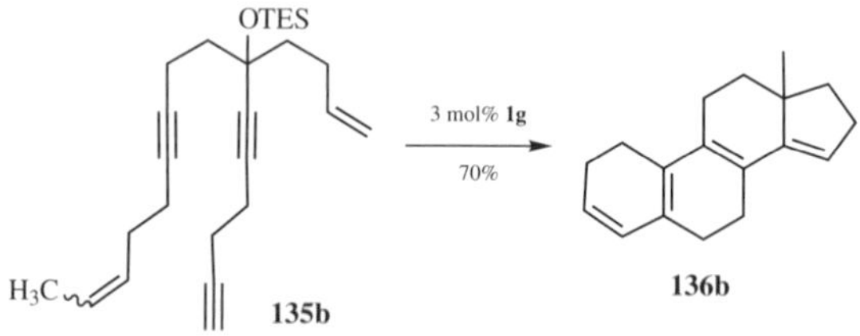

135a   136a

$$(6.106)$$

135b   136b

Trimerization of the triynes **135c** also occurs by ruthenium carbene complex **1c** to construct a benzene ring:[75]

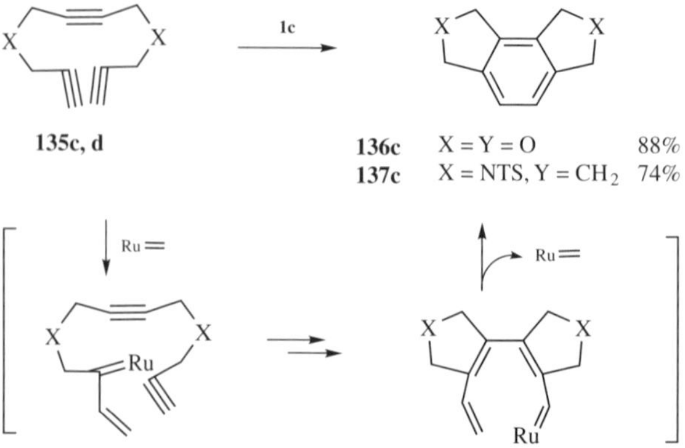

135c, d

136c X = Y = O   88%
137c X = NTS, Y = CH₂ 74%

$$(6.107)$$

6.3.4   Cross-Metathesis of Enyne

It is difficult to realize cross-metathesis of enyne because cross-alkene metathesis and cross-alkyne metathesis occur simultaneously along with cross-enyne metathesis. Thus, various products should be formed in cross-enyne metathesis. Mori succeeded in the cross-metathesis of alkyne **137** and ethylene using ruthenium carbene complex **1c**. When a CH_2Cl_2 solution of alkyne **137a** and a catalytic amount of **1c** is stirred under an atmosphere of ethylene (1 atm), 1,3-diene **138a** is obtained in good yield [Eq. (6.108)]. However, the reaction rate of alkyne **138c** having no heteroatom at the propargyl position is decreased, although alkyne **137b** gives 1,3-diene **138b** in high yield [Eq. (6.109)]: [76a,b]

(6.108)

(6.109)

138b *n*=1   81%
138c *n*=2   11%

Cross-metathesis is therefore reinvestigated using a second-generation ruthenium carbene complex, and 1,3-diene **138** is obtained in high yield from alkyne **137**.[76c] Even alkynes having no heteroatom at the propargylic position gave 1,3-dienes in high yields (Table 6.3). In this reaction, terminal alkyne **137d**, silyl alkyne **137g**, and alkyne **137h** bearing an electron-withdrawing group can be converted into the corresponding 1,3-dienes in high yields. The reaction procedure is very simple; a toluene solution of alkyne **137** is warmed at 80°C under ethylene gas in the presence of a catalytic amount of **1g**.[76d]

Using this procedure, total synthesis of anolignane A is achieved. It is interesting that two methylene groups generated from ethylene are introduced on the alkyne carbons of **137i**, respectively, to form 1,3-diene **138i**:[77]

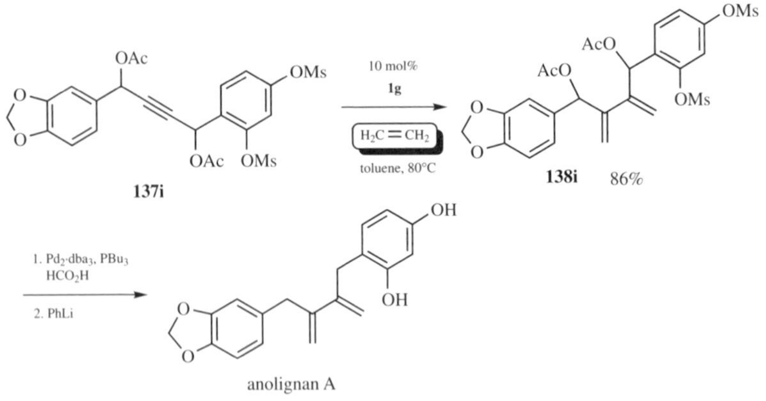

(6.110)

anolignan A

TABLE 6.3 Synthesis of Various 1,3-Dienes

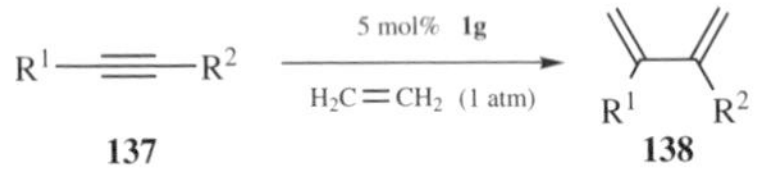

Entry	Alkyne	Diene	Yield (%)
1	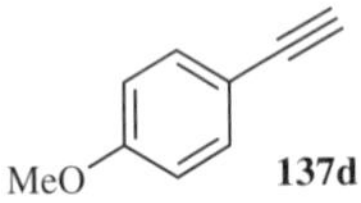 **137d**	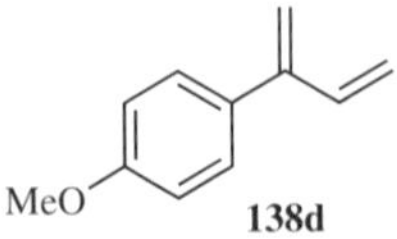 **138d**	88
2	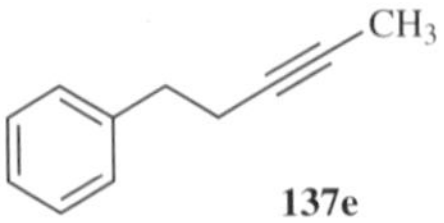 **137e**	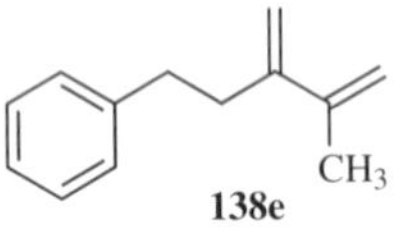 **138e**	85
3	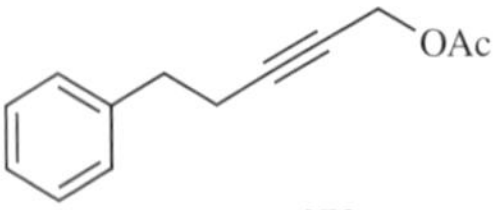 **137f**	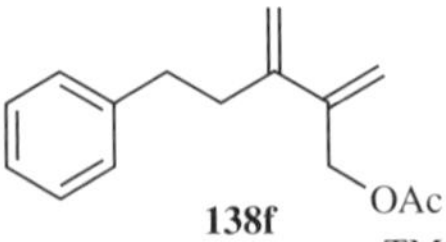 **138f**	100
4	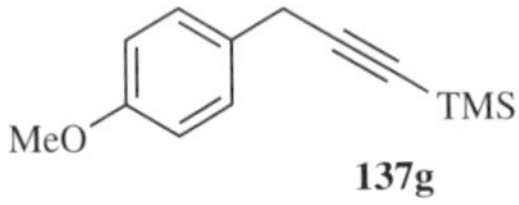 **137g**	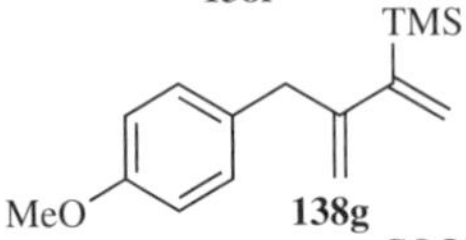 **138g**	87
5[a]	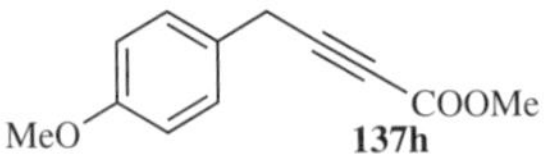 **137h**	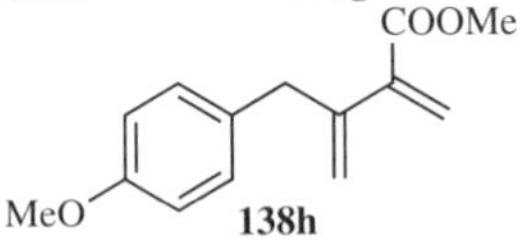 **138h**	43

[a]SM was recovered in a 34% yield.

Blechert reported a skillful method of cross–enyne metathesis. Solid-supported alkyne **139** is reacted with alkene in the presence of **1c** to give **140**. For cleavage of 1,3-diene from solid-supported product **140** having an allyl acetate moiety, palladium-catalyzed allylic substitution is used. Thus, **140** is treated with Pd(PPh$_3$)$_4$ in the presence of methyl malonate to afford three-component coupling product **141** in good yield:[78]

$$(6.111)$$

Cross-metathesis of enynes having various functional groups on the alkyne and an alkene gives dienes having useful functional groups such as vinyl silane[79a] or enol ether[79b] as the sole product:

$$(6.112)$$

$$(6.113)$$

Cross-metathesis of terminal alkyne **142** and cyclopentene gives cyclic compound **143** having a diene moiety [Eq. (6.114)].[80] Terminal ruthenium carbene generated from an alkyne and methylidene ruthenium carbene complex reacts with cyclopentene to afford two-carbon elongated cycloheptadiene **143**:

$$(6.114)$$

When enyne **144** is treated with ruthenium carbene complex **1i**[81] in the presence of methyl acrylate, RCM-CM occurs to afford cyclic compound **145** in good yield:[82]

$$(6.115)$$

6.4   ALKYNE METATHESIS

In alkyne metathesis, the triple bond of an alkyne is cleaved and triple-bond construction occurs simultaneously.[83] The first effective catalyst for alkyne metathesis was a heterogeneous mixture of tungsten oxide and silica.[84] Then Mortreux found that $Mo(CO)_6$ and phenol is effective for alkyne metathesis. When an alkyne is heated with $Mo(CO)_6$ and resorcinol, a mixture of three alkynes is formed because the reaction is in a state of equilibrium under these conditions [Eq. (6.116)].[85] Although the reaction mechanism is not clear yet, it could proceed by [2+2] cycloaddition of metal alkylidyne complex and alkyne:[86]

$$(6.116)$$

This catalytic system for alkyne metathesis is attractive since the catalyst is commercially available and strict reaction conditions are not required. Thus, a novel method for synthesis of alkynes was developed using this catalytic system.[87] When alkyne **146a** is reacted with an excess amount (3 equiv) of diphenylacetylene in the presence of 5 mol% of $Mo(CO)_6$ and p-Cl-C_6H_4OH (1 equiv) in toluene on heating, cross-metathesis product **147a** is obtained in 74% yield [Eq. (6.118)]. In a similar manner, the reaction of **146b** and 4-octyne gives cross-metathesis product **147b** in 89% yield [Eq. (6.119)]:

$$(6.117)$$

$$(6.118)$$

$$(6.119)$$

Fürstner succeeded in the synthesis of ambrettolide from a diyne using this catalyst system.[88] Treatment of diyne **148a** with $Mo(CO)_6$ (5 mol%) and p-chlorophenol (1 equiv) in chlorobenzene at 140°C gave cycloalkyne **149a** in 69% yield. Subsequent Lindlar reduction proceeded smoothly in a stereoselective manner to afford ambrettolide:

$$(6.120)$$

TABLE 6.4. Cyclization Using Mo(CO)$_6$/p-ClC$_6$H$_4$OH or (^{t}BuO)$_3$ W≡CCMe$_3$

Entry	Product		(^{t}BuO)$_3$ W≡CCMe$_3$[a]	Mo(CO)$_6$/p-ClC$_6$H$_4$OH[b]
1		**149a**	73%	64%
2		**149b**	68%	0%
3		**149c** R = H **149d** R = Me	62% 72%	0% 64%
4		**149e**	53%	70%

[a]Using 1 (5 mol%) in chlorobenzene at 80°C.
[b]Mo(CO)$_6$ (5 mol%) and p-chlorophenol (1 equiv) at 140°C.

At the same time, Fürstner used tungsten alkylidene complex **150** developed by Schrock[89] for ring-closing alkyne metathesis. He compared the reactivities of tungsten alkylidyne complex **150** and Mo(CO)$_6$-p-ClC$_6$H$_4$OH (Table 6.4) and showed that both catalysts work well, although a higher reaction temperature is required in the case of Mo(CO)$_6$-p-chlorophenol.

Ring-closing metathesis of dienes to cycloalkenes provides useful methods for the synthesis of carbo- and heterocycles and has proved to be effective in numerous syntheses of natural products. However, it usually affords a mixture of E- and Z-isomers. Ring-closing alkyne metathesis provides a novel synthetic route of E- and Z-macrocycloalkenes in a stereoselective manner using partial reduction of the cycloalkyne obtained by ring-closing alkyne metathesis. A Lindlar reduction gives Z-cycloalkene

Z-**151**, and hydroboration/protonation affords E-cycloalkene. In 2002, Trost reported an alternative procedure for the synthesis of the E-olefin from an alkyne by hydrosililation using a ruthenium catalyst.[90a] Using this procedure followed by desilylation, cycloalkyne **149f** afforded E-cycloalkene E-**151f** in a stereoselective manner:[90b]

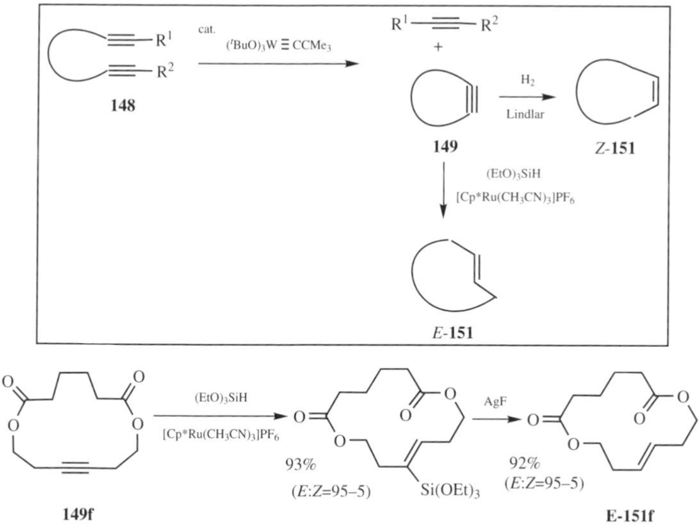

$$(6.121)$$

On the other hand, Fürstner discovered a novel molybdenum catalyst for alkyne metathesis. They investigated the reactivity of Mo[N(^{t}Bu)(Ar)]$_3$ **152** toward alkyne metathesis.[91] Although **152** itself does not affect alkyne metathesis, a strongly endothermic process is observed when **152** is dissolved in CH_2Cl_2. The resulting solution catalyzes the metathesis coupling of aromatic alkynes **146c** and **146d** and aliphatic alkynes **146e** and **146f** [Eqs. (6.123) and (6.124)]. Evaporation of the solvent of the CH_2Cl_2 solution of **152** gave different molybdenum species **153a** and **153c**. Although they could not separate the complexes, **153a** can be synthesized from **152** and Cl_2. Molybdenum alkylidyne complex **153c** is assumed to be a catalytic species toward alkyne metathesis, but it shows low reactivity.[91b] On the other

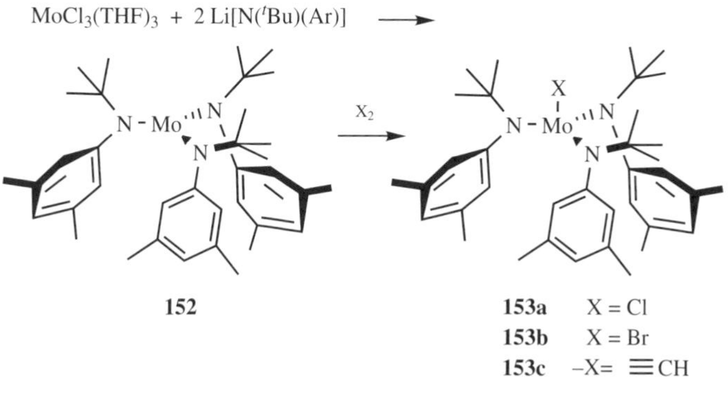

$$MoCl_3(THF)_3 + 2 Li[N(^tBu)(Ar)] \longrightarrow$$

$$(6.122)$$

152

153a   X = Cl
153b   X = Br
153c   –X= $\equiv$CH

hand, **153a** shows effective catalytic activity toward alkyne metathesis.

$$(6.123)$$

154c	R=H	60%
154d	R=CN	58%

154e	R=Me	59%
154f	R=THP	55%

$$(6.124)$$

Although the structure of the complex arising from **152**/CH_2Cl_2 is not clear, this catalyst is excellent in terms of ease of preparation. The catalyst is very active for formation of cycloalkynes with ring sizes different from those of diynes (Table 6.5).[91b] In contrast to tungsten alkylidyne complex **150**, catalyst **152**/CH_2Cl_2 is sensitive toward an acidic proton such as amide proton and exhibited remarkable tolerance towards many polar functional groups (Table 6.5).

TABLE 6.5. Diyne Metathesis Using Tungsten and Molybdenum Catalysts

Entry	Product		$(^tBuO)_3W\equiv$ —	
			150 (%)	**152**/CH_2Cl_2 (%)
1		**149g**	73	91
2		**149h**	0	84
3		**149i**	0	88

Alkyne cross-metathesis is also achieved using **152**/CH_2Cl_2. Even in the case of a 1 : 1 mixture of two different alkynes, cross-metathesis product **147j** is produced in 71% yield using **152**/CH_2Cl_2 catalyst:[92]

$$(6.125)$$

For construction of the α-chain of PGE_2 methyl ester, alkyne **155** and a slight excess of symmetric alkyne **156** in the presence of **152**/CH_2Cl_2 provided the desired cross-metathesis product **157** in 51% yield. Hydrogenation of **157** with a Lindlar catalyst afforded Z-olefin **158**, whose protecting groups were removed to give PGE_2–methyl ester:[92]

$$(6.126)$$

Tribenzocyclyne **160** is synthesized using alkyne metathesis of o-dipropynylated arene **159** by tungsten alkylidiyne complex **150** [Eq. (6.127)].[93] Superpersistent arylene ethynylene macrocycles have received much attention in the fields of supramolecular chemistry and material sciences. Phenylene ethynylene macrocycles **163a** are obtained from monomer **161a** in good yield using molybdenum catalyst **162** on a small scale. Multigram synthesis of **163b** is accomplished from **161b** in one step [Eq. (6.128)]:[94]

$$(6.127)$$

$$(6.128)$$

		product	mg-scale	g-scale
161a	R′=Me R=^{t}Bu	**163a**	61%	—
161b	R′= R=COO(CH$_2$CH$_2$O)$_3$CH$_3$	**163b**	81%	77%

6.5 CONCLUSIONS

Since the discovery of a stable and isolable catalyst for metathesis by Schrock and Grubbs, a wide range of olefin metatheses have been reported, and enyne and alkyne metatheses have been developed. The remarkable features of these reactions are that double and triple bonds are cleaved and double and triple bonds are formed simultaneously. These reactions are very interesting, and there are many reports on applications of these metatheses in synthetic organic chemistry. Olefin metathesis now occupies an important position in synthetic organic chemistry, and many reactions have been replaced by metathesis. A medium-sized ring, a macrocyclic ring, and even five- and six-membered rings are constructed by RCM instead of traditional methods. In syntheses of natural products, novel retrosynthetic analyses have been developed and the reaction steps have been shortened. It is expected that metathesis will be used in the future for the syntheses of various complex molecules and macrocyclic compounds because complex molecules can be formed by one step, the reaction procedure is very simple, the yields are high, and transition metal carbene complexes are now commercially available.

REFERENCES

1. (a) *Handbook of Metathesis*, Grubbs, R. H. (Ed.), Wiley-VCH, **2003**; (b) *Topics in Organometallic Chemistry*, Vol. 1, Fürstner, A. (Ed.), Springer-Verlag, Berlin–Heidelberg, **1998**; (c) Trunk, T. M.; Grubss, R. H. *Acc. Chem. Res.* **2001**, *34*, 18; (d) Fürstner, A. *Angew. Chem. Int. Ed.* **2000**, *39*, 3012.

2. (a) Villemin, D. *Tetrahedron Lett.* **1980**, *21*, 1715; (b) Tsuji, J.; Hashiguchi, S. *Tetrahedron Lett.* **1980**, *21*, 2955.

3. (a) Schrock, R. R.; Murdzek, J. S.; Dimare, M.; O'Regan, M. *J. Am. Chem. Soc.* **1990**, *112*, 3875; (b) Nguyen, S. T.; Johnson, L. K.; Grubbs, R. H. *J. Am. Chem. Soc.* **1993**, *115*, 9858; (c) Schwab, P.; France, M. B.; Ziller, J. W.; Grubbs, R. H. *Angew. Chem. Int. Ed. Engl.* **1995**, *34*, 2039.

4. (a) Nguyen, S. T.; Johnson, L. K.; Grubbs, R. H. *J. Am. Chem. Soc.* **1993**, *115*, 9858; (b) Schwab, P.; France, M. B.; Ziller, J. W.; Grubbs, R. H. *Angew. Chem. Int. Ed. Engl.* **1995**, *34*, 2039.

5. (a) Weskamp, T.; Schattenmann, W. C.; Spiegler, M.; Herrmann, W. A. *Angew. Chem. Int. Ed.* **1998**, *37*, 2490; (b) Weskamp, T.; Kohl, F. J.; Hieringer, W.; Gleich, D.; Herrmann, W. A. *Angew. Chem. Int. Ed.* **1999**, *38*, 2416.

6. Huang, J.; Stevens, E. D.; Nolan, S. P.; Peterson, J. L. *J. Am. Chem. Soc.* **1999**, *121*, 2674.

7. (a) Scholl, M.; Trnka, T. M.; Morgan, J. P.; Grubbs, R. H. *Tetrahedron Lett.* **1999**, *40*, 2247; (b) Scholl, M.; Ding, S.; Lee, C. W.; Grubbs, R. H. *Org. Lett.* **1999**, *1*, 953.

8. Fu, G. C.; Grubbs, R. H. *J. Am. Chem. Soc.* **1992**, *114*, 5426.

9. (a) Fu, G. C.; Grubbs, R. H. *J. Am. Chem. Soc.* **1992**, *114*, 7324; (b) Fu, G. C.; Grubbs, R. H. *J. Am. Chem. Soc.* **1993**, *115*, 3800; (c) Fu, G. C.; Nguyen, S.-B. T.; Grubbs, R. H. *J. Am. Chem. Soc.* **1993**, *115*, 9856.

10. Morehead, A.; Grubbs, R. H. *Chem. Commun.* **1998**, 275.

11. (a) Crimmins, M. T.; King, B. W. *J. Org. Chem.* **1996**, 61, 4192; (b) Crimmins, M. T.; Zuercher, W. *Org. Lett.* **2000**, 2, 1065.

12. Miller, S. J.; Kim, S.-H.; Chen, Z. R.; Grubbs, R. H. *J. Am. Chem. Soc.* **1995**, *117*, 2108.

13. Boyer, F.-D.; Hanna, I.; Nolan, S. P. *J. Org. Chem.* **2001**, *66*, 4094.

14. Fürstner, A.; Langemann, K. *J. J. Org. Chem.* **1996**, *61*, 8746.

15. (a) Crimmins, M. T.; Choy, A. L. *J. Am. Chem. Soc.* **1999**, *121*, 5653; (b) Crimmins, M. T.; Tabet, E. A. *J. Am. Chem. Soc.* **2000**, *122*, 5473.

16. Evans, P. A.; Murthy, V. S. *J. Org. Chem.* **1998**, *63*, 6768.

17. (a) Forbes, M. D.; Patton, J. T.; Myers, T. L.; Maynard, H. D.; Smith Jr. D. W.; Schuiz, G. R.; Wagener, K. B. *J. Am. Chem. Soc.* **1992**, *114*, 10978; (b) Chang, S. B.; Grubbs, R. H. *Tetrahedron Lett.* **1997**, 38, 4757.

18. Scholl, M.; Ding, S.; Lee, C. W.; Grubbs, R. H. *Org. Lett.* **1999**, *1*, 953.

19. (a) Sanford, M. S.; Ulman, M.; Grubbs, R. H. *J. Am. Chem. Soc.* **2001**, *123*, 749; (b) Sanford, M. S.; Love, A. L.; Grubbs, R. H. *J. Am. Chem. Soc.* **2001**, *123*, 6543; (c) Love, J. A.; Sanford, M. S.; Day, M. W.; Grubbs, R. H. *J. Am. Chem. Soc.* **2003**, *125*, 10103.

20. (a) Arisawa, M.; Theeraladanon, C.; Nishida, A.; Nakagawa, M. *Tetrahedron Lett.* **2001**, 42, 8027; (b) Okada, A.; Oshima, T.; Shibasaki, M. *Tetrahedron, Lett.* **2001**, 42, 8023.

21. Conrad, R. M.; Grogan, M. J.; Bertozzi, C. R. *Org. Lett.* **2002**, *4*, 1359.

22. (a) M-Contelles, J.; de Opazo, E. J. *J. Org. Chem.* **2000**, *65*, 5416; (b) M-Contelles, J.; de Opazo, E. J. *J. Org. Chem.* **2002**, *67*, 3705.

23. (a) Ghosh, A. K.; Liu, C.; *Chem. Commun.* **1999**, 1743; (b) Ghosh, A. K.; Wang, Y. *J. Am. Chem. Soc.* **2000**, 122, 11027.

24. Boiteau, J.-G.; Van de Weghe, P.; Eustache, *J. Org. Lett.* **2001**, *3*, 2737.

25. Nicolaou, K. C.; Jennings, M. P.; Dagneau, P. *Chem. Commun.* **2002**, 2480.

26. (a) Maeda, K.; Oishi, T.; Oguri, H.; Hirama, M. *Chem. Commun.* **1999**, 1063; (b) Hirama, M.; Oishi, T.; Uehara, H.; Inoue, M.; Maruyama, M.; Guri, H.; Satake, M. *Science*, **2001**, 294, 1904.

27. (a) Fürstner, A.; Langemann, K. *J. Org. Chem.* **1996**, *61*, 3942; (b) Lee, C. W.; Grubbs, R. H. *Org. Lett.* **2000**, *2*, 2145.

28. (a) Martin, S. F.; Liao Y.; Wong, Y.; Rein, T. *Tetrahedron Lett.* **1994**, *35*, 691; (b) Humphrey, J. M.; Liao, Y. S.; Ali, A.; Rein, T.; Wong, Y. L.; Chen, H.-J.; Courtney, A. K.; Martin, S. F. *J. Am. Chem. Soc.* **2002**, *124*, 8584.

29. Tang, H.; Yusuff, N.; Wood, J. L. *Org. Lett.* **2001**, *3*, 1563.

30. Marsella, M. J.; Maynard, H. D.; Grubbs, R. H. *Angew. Chem. Int. Ed.* **1997**, *36*, 1101.

31. (a) Mohr, B.; Weck, M.; Sauvage, J.-P.; Grubbs, R. H. *Angew, Chem. Int. Ed.* **1997**, *36*, 1308; (b) Weck, M.; Mohr, B.; Sauvage, J.-P.; Grubbs, R. H. *J. Org. Chem.* **1999**, *64*, 5463.

32. Choi, T.-L.; Lee, C. W.; Chaterjee, A. K.; Grubbs, R. H. *J. Am. Chem. Soc.* **2001**, *123*, 10417.

33. Choi, T.-L.; Chatterjee, A. K.; Grubbs, R. H. *Angew. Chem. Int. Ed.* **2000**, *39*, 1277.

34. Kitamura, T.; Sato, Y.; Mori, M. *Tetrahedron* **2004**, *60*, 9649.

35. Wakamatsu, H.; Blechert, S.; *Angew. Chem. Int. Ed.* **2002**, *41*, 2403.

36. Zuercher, W. J.; Hashimoto, M.; Grubbs, R. H. *J. Am. Chem. Soc.* **1996**, 118, 6634.

37. (a) Randall, M. L.; Tallarico, J. A.; Snapper, M. L. *J. Am. Chem. Soc.* **1995**, *117*, 9610; (b) White, B. H.; Snapper, M. L. *J. Am. Chem. Soc.* **2003**, 125, 14901; (c) Limanto, J.; Snapper, M. L. *J. Am. Chem. Soc.* **2000**, *122*, 8071.

38. Stragies, R.; Blechert, S. *Synlett*, **1998**, 169.

39. (a) Voligtmann, U.; Blechert, S. *Org. Lett.* **2000**, 2, 3971; (b) Stragies, R.; Blechert, S. *Tetrahedron*, **1999**, *55*, 8179.

40. (a) Harrity, J. P. A.; Visser, M. S.; Gleson, J. D.; Hoveyda, A. H. *J. Am. Chem. Soc.* **1997**, *119*, 1488; (b) Harrity, J. P. A.; La, D. S.; Cefalo, D. R.; Visser, M. S.; Hoveyda, A. H. *J. Am. Chem. Soc.* **1998**, *120*, 2343.

41. Usher, L. C.; E-Jimenez, M.; Ghibiriga, I.; Wright, D. L. *Angew. Chem. Int. Ed.* **2002**, *41*, 4560.

42. (a) Fujimura, O.; Grubbs, R. H. *J. Am. Chem. Soc.* **1996**, *118*, 2499; (b) Fujimura, O.; Grubbs, R. H. *J. Org. Chem.* **1998**, *63*, 824.

43. Alexander, J. B.; La, D. S.; Cefalo, D. R.; Hoveyda, A. H.; Schrock, R. R. *J. Am. Chem. Soc.* **1998**, *120*, 4041.

44. Zhu, S. S.; Cefalo, D. R.; La, D. S.; Jamieson, J. Y.; Davis, W. M.; Hoveyda, A. H.; Schrock, R. R. *J. Am. Chem. Soc.* **1999**, *121*, 8251.

45. La, D. S.; Alexander, J. B.; Cefalo, D. R.; Graf, D. D.; Hoveyda, A. H.; Schrock, R. R. *J. Am. Chem. Soc.* **1998**, *120*, 9720.

46. La, D. S.; Ford, J. G.; Sattely, E. S.; Bonitatebus, P. J.; Schrock, R. R.; Hoveyda, A. H. *J. Am. Chem. Soc.* **1999**, *121*, 11603.

47. Weatherhead, G. S Ford,, J. G.; Alexanian, E. J.; Schrock, R. R.; Hoveyda, A. H. *J. Am. Chem. Soc.* **2000**, *122*, 1828.

48. Cefalo, D. R.; Kiely, A. F.; Wuchrer, M.; Jamieson, J. Y.; Schrock, R. R.; Hoveyda, A. H. *J. Am. Chem. Soc.*, **2001**, *123*, 3139.

49. Kiery, A. F.; Jernelius, J. A.; Schrock, R. R.; Hoveyda, A. H. *J. Am. Chem. Soc.* **2002**, *124*, 2868.

50. Dolman, S. J.; Sattery, E. S.; Hoveyda, A. H.; Schrock, R. R. *J. Am. Chem. Soc.* **2002**, *124*, 6991.

51. Weatherhead, G. S.; Houser, J. H.; Ford, G. J.; Jamieson, J. Y.; Schrock, R. R.; Hoveyda, A. H. *Tetrahedron Lett.* **2000**, *41*, 9553.

52. Burke, S. D.; Mu1ller, N.; Beaudry, C. M. *Org. Lett.* **1999**, *1*, 1827.

53. (a) Teng, X.; Cefalo, D.; Schrock, R. R.; Hoveyda, A. H. *J. Am. Chem. Soc.* **2002**, *124*, 10779; (b) Aelits, S. L.; Cefalo, D. R.; Bonitatebus Jr, P. J.; Houser, J. H. *Angew. Chem. Int. Ed.* **2001**, *40*, 1452.

54. Hultzsch, K. C.; Jernelius, J. A.; Hoveyda, A. H.; Schrock, R. R. *Angew. Chem. Int. Ed.* **2002**, 41, 589.

55. Seiders, T. J.; Ward, D. W.; Grubbs, R. H. *Org, Lett.* **2001**, *3*, 3225.

56. Van Veldhuizen, J. J.; Garber, S. B.; Kingsbury, J. S.; Hoveyda, A. H. *J. Am. Chem. Soc.*, **2002**, *124*, 4954.

57. For reviews on enyne metathesis, see (a) Mori, M. *Top. Organomet. Chem.*, Furthner, A. (Ed.), Springer, Berlin–Heidelberg, **1998**, *1*, 133; (b) Poulsen, C. S.; Madsen, R. *Synthesis* **2003**, 1; (c) Mori, M. *Handbook of Metathesis*; Grubbs, R. H. (Ed.), Wiley-VCH, **2003**, Vol. 2, p. 176; (d) Giessert, A. J.; Diver, S. T. *Chem. Rev.* **2004**, *104*, 1317; For mechanistic study, see: (e) Lippstreu, J. J.; Straub, B. F. *J. Am. Chem. Soc.* **2005**, *127*, 7444.

58. (a) Kinoshita, A.; Mori, M. *Synlett* **1994**, 1020; (b) Mori, M.; Kitamura, T.; Sakakibara, N. Y. Sato, *Org. Lett.* **2000**, *2*, 543; (c) Mori, M.; Kitamura, T.; Sato, Y. *Synthesis* **2001**, 654.

59. Kinoshita, A.; Mori, M. *J. Org. Chem.* **1996**, *61*, 8356; (b) Kinoshita, A.; Mori, M. *Heterocycles* **1997**, *46*, 287.

60. Mori, M.; Sakakibara, N.; Kinoshita, A. *J. Org. Chem.* **1998**, *63*, 6082.

61. (a) Kitamura, T.; Sato, Y.; Mori, M. *Chem. Commun.*, **2001**, 1258; (b) Kitamura, T.; Sato, Y.; Mori, M. *Adv. Synth. Catal.* **2002**, *344*, 678.

62. Renaud, J.; Graf, C.-D.; Oberer, L. *Angew. Chem. Int. Ed.* **2000**, *39*, 3101.

63. (a) Saito, N.; Sato, Y.; Mori, M. *Org. Lett.* **2002**, *4*, 803; (b) Schramm, M. P.; Reddy, D. S.; Kozmin, S. A. *Angew. Chem. Int. Ed.* **2001**, *40*, 4274.

64. (a) Kitamura, T.; Mori, M. *Org. Lett.* **2001**, *3*, 1161; (b) Kitamura, T.; Kuzuba, Y.; Sato, Y.; Wakamatsu, H.; Fujita, R.; Mori, M. *Tetrahedon*, **2004**, *60*, 7375.

65. (a) Randl, S.; Lucas, N.; Connon, S. J.; Blechert, S. *Adv. Synth. Catal.* **2002**, *344*, 631; (b) Rückert, A.; Eisele, D.; Blechert, S. *Tetrahedron Lett.* **2001**, *42*, 5245.

66. Mori, M.; Kuzuba, Y.; Kitamura, T.; Sato, Y. *Org. Lett.* **2002**, *4*, 3855.

67. Mori, M.; Wakamatsu, H.; Tonogaki, K.; Fujita, R.; Sato, Y. *J. Org. Chem.* **2005**, *70*, 1066.

68. (a) Kim, S.-H.; Bowden, N.; Grubbs, R. H. *J. Am. Chem. Soc.* **1994**, *116*, 10801; (b) Kim, S.-H.; Zuercher, W. J.; Bowden, N.; Grubbs, R. H. *J. Org. Chem.* **1996**, *61*, 1073.

69. Boyer, F.-D.; Hanna, I.; Ricard, L. *Org. Lett.* **2001**, 3, 3095.

70. (a) Shimizu, K.; Takimoto, M.; Mori, M. *Org. Lett.* **2003**, *5*, 2323; (b) Fukumoto, H.; Esumi, T.; Ishihara, J.; Hatakeyama, S. *Tetrahedron Lett.* **2003**, *44*, 8047.

71. Maifeld, S. V.; Miller, R. L.; Lee, D. *J. Am. Chem. Soc.* **2004**, *126*, 12228.

72. Stragies, R.; Schuster, M.; Blechert, S. *Chem. Commun.* **1999**, 237.

73. (a) Banti, D.; North, M. *Tetrahedron Lett.* **2002**, *43*, 1561; (b) Banti, D.; North, M. *Adv. Synth. Catal.* **2002**, *344*, 694.

74. Zuercher, W. J.; Scholl, M.; Grubbs, R. H. *J. Org. Chem.* **1998**, 63, 4291.

75. Peters, J.-U.; Blechert, S. *Chem. Commun.* **1997**, 1983.

76. (a) Kinoshita, A.; Sakakibara, N.; Mori, M. *J. Am. Chem. Soc.* **1997**, *119*, 12388; (b) Kinoshita, A.; Sakakibara, N.; Mori, M. *Tetrahedron* **1999**, *55*, 8155; (c) Tonogaki, K.; Mori, M. *Tetraheron Lett.* **2002**, *43*, 2235; (d) Mori, M.; Tonogaki, K.; Kinoshita, A. *Org. Synth.* **2004**, *81*, 1; (e) Smulik, J. A.; Diver, S. T. *J. Org. Chem.* **2000**, *65*, 1788; (f) Smulik, J. A.; Diver, S. T. *Org. Lett.* **2000**, *2*, 2271; (g) Smulik, J. A.; Giessert, A. J.; Diver, S. T. *Tetrahedron Lett.* **2002**, *43*, 209.

77. Mori, M.; Tonogaki, K.; Nishiguchi, N. *J. Org. Chem.* **2002**, *67*, 224.

78. Schuter, S. C.; Blechert, S. *Synlett* **1998**, 166.

79. (a) Kim, M.; Park, S.; Mafeld, S. V.; Lee, D. *J. Am. Chem. Soc.* **2004**, *126*, 10242; (b) Giessert, A. J.; Brazis, N. J.; Diver, S. T. *Org. Lett.* **2003**, 5, 3819.

80. Kulkami, A. K., Diver, S. T. *Org. Lett.* **2003**, 5, 3463.

81. Garber, S. B.; Kingsbury, J. S.; Gray, B. L.; Hoveyda, A. H. *J. Am. Chem. Soc.* **2000**, *122*, 8168

82. Royer, F.; Vilain, C.; Elkaim, L.; Grimaud, L. *Org. Lett.* **2003**, 5, 2007.

83. For recent reviews, see: (a) Bunz, U. H. F.; Kloppenburg, L. *Angew. Chem. Int. Ed.* **1999**, *38*, 478; (b) Fürstner, A.; Davies, P. W. *Chem. Commun.* **2005**, 2307.

84. Pennella, E.; Banks, R. L.; Bailey, G. C. *Chem. Commun.* **1968**, 1548.

85. (a) Mortreux, A.; Blanchard, M. J. *J. Chem. Commun.* **1974**, 786; (b) Bencheik, A.; Petit, M.; Mortreux, A. *J. Mol. Catal.* **1982**, *15*, 93; (c) Villemin, D.; Cadiot, P.; *Tetrahedron Lett.* **1982**, *97*, 1592.

86. Katz, T. J. *J. Am. Chem. Soc.* **1975**, *97*, 1592.

87. (a) Kaneta, N.; Hirai, T.; Mori, M. *Chem. Lett.* **1995**, 627; (b) Kaneta, N.; Hikichi, K.; Asaka, S.; Uemura, M.; Mori, M. *Chem. Lett.* **1995**, 1055.

88. (a) Fürstner, A.; Guth, O.; Rumbo, A.; Seidel, G. *J. Am. Chem. Soc.* **1999**, *121*, 11108; (b) Fürstner, A.; Seidel, G. *Angew. Chem. Int. Ed.* **1998**, *37*, 1734.

89. (a) Wengrovins, J. H.; Sabcho, J.; Schrock, R. R. *J. Am. Chem. Soc.* **1981**, *103*, 3932; (b) McCullough, L. G.; Schrock, R. R. *J. Am. Chem. Soc.* **1984**, *106*, 4067.

90. (a) Trost, B. M.; Ball, Z. T.; Joge, T. *J. Am. Chem. Soc.* **2002**, *124*, 7922; (b) Fürstner, A.; Radkowski, K. *Chem. Commun.* **2002**, 2182.

91. (a) Fürstner, A.; Mathes, C.; Lehmann, C. W. *J. Am. Chem. Soc.* **1999**, *121*, 9453; (b) Fürstner, A.; Mathes, C.;. Lehmann, C. W. *Chem. Eur. J.* **2001**, *7*, 5299.

92. Fürstner, A.; Mathes, C. *Org. Lett.* **2001**, *2*, 221.

93. Milanic, O. S.; Vollhardt, K. P. C.; Whitener, G. D. *Synlett* **2003**, 29.

94. (a) Zhang W.; Moore, J. S. *J. Am. Chem. Soc.* **2004**, *126*, 12796. (b) Ge, P.-H.; Fu, W.; Herrmann, W. A.; Herdtweck, E.; Campana, C.; Adams, R. D.; Bunz, U. H. F. *Angew. Chem. Int. Ed.* **2000**, *39*, 3607.

7

NONLINEAR EFFECTS IN ASYMMETRIC CATALYSIS

HENRI B. KAGAN

Laboratoire de Catalyse Moléculaire, Institut de Chimie Moléculaire et des Matériaux d'Orsay (CNRS UMR 8182), Université Paris-Sud, Orsay, France

7.1   INTRODUCTION

Since the early times of stereochemistry, the phenomena related to chirality ("dissymétrie moléculaire," as originally stated by Pasteur)[1] have been treated or referred to as enantiomerically pure compounds. For a long time the measurement of specific rotations has been the only tool to evaluate the enantiomer distribution of an enantioimpure sample; hence the expressions *optical purity* and *optical antipodes*. The usefulness of chiral assistance (natural products, circularly polarized light, etc.) for the preparation of optically active compounds, by either resolution or asymmetric synthesis, has been recognized by Pasteur,[2] Le Bel,[3] and van't Hoff.[4] The first chiral auxiliaries selected for asymmetric synthesis were alkaloids such as quinine or some terpenes.[5] Natural products with several asymmetric centers are usually enantiopure or close to 100% ee.[6,7] With the necessity to devise new routes to enantiopure compounds, many simple or complex auxiliaries have been prepared from natural products or from resolved materials. Often the authors tried to get the highest enantiomeric excess values possible for the chiral auxiliaries before using them for asymmetric reactions. When a chiral reagent or catalyst could not be prepared enantiomerically pure, the enantiomeric excess (ee) of the product was assumed to be a minimum value[8,9] or was corrected by the ee of the chiral auxiliary.[10,11] The experimental data measured by polarimetry or spectroscopic methods are conveniently expressed by enantiomeric excess and enantiomeric

New Frontiers in Asymmetric Catalysis, Edited by Koichi Mikami and Mark Lautens
Copyright © 2007 John Wiley & Sons, Inc.

ratios (er), respectively. The compared advantages of ee and er have been discussed.[12] In this chapter the use of nonenantiopure auxiliaries will be summarized and discussed. The deviations of the observed ee of the product (ee_{prod}) with respect to the calculated value (assuming a proportionality with the ee of the auxiliary (ee_{aux}) are especially informative and may give rise to synthetic applications. This aspect will be developed as well as the possible ramifications in other areas of chemistry.

7.2 PROPERTIES OF ENANTIOMER MIXTURES

7.2.1 Physical Properties

A solution of chiral molecules of the same configuration ("homochiral assembly")[13] in an achiral solvent have some physical properties related to chirality. For example, in a given set of experimental conditions one can measure a specific rotation, a circular dichroism, or a circularly polarized luminescence. By inverting the configuration of the molecules, one gets a mirror-image situation, allowing exact prediction of the values of the chiral properties; they remain the same as above, except a change of the sign. With the racemic composition, the chiral properties vanish, and give zero values.

What happens for a nonracemic mixture of enantiomers? Is it possible to calculate the values of the chiral properties of the solution from knowledge of the properties of the enantiopure compound? In principle, yes, on the condition that there is no autoassociation or aggregation in solution. Then, the observed properties will be simply the weighted combination of the properties of two enantiomers. A nice example of where this normal law may be broken was discovered by Horeau in 1967; it is the nonequivalence between enantiomeric excess (ee) and optical purity (op, with op $= [\alpha]_{exp}/[\alpha]_{max}$) for 2,2-methylethyl–succinic acid.[14,15] In chloroform op is inferior to ee, while in methanol op $=$ ee. This was explained by the formation of diastereomeric aggregates in chloroform, while the solvation by methanol suppresses the autoassociation.

The formation of diastereomeric aggregates may perturb the achiral properties as well, when compared to homochiral solutions, since the aggregation state will be not necessarily the same. It has been observed that the ^1H NMR spectra of racemic and enantiopure dihydroquinine in chloroform are significantly different.[16]

7.2.2 Chemical Properties

Chiral compounds are of wide interest as auxiliaries in asymmetric synthesis. The efficiency of the auxiliary in the process is usually expressed as the ee of the product (ee_{prod}). What happens if the chiral auxiliary is not enantiomerically pure? One expects a lower ee for the product. From this value one can calculate the maximum ee of the product (ee_{max}) by taking into account the enantiomeric excess of the auxiliary, assuming a proportionality between ee_{prod} and ee_{aux}:

$$ee_{prod} = ee_{max}ee_{aux} \qquad (7.1)$$

This equation can be easily established by calculations involving a weighted mixing of the effects of the (R) and (S) auxiliaries. The only assumption is that the auxiliaries are reacting independently of each other, and that they don't give aggregation phenomena. In catalytic reactions where there is a chiral catalyst, Eq. (7.1) can also be established for simple kinetic schemes where there is no aggregation of the catalyst or autoinduction (modification of the catalyst by the product).[17,18] It was anticipated some complications for oganometallic catalysts where the metal is coordinated to several chiral ligands (Ref. 17, p. 244).

Another aspect of the chemical properties of mixtures of enantiomers has been reported by Wynberg and Feringa in 1976.[19] These authors have studied some diastereoselective reactions on chiral molecules (such as the LiAlH$_4$ reduction of camphor) in the absence of chiral auxiliaries. They found that the product distribution was significantly different if the substrate was enantiopure or racemic. Similarly, it is known that reduction of enantiopure or racemic camphor by K/liquid NH$_3$ gives rise to different isoborneol/borneol ratios, a detailed mechanistic analysis has been done by Rautenstrauch.[20]

In conclusion, for a long time Eq. (7.1) described correctly many experimental data in asymmetric synthesis.

7.3 NONLINEAR EFFECT IN ASYMMETRIC CATALYSIS

7.3.1 The First Evidences

The traditional way to treat reactions involving enantioimpure auxiliaries was questioned in 1986 by us.[21] We studied several asymmetric reactions and plotted ee$_{prod}$ versus ee$_{aux}$. Instead of a straight line, some curves were found. Sharpless epoxidation of geraniol by t-BuOOH gave 97% ee for the combination Ti(O^i-Pr)$_4$/diethyl tartrate 100% ee. Decreasing the enantiomeric excess of diethyl tartrate (DET) gave a curve located above the straight line (linearity) describing the proportionality between ee of product and ee of auxiliarity. The term "positive nonlinear effect" [abbreviated as (+)-NLE] was proposed to reflect the fact that the observed ee$_{prod}$ (in absolute value) is higher than expected by the linear correlation.[22,23] It was also qualified of "asymmetric amplification."[24] A water-modified Sharpless reagent Ti(O^i-Pr)$_4$/diethyl tartrate/H$_2$O $= 1:2:1$ was very successful in asymmetric oxidation by hydroperoxides of some sulfides into sulfoxides.[25] A change of the enantiomeric excess of DET gave rise to a curve located mostly below the straight line **B** of Figure 7.1. Since ee$_{prod}$ was lower than expected, such cases were termed "negative nonlinear effects" [(−)-NLE][22,23] or *asymmetric depletion*.[18] The intramolecular aldol cyclization of a triketone into a hydrindanone system catalyzed by (S)-proline was also investigated, giving a very weak (−)-NLE (based on polarimetric measurements).[21] It was recently found that indeed the system displays a linear behavior (ee measured by HPLC methods).[26] The initial 1986 paper on NLE prompted the publication of many reports on nonlinear effects, especially those with asymmetric amplification. The catalyzed addition of dialkylzincs on aldehydes soon afforded spectacular examples of asymmetric amplifications.[24,27,28] Examples of

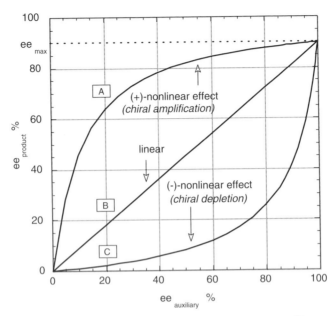

Figure 7.1. Plots of $ee_{product} = f (ee_{aux})$: the nonlinear effects.

nonlinear effects for various types of enantioselective catalytic reactions were also published.[29]

An index of asymmetric amplification, at a given value of ee_{aux} has been expressed as the ratio ee_{prod}/ee_{linear}, where ee_{linear} is the calculated value using Eq. (7.1). We proposed using also an amplification index er_{prod}/er_{linear} involving the enantiomeric ratio er.[18] This index better reflects the amplification when the chiral auxiliary is in the range 80–95% ee. Since it takes the values 1 for $ee_{aux} = 0$ or 100%, usually a maximum value occurs around $ee_{aux} = 30$–40%. This value is a convenient parameter for the comparison of various curves of asymmetric amplifications.[18]

7.3.2 Origin of Nonlinear Effects: Some Models

It is difficult to give general rules predicting the possibility of NLE in a given chemical system. One statement that we proposed is the following:

> If several molecules of the chiral auxiliary are involved in autoassociation or in any type of molecular species, then the passage from enantiopure auxiliary to nonenantiopure auxiliary may produce new diastereomeric entities. Consequently, the enantioimpure system may have new stereochemical properties.

This is well evidenced by simple kinetic models in organometallic catalysis. We first developed the now so-called ML_2 model, where M is a metalllic center and L a chiral ligand (Figure 7.2). In this model we assumed that the catalytic reaction

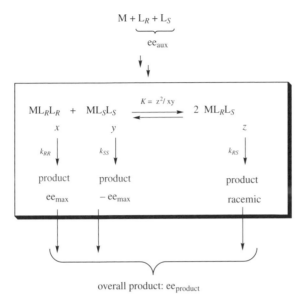

Figure 7.2. The ML_2 model with competition between homochiral and meso catalysts.

occurs through a mixture of three stereoisomeric complexes symbolized by ML_RL_R, ML_SL_S (homochiral catalysts) and ML_RL_S (meso catalyst) staying in respective amounts x, y and z. K is the equilibrium constant if the complexes are in fast equilibrium between them.

An easy kinetic treatment has been done by taking as parameters the relative rate constants and the relative amounts of the meso versus homochiral complexes: $g = k_{meso}/k_{homo}$, $\beta = z/(x + y)$.

Equation (7.2) has been established, which is equal to Eq. (7.1) multiplied by a fraction where g and β are involved. If $\beta = 0$ (no meso complex) or $g = 1$ (same reactivities of meso and homochiral complexes), then Eq. (7.2) gives back to Eq. (7.1). When $g < 1$ (meso complex slower than homochiral catalyst) the fraction will be a number >1, meaning that ee_{prod} will be higher than the value predicted by Eq. (7.1):

$$ee_{prod} = ee_{max}ee_{aux}\frac{1+\beta}{1+g\beta} \qquad (7.2)$$

Figure 7.1 shows curves ee_{prod} (%) $= f(ee_{aux}(\%))$ depicting the three main correlations between these two quantities. When Eq. (7.1) is obeyed through the whole range of values of ee_{aux}, it gives curve **B** (linear correlation). If the experimental points (curve **A**) are located above line **B**, it defines a positive nonlinear effect. Finally, curve **C** is an example of negative nonlinear effect.

An alternate model to the ML_2 model is the "reservoir model," where the catalytic species ML_R and ML_S are monomeric and in equilibrium with an inactive dimeric meso complex $(ML_R)(ML_S)$ (Figure 7.3).[29] In this model the inactive meso complex acts as a racemic trap, consequently decreasing the amount of active catalyst.

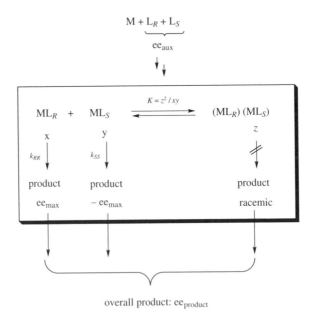

Figure 7.3. The reservoir effect. Examples of a ML catalyst with partial heterochiral dimerization into an inactive meso complex, giving rise to a (+)-NLE.

Blackmond pointed out that asymmetric amplification always has, as a consequence, a decrease in reactivity when compared to the enantiopure catalyst. This can be calculated on the various models proposed for the interpretation of nonlinear effects.[30] It is qualitatively visible in the "reservoir model" above as well as in the ML_2 model, where the asymmetric amplification given by $g < 1$ (low reactivity of the meso catalyst) has as consequence the overall slowdown in reaction rate. The generalized model ML_n has been discussed (for $n = 2,3,4$) when the various species are in equilibrium.[22a] The complexity of the curve can increase sharply as soon as $n > 2$.

From the simple models of Figures 7.2 and 7.3, one may go to more complicated situations, for example, where there is a preequilibrium between ligands and the catalyst precursor. Some of these cases have been modelized.[31]

In all these calculations, the basic assumption is the fixed relative composition of the catalytic species and the absence of modification of the catalyst by the products (autoinduction). On the contrary, the interpretation is more delicate and must take into account these perturbations.

7.4 MAIN CLASSES OF REACTIONS

There are several reviews covering many aspects of nonlinear effects.[29,31–35] The early examples of nonlinear effects almost exclusively dealt with organometallic catalysts. It is only very recently that some organocatalysts have been found to display asymmetric amplification. The two classes of catalysts are briefly presented below.

7.4.1 Organometallic Catalysts

Reactions where NLE have been discovered include Sharpless asymmetric epoxidation of allylic alcohols, enantioselective oxidation of sulfides to sulfoxides, Diels–Alder and hetero-Diels–Alder reactions, carbonyl-ene reactions, addition of Me_3SiCN or organometallics on aldehydes, conjugated additions of organometallics on enones, enantioselective hydrogenations, copolymerization, and the Henry reaction. Because of the diversity of the reactions, it is more convenient to classify the examples according to the types of catalyst involved.

Table 7.1 lists examples of NLEs induced by organometallic catalysts. *Titanium complexes* have been involved in a wide range of catalytic reactions, with the chiral ligand derived from a diol, a diphenol, a bis-sulfonamide, an aminoalcohol, and so on. The common feature is the possibility for a monomeric catalytic complex to dimerize into less active dimers or oligomers, providing a mechanism for a "reservoir model" and an asymmetric amplification. *Organozinc complexes* are another important class of catalysts, which are involved in addition of organozincs on various aldehydes. Spectacular asymmetric amplifications have been achieved with 10% ee DAIB ligand or its analog MIB; around 90% ee was oberved in the addition of diethylzinc on benzaldehyde, against 98% ee for the enantiopure ligands.[27,28] *Rare-earth complexes* are becoming increasingly important in asymmetric catalysis[36] and may generate NLEs.[37]

7.4.2 Organocatalysts

Enantioselective organocatalysts were the first to be used as nonenzymatic catalysts in the early twentieth century.[61] The literature on organocatalysis was reviewed in 2001.[62]

Only a few cases of nonlinear effects have been reported; some are listed in Table 7.2. Most of the examples involve proline as the catalyst. Chiral phosphoramides and some phase transfer catalysts were also reported to give NLEs.

7.5 ASYMMETRIC AMPLIFICATION

The expression "positive nonlinear effect" reflects the fact that the observed ee (ee_{prod}) is higher then the expected ee (ee_{linear}) calculated on the basis of Eq. (7.1). For example, let us consider an enantiopure catalyst that generates a product of 60% ee (ee_{max}). If the ligand is of 50% ee (ee_{aux}), one now calculates $ee_{prod} = 30\%$. If instead the reaction provides a product of 58% ee, this can be considered as an excellent case of asymmetric amplification.

The positive nonlinear effect is of great current interest. The possibility of using an enantioimpure chiral auxiliary for the preparation of a desired product in high ee

TABLE 7.1. Nonlinear Effects Displayed by Some Organometallic Catalysts

Catalyst	Reaction	Nonlinear Effect	Reference
Ti(O^{i}Pr)$_4$/(+)-DET	Sharpless epoxidation of geraniol	Weak (+)NLE	21
Ti(O^{i}Pr)$_4$/(+)-DET	Sulfide oxidation by ROOH/additives	(+)- or (−)-NLE	38
Ti(O^{i}Pr)$_4$/(−)-DIPT	TMSCN addition on benzaldehyde	Weak (+)-NLE	39
Ti(O^{i}Pr)$_4$/(R)-BINOL	Glyoxylate-ene reaction	Strong (+)-NLE	40
Ti(O^{i}Pr)$_4$/(R)-BINOL	Allylstannane on benzaldehyde/MS	Good (+)-NLE	41
(Ti-O-Ti)/(S)-BINOL	Allylstannane on PhCH$_2$CH$_2$CHO	Strong (+)-NLE	42
Ti(O^{i}Pr)$_2$Cl$_2$/Taddol	Diels–Alder reaction	Good (+)-NLE	43
Ti(O^{i}Pr)$_2$Cl$_2$/BINOL	Diels–Alder reaction/ MS or not	(+)-NLE or not	44
Et$_2$Zn/(−)-DAIB	Et$_2$Zn adition on benzaldehyde	High (+)-NLE	27,28
Et$_2$Zn/DIPT	1,3-Dipolar cycloaddition of nitrone on allylic alcohol	Good (+)-NLE	45
Et$_2$Zn/(R)- diphenylprolinol	Copolymerization meso-epoxide/CO$_2$	Strong (+)-NLE	46
Zn(OTf)$_2$/(+)-NME	Henry reaction CH$_3$NO$_2$ /PhCH$_2$CH$_2$CHO	Weak (+)-NLE	47
Cu-oxazolinethiolate	1,4-Addition RCu on cycloheptenone	Weak (−)-NLE	48
Cu-aminothiolate	1,4-Addition MeMgI on chalcone	Weak (+)-then (−)-NLE	49
Cu(OAc)$_2$/(S)-proline	Peracid hydroxylation of cyclohexenone	Weak (+) or (−)-NLE	50
Ni(acac)$_2$/L*	1,4-Addition of Et$_2$Zn on chalcone	Good (+)-NLE	51
Ni/bisoxazoline	Diels–Alder reaction	High (+)-NLE	52
CrN$_3$/salen*	Ring opening of meso-epoxide by N$_3$SiMe$_3$	Good (+)-NLE	53
Yb(O^{i}Pr)$_3$/(R)- BINOL	Epoxidation of chalcone by t-BuOOH	Good (+)-NLE	54
Rare earth–Li– binaphthooxides	Aza–Michael reaction on chalcone	High (+)-NLE	55
Sc-BINOLphosphate	Fluorination of a β-ketoester	Modest (+)-NLE	56
Yb(OTf)$_3$/(R)-BINOL	Diels–Alder reaction	Large (−)-NLE	57
[Rh]/2 monophosphites	Hydrogenation of dimethyl itaconate	Good (+)-NLE	58
RuBr$_2$binap	Hydrogenation of β-ketoesters	Good (+)-NLE	59
RhOH(binap)	1,4-Addition of PhB (OH)$_2$ on enone	Weak (−)-NLE	60

TABLE 7.2. Nonlinear Effects Displayed by Some Organocatalysts

Catalyst	Reaction	Nonlinear Effect	Reference
(S)-Proline	Aldol reaction of α-benzyloxy acetaldehyde	Strong (+)NLE	63
(S)-Proline	1,4-add. 2-nitropropane to 2-cyclohexenone	(+) and (−)-NLE	64
Polyleucine	Epoxidation of chalcone	Good (+)-NLE	65
(S)-Proline	α-aminooxylation of propionaldehyde	Time-dependent (+)-NLE	66
(S)-Proline	α-amination of propionaldehyde	Time-dependent (+)-NLE	67
$O=PR^1R^2R^3$	Trichlorosilyl enol ether/benzaldehyde	Weak (+)-NLE	68
NOBIN	PTC alkylation of aminoesters Schiff bases	High (+)-NLE	69

is especially attractive for fast reactions, since the simultaneous decrease in reactivity is not troubling. Two conditions have to be fulfilled in order to obtain useful (+)-NLEs: (1) the enantiopure catalyst must give a product of high ee (typically in the range of 95% ee) and (2) the curve $ee_{prod} = f(ee_{au})$ must be very steep in the range of low ee_{aux}. A descriptive way to compare the size of the asymmetric amplifications given by several catalysts is to calculate the ratio of enantiomeric ratios between the experimental results and values predicted in the linear case by Eq. (7.1). The index of amplification thus defined (er_{exp} / er_{linear}) for a given ee_{aux} takes the values 1 for $ee_{aux} = 0$ and 1, meaning that there is an intermediate optimum of the index, often located around a value of 30% ee_{aux}.

The interest of asymmetric amplification is evident in synthesis, when the chiral auxiliary is difficult to prepare as a single enantiomer. This can arise when it is prepared from enantioimpure natural products such as some terpenes or if the preparation involves the incomplete resolution of racemic materials. There are many examples where a chiral auxiliary of 80–90% ee gives a product with ee_{prod} almost identical to ee_{max} (obtained with enantiopure auxiliary). Another attractive feature of asymmetric amplification is its potential usefulness in the area of prebiotic chemistry, where generation of optical activity and amplification of ee are the two facets of the problem. Asymmetric amplification is also of interest when acting in autocatalytic reactions. *Autocatalysis* is defined as a process where the product of the reaction is the catalyst of its own formation. In a perfect asymmetric autocatalysis the R-product catalyzes only the formation of R-product, while the S-product is catalyst of the formation of S-product. Moreover, the reaction must be fully enantioselective each time the R- or the S-catalyst is acting. This is a difficult condition to fulfill. If not, there will inevitably be an erosion of the ee of the product when the reaction develops. A way to overcome the erosion of ee_{prod} with time is to compensate by incorporating a (+)-NLE in the system. The probability of finding such a system seemed very low. Fortuneously, Soai et al. discovered in 1995 an extraordinary autocatalytic system, the catalyzed addition of diisopropyl zinc on a pyrimidyl carboxaldehyde.[70] This reaction is developed by the authors in Chapter 9 of this book.

7.6 CURRENT TRENDS

The plot $ee_{prod} = f(ee_{aux})$ in an enantioselective reaction is a simple operation that can sometimes be very informative from both synthetic and mechanistic perspectives. This plot is now widely used in mechanistic discussions concerning enantioselective catalysis. However, some cautions are needed, since this approach has to be combined with additional studies in order to get firm conclusions. If linearity is observed, one cannot reach conclusions on the mechanism, since even with species involving several chiral auxiliaries one may remain linear, as in the ML_2 model with $g = 1$ in Eq. (7.3). If there are deviations from linearity, this could be a piece of information on the mechanism, for example, aggregations at some level of the chemical system,[27,44] or some competitive mechanisms.[68] References 27, 44, and 68 are only three among many published examples.

The sensitivity of nonlinear effects to *additives*, *catalyst preparation*, or *substrate structure* have been noted, giving some mechanistic informations (see, e.g., Refs. 50, 38, and 71 respectively).

Asymmetric depletion ((−)-NLE) is less frequently reported than (+)-NLE, perhaps because it seems less worthy of investigation. However, it can be informative with respect to the reaction mechanism, as demonstrated by Hayashi et al. in the catalyzed enantioselective 1,4-addition of arylboronic acids on enones.[60]

7.7 CONCLUSION

The concept of nonlinear effects has evolved since its recognition in the mid-1980s only. The plot $ee_{prod} = f(ee_{au})$ is now widely used as a tool to learn more about a catalytic system. The concepts and basic principles can been extended with some caution to kinetic resolution and to chiral reagents. Nonlinear effects that have been described thus far overwhelmingly for organometallic catalysts where aggregation processes or formation of multiligands species are easily understood. The renewal of organocatalysis is certainly expected to provide many examples of nonlinear effects in the future. Asymmetric amplification is an important aspect of nonlinear effects, in mechanistic studies as well as in autocatalysis and in synthetic organic chemistry.

ACKNOWLEDGMENT

I thank University Paris-Sud and CNRS for financial support and acknowledge the coauthors whose names appear in the papers. I am also grateful to Dr. Martial Vallet for his technical assistance.

REFERENCES AND NOTES

1. (a) Pasteur, L. *Compt. Rend. Acad. Sci.* **1848**, *26*, 535–538; (b) Pasteur, L. *Researches on the Molecular Asymmetry of Natural Organic Products*, Alembic Club Reprints, 14, Edinburgh, **1948**.

2. Pasteur, L. *Rev. Sci. Paris* (Ser. 3), **1884**, 7, 2–6.

3. Le Bel, A. *Bull. Soc. Chim. Fr.* **1874**, *22*, 337–342.

4. van't Hoff, J. H. *Bull. Soc. Chim. Fr.* **1875**, *23*, 299–301.

5. Compilation of the early examples of asymmetric synthesis up to 1970 in *Asymmetric Organic Reactions,* Morrison, J. D.; Mosher, H. S. (Eds.), Prentice-Hall, Englewoods Cliffs, NJ, **1971**.

6. Eliel, E. N.; Wylen, S.; Mander, L. N. *Stereochemistry of Organic Compounds*, Wiley, New York, **1994**.

7. The enantiomeric excess of (+)-camphor fluctuates between 99.76% and 99.52% depending on the origin of the plant: Rautenstrauch, V.; Lindström, M.; Bourdin, B.; Currie, J.; Oliveros, E. *Helv. Chim. Acta* **1993**, *76*, 607–615. Natural pinenes can be of much lower ee (70–90%).

8. (a) Brown, H. C.; Zweifel, G. *J. Am. Chem. Soc.* **1961**, *83*, 486–487; (b) Brown, H. C.; Ayyangar, N. R.; Zweifel, G. *J. Am. Chem. Soc.* **1964**, *86*, 397–403.

9. Knowles, W. S.; Sabacky,J.; Vineyard, B. D. *J. Chem. Soc. Chem. Commun.* **1972**, 10–11.

10. Tanaka, M.; Watanabe, Y.; Mitsudo, T.-A.; Yasunori, Y.; Takegami, Y. *Chem. Lett.* **1974**, 137–140.

11. Giacomelli, G.; Manicagli, R.; Lardicci, L. *J. Org. Chem.* **1973**, *13*, 2370–2376.

12. (a) Kagan, H. B. *Recl. Trav. Chim. Pays-Bas* **1995**, *114*, 203–205; (b) Gawley, R. E. *J. Org. Chem.* **2006**, *71*, 2411–2416.

13. The expression "homochiral" in its original meaning concerns isometric molecules of various kinds with the same configuration, for example, the (*L*)-aminoacids. There were propositions to use this word in a narrow sense, as the equivalent of *enantiopure*. For a discussion, see Ref. 6, p. 215.

14. Horeau, A. *Tetrahedron Lett.* **1969**, *36*, 3121–3124.

15. Horeau, A.; Guetté J.-P. *Tetrahedron* **1974**, *30*, 1923–1931.

16. Williams, T.; Pitcher, R. G.; Bommer, P.; Gutzwiller, J.; Uskokovic, M. *J. Am. Chem. Soc.* **1969**, *91*, 1871–1872.

17. Izumi, Y.; Tai, A. *Stereo-differentiating Reactions*, Academic Press, New York, **1977**, p 244.

18. Kagan, H. B.; Fenwick, D. *Topics in Stereochem*, Denmark, S. (Ed.), **1999**, *22*, 257–296.

19. Wynberg, H.; Feringa, B. *Tetrahedron* **1976**, *32*, 2831–2834.

20. (a) Rautenstrauch, V. *Helv. Chim. Acta* **1982**, *65*, 402–406; (b) Rautenstrauch, V.; Mégard, P.; Bourdin, B.; Furrer, A. *J. Am. Chem. Soc.* **1992**, *114*, 1418–1428.

21. Puchot, C.; Samuel, O.; Dunach, E.; Zhao, S.; Agami, C.; Kagan, H. B. *J. Am. Chem. Soc.* **1986**, *108*, 2353–2357.

22. (a) Guillaneux, D.; Zhao, S. H.; Samuel, O.; Rainford, D.; Kagan, H. B. *J. Am. Chem. Soc.* **1994**, *116*, 9430–9439; (b) Zhao, S. H. *Thesis*, Orsay, **1987**.

23. Mikami, K.; Terada, M. *Tetrahedron* **1992**, *48*, 5671–5680.

24. Oguni, N.; Matsuda, Y.; Kaneko, T. *J. Am. Chem. Soc.* **1988**, *110*, 7877–7878.

25. Pitchen, P.; Dunach, E.; Deshmukh, M. N.; Kagan, H. B. *J. Am. Chem. Soc*, **1984**, *106*, 8188–8194.

26. Hoang, L.; Bahmanayar, S.; Houk, K. N.; List, B. *J. Am. Chem. Soc.* **2003**, *125*, 16–17.

27. Kitamura, M.; Okada, S.; Suga, S.; Noyori, R. *J. Am. Chem. Soc.* **1989**, *111*, 4028–4036.

28. Noyori, R.; Kitamura, M. *Angew. Chem. Int. Ed. Engl.* **1991**, *30*, 49–69.

29. Girard, C.; Kagan, H. B. *Angew. Chem. Int. Ed.* **1998**, *37*, 2922–2959.

30. Blackmond, D. G. *J. Am. Chem. Soc.* **1997**, *119*, 12934–12939.

31. Blackmond, D. G. *Acc. Chem. Res.* **2000**, 33, 402–411.

32. Bolm, C. In *Advanced Asymmetric Catalysis,* Stephenson, G. R. (Ed.), Chapman & Hall, London, **1996**, pp. 9–26.

33. Avalos, M.; Babiano, R.; Cintas, P.; Jimenez, J. L.; Palacios, J. C. *Tetrahedron Asym.*, **1997**, *8*, 2997–3017.

34. Luukas, T.; Kagan, H. B. In *Comprehensive Asymmetric Catalysis,* Jacobsen, E. N.; Pfaltz, A.; Yamamoto H. (Eds.), Springer-Verlag, Berlin, **1999**, Vol. 1, pp. 101–118.

35. Mikami, K.; Terada, M.; Korenaga, T.; Matsumoto, Y.; Ueki, M.; Angelaud, R. *Angew. Chem. Int. Ed.* **2000**, *39*, 3532–3556.

36. Shibasaki, M.; Yoshikawa, N. *Chem. Rev.* **2002**, *102*, 2187–2209.

37. Inanaga, J.; Furuno, H.; Hayano, T. *Chem. Rev.* **2002**, *102*, 2211–2225.

38. Brunel, J.-M.; Luukas, T. O.; Kagan, H. B. *Tetrahedron Asym.* **1998**, *9*, 1941–1946.

39. Hayashi, M.; Matsuda, T.; Oguni, N. *J. Chem. Soc. Perkin Trans 1* **1992**, 3135–3140.

40. Terada, M.; Mikami, K.; Nakai, T. *J. Chem. Soc. Chem. Commun.* **1990**, 1623–1624.

41. (a) Keck, G. E.; Krishnamurthy, D. *J. Am. Chem. Soc.* **1995**, *117*, 2363–2364; (b) Faller, J. W.; Sams, D. W. I. *J. Am. Chem. Soc.* **1996**, *118*, 1217–1218.

42. Hanawa, H.; Hashimoto, T.; Maruoka, K. *J. Am. Chem. Soc.* **2003**, *125*, 1708–1709.

43. Iwasawa, N.; Haysahi, Y.; Sakurai, H.; Narasaka, K. *Chem. Lett.* **1989**, 1581–1584.

44. Mikami, K.; Motoyama, Y.; Terada, M. *J. Am. Chem. Soc.* **1994**, *116*, 2812–2820.

45. Shimizu, M.; Ukaji, Y.; Inomata, K. *Chem. Lett.* **1996**, 455–456.

46. Nakano, K.; Hiyama, T.; Nozaki, K. *Chem. Commun.* **2005**, 1871–1873.

47. Palomo, C.; Oiarbide, M.; Laso, A. *Angew. Chem. Int. Ed.* **2005**, *44*, 3881–3884.

48. Zhou, Q.-L.; Pfaltz, A. *Tetrahedron* **1994**, *50*, 4467–4478.

49. van Koten, G. *Pure Applied Chem.* **1994**, *66*, 1455.

50. Zondervan, C.; Feringa, B. L. *Tetrahedron Asym.* **1996**, *7*, 1895–1898.

51. Bolm, C.; Felder, M.; Müller, J. *Synlett* **1992**, 439–441.

52. Kanemasa, S.; Oderaotoshi, Y.; Sagakuchi, S.; Yamamoto, H.; Tanaka, J.; Wada, E.; Curran, D. P. *J. Am. Chem. Soc.* **1998**, *120*, 3074–3088.

53. Hansen, K. B.; Leighton, J. L.; Jacobsen, E. N. *J. Am. Chem. Soc.* **1996**, *118*, 10924–10925.

54. Bougauchi, M.; Watanabe, S.; Arai, T.; Sasai, H.; Shibasaki, M. *J. Am. Chem. Soc.* **1997**, *119*, 2329–2330.

55. Yamagiwa, N.; Qin, H.; Matsunaga, S.; Shibasaki, M. *J. Am. Chem. Soc.* **2005**, *127*, 13419–13427.

56. Suzuki, S.; Furuno, H.; Yokoyama, Y.; Inanaga, J. *Tetrahedron Asym.* **2006**, *17*, 504–507.

57. Kobayashi, S.; Ishitani, H.; Araki, M.; Hachiya, M. *J. Tetrahedron Lett.* **1994**, *35*, 6325–6328.

58. Reetz, M. T.; Meiswinkel, A.; Mehler, G.; Angermund, K.; Graf, M.; Thiel, W.; Mynott, R.; Blackmond, D. G. *J. Am. Chem. Soc.* **2005**, *127*, 10305–10313.

59. Girard, C.; Genet, J.-P.; Bulliard, M. *Eur. J. Org. Chem.* **1999**, *11*, 2937–2942.

60. Kina, A.; Iwamura, H.; Hayashi, T. *J. Am. Chem. Soc.* **2006**, *128*, 3904–3905.

61. History of asymmetric catalysis: Kagan, H. B. in *Comprehensive Asymmetric Catalysis*, Jacobsen, E. N.; Pfaltz, A.; Yamamoto H. (Eds.), Springer-Verlag, Berlin, **1999**, Vol. 1, pp. 9–30.

62. (a) Dalko, P. L.; Moisan, L. *Angew. Chem. Int. Ed.* **2001**, *40*, 3726–3748; (b) Dalko, P. L.; Moisan, L. *Angew. Chem. Int. Ed.* **2004**, *43*, 5138–5175.

63. Cordova A.; Engqvist, M.; Ibrahem, I.; Casas, J.; Sundén, H. *Chem. Commun.* **2005**, 2047–2049.

64. Hanessian, S.; Pham, V. *Org. Lett.* **2000**, *2*, 2975–2978.

65. Kelly, D. R.; Meek, A.; Roberts, S. M. *Chem. Commun.* **2004**, 2021–2022.

66. Mathew, S. P.; Iwamura, H.; Blackmond, D. G. *Angew. Chem Int. Ed.* **2004**, *43*, 3317–3321.

67. Iwamura, H.; Mathew, S. P.; Blackmond, D. G. *J. Am. Chem. Soc.* **2004**, *126*, 11770–11771.

68. Denmark, S. E.; Su, X.; Nishigaichi, Y., *J. Am. Chem. Soc.* **1998**, *120*, 12990–12991.

69. Belokon, Y. N.; Bespalova, N. B.; Churkina, T. D.; Cisarova, I.; Ezernitskaya, M. G.; Harutyunyan, S. R; Hrdina, R.; Kagan, H. B.; Kocovsky, P.; Kochetkov, K. A.; Larionov, O. V.; Lyssenko, K. A.; North, M.; Polasek, M.; Peregudov, A. S.; Prisyazhnyuk, V. V.; Vyskocil, S. *J. Am. Chem. Soc.* **2003**, *125*, 12860–12871.

70. Soai, K.; Shibata, T.; Morioka, H.; Choji, K. *Nature* **1995**, *378*, 767–768.

71. Chen, Y. K.; Costa, A. M.; Walsh, P. J. *J. Am. Chem. Soc.* **2001**, *123*, 5378–5379.

8

ASYMMETRIC ACTIVATION AND DEACTIVATION OF RACEMIC CATALYSTS

KOICHI MIKAMI AND KOHSUKE AIKAWA

Department of Applied Chemistry, Tokyo Institute of Technology, Meguro-ku, Tokyo

8.1 INTRODUCTION

The development of enantioselective catalysts, which provide enantiomerically enriched products, has attracted a crucial interest in modern asymmetric synthesis[1] in terms of "chiral catalyst economy," namely, the ratio of (product % ee × product % yield)/(catalyst % ee × catalyst mol%). Therefore racemic catalysis using unresolved racemic catalysts should be *an infinitely* efficient and environmentally "green" process.[2] Standard methods of such enantioselective catalysis employ metal complexes bearing chiral and nonracemic organic ligands, normally in enantiopure form (Scheme 8.1a). The key to developing efficient asymmetric catalytic reactions is to create chiral catalysts with suitable chiral ligands. These chiral ligands can be atropisomeric (originated from the Greek, *atropos*: *a* meaning "not" and *tropos* meaning "turn")[3] ligands such as BINAP and BINOL, which usually produce a C_2-symmetric reaction center. However, studies on the origin of chirality from achiral or racemic substrates in nature provide the basis for a new paradigm in a symmetric catalysis starting from racemic catalysts. Racemic catalysts inherently produce only racemic products (Scheme 8.1b), while nonracemic catalysts can generate nonracemic products (Scheme 8.1a). Asymmetric catalysis through enantiomeric fluctuation (or discrimination) by an external chiral bias and subsequent

New Frontiers in Asymmetric Catalysis, Edited by Koichi Mikami and Mark Lautens

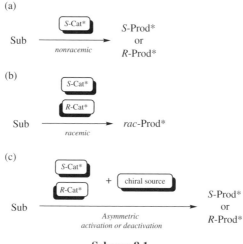

Scheme 8.1

amplification of chirality can be developed through autocatalysis. One strategy for achieving this enantiomeric discrimination is the addition of a chiral source (Scheme 8.1c), which selectively transforms one catalyst enantiomer into a highly activated or deactivated catalyst enantiomer. In this chapter, recent progress on chirally economical racemic catalysis will be highlighted on the basis of asymmetric activation and deactivation (chiral poisoning) of racemic catalysts.

8.2 RACEMIC CATALYSIS

Generally racemic catalysts give only a racemic mixture of chiral products. However, the development of an asymmetric reaction through preferential discrimination of a racemic catalyst via addition of a chemical or physical source can be expected in view of the fluctuation of the racemic environment in nature. Typically, asymmetric synthesis using racemic catalysts in the presence of a chiral source can be classified into two ways:[4] (1) asymmetric deactivation and (2) asymmetric activation. In *asymmetric deactivation*, enantiomer-selective deactivation of a racemic catalyst by a chiral deactivator (e.g., chiral poison) provides the desired catalyst enatiomer remained, yielding nonracemic products. As a conceptually opposite strategy to asymmetric deactivation of racemic catalysis, a strategy of *asymmetric activation* has also been developed. In this process, a chiral activator selectively activates one enantiomer of a racemic catalyst, thereby yielding nonracemic products. Racemic catalysts can thus be evolved into chirally activated catalysts leading to enantiomerically enriched products by association with chiral activators.

(a) Selective formation of deactivated catalyst

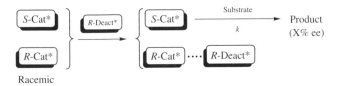

(b) Nonselective formation of deactivated catalyst

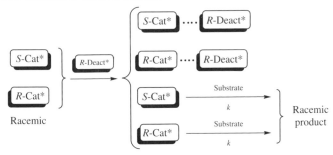

Scheme 8.2

8.2.1 Asymmetric Deactivation

Enantiomer-selective deactivation of racemic catalyts by a chiral deactivator affects the enantiomer-selective formation of a deactivated catalyst with low catalytic activity (Scheme 8.2). Therefore, it is crucial for a chiral deactivator to interact with one enantiomer of a racemic catalyst (Scheme 8.2a). As the chiral deactivator does not interact with the other enantiomer of racemic catalyst, the enantiomerically enriched product can be obtained. Therefore, the level of enantiomeric excess (% ee) could not exceed that attained by the enantiopure catalyst. On the other hand, nonselective complexation of a chiral deactivator would equally and simultaneously deactivate both catalyst enantiomers, thereby yielding a racemic product (Scheme 8.2b). Although this strategy tends to use excess chiral poison relative to the amount of catalyst, it offers a significant advantage in reducing cost and synthetic difficulty since readily available racemic catalysts and often inexpensive chiral poisons are used.

A resolution of racemic CHIRAPHOS ligand has been achieved using a chiral iridium amide complex (Scheme 8.3).[5] The chiral iridium complex (+)-**1** reacts selectively with (*S,S*)-CHIRAPHOS to form the inactive iridium complex **2**. The remaining (*R,R*)-CHIRAPHOS affords the catalytically active chiral rhodium complex **3**. The system catalyzes asymmetric hydrogenation to give the (*S*)-product with 87% ee. The opposite enantiomer (−)-**1** gives the (*R*)-product with 89.5% ee, which is almost the same level of enantioselectivity obtained by using optically pure (*S,S*)-CHIRAPHOS.

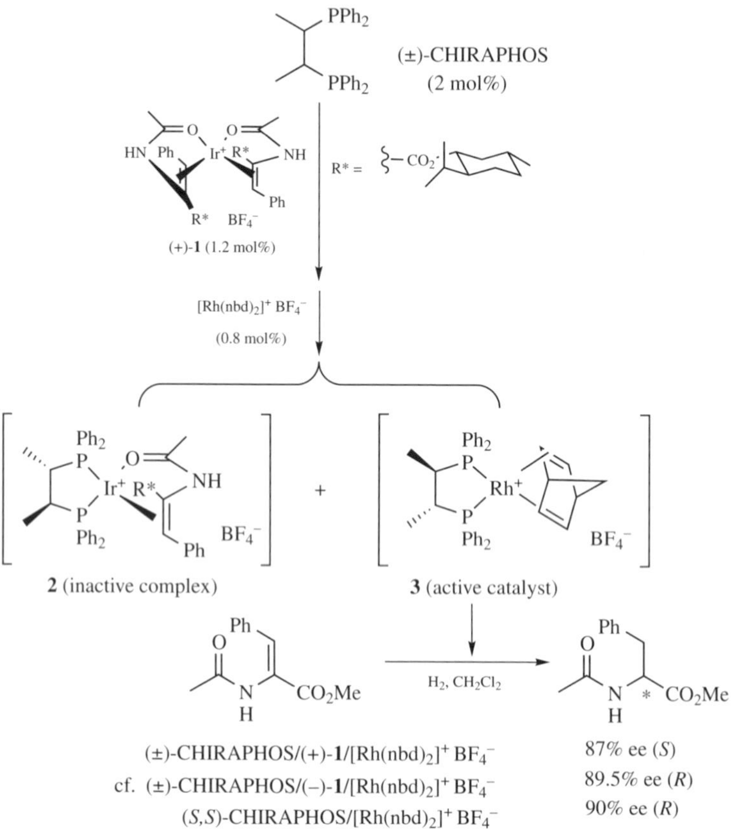

Scheme 8.3

Discrimination of the racemic aluminum reagent **4** can be carried out using chiral ketone **5**, which deactivates one enantiomer of racemic **4**. The hetero-Diels–Alder reaction is then catalyzed by the remaining opposite enantiomer of racemic **4** (Scheme 8.4).[6] The combination of racemic **4** and chiral ketone **5** in a 1:1 ratio gives better enantiomeric excess than in a 2:1 ratio, implying that one diastereomer of the **4/5** complex readily dissociates to yield optically pure **4** and the chiral ketone **5**.

The term "chiral poisoning" as a deactivating strategy has been proposed for the asymmetric hydrogenation reaction of dimethyl itaconate catalyzed by CHIRAPHOS-Rh complex (Scheme 8.5).[7] The combination of racemic CHIRAPHOS-Rh complex and (S)-METHOPHOS **6** as a catalyst poison yields the hydrogenated product in 49% ee. (S)-METHOPHOS is believed to bind to the (S,S)-CHIRAPHOS-Rh complex preferentially, as the use of enantiopure (R,R)-CHIRAPHOS-Rh complex affords the product with 98% ee.

(1R,2S)-Ephedrine as a chiral poison is also employed in the kinetic resolution of cyclic allylic alcohols using racemic BINAP-RuCl$_2$(dmf)$_n$ **7** (Scheme 8.6).[8] Chiral

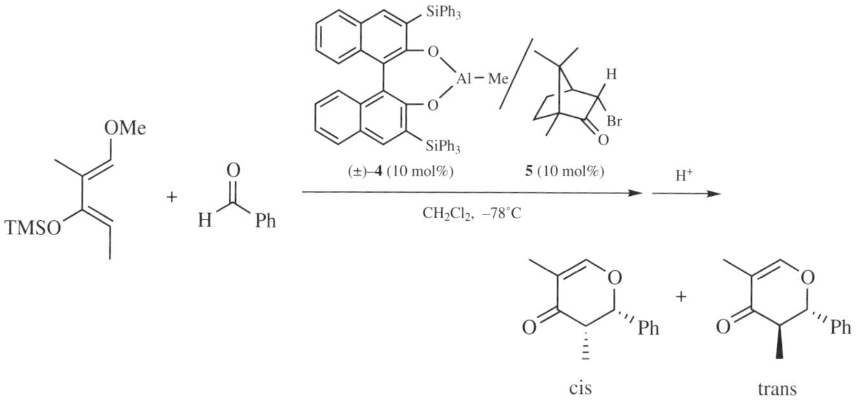

	80 (82% ee)	:	20
cf. (S)-**4** catalyst	90 (95% ee)	:	10

Scheme 8.4

poisoning of the racemic **7** with (1*R*,2*S*)-ephedrine provides (*R*)-2-cyclohexenol in >95% ee at 77% conversion (the relative rate k_f/k_s => 6.4). Enantiopure (*R*)-**7** is effective for the kinetic resolution of racemic 2-cyclohexenol to afford (*S*)-2-cyclohexenol in >95% ee at 60% conversion (k_f/k_s => 15); (*R*)-**7** hydrogenates (*R*)-2-cyclohexenol much faster than (*S*)-2-cyclohexenol. Hence, addition of

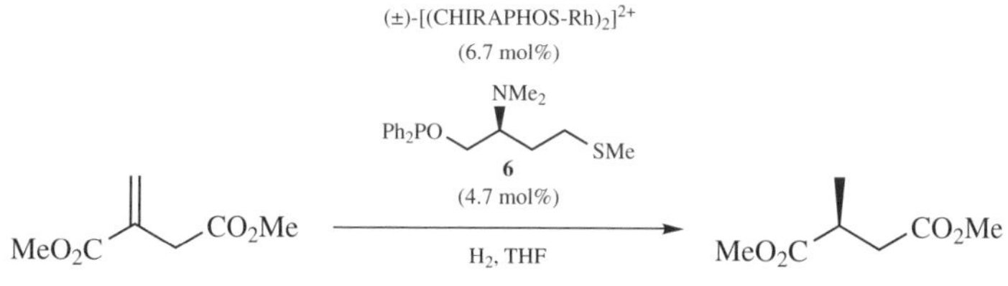

49% ee (*S*)

cf. (*R*,*R*)-[(CHIRAPHOS-Rh)$_2$]$^{2+}$ >98% ee (*S*)

Scheme 8.5

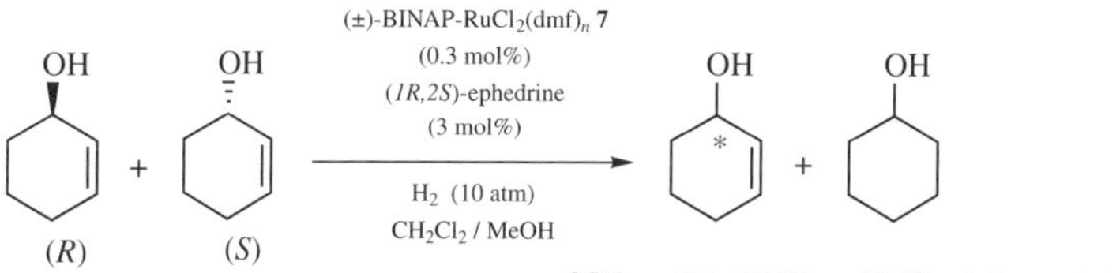

	>95% ee (*R*)	(23%)	(77%)	k_f/k_s = >6.4
cf. (*R*)-BINAP-RuCl$_2$(dmf)$_n$	>95% ee (*S*)	(40%)	(60%)	k_f/k_s = >15

Scheme 8.6

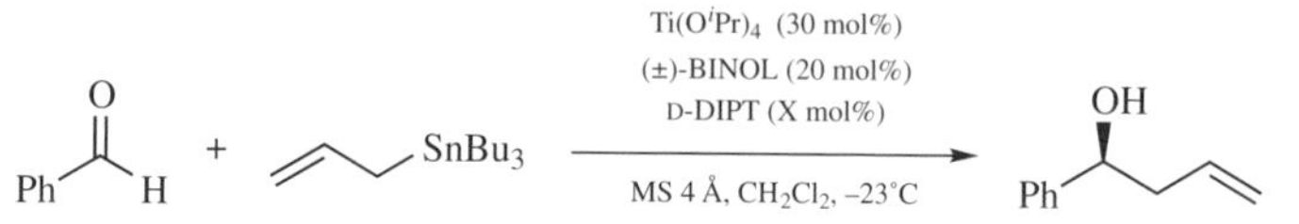

D-DIPT	
12 mol%	19% ee (44%)
15 mol%	39% ee (40%)
20 mol%	81% ee (47%)
30 mol%	91% ee (63%)

Scheme 8.7

(1*R*,2*S*)-ephedrine remarkably retards the reaction, indicating that (1*R*,2*S*)-ephedrine selectively deactivates the (*R*) component of the racemic catalyst **7**.

Enantiomerically pure diisopropyl D-tartrate (D-DIPT) can be used effectively as a chiral poison for racemic BINOLato-Ti(O^{i}Pr)$_2$- catalyzed addition of allyltributyltin to aldehydes (Scheme 8.7).[9] The enantioselectivity of the product increases with an increase in the amount of D-DIPT employed. When Ti(O^{i}Pr)$_4$ and D-DIPT are employed in the ratio of 1 : 3, the enantioselectivity and yield of the homoallylic alcohol product increase to 91% ee and 63% from 19% ee and 44% in a ratio of 1 : 1.

The Ti(O^{i}Pr)$_2$Cl$_2$/D-DIPT poison has also been used for the Ti(O^{i}Pr)$_2$Cl$_2$/BINOL-catalyzed asymmetric carbonyl-ene reaction with chloral (Scheme 8.8).[10] With the Ti(O^{i}Pr)$_4$/D-DIPT poison in a 1 : 3 ratio, both the regioselectivity and the enantioselectivity of the ene product are improved.

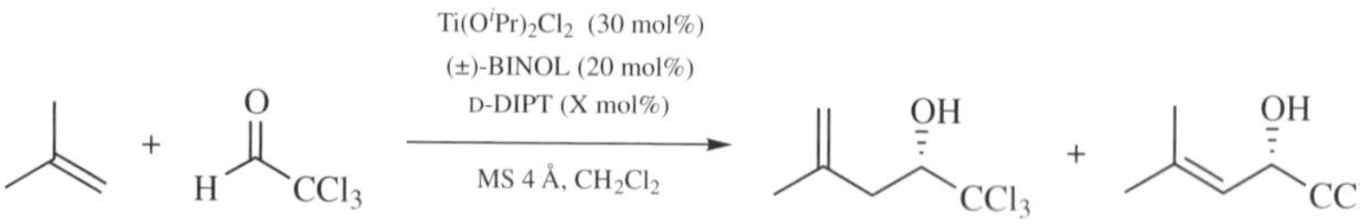

D-DIPT		
10 mol%	14% ee (55%)	14% ee (11%)
30 mol%	48% ee (37%)	35% ee (6%)
cf. Ti(O^{i}Pr)$_2$Cl$_2$/(*S*)-BINOL (10 mol%)	24% ee (58%)	66% ee (29%)

Scheme 8.8

A chiral poisoning strategy has been used with the racemic [*p*-cymeneRu(H_2O)(BINPO)](SbF_6)$_2$ complex **8**-catalyzed asymmetric Diels–Alder reaction of methacrolein with cyclopentadiene (Table 8.1).[11] The racemic [*p*-cymeneRuCl-(BINPO)]SbF_6 with 1.0 equiv of $AgSbF_6$ leads to racemic **8**. Subsequent addition of L-proline or L-prolinamide as a chiral poison (**P***) attains an equilibrium between the active aqua-Ru complex **8** and the deactivated Ru complex **8-P***. Enantioselectivity of the Diels–Alder product is in the range of 8–60% ee when the chiral poison is added in greater or stoichiometric amounts with respect to (±)-**8**. The chiral poisons, L-proline and L-prolinamide, selectively deactivate (S_{Ru},*R*)-**8** via different deactivation modes, selective displacement of BINPO and aqua ligands by

TABLE 8.1

Entry	($\pm$)-**8** (equiv)	**P*** (equiv)	Time (h)	Conv. (%)	% de (exo:endo)	% ee (Config.)
1	0.2	L-Proline (0.1)	16	76	98	8 (S)
2	0.1	L-Proline (0.1)	140	28	95	54 (S)
3	0.2	L-Proline (1.0)	120	35	95	59 (S)
4	0.2	L-Prolinamide (0.1)[a]	16	48	95	30 (S)
5	1.25	L-Prolinamide (1.0)[a]	140	92	96	60 (S)

[a]The reaction was carried out at $-78°C$.

L-proline and L-prolinamide, respectively. In both cases, the remaining (R_{Ru},S)-**8** is free to catalyze the asymmetric Diels–Alder reaction.

As an excellent chiral poison for BINAPs-RuCl$_2$(dmf)$_n$ **7**, 3,3'-dimethyl-2,2'-diamino-1,1'-binaphthyl (DM-DABN) has been designed in which methyl groups in (R)-2,2'-diamino-1,1'-binaphthyl (DABN) would interact with equatorial Ar groups in (R)-BINAP but not in (S)-BINAP (Figure 8.1).

Complete enantiomer discrimination and asymmetric deactivation of the racemic XylBINAP-RuCl$_2$(dmf)$_n$ ($\pm$)-**7b** using DM-DABN as a chiral poison are shown to be effective in the kinetic resolution of 2-cyclohexenol (Scheme 8.9).[12] Use of just a 0.5 molar amount of (S)-DM-DABN relative to ($\pm$)-**7b** gives enantiopure (S)-2-cyclohexenol, which is kinetically resolved in the same conversion as enantiopure **7b**. Indeed, the relative rate of hydrogenation of (R)- versus (S)-2-cyclohexenol in the presence of only a 0.5 molar amount of (S)-DM-DABN relative to ($\pm$)-**7b** is significantly large ($k_f/k_s = 102$). The combination of ($\pm$)-**7b** with (S)-DM-DABN also gives 99.3% ee of (R)-methyl 3-hydroxybutanoate quantitatively

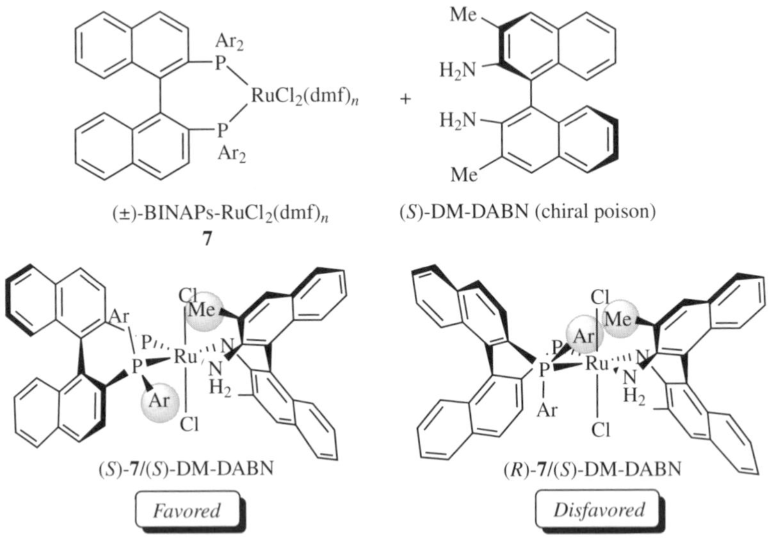

(±)-BINAPs-RuCl$_2$(dmf)$_n$
7

(S)-DM-DABN (chiral poison)

(S)-**7**/(S)-DM-DABN

Favored

(R)-**7**/(S)-DM-DABN

Disfavored

Figure 8.1

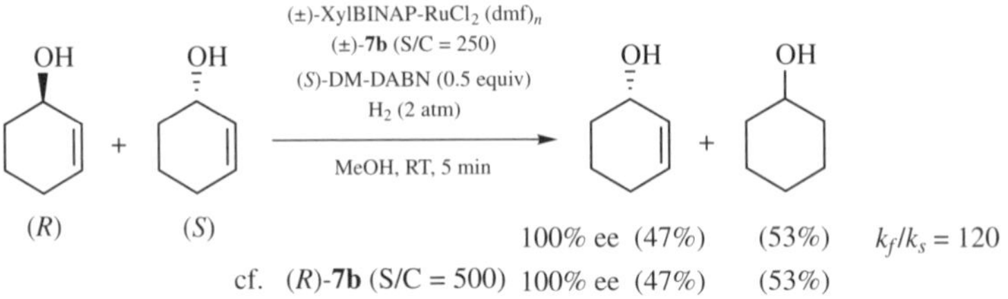

(±)-XylBINAP-RuCl$_2$ (dmf)$_n$
(±)-**7b** (S/C = 250)
(S)-DM-DABN (0.5 equiv)
H$_2$ (2 atm)
─────────────
MeOH, RT, 5 min

(R)   (S)   100% ee (47%)   (53%) $k_f/k_s = 120$

cf. (R)-**7b** (S/C = 500) 100% ee (47%)   (53%)

Scheme 8.9

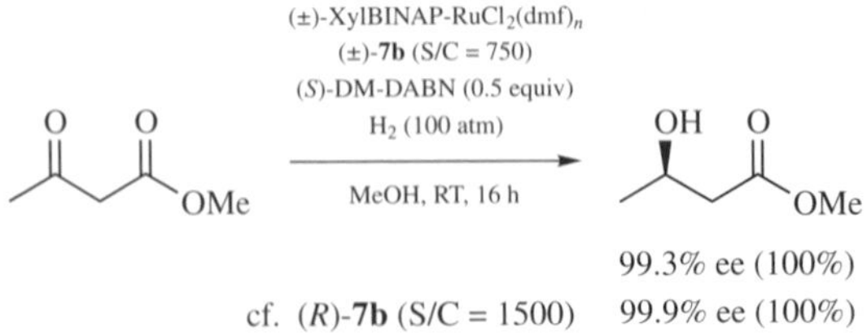

(±)-XylBINAP-RuCl$_2$(dmf)$_n$
(±)-**7b** (S/C = 750)
(S)-DM-DABN (0.5 equiv)
H$_2$ (100 atm)
─────────────
MeOH, RT, 16 h

99.3% ee (100%)

cf. (R)-**7b** (S/C = 1500)   99.9% ee (100%)

Scheme 8.10

from the corresponding β-ketoester. The enantioselectivity is equally high to 99.9% ee which is actually obtained using the enantiopure (R)-catalyst (Scheme 8.10).

8.2.2 Asymmetric Activation of Chirally Rigid (Atropos) Catalysts

In contrast to asymmetric deactivation, Mikami[13] has reported a conceptually opposite strategy, asymmetric activation. A highly activated chiral catalyst can be produced by addition of a chiral activator (Scheme 8.11). This strategy has the advantage that the activated catalyst can afford products with a higher enantiomeric

(a) Selective formation of activated catalyst

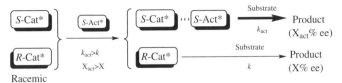

(b) Nonselective formation of activated catalyst

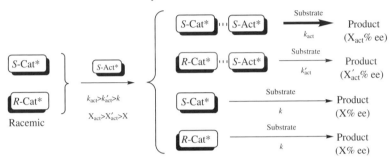

Scheme 8.11

excess (X_{act}% ee) than the enantiomerically pure catalyst (X% ee). Therefore, the complexation of the activated catalyst with the substrate is the most decisive step in determining the overall enantioselectivity. Asymmetric activation can also be attained via a nonselective complexation giving activated diastereomeric catalysts (Scheme 8.11b), as a result of different turnover frequencies (catalytic activities) between the diastereomers ($k_{act} > k'_{act}$). The use of 1.0 equiv of the activator to a parent catalyst may provide a 1 : 1 diastereomeric mixture; a higher level of enantioselectivity than that of an enantiomerically pure catalyst (X_{act}% ee > X% ee) can be obtained by a difference of more than two orders of magnitude in the catalytic efficiency ($k_{act}/k > 10^2$). Figure 8.2 indicates the effect of the variation of the relative catalytic efficiency ($k_{rel} = k'_{act}/k_{act}$ ranges from 0.01 to 100). In one case, an activated diastereomeric complex ((R)-ML$_n$/(R)-Act*) leads to the product in 100% ee (R) and the other diastereomer ((S)-ML$_n$/(R)-Act*) provides the opposite enantiomeric product in 50% ee (S). When the two diastereomeric activated complexes are formed in a 1 : 1 ratio, more than 99% ee can be obtained if the relative catalytic activity of the two activated diastereomers is 10^2 (log $k_{rel} = 2$).

Mikami and co-workers have reported the asymmetric activation in ene, aldol, and Diels–Alder reactions catalyzed by BINOLato-Ti(O^{i}Pr)$_2$ complex **9**. The racemic BINOLato-Ti(O^{i}Pr)$_2$ ($\pm$)-**9**-catalyzed ene reactions, provides extremely high enantioselectivity through the enantiomer-selective activation by chiral diols (Table 8.2).[13] The high level of enantioselectivity (89.8% ee) can be achieved by using only 5 mol% of (R)-BINOL as a chiral activator per 10 mol% of ($\pm$)-**9**. The advantage of asymmetric activation is highlighted in the case of a catalytic version. Significantly high enantioselectivity (80.0% ee)

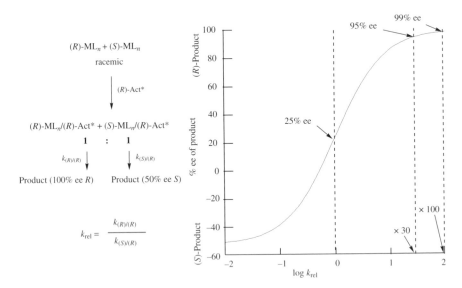

Figure 8.2

is obtained by adding a catalytic amount (0.25 equiv to (±)-**9**) of (R)-BINOL (entry 5). This result indicates the advantage of the catalytic version in asymmetric activation concept.

Ene reactions catalyzed by enantiopure (R)-**9** can also be achieved with (R)-BINOL as a chiral activator (Table 8.3). The reaction proceeds to give higher chemical yield (82.1%) and enantioselectivity (96.8% ee) than those attained without the additional BINOL activator (19.8%, 94.5% ee) (entry 1 vs. 2). Kinetic studies indicate that the reaction catalyzed by (R)-BINOLato-Ti(O^iPr)$_2$/(R)-BINOL complex ((R,R_{act})-**9**) is 25.6 times as fast as that catalyzed by (R)-**9**. These results imply that (±)-**9** and a half-molar amount of (R)-BINOL complex give (R,R_{act})-**9**, leaving (S)-**9** uncomplexed. In contrast, (S)-BINOL activates (R)-**9** to a smaller degree (entry 3), giving lower optical (86.0% ee) and chemical (48.0%) yields than with (R)-BINOL.

TABLE 8.2

Entry	Act*	Yield (%)	ee (%)
1	–	5.9	0
2	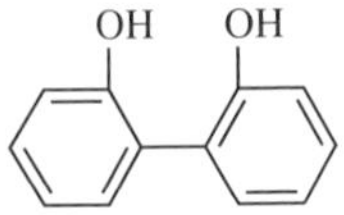	20	0
3	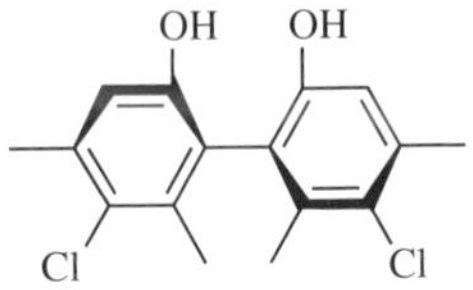	38	80.8
4	(R)-BINOL	52	89.8
5	(R)-BINOL	35	80.0[a]

[a] 2.5 mol% of (R)-BINOL was used per 10 mol% ($\pm$)-BINOLate-Ti(O^{i}Pr)$_2$.

A similar enantiomer-selective activation has been observed for aldol[14] and hetero-Diels–Alder reactions.[15] Asymmetric activation of (R)-**9** by (R)-BINOL is also effective in giving higher enantioselectivity (97% ee) than those by the parent (R)-**9** (91% ee) in the aldol reaction of silyl enol ethers (Scheme 8.12a). Asymmetric activation of (R)-**9** by (R)-BINOL is the key to provide higher enantioselectivity (84% ee) than those obtained by (R)-**9** (5% ee) in the hetero-Diels–Alder reaction with Danishefsky's diene (Scheme 8.12b). Activation with (R)-6-Br-BINOL gives lower yield (25%) and enantioselectivity (43% ee) than the one using (R)-BINOL (50%, 84% ee). One can see that not only steric but also electronic factors are important in a chiral activator.

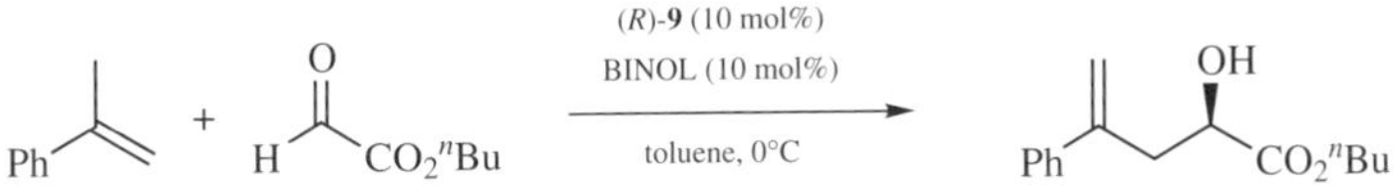

TABLE 8.3

Entry	BINOL	Yield (%)	ee (%)
1	—	19.8	94.5
2	(R)-BINOL	82.1	96.8
3	(S)-BINOL	48.0	86.0
4	($\pm$)-BINOL	69.2	95.7

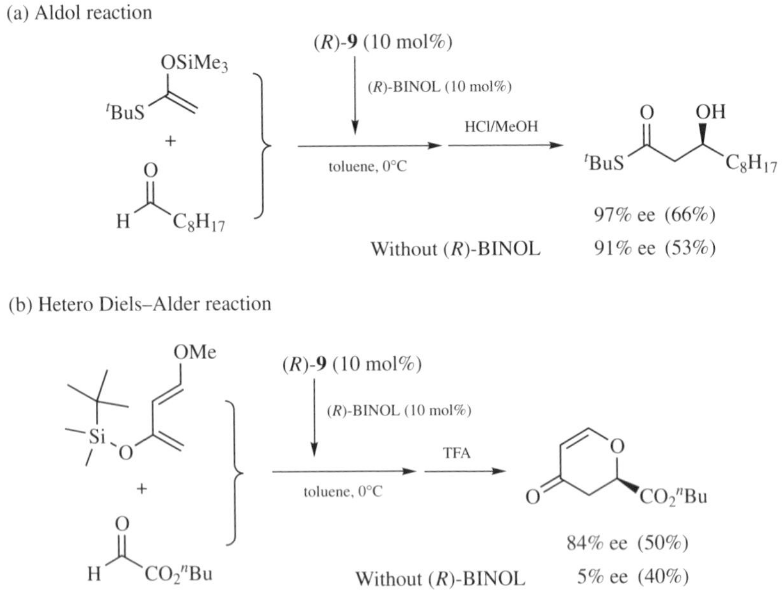

(a) Aldol reaction

(b) Hetero Diels–Alder reaction

Scheme 8.12

A symmetric activation is also observed in the combination of (R)-BINOL and $Zr(O^tBu)_4$, which promotes enantioselective synthesis of homoallylic alcohols (Scheme 8.13).[16] A $2:1$ ratio of (R)-BINOL and $Zr(O^tBu)_4$ without any other chiral source affords the homoallylic alcohol product in 27% ee and 44% yield. Addition of (R)-(+)-α-methyl-2-naphthalenemethanol $((R)$-MNM) leads to higher enantiomeric excess (53% ee) than those using only (R)-BINOL. Therefore, (R)-MNM can act as a chiral activator; a higher ee can be achieved via activation of the allylation of benzaldehyde by addition of (R)-MNM as a product-like activator.

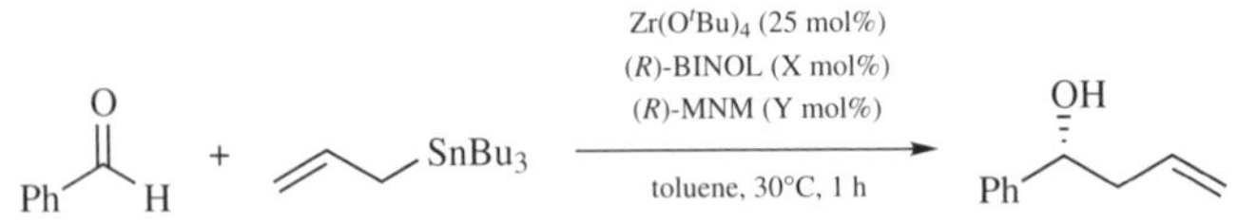

(R)-BINOL	(R)-MNM	
50 mol%	none	27% ee (44%)
50 mol%	25 mol%	53% ee (34%)
25 mol%	25 mol%	43% ee (20%)
25 mol%	50 mol%	57% ee (17%)

Scheme 8.13

X = H: (R)-BINOL
X = Ph: (R)-Ph$_2$-BINOL

10a: Ar = Ph (R,R)
10b: Ar = Ph (S,S)
10c: Ar = 2,4,6-Me$_3$Ph (R,R)

TABLE 8.4

Entry	BINOLs	Diamine	Yield (%)	ee (%)
1	(R)-BINOL	—	54	8.2
2	—	10a	64	1.1
3	(R)-BINOL	10a	100	37
4	(R)-Ph$_2$-BINOL	10a	100	65
5	(R)-Ph$_2$-BINOL	10b	100	65
6	(R)-Ph$_2$-BINOL	10c	100	90
7[a]	(R)-Ph$_2$-BINOL	10c	100	99

[a]The reaction was carried out at −78°C.

Ternary combination of diethylzinc, BINOLs, and chiral diamine activators **10** has been found to enhance the activity and enantioselectivity in asymmetric alkylation (Table 8.4).[17] By employing either (R)-BINOL or the chiral amine **10a**, (S)-1-phenylpropanol is obtained with only 8.2% or 1.1% ee (entries 1 and 2). However, the combined use of (R)-BINOL and **10a** affords the product quantitatively with 37.4% ee (entry 3). Enantioselectivity is increased by combination of a modified BINOL and matched chiral diamine activators. (R)-Ph$_2$-BINOL gives the product in up to 65% ee and quantitative yields in the presence of either chiral amine **10a** or **10b** (entries 4 and 5). The steric hindrance of the chiral activators is also crucial. The chiral activator **10c** is found to lead to the best results (entry 6). The best combination of (R)-Ph$_2$-BINOL/**10c** is further optimized by carrying out the reaction at −78°C, affording the product in 99% ee and quantitative yield (entry 7). All amine activators are found to significantly activate the BINOLs-Zn catalyst complex. Higher yields and enantioselectivities of 1-phenylpropanol are obtained in the presence of the amine activator than using the ligands themselves. The best combination of chiral ligands and activators can easily be found in an efficient manner by a super-high-throughput screening to find the most enantiose-lective and activated catalyst.

It has been also reported that an asymmetric activation approach for the enantiose-lective hydrosilylation of sterically demanding benzophenones catalyzed by the com-bination of diethylzinc, a diol, and a chiral diamine activator (Table 8.5).[18] The (S)- or (R)-absolute configuration of BINOL has little effect on the enantioselectivity of the reduction product (entries 1 and 2). Furthermore, achiral diols such as 1,3-propanediol and ethylene glycol give almost the same degree of enantioselectivity. However, use of 1,3-propanediol enhances the activity of the Zn catalyst compared to the ethylene glycol-Zn complex (entries 3 and 4). Screening of a variety of chiral diamine acti-vators finds that treatment of diol-Zn complexes and p-F diamine **11d** achieves the highest level of enantioselectivity (96% ee) (entry 7).

TABLE 8.5

Entry	Diol	Diamine	Time (h)	Conv. (%)	ee (%)
1	(S)-BINOL	**11a**	48	97	74
2	(R)-BINOL	**11a**	48	45	77
3	HO⌒⌒OH	**11a**	9	>99	76
4	HO⌒⌒OH	**11a**	24	74	74
5	HO⌒⌒OH	**11b**	24	68	85
6[a]	HO⌒⌒OH	**11c**	24	98	90
7[a]	HO⌒⌒OH	**11d**	24	97	96

[a]2.5 equiv of PMHS was used.

TABLE 8.6

Entry	Act*	Yield (%)	ee (%)
1	(S,S)-12	91	20
2[a]	(S,S)-12	99	79
3	(S)-13	99	23
4[a]	(S)-13	99	76
5	(S)-14	53	4
6[a]	(S)-14	47	30
7	(S)-15	3	1
8	(S)-16	96	4
9	(1R,2S)-17	5	5

[a]0.4 mol% of enantiopure catalyst (S)-7a was used with 0.4 mol% of a chiral activator.

Noyori has reported the catalytic enantioselective hydrogenation of simple ketones by enantiomerically pure BINAPs-RuCl$_2$(dmf)$_n$ complex (see also Chapter 1 in this book) (7a: BINAP, 7b: XylBINAP,[19] 7c: TolBINAP[20]) with an enantiopure diamine such as the (S,S)- or (R,R)-1,2-diphenylethylenediamine (DPEN).[21–23] The asymmetric activation strategy can also be applied to the racemic BINAPs-RuCl$_2$(dmf)$_n$ catalysts 7 for enantioselective carbonyl reduction (Table 8.6).[24b] Use of a chiral diamine 12-17 provides nonracemic hydrogenation products. By screening various chiral diamines, (S,S)-DPEN 12 is found to be the most effective chiral activator, giving the highest enantioselectivity (91%, 80% ee) in contrast to 2,2'-diamino-1,1'-binaphthyl (DABN) 15 (3%, 1% ee).

TABLE 8.7

Entry	Catalyst	Ketone	Temp. (°C)	Time (h)	Yield (%)	ee (%)
1	(±)-**7c**	**18**	80	10	99	80 (R)
2	(R)-**7c**	**18**	80	10	99	81 (R)
3	(S)-**7c**	**18**	80	10	91	41 (R)
4	(±)-**7b**	**19**	−35	7	95	90 (R)
5[a]	(±)-**7b**	**19**	−35	7	90	90 (R)
6	(R)-**7b**	**19**	28	4	99	56 (S)
7	(S)-**7b**	**19**	28	4	99	>99 (R)

[a]Only 0.2 mol% of (S, S)-DPEN **12** was used as a chiral activator.

Hydrogenation of ketones with (±)-**7b** or (±)-**7c**, and the enantiomerically pure diamine (S,S)-DPEN **12**, have also been carried out (Table 8.7).[24b] The use of (S,S)-DPEN **12** affords the non-racemic hydrogenated products. The asymmetric activation of (±)-**7c** leads to almost the same enantioselectivity and catalytic activity as those attained by the matched (R)-**7c**/(S,S)-DPEN complex (entries 1 and 2). The mismatched (S)-**7c**/(S,S)-DPEN complex gives lower enantioselectivity to suggest the importance of chirality of the diamine activator for selective activation of one enantiomer (entry 3). It is noted that the matched pair is dramatically changed going from the 9-acetylanthracene **18** to 1′-acetonaphthone **19**. For sterically demanding **18**, the (S)-**7b**/(S,S)-DPEN complex is a more enantioselective than the (R)-**7b**/(S,S)-DPEN one. Thus, the catalytic activity critically depends on the nature of the carbonyl substrates. Significantly, a catalytic amount of DPEN affords an equally high enantioselectivity to the case in which an equimolar amount of DPEN is used (entries 4 and 5).

The ^{1}H and ^{31}P NMR spectra of a mixture of (±)-**7c** and a 0.5 molar or 1.0 molar amount of (S,S)-DPEN are identical to that of the 1 : 1 mixture; racemic BINAPs-RuCl$_2$ even with a 0.5 equiv of the enantiopure diamine, DPEN provides a 1 : 1 mixture of two diastereomeric BINAPs-RuCl$_2$/DPEN complexes. Computational modeling studies indicate that the two diastereomeric complexes have almost the same steric energies, and that the structures are in close analogy to those reported on the basis of X-ray analysis (Figure 8.3). Therefore, the matched pair with higher enantioselectivity (R/S,S cycle for **18** and S/S,S cycle for **19** case,

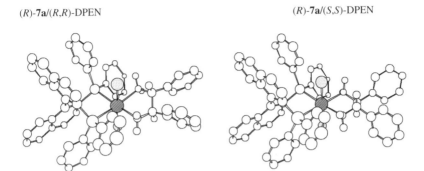

(R)-**7a**/(R,R)-DPEN   (R)-**7a**/(S,S)-DPEN

Figure 8.3

respectively) can be determined by the ratio and catalytic activity (turnover frequency) of the mono-or dihydrido-RuHX(BINAPs)/DPEN complexes (X = H or Cl)[25] generated from the diastereomeric complexes of BINAPs-RuCl₂/DPEN (Scheme 8.14).[24a]

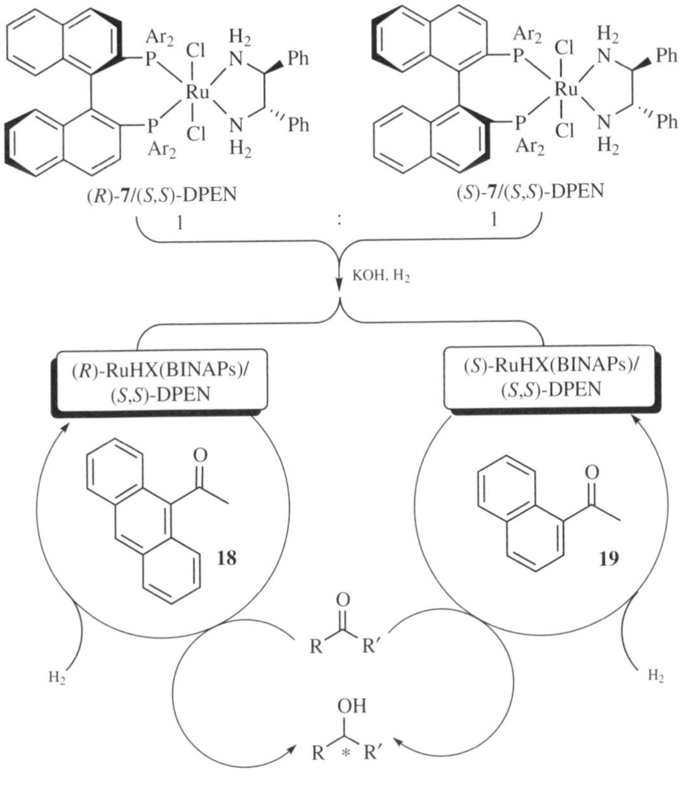

Scheme 8.14

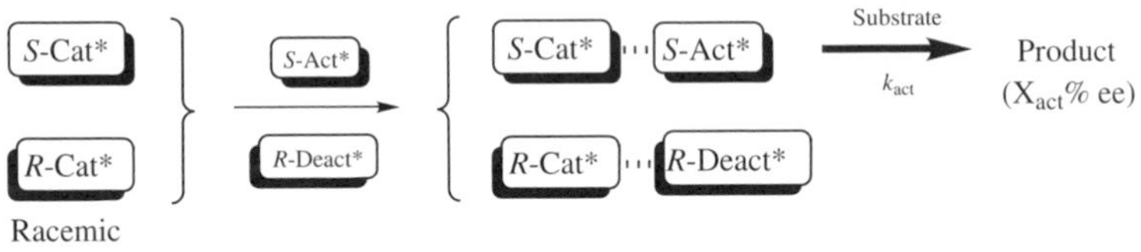

Scheme 8.15

8.2.3 Asymmetric Activation/Deactivation of Chirally Rigid (Atropos) Catalysts

Combination of the asymmetric activation and asymmetric deactivation protocols as asymmetric activation/deactivation can be achieve; the difference in catalytic activity between the two enantiomers of racemic catalysts can be maximized through selective activation and deactivation of enantiomeric catalyst, respectively (Scheme 8.15).

Since BINAPs-RuCl$_2$/DM-DABN complexes are catalytically less active than BINAPs-RuCl$_2$/DPEN complexes, ($\pm$)-XylBINAPs-RuCl$_2$(dmf)$_n$ ($\pm$)-**7b** with DM-DABN and DPEN as a chiral deactivator and activator, respectively can afford higher enantioselectivities than those attained by simple activation. By addition of first DM-DABN and then DPEN, the two enantiomers ($\pm$)-**7b** can be completely discriminated to afford an equally effective catalyst to the enantiopure **7b**, activated by DPEN (Table 8.8).[26] All ketones are readily hydrogenated at room temperature with high enantioselectivity ($>$90% ee) in quantitative yield. Thus, the asymmetric activation/deactivation strategy affords a higher level of enantioselectivity than those obtained using the ($\pm$)-**7b**/(S,S)-DPEN complexes even at the same temperature and pressure. A superiority of the asymmetric activation/deactivation concept is exemplified for 2′-acetonaphthone **20**; the enantioselectivity of (R)-1-(2-naphthyl)-ethanol using (R)-DM-DABN is increased to 91% ee from only 45% ee by simple asymmetric activation without deactivation (entries 3 and 4). 2,4,4-Trimethyl-2-cyclohexenone **23** is also hydrogenated in 92% ee by changing the chirality of DPEN from S to R (entries 13 and 14).

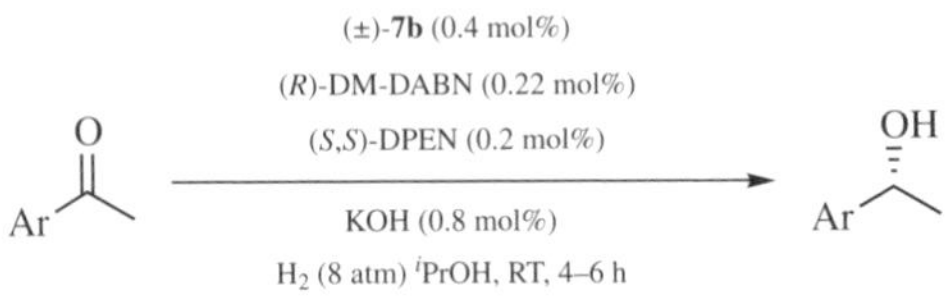

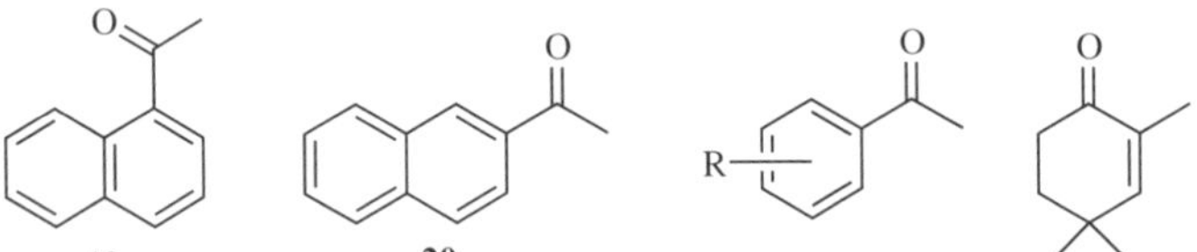

19   20   21: R = H
22a: R = o-Me
22b: R = m-Me
22c: R = p-Me
23

TABLE 8.8

Entry	Ketone	(R)-DM-DABN[a]	Yield (%)	ee (%)
1	**19**	+	>99	96
2	**19**	−	>99	80
3	**20**	+	>99	91
4	**20**	−	>99	45
5	**21**	+	>99	95
6	**21**	−	>99	70
7	**22a**	+	>99	95
8	**22a**	−	>99	82
9	**22b**	+	>99	95
10	**22b**	−	>99	60
11	**22c**	+	>99	93
12	**22c**	−	>99	60
13[b]	**23**	+	>99	92
14[b]	**23**	−	>99	84

[a]"+"denotes the presence of (R)-DM-DABN.
[b](±)-**7c** and (R,R)-DPEN instead of (±)-**7c** and (S,S)-DPEN were used.

8.2.4 Self-Assembly into the Most Enantioselective Catalyst

Sharpless et al. coined the word "ligand-accelerated catalysis" (LAC), which means the construction of an active chiral catalyst[27] from an achiral precatalyst via ligand exchange with a chiral ligand. By contrast, a combinatorial library approach in which an achiral pre-catalyst combined with several chiral ligand components (L^1*, L^2*, —) may selectively assemble in the presence of several chiral activators (A^1*, A^2*, —) into the most catalytically active and enantioselective activated catalyst ($ML^{m}*A^{n}*$) (Scheme 8.16).[28]

Two ways of self-assembly can be seen for an achiral precatalyst, $Ti(O^iPr)_4$ with chiral diol components into a single chiral titanium complex (Scheme 8.17).[29] In one case (Scheme 8.17a), a combination of $Ti(O^iPr)_4$, the acidic (R)-BINOL and a relatively basic diol such as TADDOL[30] in a molar ratio of 1 : 1 : 1 could assemble into a single (R)-BINOLato-Ti-(R,R)-TADDOLato complex **25**: In the case of matched chirality, complex **25** is obtained from either (R,R)-TADDO-Lato-Ti(O^iPr)_2 **24** and (R)-BINOL, or from (R)-BINOLato-Ti(O^iPr)_2 **9** and (R,R)-TADDOL. In the other case (Scheme 8.17b), on addition of (R)-BINOL and a more acidic diol such as (R)-5-Cl-BIPOL to $Ti(O^iPr)_4$, (R)-BINOLato-Ti(O^iPr)_2 **9**/(R)-5-Cl-BIPOL complex **27** is observed. This complex could also be obtained using BINOLato-Ti(O^iPr)_2 **9** [31] and 5-Cl-BIPOL[13a], or alternatively from 5-Cl-BIPO-Lato-Ti(O^iPr)_2 **26** and BINOL.

The role of multicomponent ligand assembly into a highly enantioselective catalyst is shown in the enantioselective catalysis for the carbonyl-ene reaction (Table 8.9). The catalyst is prepared from an achiral precatalyst, $Ti(O^iPr)_4$ and a combination of BINOL with various chiral diols such as TADDOL and 5-Cl-BIPOL in a molar ratio of 1 : 1 : 1 (10 mol% with respect to the olefin and glyoxylate) in

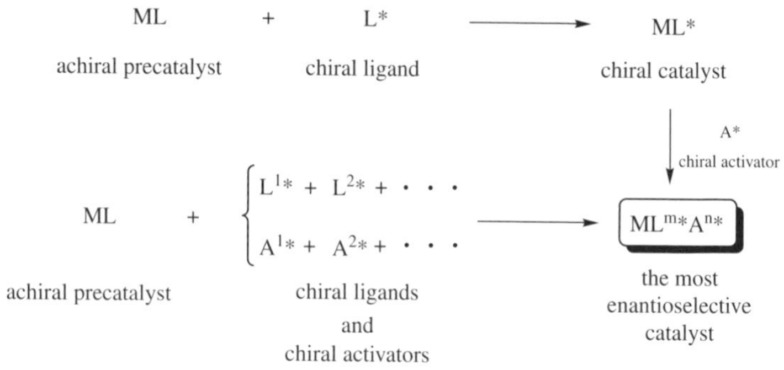

ML   +   L* $\longrightarrow$ ML*

achiral precatalyst   chiral ligand   chiral catalyst

$\downarrow$ A*

chiral activator

ML   + $\begin{cases} L^{1*} + L^{2*} + \cdots \\ A^{1*} + A^{2*} + \cdots \end{cases}$ $\longrightarrow$ $\boxed{ML^{m*}A^{n*}}$

achiral precatalyst   chiral ligands   the most
and   enantioselective
chiral activators   catalyst

Scheme 8.16

toluene. A significant leap in chemical yield from 0% to 50% is obtained with the high enantioselectivity (91% ee, R) in a combination of (R,R)-TADDOL and (R)-BINOL (enties 1 vs. 2). In a combination of (R)-5-Cl-BIPOL and (R)-BINOL, the reaction proceeds quite smoothly with the highest chemical yield and enantioselectivity (entry 3). This is in stark contrast to the lower enantioselectivity and

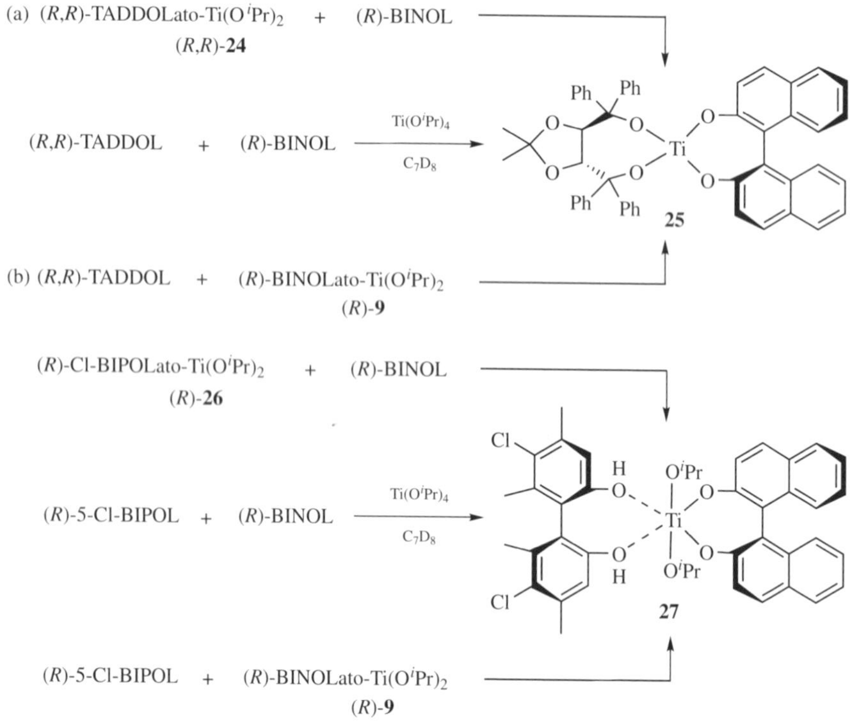

(a) (R,R)-TADDOLato-Ti(O^{i}Pr)$_2$ + (R)-BINOL
       (R,R)-**24**

(R,R)-TADDOL   + (R)-BINOL $\xrightarrow[C_7D_8]{Ti(O^iPr)_4}$ **25**

(b) (R,R)-TADDOL   + (R)-BINOLato-Ti(O^{i}Pr)$_2$
       (R)-**9**

(R)-Cl-BIPOLato-Ti(O^{i}Pr)$_2$ + (R)-BINOL
       (R)-**26**

(R)-5-Cl-BIPOL   + (R)-BINOL $\xrightarrow[C_7D_8]{Ti(O^iPr)_4}$ **27**

(R)-5-Cl-BIPOL   + (R)-BINOLato-Ti(O^{i}Pr)$_2$
       (R)-**9**

Scheme 8.17

chemical yields using the (R)-5-Cl-BIPOLato-Ti catalyst **26** or (R)-BINOLato-Ti(O^{i}Pr)$_2$ catalyst **9** (entries 4 and 5).

TABLE 8.9

Entry	R^1*(OH)$_2$	R^1*(OH)$_2$	Yield (%)	ee (%)
1	(R, R)-TADDOL	(R)-BINOL	50	91
2	(R, R)-TADDOL	—	0	—
3	(R)-5-Cl-BIPOL	(R)-BINOL	66	97
4	(R)-5-Cl-BIPOL	—	13	75
5	(R)-BINOL	—	20	95

The asymmetric hetero-Diels–Alder reaction of aldehydes with Danishefsky's diene catalyzed by Ti catalysts generated from a library of 13 chiral ligands or activators has also been reported (Scheme 8.18).[32] The catalyst library contains 104 members. The Ti catalysts bearing $\mathbf{L^4}$, $\mathbf{L^5}$, $\mathbf{L^6}$, and $\mathbf{L^7}$ are found to have a remarkable effect on both enantioselectivity (76.7–95.7% ee) and yield (63–100%). On the other hand, ligands bearing sterically demanding substituents at the 3,3′-positions are found to be detrimental to the reaction. The optimized catalysts, both $\mathbf{L^5}$/Ti/$\mathbf{L^5}$ and $\mathbf{L^5}$/Ti/$\mathbf{L^6}$, are the most efficient for the reaction of a variety of aldehydes, including aromatic, olefinic, and aliphatic derivatives.

Assembled polymeric chiral Ti catalysts using linked bis-BINOL also show high enantioselectivity for carbonyl-ene reactions under heterogeneous conditions (Schemes 8.18 and 8.19).[33] Combination of the linked bis-BINOLs **28** and Ti(O^{i}Pr)$_4$ provides the assembled catalysts **29** for the reaction (Scheme 8.19). Using the heterogeneous catalyst **29a**, the carbonyl-ene reaction proceeds smoothly at room temperature to give 94.4% ee of the (S)-α-hydroxy ester in 91% yield. On the other hand, catalyst **29b** bearing a meta-phenylene linker decreases the catalytic activity and enantioselectivity under the same conditions. These results indicate that the position of linker dramatically alters the supramolecular structure of the catalyst. The catalyst **29c** (Scheme 8.20) having simple dimer of BINOL enhances both the yield and enantioselectivity (96% ee, >99%).

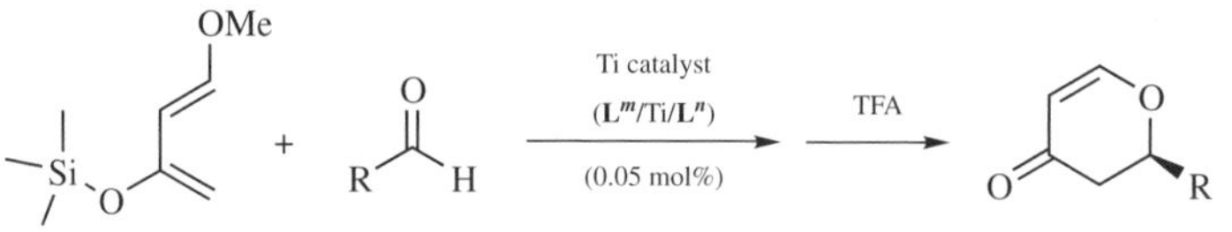

aldehyde	$L^5/Ti/L^5$	$L^5/Ti/L^6$
benzaldehyde	99.3% ee (>99%)	99.4% ee (82%)
p-nitrobenzaldehyde	97.3% ee (>99%)	99.4% ee (>99%)
furfural	99.2% ee (>99%)	99.7% ee (>99%)

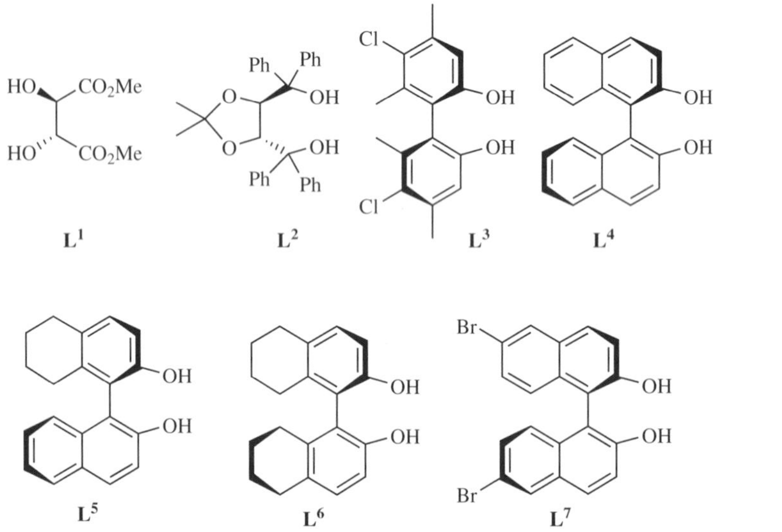

L^1 L^2 L^3 L^4

L^5 L^6 L^7

Scheme 8.18

28   **29**

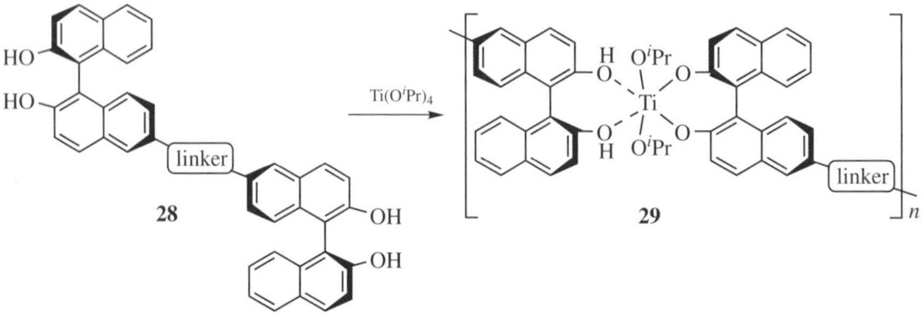

Scheme 8.19

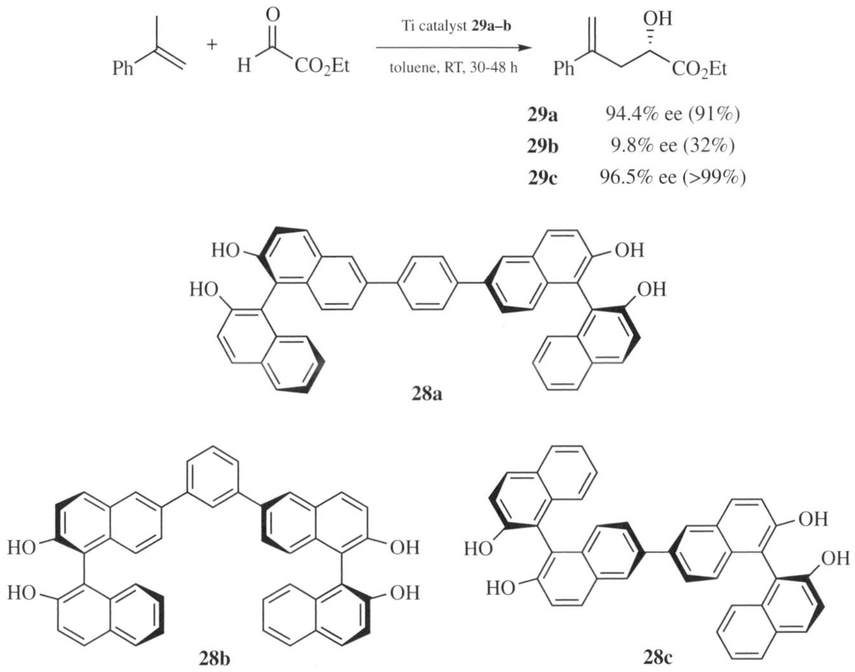

29a   94.4% ee (91%)
29b   9.8% ee (32%)
29c   96.5% ee (>99%)

28a

28b   28c

Scheme 8.20

8.2.5 Asymmetric Activation of Chirally Flexible (Tropos) Catalysts

All enantiopure atropisomeric ligands essentially require the enantioresolution or synthetic transformation from a chiral pool. Since the word *atropos* consists of *a*, meaning "not," and *tropos*, meaning "turn" in Greek, the chirally flexible dynamic behavior of a ligand with chirality (i.e., axis, planar, center, helicity), can be called *tropos*.[3] The *tropos* ligands can be used as enantiopure ligands without their asymmetric synthesis or resolution . Thus, a further advanced strategy for "asymmetric activation" can be highlighted in the combination of tropos and racemic ligands.[3b] As shown in Section 8.2.2, employing a chirally rigid (atropos) ligand with a chiral activator usually produces a diastereomeric mixture of a resulting complex, and hence the ee of the product much depends on the difference of the reactivity between the diastereomers. However, combination of a tropos ligand, such as a biphenyl ligand with axial chirality and a chiral activator can produce a certain single diastereomeric complex. By controlling the chirality through diastereomer interconversion (e.g., tropo-inversion of the chiral axis), the catalytic activity and the enantioselectivity of the resulting catalyst can be increased for effective asymmetric catalysis (Scheme 8.21).

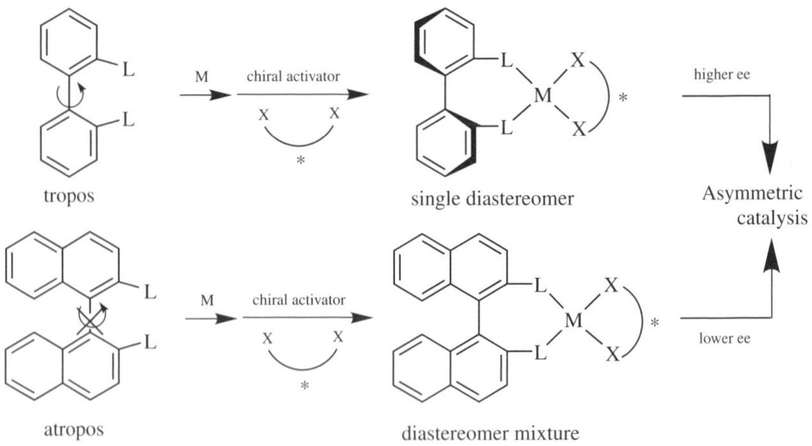

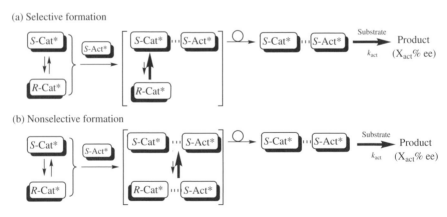

Scheme 8.21

The asymmetric activation can be done by a chiral activator through in situ diastereomer interconversion of the tropos ligand of racemic catalysts (Scheme 8.22). One possible case is that selective complexation of a chiral activator with one enantiomer of a racemic catalyst occurs. The remaining enantiomeric catalyst may then isomerize and complex with the chiral activator leading to a single diastereomer (Scheme 8.22a). The other case is that nonselective complexation of a tropos catalyst with a chiral activator initially provides a 1:1 ratio of activated diastereomers, which would isomerize to the single diastereomeric activated complex (Scheme 8.22b).

The dynamic asymmetric activation can be rationalized by a continuum from the nonselective to selective complexation with a chiral activator. Figure 8.4 shows the difference in the relative rate ($k_{rel} = k_{act}/k'_{act}$ ranges from 0.01 to 100) in the ratio of the diastereomeric catalysts (ranges from a 1:1 to 99:1);

(a) Selective formation

(b) Nonselective formation

Scheme 8.22

one activated diastereomer $((R)\text{-}ML_n/(R)\text{-}Act^*)$ leads to the product in 100% ee (R), while the other diastereomer $[(S)\text{-}ML_n/(R)\text{-}Act^*]$ provides the opposite enantiomeric product in 50% ee (S). A relative rate of 100 (log $k_{rel} = 2$) is big enough to achieve a high enantioselectivity (e.g., 99% ee), even when the diastereomeric catalysts are formed in a 1 : 1 ratio. In the case of a 20 : 1 ratio, a product with more than 99% ee can be obtained with more activated catalyst (log $k_{rel} = 1$). Even in the case that the relative rate of the diastereomeric catalysts is the same (log $k_{rel} = 0$), a high level of enantioselectivity (e.g., 99% ee) can be obtained, when the diastereomeric catalysts are formed in a 99 : 1 ratio.

Davies, Renaud, and Sibi independently reported the "chiral relay"[34] approach to control the enhanced steric extension inside a substrate to achieve increased asymmetric induction. However, as our study proves, the asymmetric activation of a tropos catalyst clearly differs from the chiral relay approach, in which substrate conformational control is utilized, since asymmetric activation controls the chiral environment of a tropos catalyst by the addition of a chiral external source (a chiral activator).

Atropisomerism was coined to describe isomerism caused by stopping the internal turn around a single bond.[3a] In view of the prime importance of the atropisomerism of binaphthol (BINOL), the tropos biphenol (BIPOL) could replace the enantiopure form of atropos BINOL obtained via the enantioresolution. The chiral Ti catalysts prepared by self-assembly of TADDOL-BINOL exhibit higher catalytic activity and enantioselectivity through asymmetric activation (Table 8.9 and Scheme 8.23a). It is expected that the enantiopure TADDOL combined with the BIPOLato-Ti(O^{i}Pr)$_2$ catalyst **30** would control the chirality of the *tropos* BIPOL moiety and increase the catalytic activity of **30** (Scheme 8.23b). Molecular mechanics (MM2) calculations of the two diastereomeric complexes 3,3'-(MeO)$_2$-BIPOLato-Ti-TADDOLato show that the $(R)/(R,R)$-diastereomer is more stable than the $(S)/(R,R)$-diastereomer by 3.60 kcal/mol (Scheme 8.24). The steric effect of the 3,3'-methoxy group is significant in maximizing the difference of relative thermodynamic stability between the diastereomeric complexes obtained from the parent BIPOLato diastereomers.

The BIPOLato-Ti-TADDOLato catalysts prepared by addition of BIPOL and TADDOL to Ti(O^{i}Pr)$_4$, catalyze methylation with an achiral methyltitanium reagent to give highly enantiomerically pure methylcarbinol. Since the sterically bulky 3,3'-substituents leads to an increase in enantioselectivity, the chirality of BIPOLato-Ti(O^{i}Pr)$_2$ catalyst **30** can be dynamically controlled by the chiral TADDOL moiety (Scheme 8.25).[35] 3,3'-Dimethoxy derivative affords complete enantioselectivity (100% ee), while the moderate enantioselectivity is obtained with the parent BIPOL (73% ee).

The atropos BINAP can also be replaced by the tropos bis(phosphanyl)biphenyl (BIPHEP).[36, 37] The tropos metal catalysts with the BIPHEP ligands can be used as activated diastereomeric complexes with the chiral diamines. On addition of (S,S)-DPEN to racemic BIPHEPs-RuCl$_2$(dmf)$_n$ (**31a**: BIPHEP, **31b**: XylBIPHEP), diastereomeric complexes are formed in equal amounts. However, the mixture of (S)- and (R)-**31b**/(S,S)-DPEN in (CD$_3$)$_2$CDOD/CDCl$_3$ (2/1) is found to give a 3 : 1 mixture

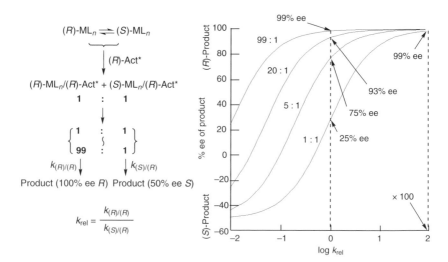

Figure 8.4

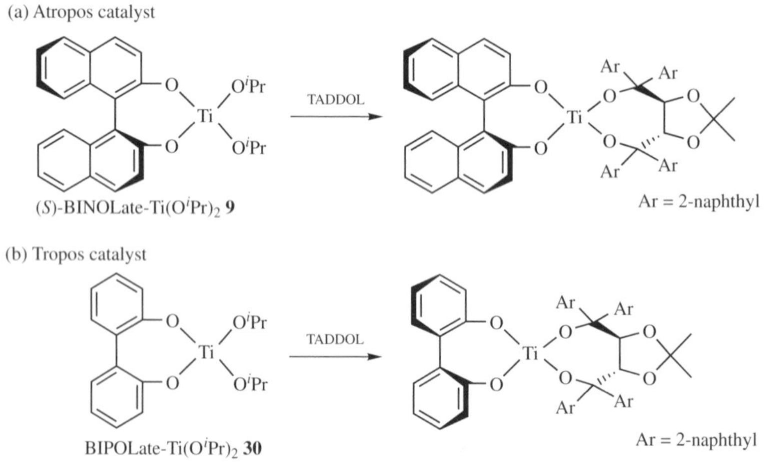

(a) Atropos catalyst

(S)-BINOLate-Ti(O^{i}Pr)$_2$ **9** $\xrightarrow{\text{TADDOL}}$ Ar = 2-naphthyl

(b) Tropos catalyst

BIPOLate-Ti(O^{i}Pr)$_2$ **30** $\xrightarrow{\text{TADDOL}}$ Ar = 2-naphthyl

Scheme 8.23

with the (S)-**31b**/(S,S)-DPEN major diastereomer on standing at room temperature (Scheme 8.26). This implies the equilibrium due to the tropos nature of **31** complexes.

The effect of the tropos complex **31b** with (S,S)-DPEN can be observed in the hydrogenation of 1'-acetonaphthone **16** as compared with the results using (±)-**10b**/(S,S)-DPEN bearing racemic and atropos XylBINAP (Table 8.10).[37]

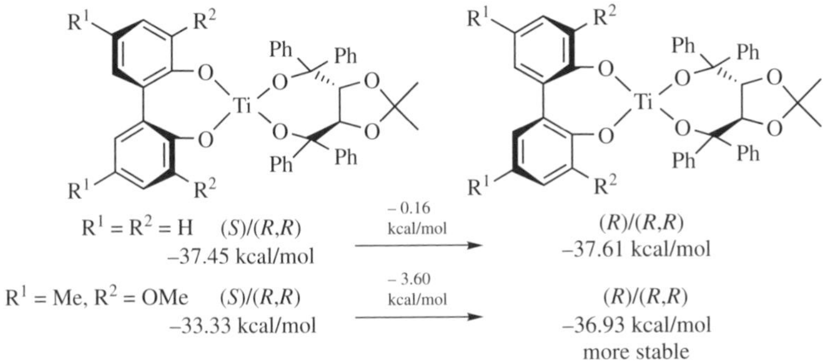

R^1 = R^2 = H (S)/(R,R)
−37.45 kcal/mol

− 0.16 kcal/mol →

(R)/(R,R)
−37.61 kcal/mol

R^1 = Me, R^2 = OMe (S)/(R,R)
−33.33 kcal/mol

− 3.60 kcal/mol →

(R)/(R,R)
−36.93 kcal/mol
more stable

Scheme 8.24

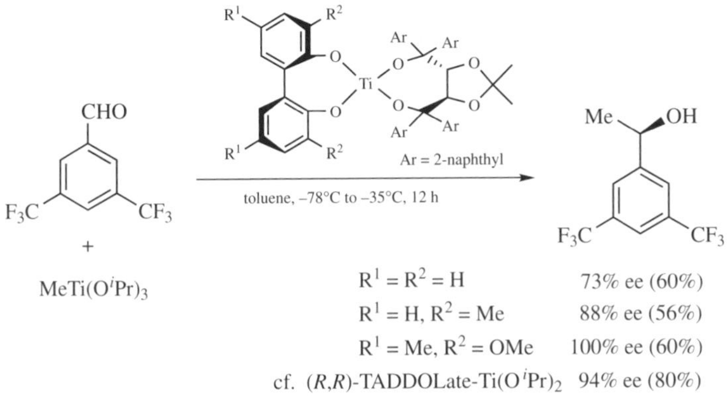

Ar = 2-naphthyl

toluene, −78°C to −35°C, 12 h

MeTi(O^{i}Pr)$_3$

R^1 = R^2 = H	73% ee (60%)
R^1 = H, R^2 = Me	88% ee (56%)
R^1 = Me, R^2 = OMe	100% ee (60%)
cf. (R,R)-TADDOLate-Ti(O^{i}Pr)$_2$	94% ee (80%)

Scheme 8.25

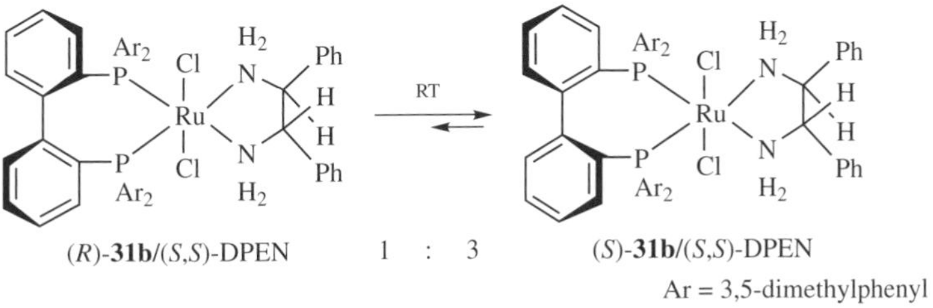

(R)-**31b**/(S,S)-DPEN   1 : 3   (S)-**31b**/(S,S)-DPEN

Ar = 3,5-dimethylphenyl

Scheme 8.26

19   22a

TABLE 8.10

Entry	Catalyst	Ketones	H_2 (atm)	Temp. (°C)	Yield (%)	ee (%)
1[a]	**31b**	**19**	8	28	>99	84
2	(±)-**10b**	**19**	8	28	>99	80
3[a]	**31b**	**19**	40	−35	>99	92
4	(±)-**10b**	**19**	40	−35	>99	89
5[a]	**31b**	**22a**	8	0	>99	88
6	(±)-**10b**	**22a**	8	0	>99	86

[a]**31b**/(S,S)-DPEN in 2-propanol was preheated at 80°C for 30 min.

A higher enantioselectivity is obtained at a lower temperature. The enantioselectivity by **31b**/(S,S)-DPEN is higher than that by (±)-**10b**/(S,S)-DPEN at the same temperature and pressure. The tropos **31b**/(S,S)-DPEN catalyst affords (R)-1-(1-naphthyl) ethanol with 92% ee in quantitative yield (entries 3 and 4). **31b**/(S,S)-DPEN is also effective for o-methylacetophenone **22a** to afford (R)-1-(1-o-methylphenyl)ethanol quantitatively with 88% ee, higher than 86% ee obtained by (±)-**10b**/(S,S)-DPEN (entries 5 and 6).

N,N-Dimethyl-1,2-diphenylethylenediamine (DM-DPEN) is thus designed as a chiral activator with a tertiary amine functionality to discriminate between the enantiomeric BIPHEPs-RuCl₂(dmf)ₙ **31**. Computational modeling studies indicate a severe steric repulsion between the N-methyl of (S,S)-DM-DPEN and the P-phenyl groups in the disfavored diastereomer (Figure 8.5). In spite of nonselective complexation of **31b** with (S,S)-DM-DPEN, (R)-**31b**/(S,S)-DM-DPEN completely isomerizes to (S)-**31b**/(S,S)-DM-DPEN in 2-propanol at 50°C after 1 h (¹H NMR in CDCl₃ at room temperature); the single diastereomer (S)-**31b**/(S,S)-DM-DPEN is quite similar to (S)-**10**/(S,S)-DM-DPEN. Indeed, the (S)-configuration of BIPHEP is confirmed by X-ray analysis of XylBIPHEP- and BIPHEP-RuCl₂ with

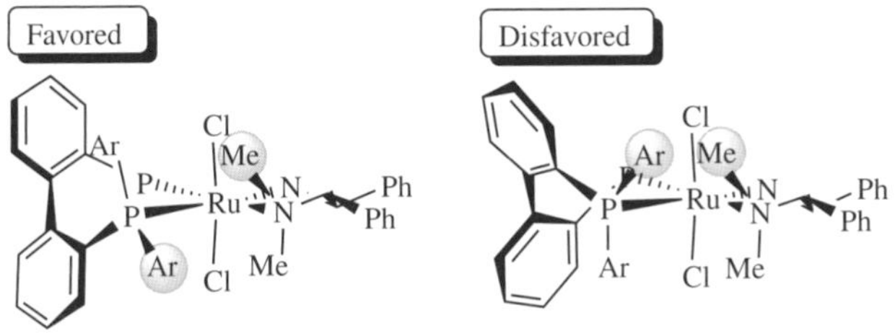

(S)-31/(S,S)-DM-DPEN   (R)-31/(S,S)-DM-DPEN

Figure 8.5

(S,S)-DM-DPEN.[38] In hydrogenation of **19**, the enantioselectivity and chemical yield (94%, 35% ee at room temperature) with the diastereomerically pure (S)-**31b**/(S,S)-DM-DPEN is lower than that (>99%, 84% ee) with the 3 : 1 diastereomeric mixture of **31b**/(S,S)-DPEN.

While the complexation of (±)-BIPHEP-Pd **32** with (R,R)-DPEN is in a nonselective manner even with 0.5 equiv. of (R,R)-DPEN, a highly selective (9 : 1) complexation of one enantiomer of (±)-**32** is observed by addition of 0.5 equiv of (R)-DABN; (R)-**32**/(R)-DABN is formed as the major diastereomer. With 1.0 equiv of (R)-DABN, however, a 1 : 1 ratio of a diastereomeric mixture of (R)-**32**/(R)-DABN and (S)-**32**/(R)-DABN is observed (Scheme 8.27). The diastereomer mixture of **32**/(R)-DABN does not isomerize at room temparature over 3 days. But tropo-inversion of the BIPHEP moiety at 80°C after 8 h leads to the favorable (R)-**32**/(R)-DABN exclusively (Scheme 8.27).[39] This shows that the BIPHEP moiety in **32**/DABN complex isomerizes at 80°C but not at room temperature.

On the other hand, a highly effective resolving agent such as DM-DABN, which would not lead to a mixture of diastereomers, can be used to clarify as to whether the remaining enantiomer of **32** isomerizes at room temperature or not (Scheme 8.28).[40] 1.0 equiv of (R)-DM-DABN leads to the single (R)-**32**/(R)-DM-DABN diastereomer along with the remaining (S)-**32** and (R)-DM-DABN. The remaining (S)-**32** does not isomerize at room temperature even after 3 days. After 2 h at 80°C, the complete isomerization of (S)-**32** to the single diastereomer (R)-**32**/(R)-DM-DABN is observed without any remaining (S)-**32** or (R)-DM-DABN. It is thus clarified that **32** could be resolved as an *atropos* metal complex at or below room temperature.

The metal complexes even with tropos ligands can thus be used as asymmetric catalysts for carbon–carbon bond-forming reactions in the same manner as atropos catalysts. The single diastereomer (R)-**32**/(R)-DABN can be employed as an activated asymmetric catalyst for the Diels–Alder reaction at room temperature (Table 8.11).[32] The high chemical yield and enantioselectivity (62%, 94% ee) in the Diels–Alder reaction of ethyl glyoxylate with 1,3-cyclohexadiene are obtained

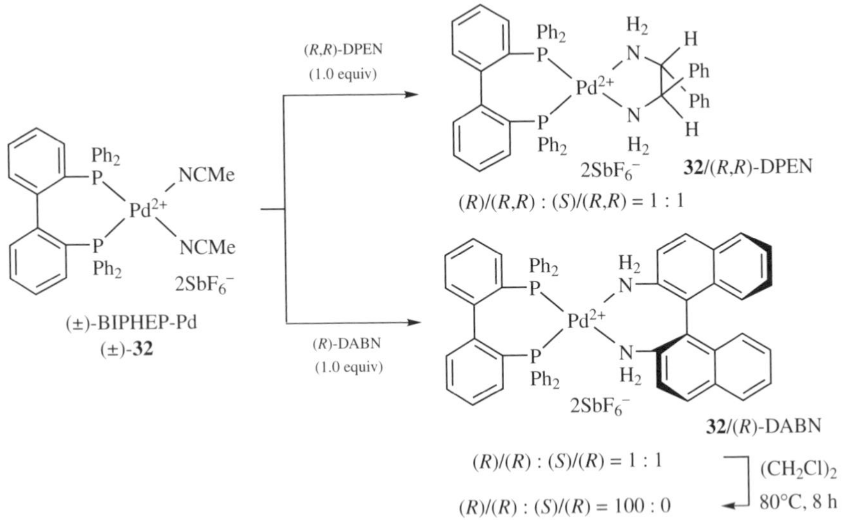

Scheme 8.27

by 0.5 mol% of the (R)-**32**/(R)-DABN catalyst (entry 2). (R)-DABN as a chiral activator is highlighted by the higher enantioselectivity and catalytic activity (94% ee, 62%) as compared with those (75% ee, 11%) using enantiopure (R)-**32** without (R)-DABN (entry 1). Higher chemical yield and enantioselectivity (75%, 92% ee) are obtained by 2.0 mol% of the (R)-**32**/(R)-DABN catalyst than those

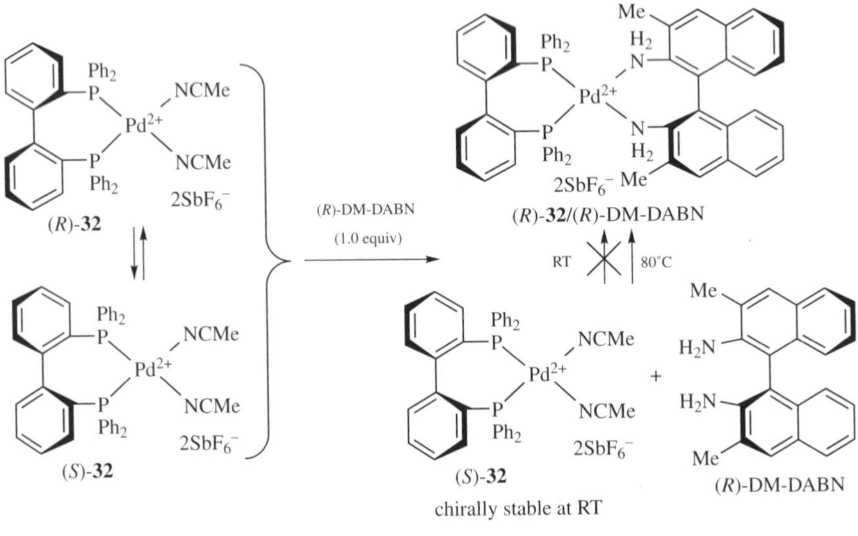

Scheme 8.28

(±)-BIPHEP: (±)-**32**/(R)-DABN
(±)-BINAP: (±)-**33**/(R)-DABN

attained by the atropos and racemic BINAP counterpart with DABN activator (61%, 7% ee) (entries 3–5).

In a similar manner, the enantiopure BIPHEP-Pt complex-catalyzed enantioselective Diels–Alder and carbonyl-ene reactions are also reported by Gagné (Scheme 8.29).[41] Racemic BIPHEP-Pt(CO₃) **34** with (S)-BINOL gives a 1:1 diastereomer mixture **34**/(S)-BINOL complexes. The (S)/(S) isomer can be seen in a 96:4 ratio at 92–122°C. The (R)/(S) isomer can be converted to a 95:5 ratio of the (S)/(S) major isomer in pyridine at lower temperature (40°C). On the other hand, the thermodynamically less favorable (R)/(S) isomer can be by recrystallization from CH₂Cl₂.

The treatment of (S)-**34**/(S)-BINOL complex (95:5) with concentrated HCl followed by recrystallization leads to the enantiopure (S)-**35** complex with retention of

TABLE 8.11

Entry	Catalyst	mol (%)	Yield (%)	ee (%)
1	(R)-**32**	0.5	11	75 (1R, 3S, 4S)
2	(R)-**32**/(R)-DABN	0.5	62	94 (1R, 3S, 4S)
3	(R)-**32**/(R)-DABN	2.0	75	92 (1R, 3S, 4S)
4	(±)-**33**/(R)-DABN	2.0	61	7 (1S, 3R, 4R)
5	(±)-**32**/(R)-DABN	2.0	64	9 (1S, 3R, 4R)

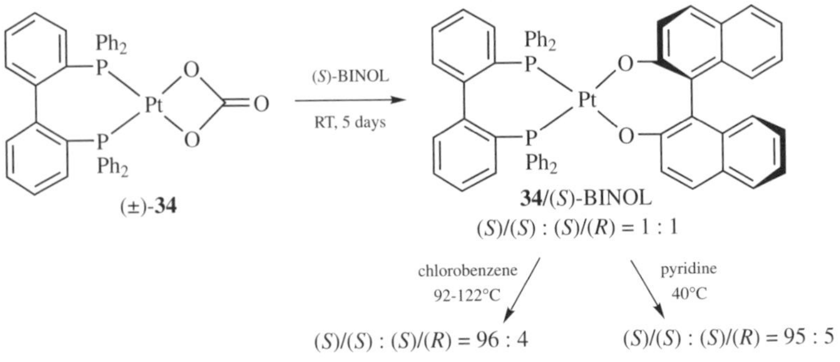

Scheme 8.29

configuration in 63% yield (Scheme 8.30a). Similarly, use of concentrated HCl or TfOH for diastereopure (*R*)-**34**/(*S*)-BINOL complex leads to the enantiopure (*R*)-**35** or (*R*)-**36** complexes, respectively (Scheme 8.30b).

The enantiopure BIPHEP-Pt complexes can act as chiral Lewis acids for the enantioselective Diels–Alder and carbonyl-ene reactions. The Diels–Alder products are obtained in 92–94% ee (93 : 7 = endo : exo) and 92–94% ee (94 : 6 = endo : exo) by (*R*)- and (*S*)-**36**, respectively (Scheme 8.31a).[41b] In the carbonyl-ene reaction catalyzed by the dication species generated from BIPHEP-PtCl$_2$ **35** and AgSbF$_6$, the (*S*)- and (*R*)-ene products are obtained with 71% ee (99% conversion at room temperature) and 70% ee (90% conversion at room temperature) from (*R*)- and (*S*)-**35**, respectively (Scheme 8.31b).

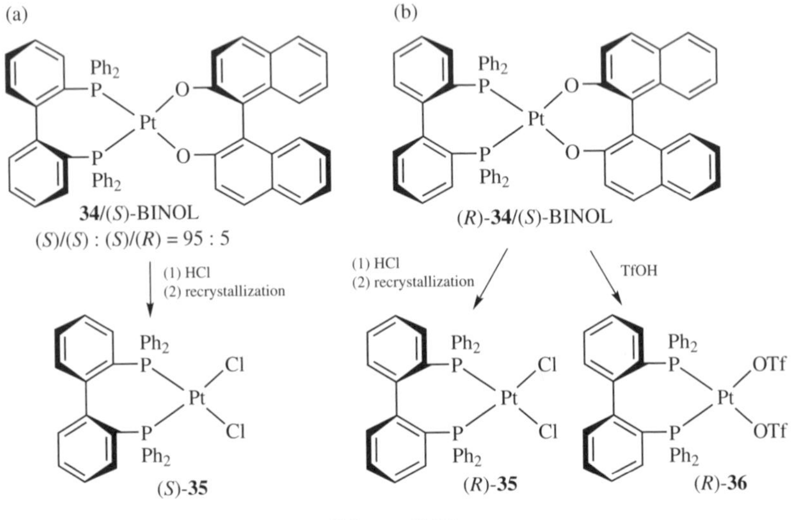

Scheme 8.30

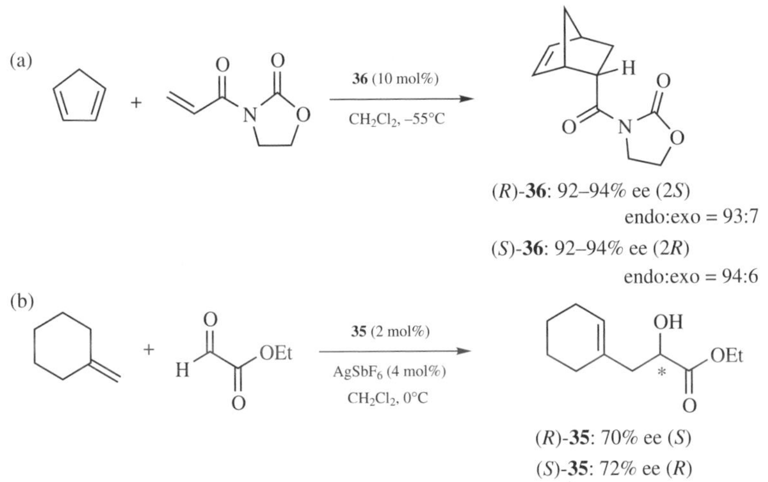

(a)

(R)-36: 92–94% ee (2S)
endo:exo = 93:7

(S)-36: 92–94% ee (2R)
endo:exo = 94:6

(b)

(R)-35: 70% ee (S)
(S)-35: 72% ee (R)

Scheme 8.31

In the neutral BIPHEP-Pt complex, the axial chirality of BIPHEP moiety is controlled by chiral diol BINOL as shown in Scheme 8.29. However, the diastereomeric purity is not high enough (95 : 5). Therefore, recrystallization is essential to obtain the single BIPHEP-Pt diastereomer and subsequent enantiomer. It has thus been required that complete chirality control of both neutral and cationic BIPHEP-Pt complexes without recrystallization and its application to asymmetric Lewis acid catalysis (Scheme 8.32).[42] Interestingly, both enantiopure (S)- and (R)-BIPHEP-Pt complexes can be obtained quantitatively through the

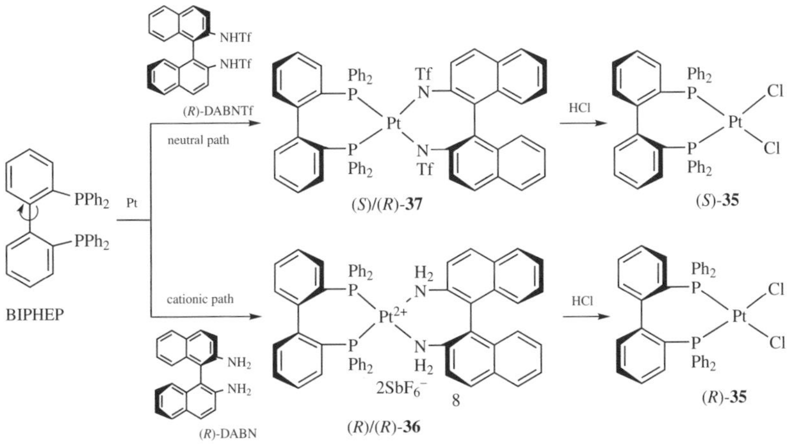

Scheme 8.32

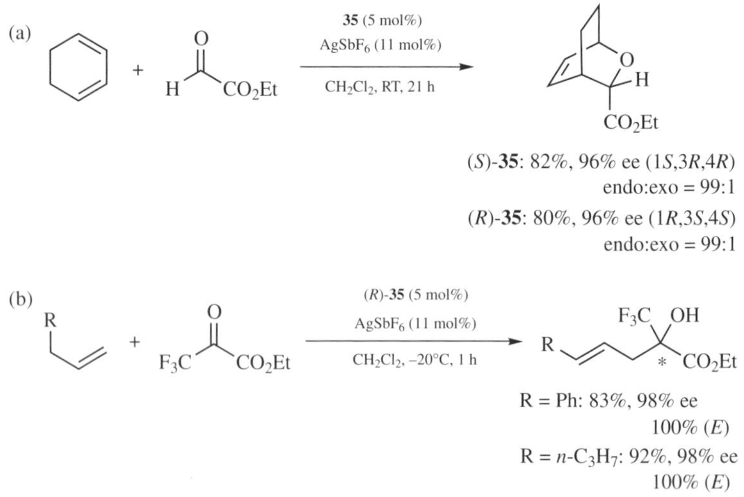
(a)
35 (5 mol%)
AgSbF6 (11 mol%)
CH2Cl2, RT, 21 h
CO2Et
(S)-35: 82%, 96% ee (1S,3R,4R)
endo:exo = 99:1
(R)-35: 80%, 96% ee (1R,3S,4S)
endo:exo = 99:1
(b)
(R)-35 (5 mol%)
AgSbF6 (11 mol%)
CH2Cl2, −20°C, 1 h
R = Ph: 83%, 98% ee
100% (E)
R = n-C3H7: 92%, 98% ee
100% (E)

Scheme 8.33

enantiodiscrimination by (*R*)-DABNTf and (*R*)- DABN with the same absolute configuration. In the neutral path through selective complexation, the (*S*)-**35** complex is obtained without recrystallization by complete control of the axial chirality followed by use of HCl. In the cationic path through nonselective complexation, the (*R*)-**35** complex is, in turn, obtained without recrystallization by complete control of the axial chirality.

The enantiopure complex **35** has since been employed as an atropos asymmetric catalyst for a variety of synthetic transformations (Scheme 8.33).[42] In addition, the hetero Diels–Alder reaction of glyoxylate could also be catalyzed by enantiopure (*S*)-**35** (5 mol%) and $AgSbF_6$ (11 mol%) as a highly efficient Lewis acid catalyst. The HDA product is obtained with high enantioselectivity (96% ee, endo : exo = 99 : 1) even at room temperature (Scheme 8.33a). Use of enantiopure (*R*)-**35** gives the HDA product with opposite absolute configuration in equally high selectivity. Additionally, the enantiopure dicationic complex obtained by (*R*)-**35** and $AgSbF_6$ gives high chemical yields and high levels of enantio- and (*E*)-selectivity in the carbonyl-ene reaction with trifluoropyruvate for less reactive mono-substituted olefins (Scheme 8.33b).

8.3 FUTURE PERSPECTIVES

The racemic catalyst systems which allow precise chiral recognition among enantiotopic atoms, groups, or faces in prochiral substrates must be created by virtue of chiral activators. "Asymmetric activation" is a novel concept that can lead to a generally effective strategy for asymmetric catalysis involving chirally economical racemic ligands without enantioresolution. Therefore, the preparation and optimization of many different catalysts can be easily achieved in situ from racemic atropos or tropos ligands. Not only the catalytic activity but also the enantioselectivity of the racemic catalyst can be increased through interaction between chiral activators and racemic ligands. Additionally, enantiopure catalysts bearing tropos ligands can be obtained through chiral recognition and control by chiral activators. The catalysts thus obtained work just like enantiopure atropos catalysts. However, so far, the stoichiometric chiral activators are needed in order to control the chirality of the tropos catalyst. Endeavors in which efficient chirality control can be achieved using enantiopure catalysts prepared in situ from their corresponding racemic catalysts and a catalytic amount of chiral activator is expected to represent a major area of future research.

REFERENCES AND NOTES

1. (a) Gawley, R. E.; Aube, J. *Principles of Asymmetric Synthesis*, Pergamon, London, **1996**; (b) Noyori, R. *Asymmetric Catalysis in Organic Synthesis*, Wiley, New York, **1994**; (c) Brunner, H.; Zettlmeier, W. *Handbook of Enantioselective Catalysis*, VCH, Weinheim, **1993**; (d) *Catalytic Asymmetric Synthesis*, 2nd ed., Ojima, I. (Ed.), VCH, New York, **2000**.

2. (a) Mikami, K. *Green Reaction Media in Organic Synthesis*, Blackwell, **2005**; (b) Anastas, P. T.; Warner, J. C. *Green Chemistry: Theory and Practice*, Oxford University Press, **1998**.

3. (a) Kuhn, R. Moleculare asymmetrie, in *Stereochemie*, Freudenberg, H. (Ed.), Franz, Deutike, Leipzig-Wien, **1933**, pp. 803–824; (b) Mikami, K.; Aikawa, K.; Yusa, Y.; Jodry, J. J.; Yamanaka, M. *Synlett* **2002**, 1561–1578.

4. (a) Mikami, K.; Terada, M.; Korenaga, T.; Matsumoto, Y.; Ueki, M.; Angelaud, R. *Angew. Chem. Int. Ed.* **2000**, *39*, 3532–3556; (b) Faller, J. W.; Lavoie, A. R.; Parr, J. *Chem. Rev.* **2003**, *103*, 3345–3367; (c) Mikami, K; Yamanaka, M. *Chem. Rev.* **2003**, *103*, 3369–3400.

5. (a) Alcock, N. W.; Brown, J. M.; Maddox, P. J. *J. Chem. Soc. Chem. Commun.* **1986**, 1532–1534; (b) Brown, J. M.; Maddox, P. J. *Chirality* **1991**, *3*, 345–354.

6. (a) Maruoka, K.; Yamamoto, H. *J. Am. Chem. Soc.* **1989**, *111*, 789–790; (b) Maruoka, K; Itoh, T; Shirasaka, T.; Yamamoto, H. *J. Am. Chem. Soc.* **1988**, *110*, 310–312.

7. (a) Faller, J. W.; Parr, J. *J. Am. Chem. Soc.* **1993**, *115*, 804–805; (b) Faller, J. W.; Mazzieri, M. R.; Nguyen, J. T.; Parr, J.; Tokunaga, M. *Pure Appl. Chem.* **1994**, *66*, 1463–1469.

8. Faller, J. W.; Tokunaga, M. *Tetrahedron Lett.* **1993**, *34*, 7359–7362.

9. Faller, J. W.; Sams, D. W. I.; Liu, X. *J. Am. Chem. Soc.* **1996**, *118*, 1217–1218.

10. Faller, J. W.; Liu, X. *Tetrahedron Lett.* **1996**, *37*, 3449–3452.

11. Faller, J. W.; Lavoie, A. R.; Grimmond, B. J. *Organometallics* **2002**, *21*, 1662–1666.

12. Mikami, K; Yusa, Y.; Korenaga, T. *Org. Lett.* **2002**, *4*, 1643–1645.

13. (a) Mikami, K.; Matsukawa, S. *Nature* **1997**, *385*, 613–615; (b) Mikami, K.; Terada M. *Comprehensive Asymmetric Catalysis*, Jacobsen, E. N.; Pfaltz, A.; Yamamoto, H. (Eds.), Springer, Heidelberg, **1999**; p. 1143.

14. Matsukawa, S.; Mikami, K. *Enantiomer* **1996**, *1*, 69–73.

15. Matsukawa, S.; Mikami, K. *Tetrahedron Asym.* **1997**, *8*, 815–816.

16. Volk, T.; Korenaga, T.; Matsukawa, S.; Terada, M.; Mikami, K. *Chirality* **1998**, *10*, 717–721.

17. (a) Ding, K.; Ishii, A.; Mikami, K. *Angew. Chem. Int. Ed.* **1999**, *38*, 497–501; (b) Mikami, K.; Ding, K.; Ishii, A.; Tanaka, A.; Sawada, N.; Kudo, K. *Chromatography* **1999**, *20*,

65–69; (c) Angelaud, L.; Matsumoto, Y.; Korenaga, T.; Kudo, K.; Senda, M.; Mikami, K. *Chirality* **2000**, *12*, 544–547; (d) Mikami, K.; Angelaud, L.; Ding, K.; Ishii, A.; Tanaka, A.; Sawada, N.; Kudo, K.; Senda, M. *Chem. Eur. J.* **2001**, *7*, 730–737. See also an excellent introductory review on high-throughput screening (HTS): Reetz, M. T. *Angew. Chem. Int. Ed.* **2001**, *40*, 284–310.

18. Ushio, H.; Mikami, K. *Tetrahedron Lett.* **2005**, *46*, 2903–2906.

19. XylBINAP = 2,2′-bis(di-3,5-xylylphosphanyl)-1,1′-binaphthyl: (a) Mashima, K.; Matsumura, Y.; Kusano, K.; Kumobayashi, H.; Sayo, N.; Hori, Y.; Ishizaki, T.; Akutagawa, S.; Takaya, H. *J. Chem. Soc. Chem. Commun.* **1991**, 609–610; (b) Ohkuma, T.; Koizumi, M.; Doucet, H.; Pham, T.; Kozawa, M.; Murata, K. Katayama, E.; Yokozawa, T.; Ikariya, T.; Noyori, R. *J. Am. Chem. Soc.* **1998**, *120*, 13529–13530.

20. TolBINAP = 2,2′-bis(di-*p*-tolylphosphanyl)-1,1′-binaphthyl: (a) Takaya, H.; Mashima, K.; Koyano, K.; Yagi, M.; Kumobayashi, H.; Taketomi, T.; Akutagawa, S.; Noyori, R. *J. Org. Chem.* **1986**, *51*, 629–635; (b) Kitamura, M.; Tokunaga, M.; Ohkuma, T.; Noyori, R. *Org. Synth.* **1992**, *71*, 1–13.

21. (a) Mangeney, P.; Tejero, T.; Alexakis, A.; Grosjean, F.; Normant, J. *Synthesis* **1988**, 255–257; (b) Pikul, S.; Corey, E. J. *Org. Synth.* **1992**, *71*, 22–29.

22. Reviews: (a) Noyori, R.; Hashiguchi, S. *Acc. Chem. Res.* **1997**, *30*, 97–102; (b) Noyori, R.; Ohkuma, T. *Angew. Chem. Int. Ed.* **2001**, *40*, 40–73.

23. (a) Ohkuma, T.; Ooka, H.; Hashiguchi, S.; Ikariya, T.; Noyori, R. *J. Am. Chem. Soc.* **1995**, *117*, 2675–2676; (b) Ohkuma, T.; Ooka, H.; Ikariya, T.; Noyori, R. *J. Am. Chem. Soc.* **1995**, *117*, 10417–10418; (c) Ohkuma, T.; Ooka, H.; Yamakawa, M.; Ikariya, T.; Noyori, R. *J. Org. Chem.* **1996**, *61*, 4872–4873; (d) Ohkuma, T.; Ikehira, H.; Ikariya, T.; Noyori, R. *Synlett* **1997**, 467–468; (e) Doucet, H.; Ohkuma, T.; Murata, K.; Yokozawa, T.; Kozawa, M.; Katayama, E.; England, A. F.; Ikariya, T.; Noyori, R. *Angew. Chem. Int. Ed.* **1998**, *37*, 1703–1707.

24. (a) Ohkuma, T.; Doucet, H.; Pham, T.; Mikami, K.; Korenaga, T.; Terada, M.; Noyori, R. *J. Am. Chem. Soc.* **1998**, *120*, 1086–1087; (b) Mikami, K.; Korenaga, T.; Matsumoto, Y.; Ueki, M.; Terada, M.; Matsukawa, S. *Pure Appl. Chem.* **2001**, *73*, 255–259.

25. The real catalyst has been suggested to be a mono- or dihydride species (X = H or Cl): (a) Abdur-Rashid, K.; Lough, A. J.; Morris, R. H. *Organometallics* **2000**, *19*, 2655–2657; (b) Abdur-Rashid, K.; Lough, A. J.; Morris, R. H. *Organometallics* **2001**, *20*, 1047–1049; (c) Abdur-Rashid, K.; Faatz, M.; Lough, A. J.; Morris, R. H. *J. Am. Chem. Soc.* **2001**, *123*, 7473–7474; (d) Hartmann, R.; Chen, P. *Angew. Chem. Int. Ed.* **2001**, *40*, 3581–3585.

26. (a) Mikami, K.; Korenaga, T.; Ohkuma, T.; Noyori, R. *Angew. Chem. Int. Ed.* **2000**, *39*, 3707–3710; (b) Mikami, K.; Korenaga, T.; Yusa, Y.; Yamanaka, M. *Adv. Synth. Catal.* **2003**, *345*, 246–254.

27. Berrisford, D. J.; Bolm, C.; Sharpless, K. B. *Angew. Chem., Int. Ed. Engl.* **1995**, *34*, 1059–1070.

28. (a) Nitschke, J. R.; Lehn, J.-M. *Proc. Natl. Acad. Sci. USA* **2003**, *100*, 11970–11974; (b) Corbett, P. T.; Leclaire, J.; Vial, L.; West, K. R.; Wietor, J.-L.; Sanders, J. K. M.; Otto, S. *Chem Rev.* **2006**, *106*, 3652–3711; (c) Mikami, K. Bristol-Myers Squibb Lecture, Colorado State University, Aug. 4, 1997.

29. Mikami, K.; Matsukawa, S.; Volk, T.; Terada, M. *Angew. Chem., Int. Ed. Engl.* **1997**, *36*, 2768–2771.

30. (a) Review: Braun, M. *Angew. Chem., Int. Ed. Engl.* **1996**, *35*, 519–522; (b) Beck, A. K.; Bastani, B.; Plattner, D. A.; Petter, W.; Seebach, D.; Braunschweiger, H.; Gysi, P.; La

Vecchia, L. *Chimia* **1991**, *45*, 238–244; (c) Seebach, D.; Plattner, D. A.; Beck, A. K.; Yang, Y. M.; Hunziker, D. *Helv. Chim. Acta* **1992**, *75*, 2171–2209; (d) Narasaka, K.; Iwasawa, N.; Inoue, M.; Yamada, T.; Nakashima, M.; Sugimori, J. *J. Am. Chem. Soc.* **1989**, *111*, 5340–5344.

31. (a) Martin, C. A. Ph.D. thesis under the supervision of Prof. K. B. Sharpless, MIT, **1988**; (b) Wang, J. T.; Fan, X.; Feng, X.; Qian, Y. M. *Synthesis* **1989**, 291–292; (c) Keck, G. E.; Tarbet, K. H.; Geraci, L. S. *J. Am. Chem. Soc.* **1993**, *115*, 8467–8468; (d) Weigand, S.; Brückner, R. *Chem. Eur. J.* **1996**, *2*, 1077–1084; (d) Boyle, T. J.; Barnes, D. L.; Heppert, J. A.; Morales, L.; Takusagawa, F. *Organometallics* **1992**, *11*, 1112–1126.

32. (a) Long, J.; Hu, J.; Shen, X.; Ji, B.; Ding, K. *J. Am. Chem. Soc.* **2002**, *124*, 10–11; (b) Yuan, Y.; Zang, X.; Ding, K. *Angew. Chem. Int. Ed.* **2003**, *42*, 5478–5480.

33. Guo, H.; Wang, X.; Ding, K. *Tetrahedron Lett.* **2004**, *45*, 2009–2012.

34. (a) Bull, S. D.; Davies, S. G.; Epstein, S. W.; Ouzman, J. V. A. *Chem. Commun.* **1998**, 659–660; (b) Corminboeuf, O.; Quaranta, L.; Renaud, P.; Liu, M.; Jasperse, C. P.; Sibi, M. P. *Chem. Eur. J.* **2003**, *9*, 28–35.

35. Ueki, M.; Matsumoto, Y.; Jodry, J. J.; Mikami, K. *Synlett* **2001**, 1889–1892.

36. BIPHEP = 2,2′-bis(diphenylphosphanyl)-1,1′-biphenyl; this ligand was also named BPBP, but contrary to what was claimed in the publication, it was unsuccessfully synthesized to instead give the monophosphine derivative: (a) Uehara, A.; Bailar Jr, J. C. *J. Organomet. Chem.* **1982**, *239*, 1–15; (b) Bennett, M. A.; Bhargava, S. K.; Griffiths, K. D.; Robertson, G. B. *Angew. Chem., Int. Ed. Engl.* **1987**, *26*, 260–261; (c) Desponds, O.; Schlosser, M. *J. Organomet. Chem.* **1996**, *507*, 257–261; (d) Desponds, O.; Schlosser, M. *Tetrahedron Lett.* **1996**, *37*, 47–48; (e) Mikami, K.; Aikawa, K.; Korenaga, T. *Org. Lett.* **2001**, *3*, 243–245.

37. Mikami, K.; Korenaga, T.; Terada, M.; Ohkuma, T.; Pham, T.; Noyori, R. *Angew. Chem. Int. Ed.* **1999**, *38*, 495–497.

38. Korenaga, T.; Aikawa, K.; Terada, M.; Kawauchi, S.; Mikami, K. *Adv. Synth. Catal.* **2001**, *343*, 284–288.

39. Mikami, K.; Aikawa, K.; Yusa, Y. *Org. Lett.* **2002**, *4*, 95–98.

40. Mikami, K.; Aikawa, K.; Yusa, Y.; Hatano, M. *Org. Lett.* **2002**, *4*, 91–94.

41. (a) Tudor, M. D.; Becker, J. J.; White, P. S.; Gagné, M. R. *Organometallics* **2000**, *19*, 4367–4484; (b) Becker, J. J.; White, P. S.; Gagné, M. R. *J. Am. Chem. Soc.* **2001**, *123*, 9478–9479.

42. Mikami, K.; Kakuno, H; Aikawa, K. *Angew. Chem. Int. Ed.* **2005**, *44*, 7257–7260.

9

ASYMMETRIC AUTOCATALYSIS WITH AMPLIFICATION OF CHIRALITY AND ORIGIN OF CHIRAL HOMOGENEITY OF BIOMOLECULES

KENSO SOAI, TSUNEOMI KAWASAKI, AND ITARU SATO

Department of Applied Chemistry, Tokyo University of Science, Kagurazaka, Shinjuku-ku, Tokyo, Japan

9.1 INTRODUCTION

Significant progress has been witnessed in asymmetric catalysis.[1] In conventional asymmetric catalysis, the asymmetric catalyst C* provides the enantioenriched product P*, whose structures are generally different from those of the asymmetric catalysts. In contrast, asymmetric autocatalysis is an automultiplication of a chiral compound P*, in which the chiral product P* acts as a chiral catalyst P* for its own production:[2,3]

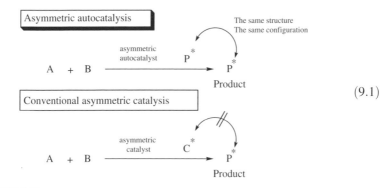

$$(9.1)$$

Asymmetric autocatalysis has the following intrinsic merits: (1) the efficiency is high because the process is automultiplication; (2) in an ideal asymmetric autocatalysis, no decrease in the amount of catalyst and no deterioration of the catalytic activity should be observed because the amount of catalyst increases during the reaction; and (3) there is no need to separate the catalyst from the product because their structures are identical. Frank proposed a kinetic model of asymmetric autocatalysis without mentioning a specific compound or reaction.[4]

All living organisms on earth are composed of highly L-enriched amino acids and D-enriched sugars. However, how and when biomolecules became highly enantiomerically enriched is a significant issue requiring further elucidation. The origin of homochirality of biomolecules has attracted much attention because the chiral homogeneity of biomolecules is considered to be closely related to the origin and evolution of life.[5] Several mechanisms have been proposed for elucidating the origins of the chirality of organic compounds, such as circularly polarized light (CPL) and quartz. However, enantiomeric excesses of organic compounds induced by the proposed mechanisms have been very low (<2% ee). An amplification process of chirality is therefore required, but reaching a very high enantiomeric enrichment of organic compounds remains an unsolved problem.

We describe highly enantioselective asymmetric autocatalysis with amplification of chirality[2] and asymmetric autocatalysis initiated by chiral triggers. Asymmetric autocatalysis correlates between the origin of chirality and the homochirality of organic compounds. We also describe spontaneous absolute asymmetric synthesis in combination with asymmetric autocatalysis.

9.2   ASYMMETRIC AUTOCATALYSIS

Enantioselective addition of dialkylzincs to aldehydes using β-amino alcohols as chiral catalysts affords chiral *sec*-alcohols.[6] During our study on the enantioselective addition of dialkylzincs to pyridine-3-carbaldehyde,[7] we found that chiral 3-pyridyl alkanol acts as an asymmetric autocatalyst in the addition of dialkylzincs to pyridine-3-carbaldehyde.[8] (S)-3-Pyridyl alkanol with 86% ee acts as an asymmetric autocatalyst in the enantioselective addition of diisopropylzinc (i-Pr$_2$Zn) to 3-pyridinecarbaldehyde to afford the same compound of the same configuration with 35% ee. This stands as the first example of asymmetric autocatalysis. After examining various systems of asymmetric autocatalysis,[9] we finally found that chiral 5-pyrimidyl alkanol **2**,[10] 3-quinolyl alkanol,[11] and 5-carbamoyl-3-pyridyl alkanol[12] serve as highly enantioselective asymmetric autocatalysts for the addition of i-Pr$_2$Zn to pyrimidine-5-carbaldehyde **1**, 3-quinolinecarbaldehyde, and 5-carbamoyl-3-pyridinecarbaldehyde, respectively. Among these, chiral 5-pyrimidyl alkanol exhibits the most significant asymmetric autocatalysis [Eqs. (9.2) and (9.3)].

$$(9.2)$$

When (S)-2-alkynyl-5-pyrimidyl alkanol **2c** with >99.5% ee was employed as an asymmetric autocatalyst, (S)-**2c** with >99.5% ee composed of both the newly formed **2c** and the initially used **2c** was obtained. The yield of the newly formed **2c** was >99%.[13] To make use of the advantage of asymmetric autocatalysis—that the structures of the asymmetric autocatalyst and the product are the same—the **2c** obtained in the first round was used as an asymmetric autocatalyst for the following round. Again, the product (S)-**2c** and the initial autocatalyst had an ee of >99.5% and the yield of the newly formed (S)-**2c** was >99%. The product **2c** was therefore used as an asymmetric autocatalyst for the following round. Even after the 10th round, the yield of **2c** was >99% and the ee was >99.5%. Thus, 2-alkynyl-5-pyrimidyl alkanol **2c** served as a virtually perfect asymmetric autocatalyst. Moreover, the amount of (S)-**2c** automultiplied by a factor of ~60 million during the 10 rounds:

1st round:	>99%, >99.5% ee
2nd round:	>99%, >99.5% ee
:	:
10th round:	>99%, >99.5% ee

Obtained alcohol was used as an asymmetric autocatalyst for the next round.

$$(9.3)$$

9.3   AMPLIFICATION OF CHIRALITY BY ASYMMETRIC AUTOCATALYSIS

If the chirality could be amplified in asymmetric autocatalysis, the process would become a very powerful method for amplifying the very tiny enantioenrichment to very high enantioenrichment. Indeed, the ee of pyrimidyl alkanol was found to incr- ease in asymmetric autocatalysis.[10a] When pyrimidyl alkanol with low ee was used as an asymmetric autocatalyst, the ee of the product (including the original autocatalyst) was higher than that of the original catalyst. To take advantage of asymmetric autoca- talysis with amplification of ee over nonautocatalytic amplification of ee,[14] the product of one round was used as an asymmetric autocatalyst for the following round. Thus, extremely low enantioenrichment of pyrimidyl alkanol was amplified to very high enantioenrichment by consecutive asymmetric autocatalysis.

Asymmetric autocatalysis using (S)-pyrimidyl alkanol **2a** with only 2% ee afforded (S)-**2a** with an increased ee of 10%, [Eq. (9.4)]. The (S)-**2a** obtained with 10% ee was then used as an asymmetric autocatalyst for the following asymmetric autocatalysis. (S)-Pyrimidyl alkanol **2a** with an increased ee of 57% was obtained. The subsequent consecutive asymmetric autocatalysis and the use of that product as an asymmetric autocatalyst for the following round gave (S)-pyrimidyl alkanol **2a** with 81% and 88% ee, respectively. Thus, the overall process was the asymmetric autocatalysis of (S)-**2a** starting from a low ee of 2% with significant amplification of chirality to 88% ee, with the increase in the amount without need for other chiral auxiliary.[10a] This stands as the first example of an asymmetric autocatalysis with amplification of ee. In addition, one-pot asymmetric autocatalysis of pyrimidyl alkanol **2b** also significantly increased the chirality from 0.28 to 87% ee.[15a]

(9.4)

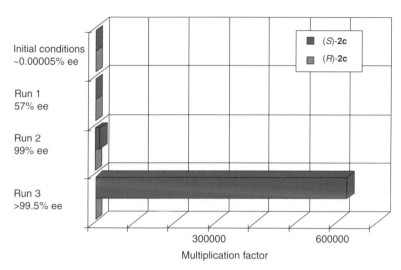

Figure 9.1. Automultiplication of (S)- and (R)-**2c** in consecutive asymmetric autocatalysis.

Moreover, we found the efficient amplification of chirality by using 2-alkynyl-5-pyrimidyl alkanol **2c** from as low as ~ 0.00005% ee to an almost enantiomerically pure (>99.5% ee) product in only three consecutive asymmetric autocatalyses, [Eq. (9.4)].[15b] The first round of asymmetric autocatalysis using (S)-**2c** with ~0.00005% ee gave (S)-**2c** in a 96% yield with an enhanced ee of 57%. The second round of asymmetric autocatalysis with the autocatalyst of 57% ee afforded (S)-**2c** with 99% ee, and the ee of (S)-**2c** finally reached >99.5% ee following the third round of asymmetric autocatalysis. During these three consecutive asymmetric autocatalyses, the initially major (S)-enantiomer of **2c** automultiplied by a factor of ~630,000, whereas the initially minor (R)-enantiomer of **2c** automultiplied by less than 1000. It is worth emphasizing how tiny the enantiomeric imbalance of ~0.00005% ee is; this extremely low ee corresponds to only several molecules of difference in the number of enantiomeric **2c** in an almost racemic mixture of ~5,000,000 molecules of (S)-**2c** and ~5,000,000 molecules of (R)-**2c** (Figure 9.1).

As described, pyrimidyl alkanols **2** act as highly efficient asymmetric autocatalysts with significant amplification of chirality. 3-Quinolyl alkanol[10,16a] and 5-carbamoyl-3-pyridyl alkanol[16b] also serve as efficient asymmetric autocatalysts with amplification of chirality.

9.4 ASYMMETRIC AUTOCATALYSIS AND ITS ROLE IN THE ORIGIN AND AMPLIFICATION OF CHIRALITY

9.4.1 Asymmetric Autocatalysis Triggered by Organic Compounds Induced by Circularly Polarized Light

Circularly polarized light (CPL) has been proposed as one of the origins of the chirality of organic compounds.[17] Asymmetric photolysis of racemic leucine by right-handed circularly polarized light (r-CPL, 213 nm) affords L-leucine with only 2% ee.[17a] Hexahelicene with low (<2%) ee is formed by asymmetric photosynthesis using CPL.[17b,17c] However, these low enantiomeric enrichments induced by CPL have not been correlated with high ees of an organic compound.

We thought that chiral organic compounds with low ee induced by CPL act as a chiral trigger in the enantioselective addition of i-Pr$_2$Zn to pyrimidine-5-carbaldehyde **1**, and that highly enantioenriched pyrimidyl alkanol **2** with an absolute configuration corresponding to that of the chirality of CPL should be formed. Indeed, in the presence of L-leucine with only 2% ee as a chiral trigger, the reaction of 2-methylpyrimidine-5-carbaldehyde **1b** with i-Pr$_2$Zn afforded (R)-pyrimidyl alkanol **2b** with an enhanced ee of 21% [Eq. (9.5)].[18a] In contrast, when D-leucine with 2% ee was used as a chiral trigger, (S)-**2b** with an increased ee (26%) was obtained. As shown in the preceding section, the ee of the obtained pyrimidyl alkanol can be further amplified significantly by the consecutive asymmetric autocatalysis.

$$(9.5)$$

In the presence of (P)-hexahelicene with very low (0.13%) ee as a chiral trigger, the reaction between aldehyde **1c** and i-Pr$_2$Zn gave (S)-pyrimidyl alkanol **2c** with 56% ee.[18b] When (M)-hexahelicene with 0.54% ee was used instead of (P)-hexahelicene, (R)-**2c** with 62% ee was formed.

Irradiation of CPL to racemic alkylidenecyclohexanone induces a small enantiomeric imbalance, which triggers the subsequent asymmetric autocatalysis to afford highly enantioenriched pyrimidyl alkanol **2c** with the absolute configuration corresponding to the chirality of CPL:[18c]

$$(9.6)$$

Thus, a slight enantiomeric imbalance in compounds induced by CPL was correlated for the first time to an organic compound with very high ee by asymmetric autocatalysis with amplification of chirality. Moreover, various chiral organic compounds such as 1,1-binaphthyl,[18d] [2.2]paracyclophanes,[18e–18g] and primary alkanols due to deuterium substitution[18h] have been found to serve as chiral triggers in asymmetric autocatalysis.

9.4.2 Asymmetric Autocatalysis Triggered Directly by Circularly Polarized Light

Both (R)- and (S)-pyrimidyl alkanols exhibit positive and negative CD spectra at 313 nm, respectively.[19] We thought that the direct irradiation of left handed (l) CPL to racemic pyrimidyl alkanol would induce the asymmetric photodegradation of (R)-pyrimidyl alkanol and leave the slightly enantioenriched (S)-pyrimidyl alkanol. Even when the enantioenrichment of (S)-pyrimidyl alkanol that remains is extremely low, the compound serves as an asymmetric autocatalyst in the subsequent asymmetric autocatalysis with amplification of chirality, as described in the preceding section, to afford itself with high enantioenrichment. Indeed, direct irradiation of l-CPL to racemic pyrimidyl alkanol and the subsequent asymmetric autocatalysis afford highly enantioenriched (S)-pyrimidyl alkanol with >99.5% ee [Eq. (9.7)].[19] On the other hand, irradiation of right-handed (r) CPL instead of l-CPL, (R)-pyrimidyl alkanol with >99.5% ee was formed. The process provides direct correlation of the chirality of CPL with that of organic compound with high enantiomeric excess:

$$(9.7)$$

We have demonstrated the enantioselective synthesis of near-enantiopure compounds by asymmetric photodegradation of racemic pyrimidyl alkanol **2c** by circularly polarized light followed by asymmetric autocatalysis. This is the first example of asymmetric autocatalysis triggered directly by a chiral physical factor: CPL.

9.4.3 Asymmetric Autocatalysis Triggered by Chiral Inorganic Crystals

Quartz is a naturally occurring chiral inorganic crystal. It exhibits either a dextrorotatory (d) or levorotatory (l) enantiomorph. Quartz has been considered as one of the origins of chirality in nature. However, no apparent asymmetric induction using quartz has been observed.[20] Only a very small asymmetric induction has been reported in an asymmetric adsorption of chiral compounds on quartz.[21]

We anticipated that asymmetric autocatalysis triggered by quartz would afford pyrimidyl alkanol with high ee [Eq. (9.8)]. When 2-alkynylpyrimidine-5-carbaldehyde **1c** was reacted with i-Pr$_2$Zn in the presence of the powder of d-quartz, (S)-pyrimidyl alkanol **2c** with 97% ee was obtained in a yield of 95%.[22] In contrast, in the presence of l-quartz, (R)-**2c** with 97% ee was obtained in a yield of 97%. These results clearly show that the absolute configurations of pyrimidyl alkanol **2c** formed were regulated by the chirality of quartz. A small enantiomeric imbalance of the initially formed (zinc alkoxide of) pyrimidyl alkanol induced by quartz was amplified significantly by the subsequent one-pot asymmetric autocatalysis to afford pyrimidyl alkanol **2c** with very high ee. A chiral organic compound with high ee has therefore been formed for the first time, using chiral inorganic crystal as the chiral trigger in conjunction with asymmetric autocatalysis with amplification of chirality.

$$(9.8)$$

Sodium chlorate (NaClO$_3$) crystal was also found to act as a chiral trigger. In the presence of d-NaClO$_3$ crystal, (S)-pyrimidyl alkanol **2c** with 98% ee was formed in a >90% yield [Eq. (9.8)].[23a] On the other hand, in the presence of l-NaClO$_3$ crystal, (R)-**2c** with 98% ee was formed. Crystallization with stirring of an achiral aqueous solution of NaClO$_3$ affords enantioenriched d- or l-crystals.[24] In contrast to d-NaClO$_3$ crystal, d-sodium bromate (NaBrO$_3$) crystal afforded (R)-**2c** and l-NaBrO$_3$ crystal (S)-**2c**[23b]. The opposite configurations of **2c** from d-NaClO$_3$ and d-NaBrO$_3$ are reasonable because the senses of the optical rotations are opposite between d-NaClO$_3$ and d-NaBrO$_3$ crystals; that is, the shapes of the enantiomorph are opposite. Thus, the results presented above show that the asymmetric autocatalysis does recognize the chirality of the enantiomorph of d-NaClO$_3$ and d-NaBrO$_3$ crystals to afford (S)-**2c** and (R)-**2c**, respectively.

In addition, helical silica[23c] and chiral organic-inorganic hybrid silica[23d,23e] were found to serve as chiral triggers of asymmetric autocatalysis:

$$(9.9)$$

As described, a chiral organic compound with high ee is formed using chiral inorganic crystals in conjunction with asymmetric autocatalysis.

9.4.4 Asymmetric Autocatalysis Triggered by Chiral Organic Crystals Composed of Achiral Organic Compounds

It has been recognized that some of the achiral organic compounds crystallize in chiral space groups to give enantiomorphous crystals. Stereospecific reactions have been reported using these chiral organic crystals of achiral compounds as reactants.[25] However, enantiomorphous crystals formed from achiral organic compounds have rarely been used as a chiral inducer (or a catalyst) in enantioselective synthesis of external compounds. From the prebiotic perspective, to investigate the highly enantioselective reaction utilizing the crystal chirality of achiral compounds is an important experimental approach to understand the origin of chirality. So we examined the possibility of these chiral crystals as chiral initiators in asymmetric autocatalysis. Two-component molecular crystals formed from tryptamine and 4-chlorobenzoic acid exhibits enantiomorph.[26] In the presence of

P-cocrystal, (*R*)-pyrimidyl alkanol **2c** with high enantioenrichment is formed, while *M*-cocrystal affords (*S*)-**2c** with high enantioenrichment [Eq. (9.10)].[27]

In addition, hippuric acid (*N*-benzoylglycine) is an achiral naturally occurring amino acid derivative, and is formed in mammals when benzoic acid is detoxified by conjunction with glycine. It has been reported that a single crystal of hippuric acid forms enantiomorphous $P2_12_12_1$ crystals, which belongs to a chiral space group.[28] When pyrimidine-5-carbaldehyde **1c** was treated with *i*-Pr$_2$Zn in the presence of [CD(+)260]-crystal which has a positive Cotton effect at 260 nm in solid CD spectrum, (*S*)-pyrimidyl alkanol **2c** with high ee was obtained. On the other hand, in the presence of [CD(−)260]-crystals, which have a negative Cotton effect at 260 nm, the opposite enantiomer (*R*)-**2c** with high ee was produced:[29]

$$(9.10)$$

In these systems, after the crystal chirality induced the chirality of asymmetric carbon in external organic compound, the subsequent asymmetric autocatalysis gives the greater amount of enantiomerically amplified product. These results clearly demonstrate that the crystal chirality of achiral organic compound is responsible for the enantioselective addition of *i*-Pr$_2$Zn to pyrimidine-5-carbaldehyde **1c**.

9.4.5 Spontaneous Absolute Asymmetric Synthesis

Spontaneous absolute asymmetric synthesis, that is the statistical formation of enantioenriched compounds from achiral reagents without the intervention of any chiral auxiliary, has been proposed as one of the origins of chirality. Without using any chiral substance, nucleophilic attack on a prochiral aldehyde on the *Re*- or *Si*-face occurs with an equal probability, providing (*S*)- and (*R*)-products with stochastic distribution, that is, so-called racemic modification. However, as Mislow described the inevitability of enantiomeric enrichment in an absolute asymmetric synthesis,[5e] it is considered that small fluctuations in the ratio of the two enantiomers are present if chiral molecules are produced from achiral starting materials under conditions where the probability of formation of the enantiomers is equal. Thus, a so-called racemic sample seldom contains the exact numbers of (*S*)- and (*R*)-enantiomers; thus, it is called *cryptochiral*, although the ee is below the detection level. Mislow introduced the term *cryptochiral* to express such very small enantioenrichment.[5a] In addition, when the total number of molecules is an odd number, the numbers of (*S*)- and (*R*)-enantiomers cannot be the same.

We thought that when *i*-Pr$_2$Zn was treated with pyrimidine-5-carbaldehyde without adding any chiral substance, extremely slight enantioenrichment would be induced statistically in the initially formed zinc alkoxide of the pyrimidyl alkanol, and that the subsequent amplification of chirality by asymmetric autocatalysis would afford the pyrimidyl alkanol with detectable enantioenrichment [Eq. (9.11)]. Indeed, we found that pyrimidyl alkanol with an ee that is above the detection level was formed.[30] Pyrimidine-5-carbaldehyde was reacted with *i*-Pr$_2$Zn, and the resulting pyrimidyl alkanol was used as an asymmetric autocatalyst for the subsequent asymmetric autocatalysis. The consecutive asymmetric autocatalysis afforded pyrimidyl alkanol of either *S* or *R* configuration with enantiomeric enrichment above the detection level.[30]

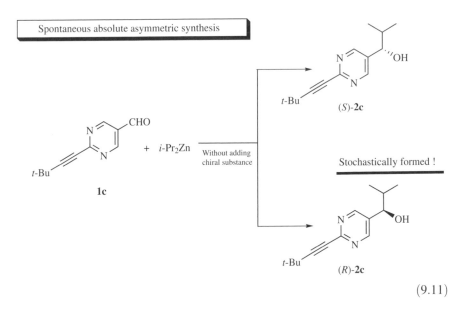

(9.11)

When 2-alkynylpyrimidine-5-carbaldehyde was reacted with *i*-Pr$_2$Zn in a mixed solvent of ether and toluene, the subsequent one-pot asymmetric autocatalysis

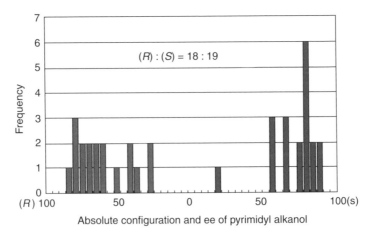

Figure 9.2. Histogram of the absolute configuration and the enantiomeric excess of pyrimidyl alkanol **2c**.

with amplification of ee gave enantiomerically enriched pyrimidyl alkanol well above the detection level.[31] The absolute configurations of the pyrimidyl alkanol formed exhibit an approximate stochastic distribution of S and R enantiomers (19 times formation of S and 18 times R), (Figure 9.2). The approximate stochastic behavior in the formation of pyrimidyl alkanols fullfils one of the conditions necessary for spontaneous absolute asymmetric synthesis.[32]

9.5 CONCLUSIONS

Chiral 5-pyrimidyl alkanol, 3-quinolyl alkanol, and 5-carbamoyl-3-pyridyl alkanol were found to act as highly enantioselective asymmetric autocatalysts for adding i-Pr$_2$Zn to pyrimidine-5-carbaldehyde, 3-quinolinecarbaldehyde, and 5-carbamoyl-3-pyridinecarbaldehyde, respectively. Among these, 2-alkenyl- and 2-alkynyl-5-pyrimidyl alkanols serve as highly efficient asymmetric autocatalysts. Asymmetric autocatalysis with amplification of ee, from extremely low ee to very high ee, was observed for the first time. Kinetic analysis of pyrimidyl alkanol suggested that the reaction is second order in the zinc monoalkoxide of pyrimidyl alkanol and first order in aldehyde and i-Pr$_2$Zn.[33] For the mechanism of the amplification from very low ee, the presence of an additional mechanism as well as the second-order mechanism of the zinc monoalkoxide of pyrimidyl alkanol is envisaged.[33b] However, the elucidation of the structure of actual reactive species awaits future investigation.

   Chiral organic compounds with low ee that are induced by circularly polarized light (CPL) serve as chiral triggers of asymmetric autocatalysis with amplification of chirality. The overall process links, for the first time, the physical chirality of CPL to organic compounds with very high ee. Chiral inorganic crystals such as quartz and sodium chlorate act as chiral triggers and regulate the sense of the asymmetric autocatalysis. The process correlates, for the first time, the chirality of inorganic crystals with highly enantioenriched organic compounds.

Spontaneous absolute asymmetric synthesis is described in the formation of enantiomerically enriched pyrimidyl alkanol from the reaction of pyrimidine-5-carbaldehyde and i-Pr$_2$Zn without adding chiral substance in combination with asymmetric autocatalysis. The approximate stochastic distribution of the absolute configurations of the product pyrimidyl alkanol strongly suggests that the reaction is a spontaneous absolute asymmetric synthesis.

As described, asymmetric autocatalysis with amplification of chirality is a powerful tool to correlate the origin of chirality with highly enantioenriched organic compounds.

ACKNOWLEDGMENT

The authors are grateful to the coworkers whose names appear in the papers. Financial support from the Ministry of Education, Culture, Sports, Science and Technology (MEXT), Japan is gratefully acknowledged.

REFERENCES

1. (a) *Comprehensive Asymmetric Catalysis*, Jacobsen, E. N.; Pfaltz, A.; Yamamoto, H. (Eds.), Springer-Verlag, Heidelberg, **1999**; (b) *Catalytic Asymmetric Synthesis, 2nd ed.,* Ojima, I. (Ed.), Wiley, New York, **2000**; (c) *Methodologies in Asymmetric Catalysis*, Malhotra, S. V. (Ed.), American Chemical Society, Washington, DC, **2004**.

2. Reviews and accounts by our group: (a) Soai, K.; Shibata, T. In *Advances in Biochirality*; Pályi, G., Zucchi, C., Caglioti, L. (Eds.), Elsevier, Amsterdam, 1999; Chap. 11; (b) Soai, K.; Shibata, T.; Sato, I. *Acc. Chem. Res.* **2000**, *33*, 382–390; (c) Soai, K. *Enantiomer* **1999**, *4*, 591–598; (d) Soai, K.; Shibata, T. *Yuki Gosei Kagaku Kyokaishi (J. Synth. Org. Chem. Jpn.)* **1997**, *55*, 994–1005; (e) Soai, K.; Shibata, T. In *Catalytic Asymmetric Synthesis, 2nd ed.*, Ojima, I. (Ed.), Wiley, New York, **2000**; Chap. 9; (f) Soai, K.; Shibata, T.; Sato, I. *Nippon Kagaku Kaishi (J. Chem. Soc. Jpn. Chem. Ind. Chem.)* **2001**, 141–149; (g) Soai, K.; Sato, I.; Shibata, T. *Chem. Rec.* **2001**, *1*, 321–332; (h) Soai, K.; Sato, I. *Chirality* **2002**, *14*, 548–554; (i) Soai, K. In *Fundamentals of Life*, Pályi, G.; Zucchi, C.; Caglioti, L. (Eds.), Elsevier, Paris, **2002**, pp. 427–435; (j) Soai, K.; Sato, I.; Shibata, T. *Yuki Gosei Kagaku Kyokaishi (J. Synth. Org. Chem. Jpn.)* **2002**, *60*, 668–678; (k) Soai, K.; Sato, I. *Viva Origino* **2002**, *30*, 186–198; (l) Soai, K.; Sato, I.; Shibata, T. In *Methodologies in Asymmetric Catalysis*, Malhotra, S. V. (Ed.), American Chemical Society, Washington, DC, **2004**; Chap. 6, pp. 85–102; (m) Soai, K. *Yuki Gosei Kagaku Kyokaishi (J. Synth. Org. Chem. Jpn.)* **2004**, *62*, 673–681; (n) Soai, K.; Shibata, T.; Sato, I. *Bull. Chem. Soc. Jpn.* **2004**, *77*, 1063–1073.

3. Reviews by other groups: (a) Bolm, C.; Bienewald, F.; Seger, A. *Angew. Chem. Int. Ed. Engl.* **1996**, *35*, 1657–1659; (b) Avalos, M.; Babiano, R.; Cintas, P.; Jiménez, J. L.; Palacios, J. C. *Chem. Commun.* **2000**, 887–892; (c) Buschmann, H.; Thede, R.; Heller, D. *Angew. Chem. Int. Ed.* **2000**, *39*, 4033–4036; (d) Todd, M. H. *Chem. Soc. Rev.* **2002**, *31*, 211–222;

(e) Blackmond, D. G. *Proc. Nat; Acad. Sci. USA* **2004**, *101*, 732–736; (f) Podlech, J.; Gehring, T. *Angew. Chem. Int. Ed.* **2005**, *44*, 5776–5777; (g) Mikami, K.; Yamanaka, M. *Chem. Rev.* **2003**, *103*, 3369–3400; (h) Gridnev, I. D. *Chem. Lett.* **2006**, *35*, 148–153; (i) Caglioti, L.; Zucchi, C.; Pályi, G. *Chemistry Today (Chimica Oggi)* **2005**, *23*, 38–43; (j) Pályi, G.; Micskei, K.; Zékány, L.; Zucchi, C.; Caglioti, L. *Magy. Kem. Lapja* **2005**, *60*, 17–24; (k) Stankiewicz, J.; Eckardt, L. H. *Angew. Chem. Int. Ed.* **2006**, *45*, 342–344.

4. Frank, F. C. *Biochim. Biophys. Acta* **1953**, *11*, 459–463.

5. (a) Mislow, K.; Bickart, P. *Isr. J. Chem.* **1976/1977**, *15*, 1–6; (b) Siegel, J. S. *Chirality* **1988**, *10*, 24–27; (c) Kondepudi, D. K., Nelson, G. W. *Nature* **1985**, *314*, 438–441; (d) Weissbuch, I.; Addadi, L.; Leiserowitz, L.; Lahav, M. *J. Am. Chem. Soc.* **1998**, *110*, 561–567. Reviews: (e) Mislow, K. *Collect. Czech. Chem. Commun.* **2003**, *68*, 849–864; (f) Eschenmoser, A. *Science* **1999**, *284*, 2118–2124; (g) Keszthelyi, L. *Q. Rev. Biophys.* **1995**, *28*, 473–507; (h) Bonner, W. A. *Orig. Life Evol. Biosphere* **1991**, *21*, 59–111; (i) Mason, S. F.; Tranter, G. E. *Proc. Roy. Soc. Lond. A* **1985**, *397*, 45–65; (j) Avetisov, V.; Goldanskii, V. I. *Proc. Natl. Acad. Sci. USA* **1996**, *93*, 11435–11442; (k) Avalos, M.; Babiano, R.; Cintas, P.; Jiménez, J. L.; Palacios, J. C. *Tetrahedron Asym.* **2000**, *11*, 2845–2874; (l) Feringa, B. L.; van Delden, R. A. *Angew. Chem., Int. Ed. Engl.* **1999**, *38*, 3418–3438; (m) Mason, S. *Chem. Soc. Rev.* **1988**, *17*, 347–359; (n) Podlech J. *Cell. Mol. Life Sci.* **2001**, *58*, 44–60.

6. Reviews: (a) Soai, K.; Niwa, S. *Chem. Rev.* **1992**, *92*, 833–856; (b) Noyori, R.; Kitamura, M. *Angew. Chem. Int. Ed. Engl.* **1991**, *30*, 49–69; (c) Pu, L.; Yu, H.-B. *Chem. Rev.* **2001**, *101*, 757–824.

7. Soai, K.; Hori, S.; Niwa, S. *Heterocycles* **1989**, *29*, 2065–2067.

8. Soai, K.; Niwa, S.; Hori, H. *J. Chem. Soc. Chem. Commun.* **1990**, 982–983.

9. (a) Soai, K.; Hayase, T.; Takai, K. *Tetrahedron Asym.* **1995**, *6*, 637–638; (b) Soai, K.; Hayase, T.; Shimada, C.; Isobe, K. *Tetrahedron Asym.* **1994**, *5*, 789–792.

10. (a) Soai, K.; Shibata, T.; Morioka, H.; Choji, K. *Nature (London)* **1995**, *378*, 767–768; (b) Shibata, T.; Morioka, H.; Hayase, T.; Choji, K.; Soai, K. *J. Am. Chem. Soc.* **1996**, *118*, 471–472.

11. Shibata, T.; Choji, K.; Morioka, H.; Hayase, T.; Soai, K. *Chem. Commun.* **1996**, 751–752.

12. Shibata, T.; Morioka, H.; Tanji, S.; Hayase, T.; Kodaka, Y.; Soai, K. *Tetrahedron Lett.* **1996**, *37*, 8783–8786.

13. Shibata, T.; Yonekubo, S.; Soai, K. *Angew. Chem. Int. Ed.* **1999**, *38*, 659–661.

14. Reviews: (a) Girard, C.; Kagan, H. B. *Angew. Chem. Int. Ed.* **1998**, *37*, 2922–2959; (b) Avalos, M.; Babiano, R.; Cintas, P.; Jiménez, J. L.; Palacios, J. C. *Tetrahedron: Asymmetry* **1997**, *8*, 2997–3017; (c) Fenwick, D. R.; Kagan, H. B. In *Topics in Stereochemistry*, Denmark S. E. (Ed.), Wiley, New York, **1999**, Vol. 22, p. 257; (d) Bolm, C. *Advanced Asymmetric Synthesis*, Stephenson, G. R. (Ed.), Blackie, Glasgow, **1996**, p. 9; (e) Green, M. M.; Park, J.-W.; Sato, T.; Lifson, S,; Selinger, R. L. B.; Seliger, J. V. *Angew. Chem. Int. Ed.* **1999**, *38*, 3138–3154.

15. (a) Shibata, T.; Hayase, T.; Yamamoto, J.; Soai, K. *Tetrahedron Asym.* **1997**, *8*, 1717–1719; (b) Sato, I.; Urabe, H.; Ishiguro, S.; Shibata, T.; Soai, K. *Angew. Chem. Int. Ed.* **2003**, *42*, 315–317.

16. (a) Sato, I.; Nakao, T.; Sugie, R.; Kawasaki, T.; Soai, K. *Synthesis* **2004**, 1419–1428; (b) Tanji, S.; Kodaka, Y.; Ohno, A.; Shibata, T.; Sato I.; Soai, K. *Tetrahedron: Asymmetry* **2000**, *11*, 4249–4253.

17. (a) Flores, J. J.; Bonner, W. A.; Massey, G. A. *J. Am. Chem. Soc.* **1977**, *99*, 3622–3625; (b) Moradpour, A.; Nicoud, J. F.; Balavoine, G.; Kagan, H.; Tsoucaris. G. *J. Am. Chem. Soc.* **1971**, *93*, 2353–2354; (c) Bernstein, W. J.; Calvin, M.; Buchardt, O. *J. Am. Chem. Soc.* **1972**, *94*, 494–498; (d) Inoue, Y.; Tsuneishi, H.; Hakushi, T.; Yagi, K.; Awazu, K.; Onuki, H. *Chem. Commun.* **1996**, 2627–2628. Review: (e) Buchardt, O. *Angew. Chem. Int. Ed. Engl.* **1974**, *13*, 179–185; (f) see also Ref. 4h; (g) Inoue, Y. *Chem. Rev.* **1992**, *92*, 741–770.

18. (a) Shibata, T.; Yamamoto, J.; Matsumoto, N.; Yonekubo, S.; Osanai, S.; Soai, K. *J. Am. Chem. Soc.* **1998**, *120*, 12157–12158; (b) Sato, I.; Yamashima, R.; Kadowaki, K.; Yamamoto, J.; Shibata, T.; Soai, K. *Angew. Chem. Int. Ed.* **2001**, *40*, 1096–1098; (c) Sato, I.; Sugie, R.; Matsueda, Y.; Furumura, Y.; Soai, K. *Angew. Chem. Int. Ed.* **2004**, *43*, 4490–4492; (d) Sato, I.; Osanai, S.; Kadowaki, K.; Sugiyama, T.; Shibata, T.; Soai, K. *Chem. Lett.* **2002**, 168–169; (e) Sato, I.; Matsueda, Y.; Kadowaki, K.; Yonekubo, S.; Shibata, T.; Soai, K. *Helv. Chim. Acta* **2002**, *85*, 3383–3387; (f) Sato, I.; Ohno, A.; Aoyama, Y.; Kasahara, T.; Soai, K. *Org. Biomol. Chem.* **2003**, *1*, 244–246; (g) Tanji, S.; Ohno, A.; Sato, I.; Soai, K. *Org. Lett.* **2001**, *3*, 287–289; (h) Sato, I.; Omiya, D.; Saito, T.; Soai, K. *J. Am. Chem. Soc.* **2000**, *122*, 11739–11740.

19. Kawasaki, T.; Sato, M.; Ishiguro, S.; Saito, T.; Morishita, Y.; Sato, I.; Nishino, H.; Inoue, Y.; Soai, K. *J. Am. Chem. Soc.* **2005**, *127*, 3274–3275.

20. (a) Schwab, G. M.; Rudolph, L. *Naturwissenschaften* **1932**, *20*, 363–364; (b) Terentev, A. P.; Klabunovskii, E. I. *Sbornik Statei Obshchei Khim.* **1953**, *2*, 1598; *Chem. Abstr.* **1955**, *49*, 5262g; (c) the reproducibility of these reports of the very low asymmetric induction using quartz was disproved in later careful experiments by Amariglio, A.; Amariglio, H.; Duval, X. *Helv. Chim. Acta* **1968**, *51*, 2110–2132.

21. Bonner, W. A.; Kavasmaneck, P. R.; Martin, F. S.; Flores, J. J. *Science* **1974**, *186*, 143–144.

22. Soai, K.; Osanai, S.; Kadowaki, K.,; Yonekubo, S.; Shibata, T.; Sato, I. *J. Am. Chem. Soc.* **1999**, *121*, 11235–11236.

23. (a) Sato, I.; Kadowaki, K.; Soai, K. *Angew. Chem. Int. Ed.* **2000**, *39*, 1510–1512; (b) Sato, I.; Kadowaki, K.; Ohgo, Y.; Soai, K. *J. Mol. Cat. A Chem.* **2004**, *216*, 209–214; (c) Sato, I.; Kadowaki, K.; Urabe, H.; Hwa Jung, J.; Ono, Y.; Shinkai, S.; Soai, K. *Tetrahedron Lett.* **2003**, *44*, 721–724; (d) Kawasaki, T.; Ishikawa, K.; Sekibata, H.; Sato, I.; Soai, K. *Tetrahedron Lett.* **2004**, *45*, 7939–7941; (e) Sato, I.; Shimizu, M.; Kawasaki, T.; Soai, K. *Bull. Chem. Soc. Jpn.* **2004**, *77*, 1587–1588.

24. Kondepudi, D. K.; Kaufman, R. J.; Singh, N. *Science* **1990**, *250*, 975–976.

25. (a) Penzien, K.; Schmidt, G. M. *Angew. Chem. Int. Ed. Engl.* **1969**, *8*, 608–609; (b) Green, B. S.; Lahav, M.; Rabinovich, D. *Acc. Chem. Res.* **1979**, *12*, 191–197; (c) Matsuura, T.; Koshima, H. *J. Photochem. Photobiol. C Photochem. Rev.* **2005**, *6*, 7–24; (d) Tanaka, K.; Toda, F. *Chem. Rev.* **2000**, *100*, 1025–1074; (e) Sakamoto, M. *Chem. Eur. J.* **1997**, *3*, 684–689.

26. (a) Koshima, H.; Hayashi, E.; Matsuura, T.; Tanaka, K.; Toda, F.; Kato, M.; Kiguchi, M. *Tetrahedron Lett.* **1997**, *38*, 5009–5012; (b) Koshima, H.; Honke, S.; Fujita, F. *J. Org. Chem.* **1999**, *64*, 3916–3921; (c) Koshima, H.; Miyauchi, M.; Shiro, M. *Supramol. Chem.* **2001**, *13*, 137–142; (d) Koshima, H.; Nagano, M.; Asahi, T. *J. Am. Chem. Soc.* **2005**, *127*, 2455–2463.

27. Kawasaki, T.; Jo, K.; Igarashi, H.; Sato, I.; Nagano, M.; Koshima, H.; Soai, K. *Angew. Chem. Int. Ed.* **2005**, *44*, 2774–2777.

28. Ringertz, H. *Acta Cryst.* **1971**, *B27*, 285–291.

29. Kawasaki, T.; Suzuki, K.; Hatase, K.; Otsuka, M.; Koshima H.; Soai, K. *Chem. Commun.* **2006**, 1869–1871.

30. Soai, K.; Shibata, T.; Kowata, Y. *Japan Kokai Tokkyo Koho* 9,268,179 (**1997**). (application dates Feb. 1, 1996 and April 18, 1996).

31. Soai, K.; Sato, I.; Shibata, T.; Komiya, S.; Hayashi, M.; Matsueda, Y.; Imamura, H.; Hayase, T.; Morioka,H.; Tabira, H.; Yamamoto, J.; Kowata, Y. *Tetrahedron: Asymmetry* **2003**, *14*, 185–188.

32. (a) For an excellent review and commentary, see Ref. 5e; (b) commentary: Siegel, J. S. *Nature* **2002**, *419*, 346–347; (c) for nonstochastic distribution of pyrimidyl alkanol, see Singleton, D. A.; Vo, L. K. *J. Am. Chem. Soc.* **2002**, *124*, 10010–10011.

33. (a) Sato, I.; Omiya, D.; Tsukiyama, K.; Ogi, Y.; Soai, K. *Tetrahedron Asym.* **2001**, *12*, 1965–1969; (b) Sato, I.; Omiya, D.; Igarashi, H.; Kato, K.; Ogi, Y.; Tsukiyama, K.; Soai, K. *Tetrahedron Asym.* **2003**, *14*, 975–979; (c) Blackmond, D. G.; McMillan, C. R.; Ramdeehul, S.; Schorm, A.; Brown, J. M. *J. Am. Chem. Soc.* **2001**, *123*, 10103–10104; (d) Gridnev, I. D.; Serafimov, J. M.; Brown, J. M. *Angew. Chem. Int. Ed.* **2004**, *43*, 4884–4887; (e) Buhse, T. *Tetrahedron Asym.* **2003**, *14*, 1055–1061; (f) Islas, J. R.; Lavabre, D.; Grevy, J.-M.; Lamoneda, R. H.; Cabrera, H. R.; Micheau, J.-C.; Buhse, T. *Proc. Natl. Acad. Sci. USA* **2005**, *102*, 13743–13748; (g) Saito, Y.; Hyuga, H. *J. Phys. Soc. Jpn.* **2004**, *73*, 33–35; (h) Buhse, T. *J. Mex. Chem. Soc.* **2005**, *49*, 118–125; (i) Lente, G. *J. Phys. Chem. A* **2004**, *108*, 9475–9478.

10

RECENT ADVANCES IN CATALYTIC ASYMMETRIC DESYMMETRIZATION REACTIONS

TOMISLAV ROVIS

Department of Chemistry, Colorado State University, Fort Collins, Colorado, USA

10.1 INTRODUCTION

Desymmetrization, the removal of elements of symmetry from a symmetric object, is readily accomplished by simple synthetic means: $NaBH_4$ reduction of a meso anhydride, monooxidation of 2-methylpropane-1,3-diol or the monohydrolysis of substituted malonate esters. Indeed, in 1984, Mislow and Siegel suggested that the "use of the term 'desymmetrization' will hereafter be restricted to the point-replacement scheme described" (the conversion of an achiral object into a chiral one by replacement of a point by a differently labeled one).[1] Eliel et al. define desymmetrization in a broader sense: "Symmetry point groups may be arranged in a hierarchy by starting with a highly symmetrical one, such as O_h or I_h, and then systematically removing elements of symmetry."[2] Worth noting is that no distinction was made by either of these authors between the conversion of achiral objects into chiral, racemic objects *versus* chiral, enantiomerically enriched objects. Indeed, citations of desymmetrizations in the organic literature of the 1980s and early 1990s consisted of a number of examples of nonasymmetric desymmetrizations, a situation that in recent years has distinctly reversed itself. Even a cursory glance at the use of the term *desymmetrization* in recent years reveals itself to be nearly always accompanied by the words *asymmetric* or *enantioselective*. What is clear, then, is that the organic community associates the word desymmetrization

New Frontiers in Asymmetric Catalysis, Edited by Koichi Mikami and Mark Lautens
Copyright © 2007 John Wiley & Sons, Inc.

with the concomitant introduction of an element of asymmetry, making the terms *asymmetric desymmetrization* and *enantioselective desymmetrization* redundant.

Whatever the nomenclature, the concept of an enantioselective desymmetrization can be a powerful strategy when executed efficiently. The ability of the reaction to simultaneously control multiple stereogenic centers in a single transformation is attractive. The caveat that must be applied is that the substrates are not always readily available and the success of the reaction may be greatly tempered by the difficulty of accessing the starting material.[3]

The purpose of this review is to cover catalytic enantioselective desymmetrizations that have appeared in the literature from early 1999 to mid-2004. Stoichiometric desymmetrizations will not be considered. Enzymatic desymmetrizations will also not be considered, largely because this method is firmly engrained as a powerful technique and the bulk of the work in this area in recent years has centered on applications of the method to specific substrates rather than developmental advances. This review will also attempt to focus on method development rather than application to synthesis. This field has recently been reviewed,[4] and the purpose of this manuscript is to complement the previous treatises of this subject.

10.2 ALLYLIC ALKYLATION

Meso allylic diesters have been extensively used in enantioselective desymmetrizations, largely through the work of Trost and coworkers.[5] Trost has recently broadened the substrate scope of this class of reactions to include biscarbamate **1**, wherein the product has a masked alkene which may be revealed by a retro-Diels–Alder reaction.[6] The use of the furan-derived bisacetal **4** as a substrate in this reaction provides a rapid route to *C-2-epi*-hygromycin A. The enantioselective desymmetrization of this system using nitronate **5**, an acyl anion equivalent, proceeded in excellent yield and selectivity and set the stage for a second palladium-catalyzed allylic alkylation. Although a chiral ligand is not necessary at this stage, better results were obtained in its presence, which the authors argue is due to the increased reactivity of these ligands in promoting the alkylation:[7]

$$(10.1)$$

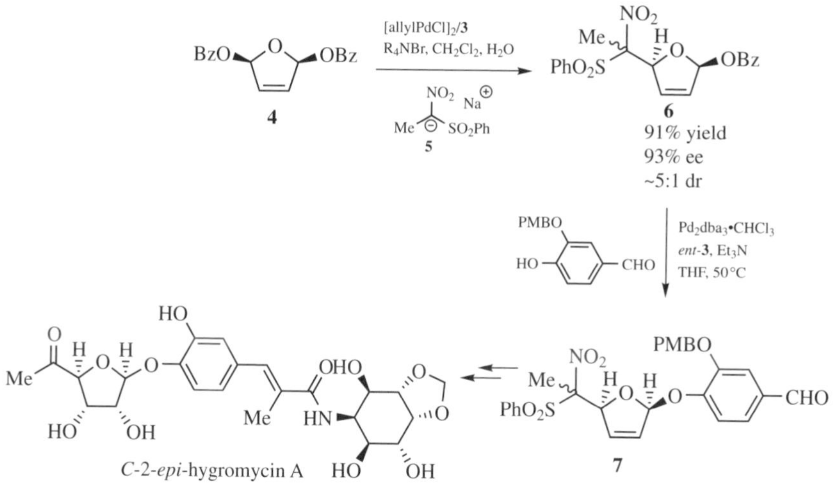

C-2-epi-hygromycin A

Burke and Jiang have utilized Trost's ligand and the asymmetric allylic alkylation to desymmetrize a tetraol diacetate as a potential route to a fragment of halichondrin B.[8] Unfortunately, enantioselectivity was not determined in this reaction, although optical rotation suggests measurable selectivity:

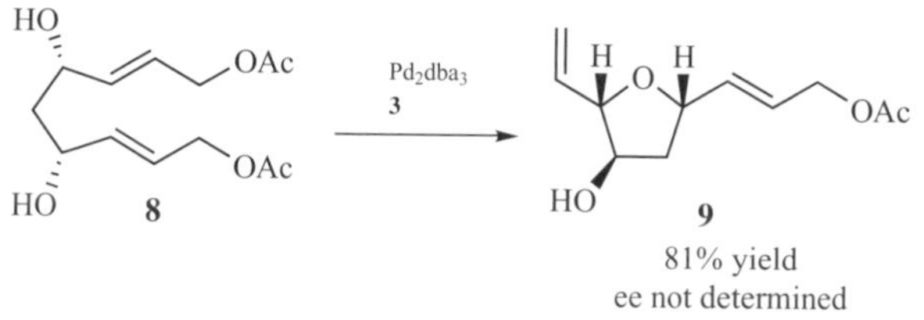

$$\tag{10.2}$$

81% yield
ee not determined

A fundamentally different allylic alkylation is the displacement of allylic leaving groups using organozinc reagents under copper catalysis.[9] Gennari and coworkers have examined meso allylic bisphosphates using diorganozinc reagents as nucleophiles.[10] A library of salicylaldimine-derived ligands was screened from which ligand **12** provided promising levels of enantioselectivity albeit moderate yields. Unlike palladium catalysis, these reactions proceed with net inversion:

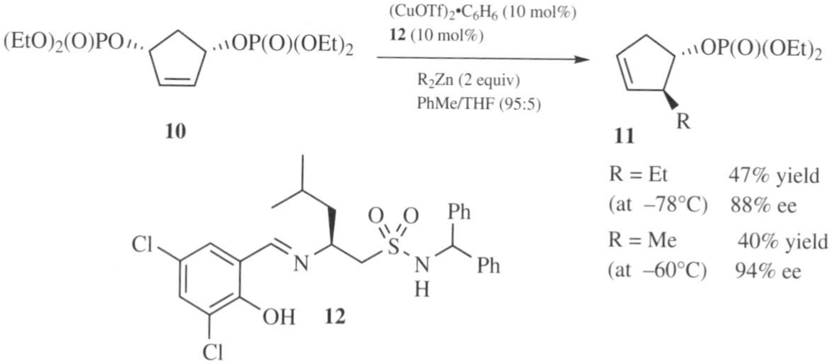

$$\tag{10.3}$$

R = Et   47% yield
(at −78°C) 88% ee

R = Me   40% yield
(at −60°C) 94% ee

In collaboration with Feringa's group, these workers applied the BINOL phos-phoramidite as a ligand in these reactions [Eq. (10.4)].[11] This system proved to be more general than the salicylaldimine system in its ability to induce good yields and high enantioselectivities across a range of cyclic bisphosphates. The salicylaldimine ligands provide only racemic product when cyclohexenyl bisphosphates are used:[9]

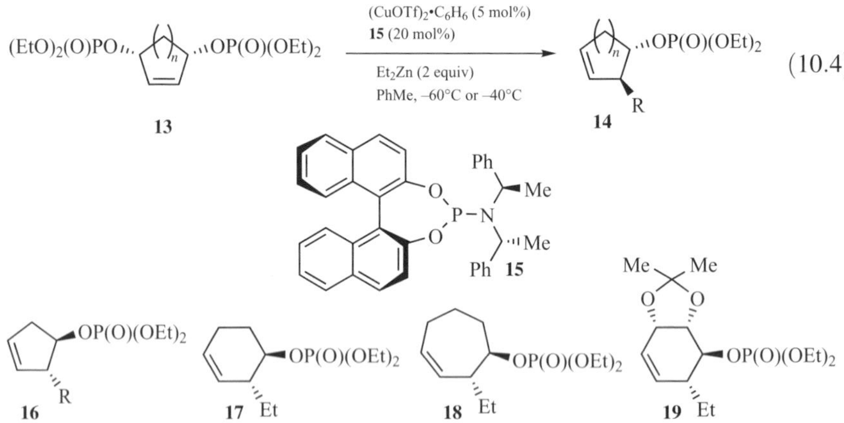

(10.4)

16
98% yield
87% ee

17 Et
77% yield
90% ee
87:13 (C_2 epimer)

18 Et
85% yield
>98% ee

19 Et
52% yield
86% ee

Pineschi and coworkers have examined the desymmetrization of cyclooctate-traene monoepoxide using Feringa's ligand in the presence of Cu(II) precatalysts.[12] Very good enantioselectivities were observed for dialkylzinc reagents with the disubstituted cyclooctatrienols isolated as single regioisomers [Eq. (10.5)]. Pineschi and Feringa also investigated the alkylation of divinyl epoxides such as **22** and found that some of these systems were excellent substrates for the Cu-phosphoramidite system, providing desymmetrized divinylcarbinols in moderate to very high enantioselectivity [Eq. (10.6)]:[13]

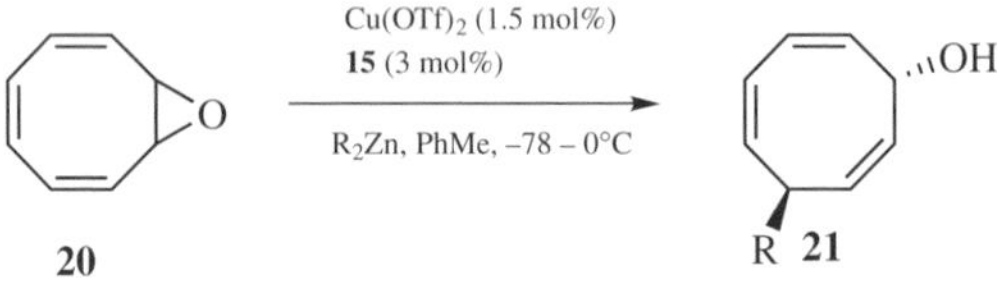

(10.5)

R = Me, Et, Bu   65 – 90% yield
82 – 90% ee

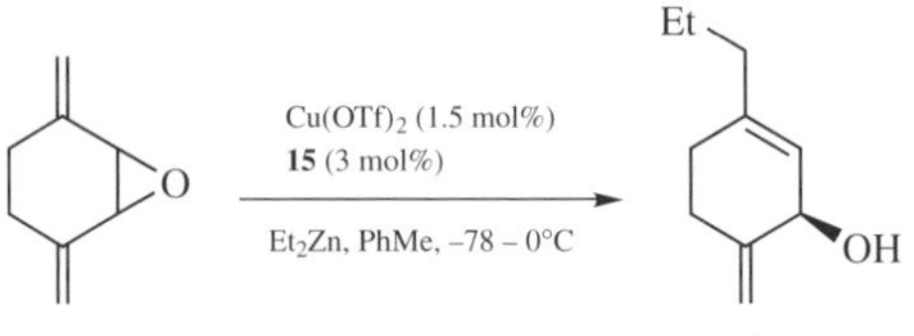

(10.6)

23
90% yield
97% ee

10.3   RING OPENING OF EPOXIDES AND AZIRIDINES

The asymmetric ring-opening of epoxides and other small strained heterocycles has been a popular target in asymmetric desymmetrizations likely because of the ease of substrate synthesis and the breadth of its reactivity. Extensive investigations by a number of workers in this area, notably Jacobsen, has led to the emergence of this strategy as a useful method of accessing valuable chiral building blocks.[14] Jacobsen has more recently extended this chemistry to include the use of cyanide as a nucleophile in epoxide ring opening. The use of $YbCl_3 \bullet PyBox$ complexes as catalysts leads to an asymmetric desymmetrization of meso epoxides in good yield and high selectivity [Eq. (10.7)].[15] Intriguingly, a report by Zhu and coworkers reveals that the use of a soft Lewis acid in this type of reaction may lead to isocyanosilylation (N attack) of meso epoxides.[16] A BINOL-Ga Lewis acid proved optimal affording moderate to high enantioselectivities and good yields [Eq. (10.8)]:

$$ (10.7) $$

$$ (10.8) $$

Shibasaki and coworkers have applied heterobimetallic gallium complexes to the desymmetrization of meso epoxides using phenolic nucleophiles.[17] Complex **35** is effective at inducing reasonable levels of enantioselectivity in these reactions, but yields are low [Eq. (10.9)]. Shibasaki argues that the reaction suffers from competitive displacement of one BINOL by excess phenol. To solve this problem, these workers introduced the linked-BINOL Ga system **36** which affords similar levels of enantioselectivity with greatly increased yields in most cases:

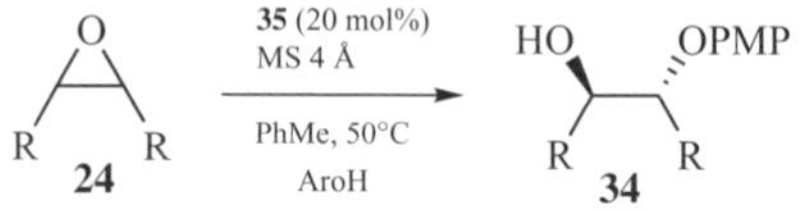

(10.9)

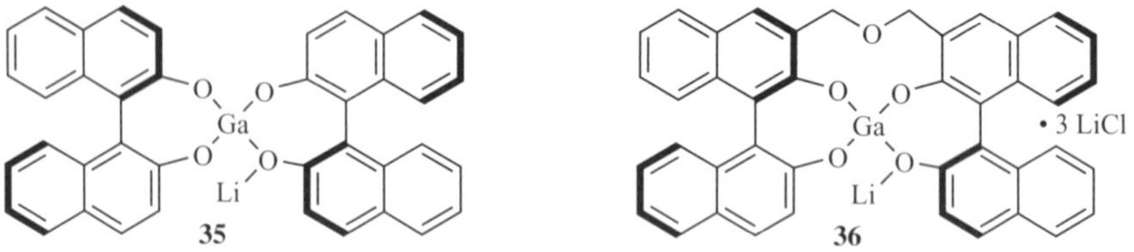

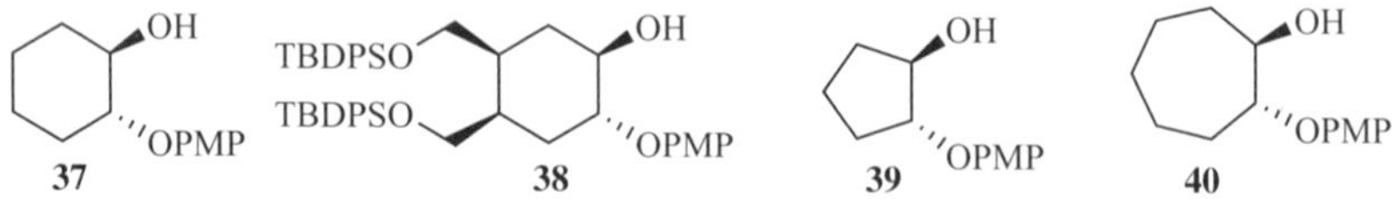

37	**38**	**39**	**40**
48% yield	34% yield	75% yield	31% yield
93% ee	80% ee	86% ee	67% ee
with 10 mol% **36**:	with 10 mol% **36**:	with 10 mol% **36**:	with 10 mol% **36**:
72% yield	72% yield	88% yield	82% yield
91% ee	79% ee	85% ee	66% ee

Zhu and coworkers have applied another Ga-based system to the desymmetrization of meso epoxides using acetylide anions as nucleophiles.[18] A complex generated from Me_3Ga and a novel salen provides modest selectivities and yields using phenylacetylide as the nucleophile:

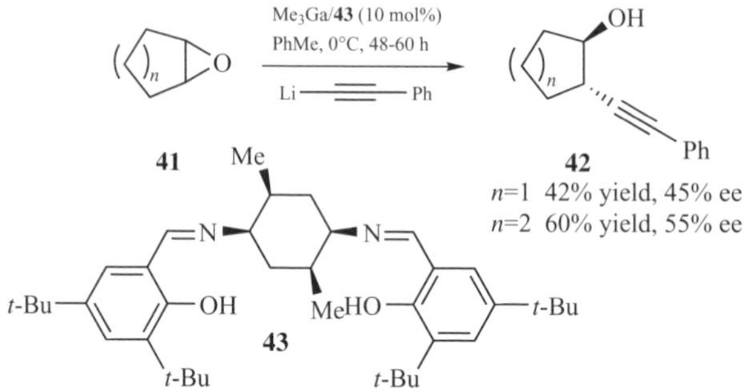

(10.10)

n=1 42% yield, 45% ee
n=2 60% yield, 55% ee

Inaba and coworkers reported that a Ti–BINOL complex is an effective catalyst for the desymmetrization of epoxide **44** using primary amines as nucleophiles.[19] Of significant note is the efficiency of this reaction, with only 1 mol% catalyst necessary to attain high yields and selectivities [Eq. (10.11)]. Unfortunately, this epoxide is uniquely effective in this reaction. Cycloheptene oxide, dihydrofuran oxide, and an acyclic version of **44** each provided negligible yields under these reaction conditions:

$$(10.11)$$

94% conversion
93% ee

Jacobsen has exploited the reactivity of Cr–salen complexes in the desymmetrization/ring opening of meso epoxides.[20] This chemistry has been extended by others. Ganem has shown that stereochemically congested *meso*-cyclopentene oxides may be desymmetrized with high enantioselectivities using the Jacobsen system, [Eq. (10.12a)],[21] while Spivey and coworkers have used the analogous Co–salen hydrolytic epoxide ring-opening chemistry in a desymmetrization of a centrosymmetric bisepoxide en route to a fragment of hemibrevetoxin B [Eq. (10.12b)].[22] Aminolysis of *cis*-stilbene oxide has been achieved using a Cr–salen complex with excellent enantioselectivity and yield [Eq. (10.12c)]. The researchers found that a cocatalytic amount of base led to improved selectivity at the expense of reactivity (20 h vs. 40 h reaction time).[23] A Cr–salen complex has been used for the ring-opening of meso epoxides using fluoride, but the reaction required 50–100 mol% catalyst for efficient conversion:[24]

$$(10.12a)$$

95% yield
98% ee

48a MX = CrN$_3$
48b MX = CrCl
49 MX = CoOAc

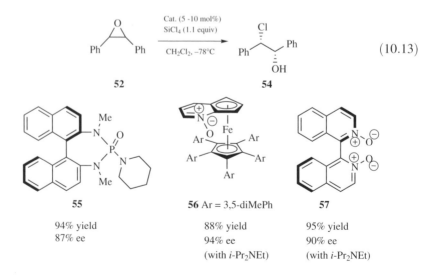

Following Denmark's discovery that nucleophilic phosphoramides catalyze the ring opening of epoxides using $SiCl_4$,[25] a number of workers have developed competent catalysts to perform this reaction enantioselectively. Denmark had applied the binaphthyl-derived phosphoramide **55** as a catalyst, which provides the chlorohydrin **54** in 87% ee.[25] Fu has developed the planar chiral ferrocene-based pyridine oxide catalyst **56** for this reaction, which provides high enantioselectivities in the ring opening of a variety of meso epoxides bearing aromatic groups as substituents [Eq. (10.13)].[26] Nakajima has developed a C_2-symmetric pyridine oxide based catalyst **57** for the ring opening of meso epoxides using $SiCl_4$.[27] While these catalysts are each fairly selective at the desymmetrization of stilbene-like epoxides, they perform somewhat less effectively with aliphatic epoxides such as **58** [Eq. (10.14)]:

$$
\text{BnO} \diagdown \overset{\text{O}}{\triangle} \diagdown \text{OBn} \quad \xrightarrow[\substack{\text{CH}_2\text{Cl}_2, -78^\circ\text{C}}]{\substack{\text{Cat. (5-10 mol\%)} \\ \text{SiCl}_4 \text{ (1.1 equiv)}}} \quad \text{BnO} \diagdown \overset{\text{Cl}}{\diagup} \diagdown \overset{}{\underset{\text{OH}}{\diagup}} \text{OBn} \quad (10.14)
$$

	58	Cat.	**59**
		55	95% yield 71% ee
		56	91% yield 50% ee
		57	98% yield 74% ee

Ring opening of meso aziridines has also been investigated. Jacobsen and coworkers found that tridentate Schiff base Cr(III) complex **62** is efficient at inducing good levels of enantioselectivity in the azidolysis of aziridines **60** [Eq. (10.15)].[28] Alkyl-substituted aziridines were found to work best with the electron-deficient dinitrophenylmethyl providing optimal results:

$$ (10.15) $$

63	**64**	**65**	**66**
95% yield 94% ee	75% yield 88% ee	87% yield 87% ee	80% yield 83% ee

Müller and Nury examined aziridine ring opening using Grignard reagents as nucleophiles. They found that the reaction proceeds in moderate selectivities in the presence of copper catalysts such as **69**:[29]

$$
\textbf{67} \xrightarrow[\substack{\text{MeMgBr} \\ \text{THF, 0}^\circ\text{C}}]{\textbf{69 (10 mol\%)}} \textbf{68} \quad (10.16)
$$

89% yield
55% ee

10.4   RING OPENING OF BRIDGED SYSTEMS

Lautens and his group have realized considerable success in the enantioselective ring opening of oxabicyclic alkenes using a variety of nucleophiles.[30] In 2000, Lautens and coworkers reported the rhodium-catalyzed asymmetric ring opening of oxabenzonorbornadiene and derivatives using alcohols as nucleophiles.[31] The reaction is remarkably efficient, proceeding in high enantioselectivity and excellent yield using only 0.25 mol% rhodium complex [Eq (10.17)]. A variety of alcohols were found to participate in this transformation, including allyl alcohol, trifluoroethanol, and trimethylsilylethanol. These workers subsequently illustrated that phenols[32] and thiols[33] participate with equal facility using the same catalyst system:

$$(10.17)$$

53-96% yield
93-99% ee

R = Me, Et, *i*-Pr, Bn, allyl, etc.

Although the initial report included amine nucleophiles, the scope was limited to activated amines such as indole (which actually undergoes *C*-alkylation at the 3-position), phthalimide, and *N*-methylaniline. Furthermore, enantioselectivities were inferior to those observed with alcohols as nucleophiles. Lautens and Fagnou subsequently discovered a profound halide effect in these reactions.[34] The exchange of the chloride for an iodide on the rhodium catalyst resulted in an increased enantioselectivity that is now comparable to levels achieved with alcoholic nucleophiles:

$$(10.18)$$

$[Rh(COD)X]_2 = [Rh(COD)Cl]_2 + 2AgOTf \text{ then } 2R_4NX$

X	yield (%)	ee (%)
OTf	93	96
F	91	96
Cl	92	74
Br	90	78
I	97	92

The halide exchange protocol also allows the use of other nucleophiles such as activated methylenes. The rhodium iodide complex was found to be the most general precatalyst, allowing the incorporation of malonates in excellent yield and enantioselectivity [Eq. (10.19)]; these conditions were also found to be effective in the asymmetric ring opening of less activated oxabicyclo[2.2.1]heptene systems [Eq. (10.20)]:

$$(10.19)$$

$$(10.20)$$

NuH = PhOH   84% yield, 93% ee
NuH = PhNHMe   93% yield, 95% ee

Lautens and coworkers have also reported the asymmetric ring opening of azabicyclic alkenes. Rhodium-catalyzed aminolysis of azabenzonorbornadiene provides access to diaminotetralins, subunits of analgesic agents. A modified aminoferrocenylphosphine was used as the optimal ligand, affording the diaminotetralin core in 77% yield and 86% ee:[35]

$$(10.21)$$

It seems logical that these reactions are mechanistically related. In every case reported to date, the relative stereochemistry between the nucleophile and alcohol moiety is trans, indicating anti addition in the ring-opening event. Lautens and coworkers suggest a syn insertion of the rhodium to form the allyl rhodium alkoxide.[36] Anti addition of the soft nucleophile, with possible initial protonation of the rhodium alkoxide,[37] results in the regeneration of the catalyst providing the observed relative stereochemistry in the product.

Enantioselective alkylative ring opening of these oxabicyclic alkenes has also been studied. Lautens and coworkers discovered that palladium complexes efficiently catalyze the addition of organozinc reagents to these activated alkenes with concomitant ring opening. In the presence of (Tol-BINAP)PdCl$_2$, diethylzinc adds to oxabenzonorbornadiene in 96% ee and 98% yield [Eq. (10.22)].[38] A matching of ligand to alkyl zinc is required for optimal results; dimethylzinc works best with (i-Pr-PHOX)PdCl$_2$. The less activated substrates **74** and **81** also undergo enantioselective alkylative ring-opening in high ee at elevated temperatures, [Eqs. (10.23) and (10.24)]:[39]

$$(10.22)$$

ligand	R	yield (%)	ee (%)
Tol-BINAP	Et	98	96
i-Pr-PHOX	Me	90	89

$$(10.23)$$

$$(10.24)$$

The syn addition of the adducts suggests a mechanism different from that observed in rhodium-catalyzed alcoholysis and aminolysis reactions. Mechanistic investigations from the Lautens laboratory have revealed that the most likely mechanism involves an enantioselective carbopalladation followed by a β-alkoxide elimination to afford the ring-opened product.[40]

Feringa and coworkers have developed a copper phosphoramidite catalyst system that accesses the diastereomeric anti products in the alkylative ring opening of oxabenzonorbornadiene:[41]

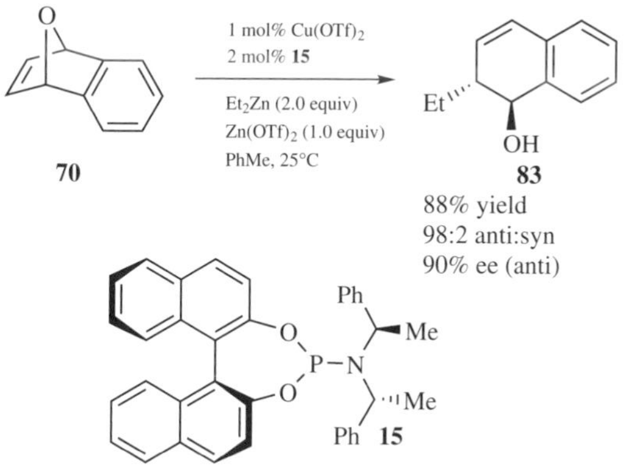

(10.25)

Lautens has extended this chemistry to include the alkylative ring opening of these systems using aryl and vinyl boronic acids as nucleophiles in the presence of rhodium catalysts [Eq. (10.26)].[42] The ferrocenyl bisphosphine proved to be the optimal ligand in these studies, affording the ring-opened product in 88–99% ee for a variety of oxabicyclo[2.2.1]heptenes. It is worth noting that the reaction is performed in the presence of aqueous base:

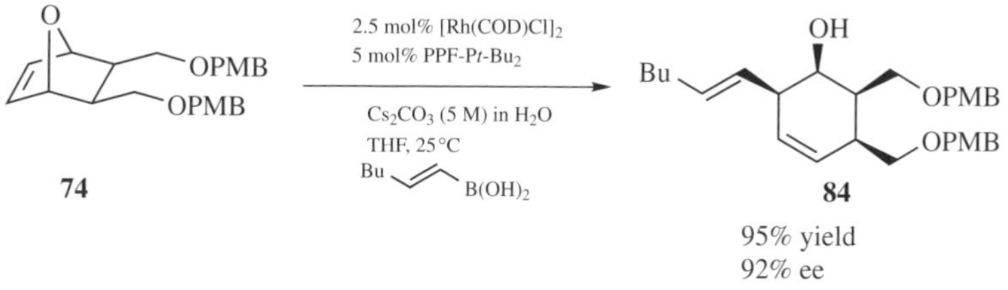

(10.26)

Waymouth and coworkers have reported that chiral zirconocene catalysts are capable of effecting the asymmetric ring opening of a number of oxabicyclic alkenes in moderate to good enantioselectivity using organoaluminum reagents as nucleophiles.[43] Intriguingly, the use of triethyl aluminum results in the ethylation/ ring opening product in high enantioselectivity while a switch to triisobutyl aluminum affords the reductive ring-opening product in much lower enantioselectivities (17% vs. 96% ee):

$$(10.27)$$

R	Nu	yield (%)	ee (%)
Et	Et	64	96
i-Bu	H	30	17

Previous efforts at asymmetric reductive ring opening of these systems had focused on metal hydrides as nucleophiles.[44] Cheng and coworkers more recently reported a fundamentally different approach, using carboxylic acids as the hydride source.[45] They found that in the presence of (BINAP)PdCl$_2$ and zinc dust, oxaben-zonorbornadiene undergoes reductive ring opening with 2-propylpentanoic acid in 90% ee and excellent chemical yield [Eq. (10.28)]. The authors suggest that the carboxylic acid serves to protonate the Pd(0) complex to generate a palladium hydride that undergoes alkene insertion and β-oxygen elimination to afford the ring-opened product.

$$(10.28)$$

Bicyclic hydrazines have also been ring-opened using π-allyl chemistry. Treatment of *meso*-bicyclic hydrazine **88** with a chiral Pd complex in the presence of phenol provides the allyl phenyl ether **89** in good yield and moderate enantios-electivity:[46]

$$(10.29)$$

10.5 OLEFIN METATHESIS

A collaborative effort between Schrock and Hoveyda has been extremely fruitful as evidenced by the advent of the first efficient enantioselective olefin metathesis catalyst.[47] The initial communication included one example of a desymmetrization. Since then, these workers have broadened the scope of this transformation and developed several new catalysts that are more effective in certain cases.[48] A selection of results is illustrated in Eqs. (10.30a)–(10.30d). These workers have also examined the corresponding tungsten-based catalysts in these reactions:[49]

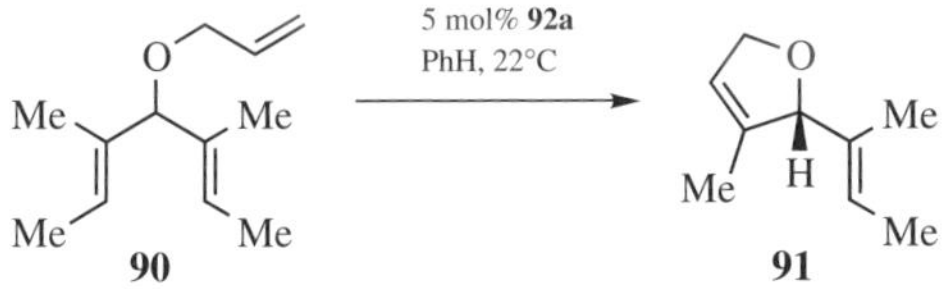

(10.30a)

83% yield
99% ee

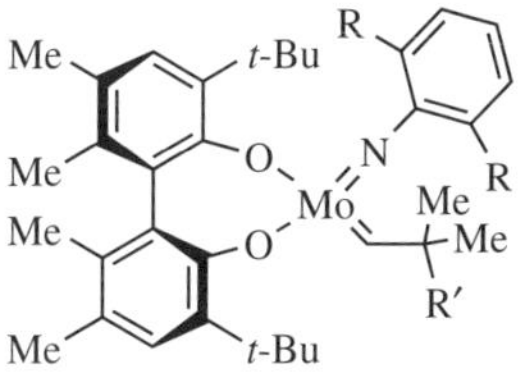

92a R=Me, R′=Ph
92b R=i-Pr, R′=Ph
92c R=Cl, R′=Me

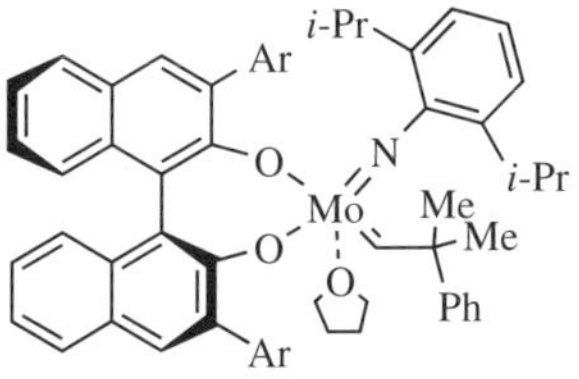

93 (Ar = 2,4,6-triisopropylphenyl)

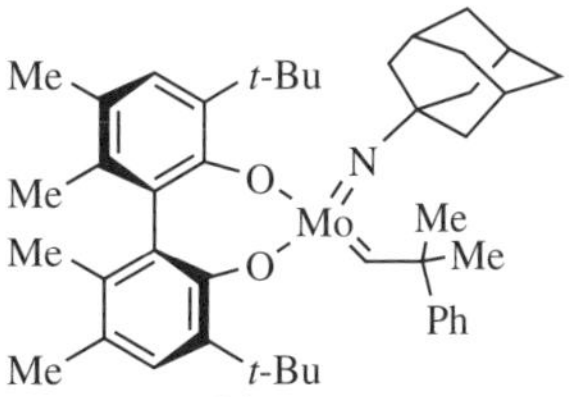

94

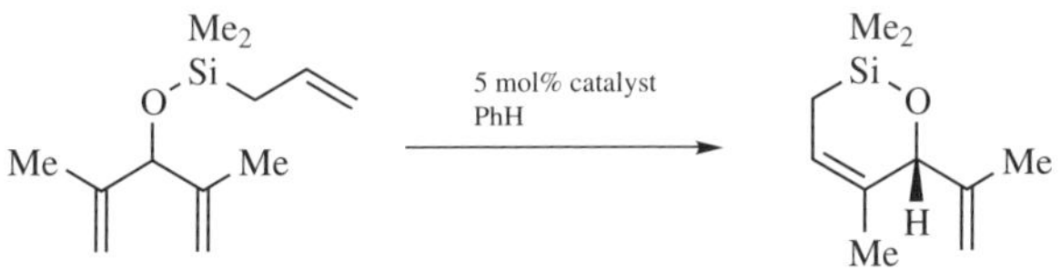

(10.30b)

Cat.	
92a (at 22°C)	20% yield 85% ee
93 (at 60°C)	98% yield 99% ee

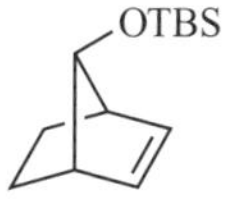

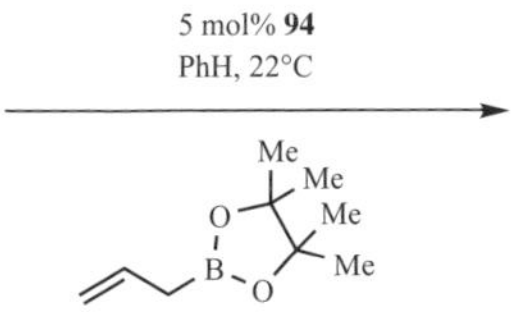

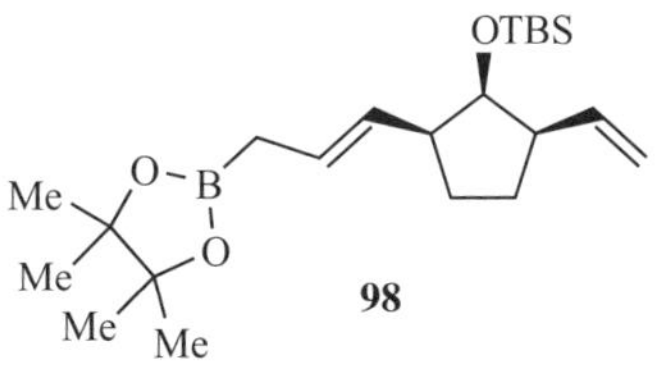

(10.30c)

88% yield
98% ee

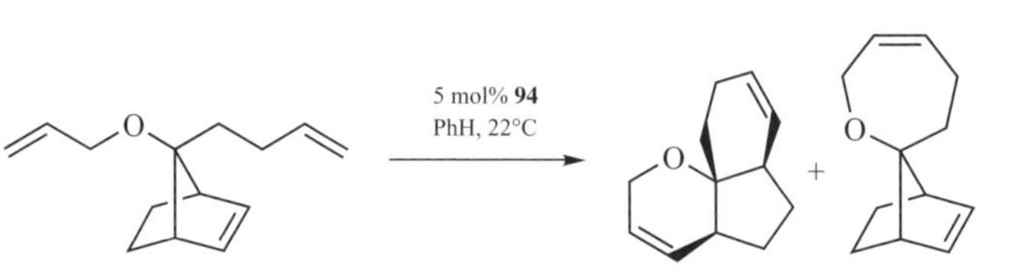

(10.30d)

99   **100**   **101**
                                          80% yield   20% yield
                                          96% ee

Some substrates require tuning of the catalyst to obtain a match in reactivity and selectivity. The authors argue that the modularity of this class of catalysts allows for ready solutions to this type of problem. Indeed, this family of catalysts is capable of effecting the asymmetric desymmetrization of substrates containing an aniline [Eq. (10.31)].[50] Particularly noteworthy here is the ability to form six-, seven-, and eight-membered rings enantioselectively:

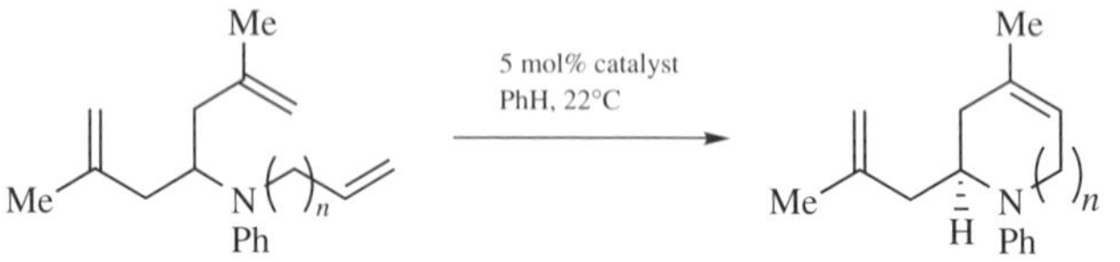

n	Cat.	
1	**92b**	78% yield
		98% ee
2	**92b**	90% yield
		95% ee
3	**92c**	93% yield
		>98% ee

(10.31)

Grubbs has described a chiral ruthenium-based catalyst and has illustrated its application to enantioselective desymmetrization. The catalyst is a derivative of Grubbs' second-generation dihydroimidazolinylidene-containing catalyst with the chirality derived from diphenylethanediamine in the carbene backbone. Substrate **90** undergoes ring-closing metathesis in 90% ee on treatment with this catalyst and NaI:[51]

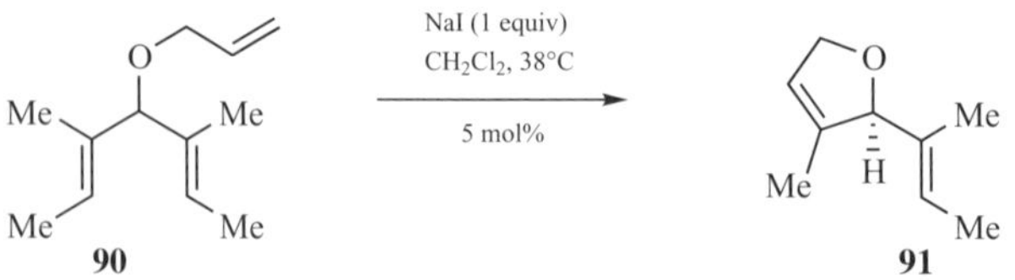

82% conversion
90% ee

(10.32)

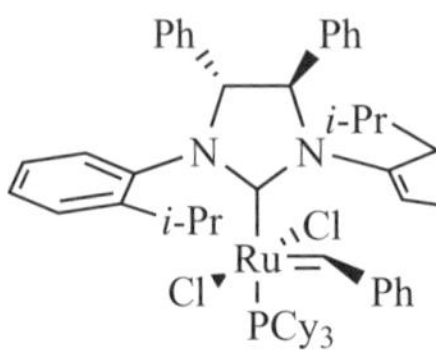

Hoveyda and coworkers have designed an analogous chiral ruthenium-based catalyst by taking advantage of bidentate binding.[52] A binaphthyl system coordinates the Ru through the phenolate of one naphthyl system and through a carbene on the other. The catalyst thus formed is active in the ring-opening metathesis/cross-metathesis of [2.2.1] systems such as **105** and **108**, providing excellent yields and selectivities [Eqs. (10.33a) and (10.33b)]. Of significance is that the catalyst may be recovered by silicagel chromatography at the end of the reaction. Hoveyda and coworkers have more recently improved the catalytic efficiency of this reaction and extended the scope by the use of a biaryl styrene ether, which, the authors argue, promotes release of the benzylidene and initiation of the catalytic cycle.[52b] Using catalyst **107b**, substrate **110** may be desymmetrized in moderate yield but excellent enantioselectivity [Eq. (10.33c)]:

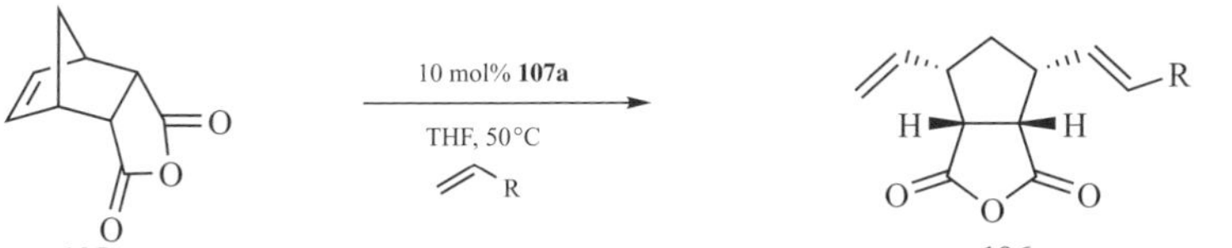

$$\text{(10.33a)}$$

R = Ph   71% yield, 80% ee
R = *n*-Pent 57% yield, 98% ee
R = *c*-Hex 60% yield, 98% ee

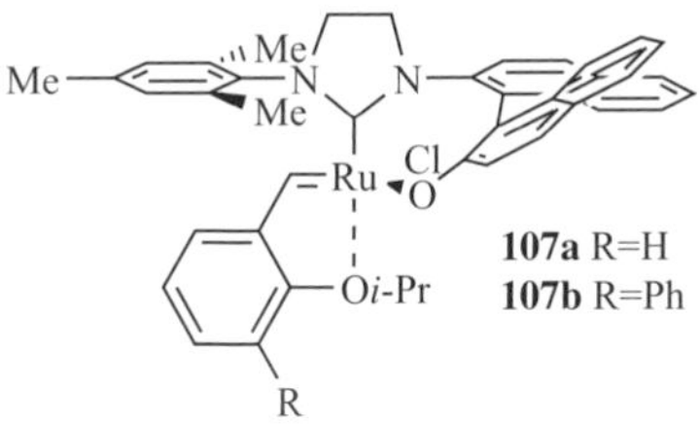

107a R=H
107b R=Ph

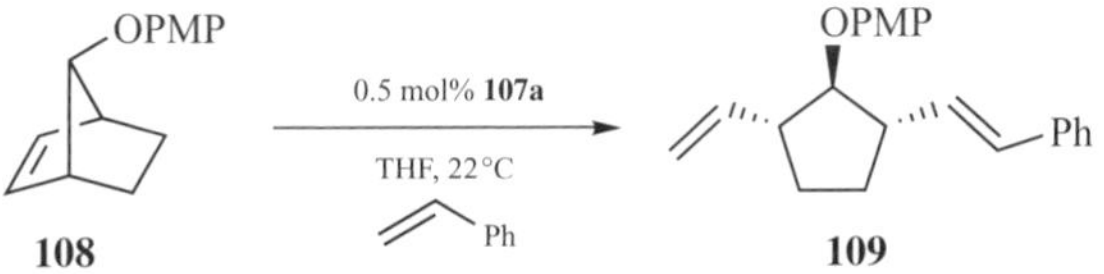

$$\text{(10.33b)}$$

76% yield, 95% ee

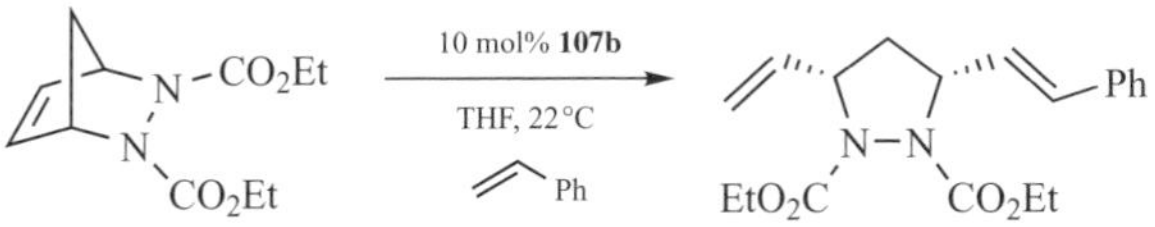

$$\text{(10.33c)}$$

65% yield, 92% ee

10.6 ACYLATION

Monoacylation of achiral and meso diols has been a popular strategy for introducing asymmetry and, in addition to kinetic resolutions of secondary alcohols, a traditional proving ground for new chiral acylation catalysts. Trost has shown that a chiral dinuclear zinc catalyst is effective at desymmetrizing 1,3- and 1,4-diols using vinyl benzoate as the acylation agent.[53] The acylation of 2-aryl propanediols using these catalysts affords selectivities comparable to the best enzymatic desymmetrizations, but Trost's dinuclear zinc catalysts stand out in their ability to desymmetrize 2-methyl propanediol and the cyclohexanedimethanol with high efficiency [Eqs. (10.34) and (10.35)]. On a related note, Oriyama and coworkers have reported that proline-derived diamine **115** effects the desymmetrization of 2-methylpropanediol in 96% ee with good turnover numbers [0.5 mol% catalyst provides 33% yield (66 TON)]; the reaction is conducted at −78°C in propionitrile in the presence of Hünig's base and molecular sieves:[54]

$$(10.34)$$

$$(10.35)$$

meso-Hydrobenzoin (**118**) has proved a popular testing ground for new chiral acylation catalysts. Vedejs has explored the use of phosphine catalysts in enantio-selective acylation reactions. Catalyst **120** was found to effect the desymmetrization of *meso*-hydrobenzoin in 83% ee and good yield, although the product was isolated as a mixture contaminated with the bis(benzoate ester).[55] Fujimoto and coworkers have described a cinchonine system for the asymmetric monoacylation of meso 1,2-diols.[56] The phosphinite derivative **121** proved to be most selective, providing the monoacylation product of a number of diols in high enantioselectivity. Surprisingly, *meso*-hydrobenzoin affords product of opposite absolute stereochemistry relative to that obtained with either cyclohexanediol or butanediol. The authors offer no comment on this observation. Matsumura has published a Lewis acid-mediated acylation using Ph-BoxCuCl$_2$ complex **122**. *meso*-Hydrobenzoin undergoes mono-acylation in excellent enantioselectivity with this system:[57]

$$(10.36)$$

In the context of our work in the area of chiral nucleophilic carbenes and their utility in organic synthesis, we have developed a conceptually distinct approach to catalyzed acylation using α-haloaldehydes as acylation precursors.[58] The use of a chiral triazolium salt in the presence of base allows an enantioselective desymmetrization of *meso*-hydrobenzoin to proceed in 83% ee and good yield:

$$(10.37)$$

A number of laboratories have targeted chiral DMAP analogs as acylation catalysts with perhaps the best success realized by Fu's planar chiral ferrocenyl pyridines.[59,60] Spivey and coworkers reported the use of a chiral biaryl-derived DMAP analog for the desymmetrization of *meso*-diols.[61] The catalytic efficiency of this system is moderate, with cyclohexanediol desymmetrized to provide the isobutyrate ester in 20% yield and 77% ee:

$$(10.38)$$

Miller has reported a "minimal kinase" for the phosphorylative enantioselective desymmetrization of a *meso*-inositol derivative.[62] Peptide **132**, identified from a small library, provided the highest selectivity using the *meso*-triol **130** as substrate. This approach was applied to a five-step synthesis of D-myoinositol-1-phosphate from readily available myoinositol:

$$(10.39)$$

10.7 ASYMMETRIC DEPROTONATION

Enantioselective deprotonations of meso substrates such as ketones or epoxides are firmly entrenched as a method in asymmetric synthesis, although the bulk of this work involves stoichiometric amounts of the chiral reagent.[63] Nevertheless, a handful of reports have appeared detailing a catalytic approach to enantioselective deprotonation. The issue that ultimately determines whether an asymmetric deprotonation may be rendered catalytic is a balance of the stoichiometric base's ability

to turn the catalyst over against its inefficiency in the deprotonation itself. Therefore, a two-pronged approach has been taken to this problem: a search for a more
effective chiral base to accelerate the productive reaction and a less effective stoichiometric base to decelerate the racemic pathway.

A number of workers have made progress on this front. Asami and coworkers
have anchored the stoichiometric base on the solid phase to realize a catalytic
desymmetrization using lithiated diamine 135.[64] Andersson has shown that slow
addition of LDA results in an improvement in enantioselectivity when using his
bicyclic base 136,[65] while Ahlberg has illustrated that a stoichiometric base such
as lithiated 1,2-dimethylimidazole results in an efficient catalytic system using diamine 137.[66] Alexakis has published a study involving a number of chiral ethane-
and propane-diamines in the catalytic deprotonation of cyclohexene oxide. Enantioselectivities observed are moderate, with diamine 138 providing the desired product in 59% ee and 80% yield.[67]

Building on Inoue's pioneering work[68] and using Andersson's versatile diamine,[69] Kozmin and coworkers have described a catalytic base-induced isomerization of silacyclopentene oxide **139** [Eq. (10.41)] as a potential route to acyclic polyol domains, important components of a number of natural products. Epoxidation of **140**, epoxide ring opening, and oxidation of the C–Si bonds provides tetraols in good overall yield and selectivity:[70]

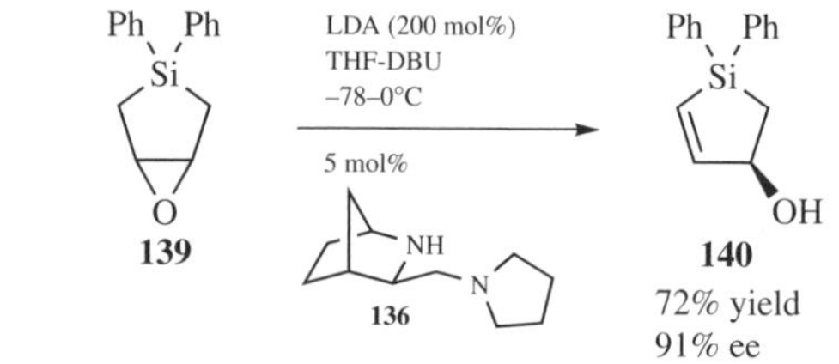

$$
\tag{10.41}
$$

139

136

140
72% yield
91% ee

10.8 OXIDATIONS

The groups of Sigman and Stoltz have concurrently published the palladium-catalyzed oxidative kinetic resolution of secondary alcohols using molecular oxygen as the stoichiometric oxidant.[71] Both communications also described a single example of a diol desymmetrization using a palladium catalyst in the presence of (−)-sparteine [Eqs. (10.42)[70a] and (10.43)[70b]]:

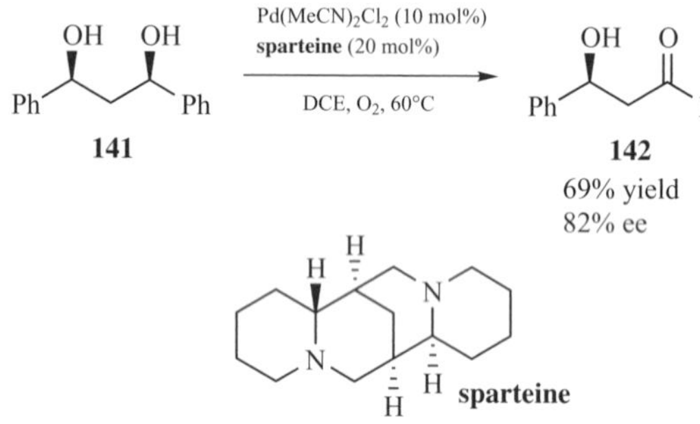

$$
\tag{10.42}
$$

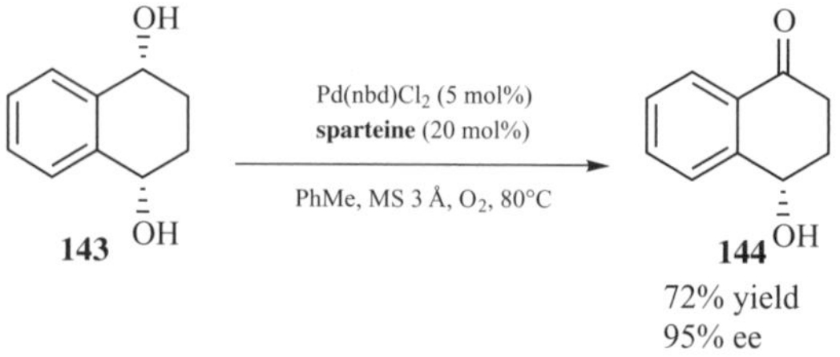

141

142
69% yield
82% ee

sparteine

$$
\tag{10.43}
$$

143

144
72% yield
95% ee

A hydroboration–oxidation sequence has been described for the desymmetrization of bicyclic hydrazino-alkenes. The use of BDPP as a chiral ligand on Rh provides the desired alcohol in 84% ee, following oxidation of the hydroborated intermediate [Eq. (10.44)].[72] Surprisingly, the use of Ir catalysts in this reaction leads to a reversal in the absolute sense of induction using the same chiral ligand:[73]

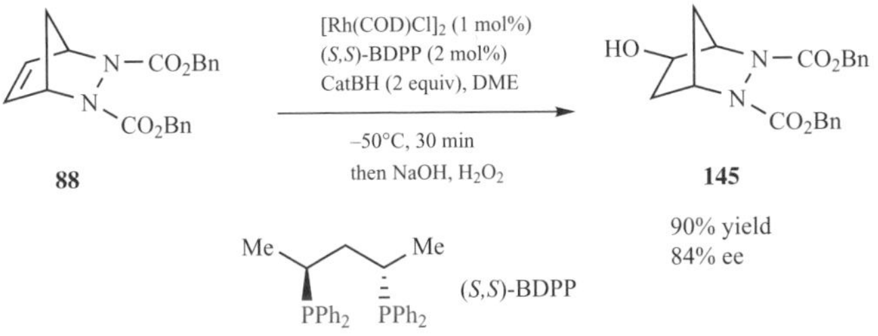

$$(10.44)$$

90% yield
84% ee

Katsuki and coworkers have developed a family of salen–metal complexes capable of effecting a C—H oxidation at activated positions. *meso*-Tetrahydrofurans may be oxidized to the lactol in good yield and excellent enantioselectivity using iodosylbenzene as the stoichiometric oxidant and a Mn–salen complex as catalyst [Eq. (10.45)].[74] Meso acylpyrrolidines behave similarly, providing slightly lower enantioselectivities using a similar catalyst [Eq. (10.46)]:[75]

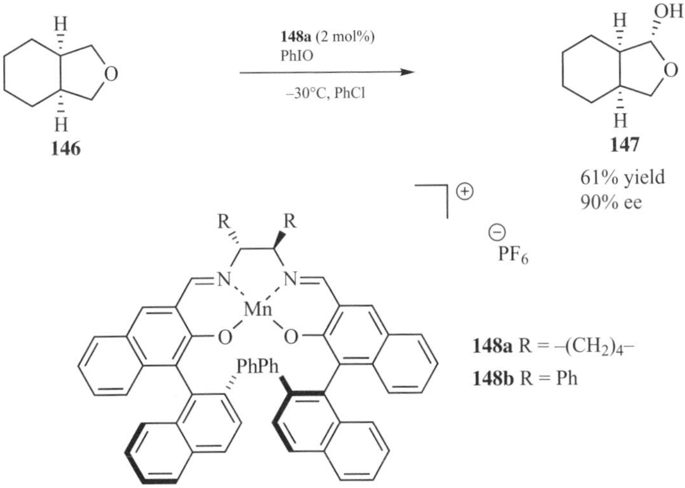

61% yield
90% ee

$$(10.45)$$

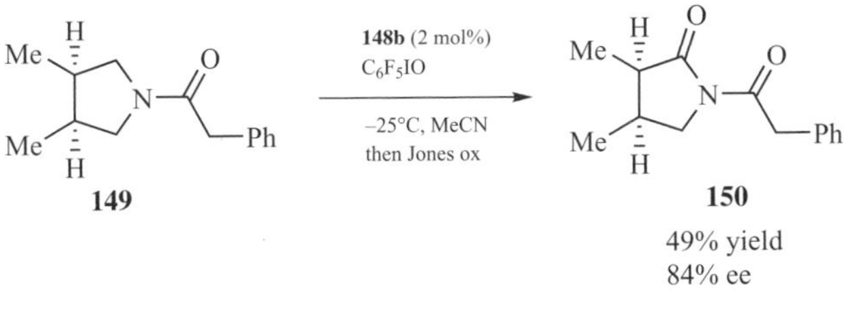

148a R = –(CH$_2$)$_4$–
148b R = Ph

$$(10.46)$$

49% yield
84% ee

These workers have also developed a Ru–salen system capable of desymmetriz-ing *meso*-diols in moderate enantiomeric excess [Eq. (10.47)].[76] It is of some interest that atmospheric oxygen suffices to induce catalyst turnover. Photolysis is presumably required to initiate nitrosyl loss from the Ru center:

$$(10.47)$$

Baeyer–Villiger oxidations of prochiral ketones have been developed by a num-ber of groups. As it stands, the asymmetric versions of these reactions are almost uniformly limited to strained cycloalkanones.[77] Nevertheless, some intriguing results have been published. Desymmetrization of 3-phenylcyclobutanone has been established as a standard in this field, and it is illustrative to compare the var-ious catalysts that have been developed for this purpose. These results are outlined in Eq. (10.48a).[78] The salen and BINOL ligand cores have been popular with a number of metals showing catalytic competence, with the unifying aspect being Lewis acidity as well as oxidative stability. Hydrogen peroxide is often used as the oxidant, although both *tert*-butyl hydroperoxide and cumene hydroperoxide have found application here. Of the metal-based catalysts, the Zr complex **157** seems to be optimal.[78b] Two intriguing systems that do not involve metals have been pub-lished. The C_2-symmetric bisflavin **158** affords moderate selectivity in this system albeit requiring long reaction times,[78e] while the selenium-based catalyst **159** in con-junction with a Lewis acid affords moderate reactivity but poor enantioselectivity.[78f] The tricyclic cyclobutanone **160** has also been a popular substrate providing excellent selectivities with a number of different catalyst systems; Katsuki's Zr complex is par-ticularly successful here [Eq. (10.48b).].[78b]

$$(10.48a)$$

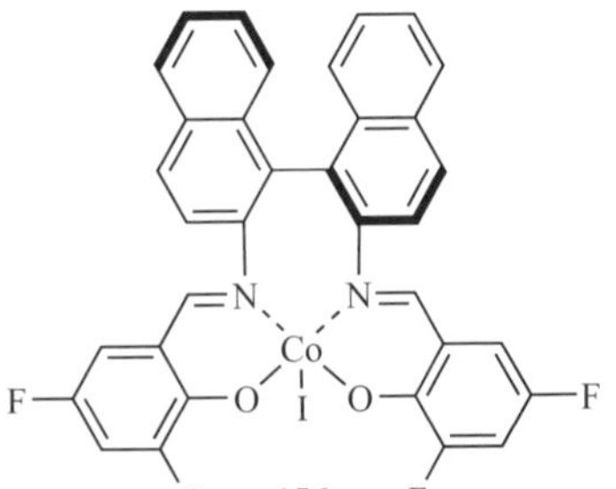

156

H₂O₂, EtOH, 0°C
96% yield, 79% ee

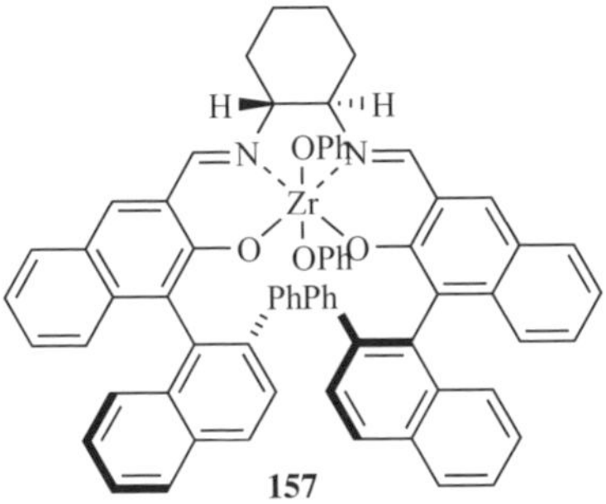

157
5 mol%
urea•H₂O₂, CH₂Cl₂, 23°C
68% yield, 87% ee

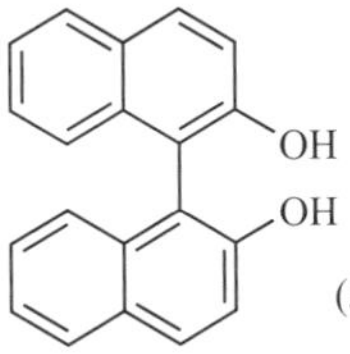

BINOL
(50 mol%)

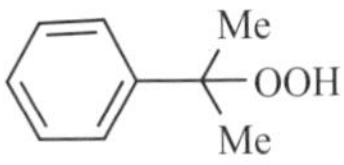

CHP
CH₂Cl₂, −25°C

MgI₂   99% conversion, 65% ee
Me₂AlCl 99% conversion, 71% ee

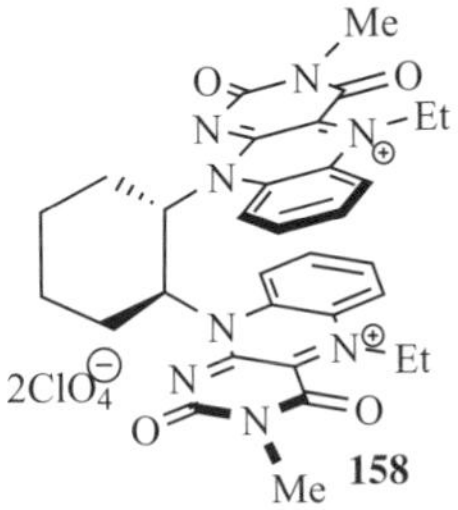

158
10 mol%
H₂O₂, NaOAc
TFE/MeOH/H₂O, −30°C
67% yield, 63% ee

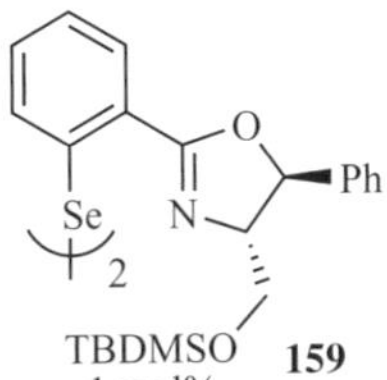

TBDMSO **159**
1 mol%
Yb(OTf)₃ (2 mol%)
H₂O₂, THF, 0°C
54% yield, 19% ee

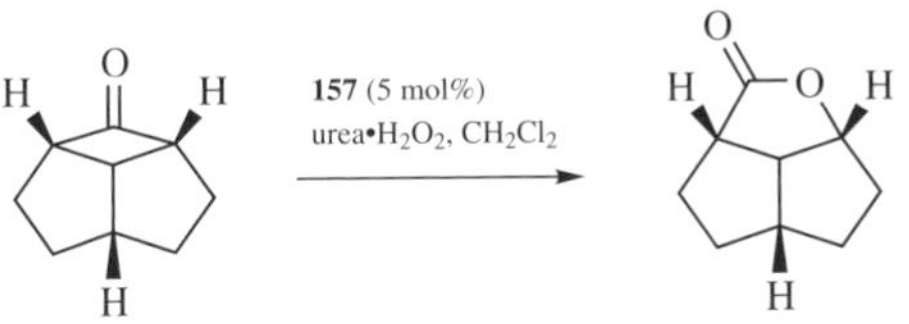

160

157 (5 mol%)
urea•H₂O₂, CH₂Cl₂

161
99% yield, 94% ee

(10.48b)

The challenge in the asymmetric Baeyer–Villiger reaction clearly seems to be reactivity. The above catalysts are incapable of inducing the reaction with larger cycloalkanones with a few exceptions. One catalyst system that is not subject to this limitation is Strukul's Pt-based catalyst **164**.[79] Although the system is still far from optimal, it is worth noting that this catalyst is capable of oxidizing cyclohexanones with some turnover and modest enantioselectivity using hydrogen peroxide as the oxidant:

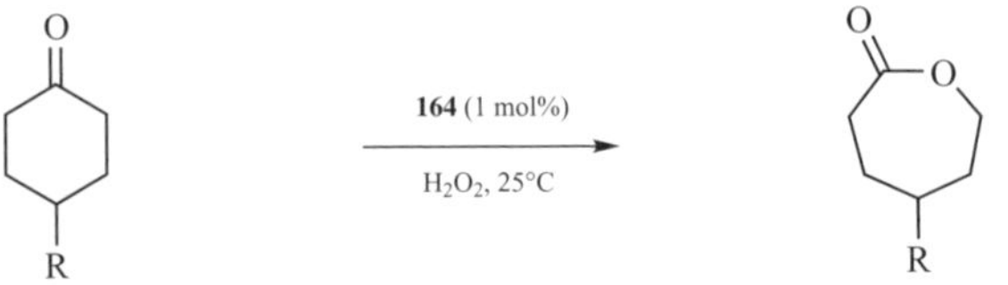

$$(10.49a)$$

162a (R=Me)
162b (R=Ph)

163a (R=Me) 17% yield, 50% ee
163b (R=Ph) 9% yield, 69% ee

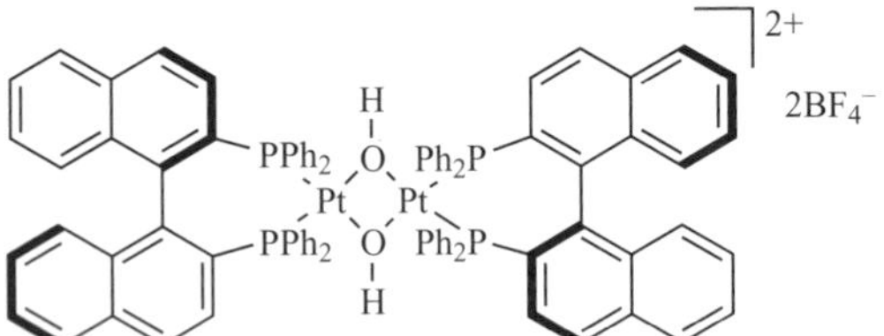

164

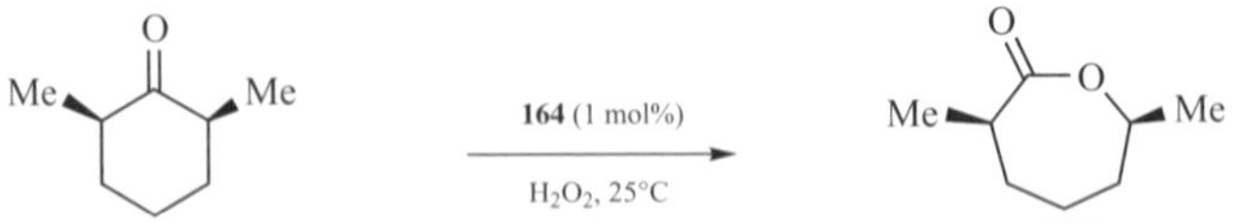

$$(10.49b)$$

165

166
18% yield, 75% ee

10.9   CYCLIC ANHYDRIDE DESYMMETRIZATION

Alcoholysis of *meso*-cyclic anhydrides offers a versatile route to succinate and glu-tarate half-esters. Although a number of stoichiometric approaches to this problem have been investigated, a successful catalytic version of this reaction appeared as recently as 2003.[80] Bolm and coworkers have developed a protocol for the metha-nolysis of a variety of succinic anhydrides using cinchona alkaloids [Eq. (10.50)]. The reaction may be made catalytic in alkaloid when pentamethylpiperidine is used as a stoichiometric additive.[81] A moderate decrease in enantioselectivity is observed in a number of cases, although excellent selectivities are still attainable. More problematic is the reaction time (6 days under catalytic conditions):

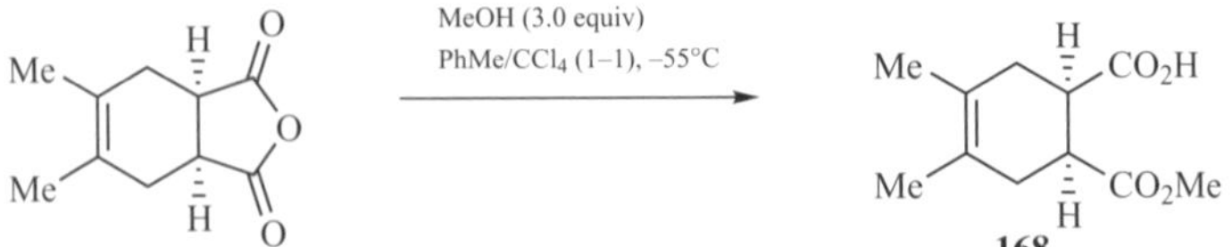

$$(10.50)$$

167

168

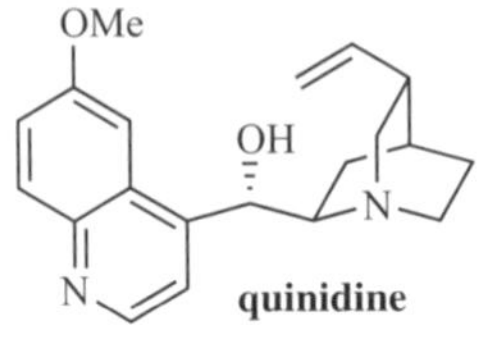

quinidine

1.1 equiv   96% yield, 97% ee

0.1 equiv +   96% yield, 92% ee
1.0 equiv

quinidine

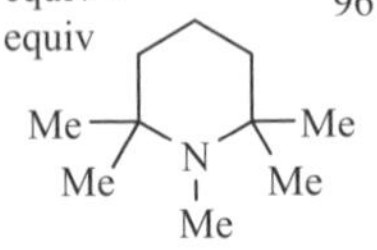

• Catalytic conditions:

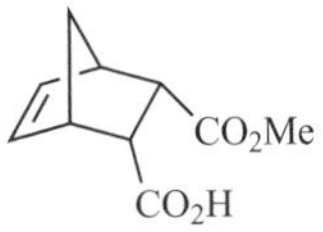

169
98% yield, 90% ee

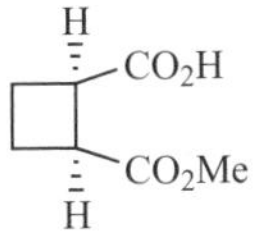

170
97% yield, 81% ee

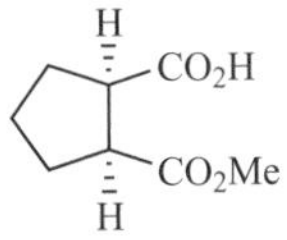

171
96% yield, 89% ee

Deng and coworkers showed that the use of the Sharpless-type ligands (DHQ)AQN and the pseudoenantiomer (DHQD)AQN provided a compact solution to this problem [Eq. (10.51)].[82] Of significance is that this protocol does not require the use of an exogenous base or other additive to achieve effective catalysis. The reaction seems to be general for succinic anhydrides and works well even for 4-substituted glutaric anhydrides. Deng has extended this concept to kinetic resolutions of various cyclic anhydrides[83] and to a formal total synthesis of biotin.[84] This catalyst has also been shown to be amenable to attachment to a surface while retaining catalytic activity:[85]

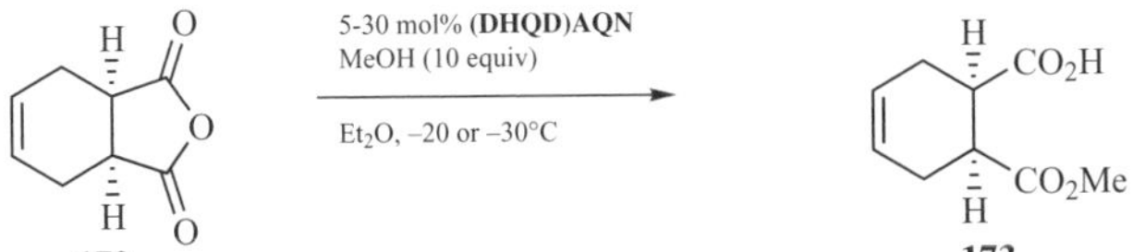

$$ \text{(10.51)} $$

5-30 mol% **(DHQD)AQN**
MeOH (10 equiv)

Et$_2$O, –20 or –30°C

172

173
95% yield, 98% ee

DHQD

AQN

DHQ

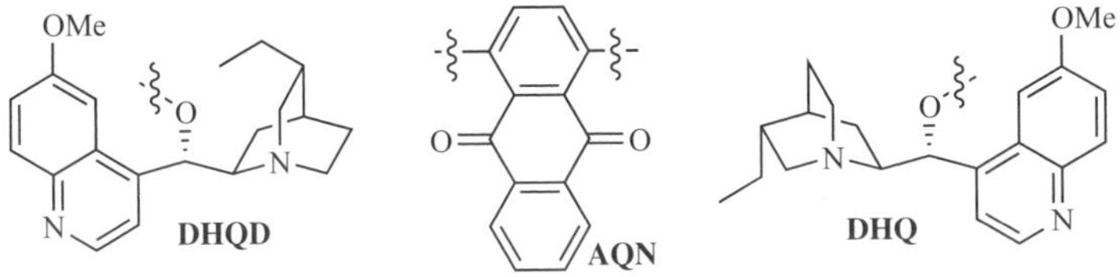

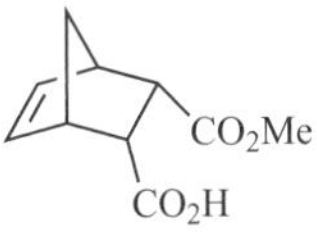

169
82% yield, 95% ee

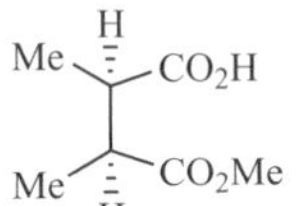

174
93% yield, 98% ee

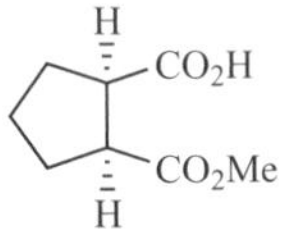

171
99% yield, 95% ee

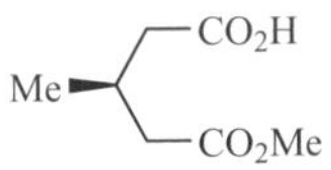

175
72% yield, 90% ee

Uozumi and coworkers have briefly investigated methanolysis of cyclic anhydrides catalyzed by chiral imidazolones.[86] Moderate reactivity and selectivity was observed:

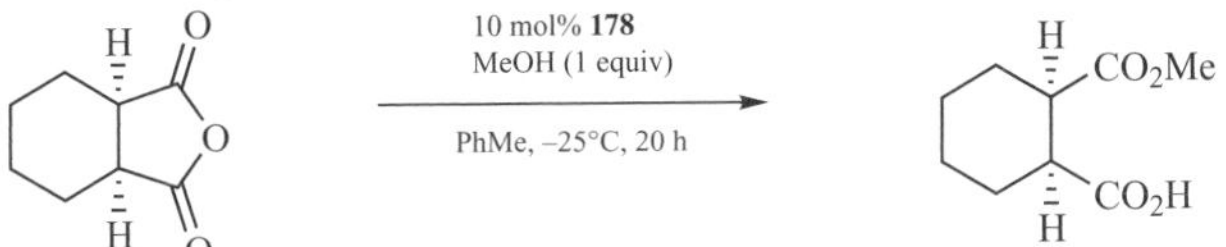

$$ \text{(10.52)} $$

10 mol% **178**
MeOH (1 equiv)

PhMe, –25°C, 20 h

176

177
33% yield, 65% ee

n-Oct

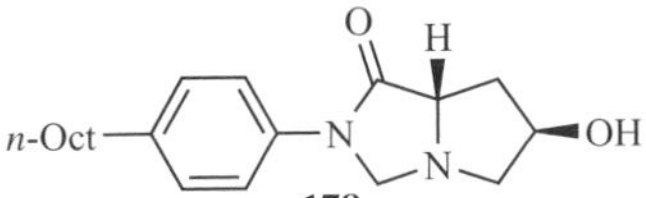

178

Alkylative desymmetrization of cyclic anhydrides would be an attractive alternative to this chemistry since the product would be a keto-acid and would allow elaboration of the molecular scaffold. The first catalyzed asymmetric alkylative desymmetrization of a cyclic anhydride was published from our laboratories in early 2002. The phosphinooxazoline ligand i-PrPHOX provided an effective catalyst when bound to nickel, affording keto-acid **179** in 85% yield and 79% ee:[87]

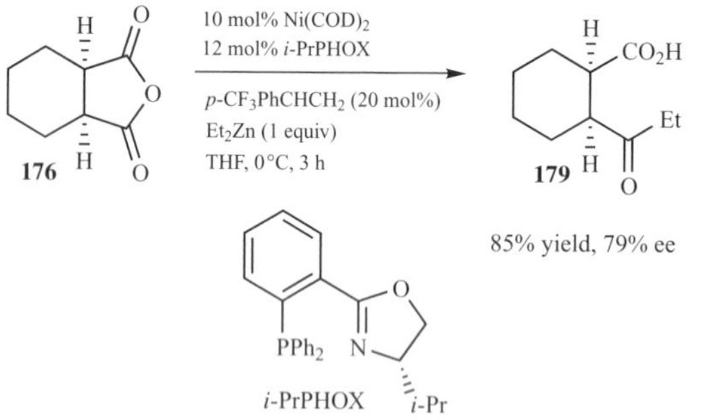

$$\begin{array}{c}\text{(scheme)}\end{array}\tag{10.53}$$

10 mol% Ni(COD)$_2$
12 mol% i-PrPHOX

p-CF$_3$PhCHCH$_2$ (20 mol%)
Et$_2$Zn (1 equiv)
THF, 0°C, 3 h

176

179

85% yield, 79% ee

PPh$_2$ N

i-PrPHOX i-Pr

More recently, we have discovered that Pd-JOSIPHOS complexes effectively desymmetrize a variety of succinic anhydrides in excellent yield and enantioselectivity [Eq. (10.54)].[88] The reaction proceeds at ambient temperature in some cases and can deliver aryl and alkylzinc reagents with equal facility. For reasons that are unclear, the latter protocol requires a styrenic additive for high enantioselectivity:

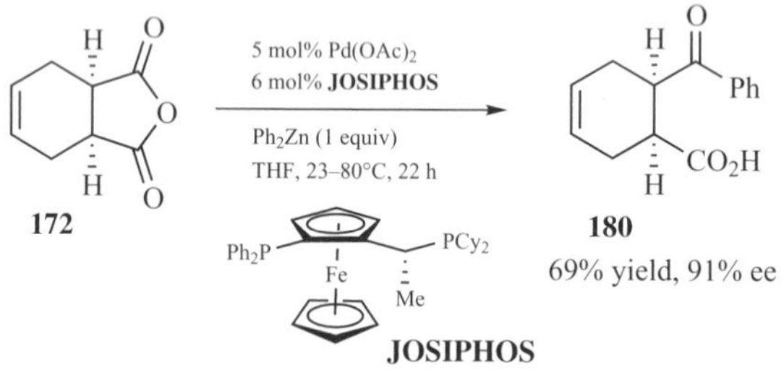

$$\begin{array}{c}\text{(scheme)}\end{array}\tag{10.54}$$

5 mol% Pd(OAc)$_2$
6 mol% **JOSIPHOS**

Ph$_2$Zn (1 equiv)
THF, 23–80°C, 22 h

172

180

69% yield, 91% ee

Ph$_2$P   Fe   PCy$_2$
Me

JOSIPHOS

181
83% yield, 97% ee

182
74% yield, 89% ee

183
72% yield,92% ee

184
61% yield, 89% ee

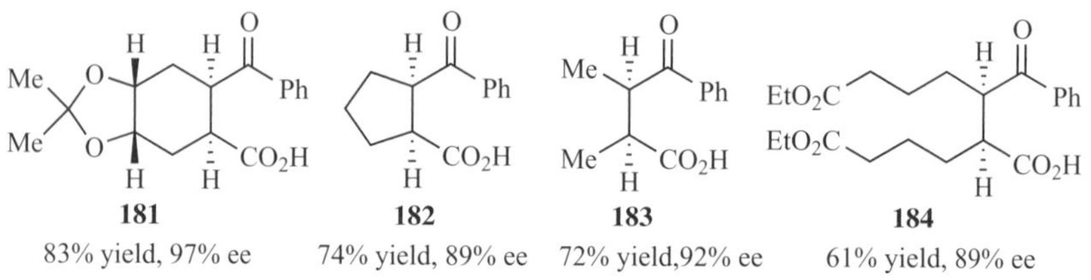

A catalyzed asymmetric alkylation of glutaric anhydrides has yet to appear. However, Fu has reported that stoichiometric amounts of sparteine efficiently mediate the addition of aryl Grignard reagents to 4-substituted glutaric anhydrides, providing the δ-ketoacids in good yields and excellent enantioselectivities:[89]

$$(10.55)$$

10.10 MISCELLANEOUS

A number of workers have examined enantioselective isomerizations as a desymmetrization strategy. Frauenrath and coworkers have been interested in the isomerization of dihydrooxepins using nickel complexes.[90] More recently, these workers reported that a Me-DUPHOS-NiI$_2$ complex in the presence of LiEt$_3$BH provides high enantioselectivity in the isomerization of dihydrooxepins:

$$(10.56)$$

Fu and coworkers have discovered an intriguing *trans*-hydroacylation of alkynals catalyzed by Rh complexes.[91] During the course of these investigations, they observed high enantioselectivities in the desymmetrization of a diyne. With

Tol-BINAP as ligand, the reaction proceeds in excellent yield and enantioselectivity to afford a series of 2-alkylcyclopentenones:

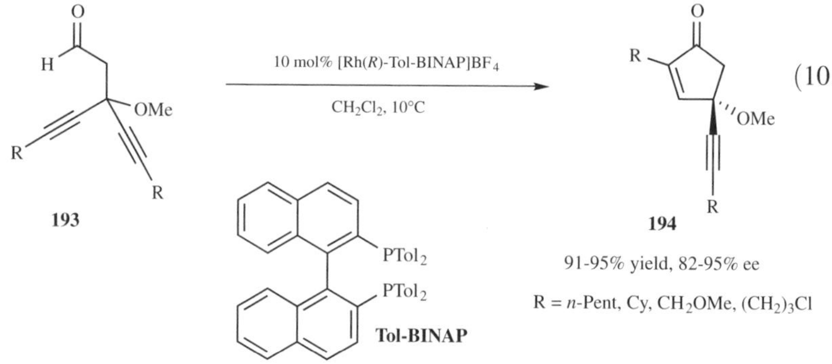

(10.57)

91-95% yield, 82-95% ee

R = n-Pent, Cy, CH$_2$OMe, (CH$_2$)$_3$Cl

Desymmetrizations by C—H insertion using diazo decomposition strategies is well established.[92] *Chiu and coworkers* have recently applied this strategy to the oxabicyclic system **195**. A *tert*-leucine-derived Rh(II) catalyst provided the best results [Eq. (10.58a)]. Wardrop and coworkers have applied a similar strategy en route to sordidin;[93] unfortunately, both yield and ee proved to be suboptimal [Eq. (10.58b)]:

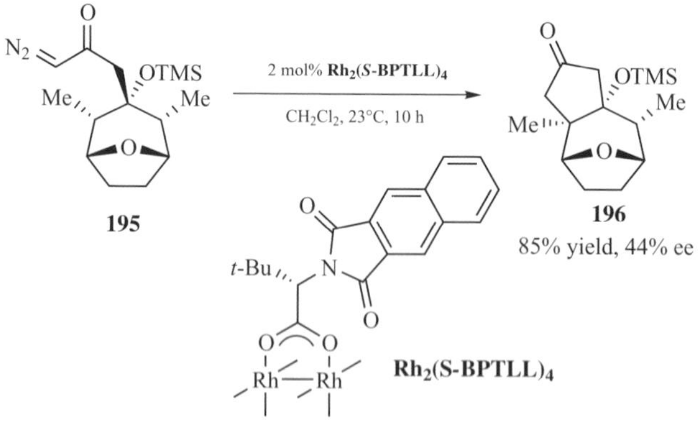

(10.58a)

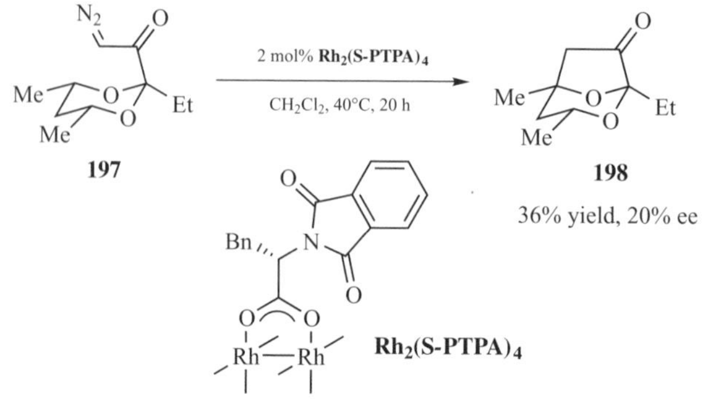

85% yield, 44% ee

(10.58b)

36% yield, 20% ee

Desymmetrization by oxidative addition has been examined in a number of contexts. Bräse reported that bisenolnonaflates undergo a desymmetrization/Heck reaction cascade with modest efficiency.[94] The use of BINAP as a ligand for Pd affords low ee's and poor yields:

$$\text{(10.59)}$$

199

Nf = $F_9C_4SO_2$

200

37% yield, 28% ee

Willis and coworkers realized better results using a similar strategy in a Suzuki coupling reaction.[94] The bistriflate of cyclopentanedione underwent a desymmetrizing cross-coupling in the presence of $Pd(OAc)_2$ and MeO-MOP as ligand to provide modest yields but good enantioselectivities of the cross-coupled product [Eq. (10.60)]. Of significance is that the desymmetrization generates a quaternary stereocenter, which is difficult to control by other means:

$$\text{(10.60)}$$

201

202

41-66% yield, 72-86% ee

Ar = Fu, indolyl, o-,m-, p-(CHO)Ph, etc.

MeO-MOP

The desymmetrizing cross-coupling reaction has been applied to setting axial chirality in bis(triflates).[95] More recently, Schmalz and coworkers have applied this strategy to controlling planar chirality.[96] The dichloroarenechromium substrate **203** was treated with a chiral Pd complex under methoxycarbonylation conditions. Under optimized conditions, the product ester was formed in moderate yield and excellent enantioselectivity [Eq. (10.61)]. Shorter reaction times resulted in higher yields of **204** but with lower enantiomeric excess, a scenario easily rationalized since the minor enantiomer is selectively activated in the second carbonylation event to afford the diester **205**, thereby enriching enantiomeric excess by a kinetic resolution:

$$(10.61)$$

Jeong and coworkers have executed a desymmetrization of a dienyne by asymmetric Pauson–Khand-type reaction.[97] Intriguingly, the use of a Rh catalyst resulted in preferential formation of one diastereomer, while a switch to the analogous Ir system provided the other diastereomer in excellent selectivity [Eq. (10.62)]; the system has been shown to be tolerant of oxygen in the linker as well as modest substitution on the alkyne (Ph):

$$(10.62)$$

5 mol% [Rh(COD)Cl]₂	1	:	11	
10 mol% **MeO-BINAP**			74% yield, 71% ee	
cinnamaldehyde				
120°C				
15 mol% [Ir(COD)Cl]₂	75	:	1	
30 mol% **MeO-BINAP**	75% yield, 96% ee			
PhMe, CO (1 atm)				
130°C				

MeO-BINAP
Ar = 4-MeOPh

Landais has extended his desymmetrization of dienes from dihydroxylation approaches[98] to a cyclopropanation reaction. A Cu-pybox complex provides the highest enantioselectivities and good diastereoselectivity in the asymmetric cyclopropanation of the silyl-substituted cyclopentadiene **210**:

$$(10.63)$$

211
48% yield, 68% ee, 84:16 dr

Spivey and coworkers have applied the CBS catalyst to the reduction of centrosymmetric imides.[99] The efficiency of this catalyst system leaves room for improvement, but the reaction is capable of forming up to four stereocenters in a single operation with good enantioselectivity:

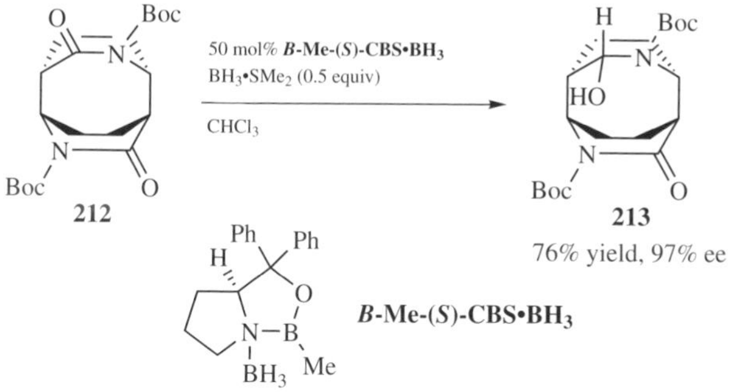
50 mol% B-Me-(S)-CBS•BH3
BH3•SMe2 (0.5 equiv)
CHCl3
212
213
76% yield, 97% ee
B-Me-(S)-CBS•BH3

(10.64)

A desymmetrizing reduction of a dicarbonyl has also been achieved as a route to *anti*-aldol adducts. Yamada and coworkers have shown that a chiral cobalt complex catalyzes the desymmetrization of diaryl-1,3-diketones in excellent yield and enantioselectivity, greatly favoring the anti isomer [Eq. (10.65)].[100] Anti selectivity is rationalized using a Felkin–Anh model:

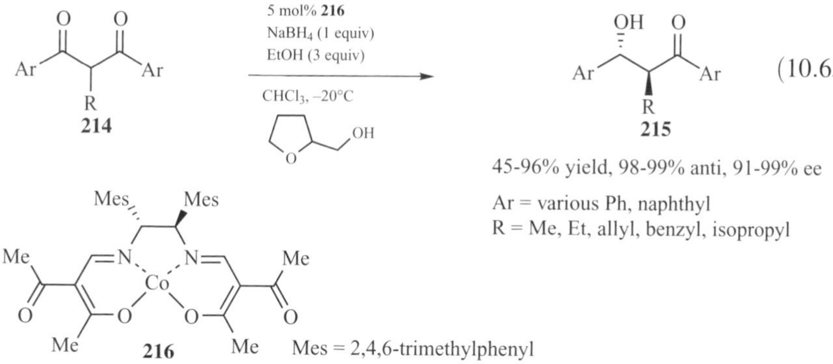
5 mol% 216
NaBH4 (1 equiv)
EtOH (3 equiv)
CHCl3, –20°C
214
215
45-96% yield, 98-99% anti, 91-99% ee
Ar = various Ph, naphthyl
R = Me, Et, allyl, benzyl, isopropyl
216
Mes = 2,4,6-trimethylphenyl

(10.65)

10.11 CONCLUDING REMARKS

Catalytic asymmetric desymmetrization as a field is still growing, with new applications appearing weekly. It is evident that advances in this subfield have kept in step with advances in catalysis as a whole. Some spectacular successes have been reported in recent years, and this strategy has been applied to many new reactions. Willis mentions in conclusion to his 1999 review of this field that desymmetrization reactions involving catalytic enantioselective construction of C–C bonds are scarce.[4a] It is of considerable interest that a little less than half of the examples shown in this review involve C—C bond formation, a number of which proceed with high efficiency. Nevertheless, much remains to be done. The substrate scope of many reactions is still limited, and the efficiency of some of the processes described above is below the threshold of synthetic utility. Given the ingenuity of organic chemists, one feels confident that these and other problems will be satisfactorily addressed in the near future.

ACKNOWLEDGMENTS

I thank my coworkers Eric Bercot, Mark Kerr, and Javier Read de Alaniz (Colorado State University) as well as Professor Keith Fagnou (University of Ottawa) for their careful reading of this manuscript.

REFERENCES

1. Mislow, K.; Siegel, J. *J. Am. Chem. Soc.* **1984**, *106*, 3319–3328.

2. Eliel, E. L.; Wilen, S. H.; Mander, L. N. *Stereochemistry of Organic Compounds*, Wiley, New York, **1994**.

3. Hoffmann, R. W. *Angew. Chem. Int. Ed.* **2003**, *42*, 1096–1109.

4. (a) Willis, M. C. *J. Chem. Soc. Perkin Trans 1* **1999**, 1765–1784; (b) Mikami, K.; Yoshida, A. *J. Syn. Org. Chem. Jpn.* **2002**, *60*, 732–739.

5. Trost, B. M. *Isr. J. Chem.* **1997**, *37*, 109–118.

6. Trost, B. M.; Patterson, D. E. *Chem. Eur. J.* **1999**, *5*, 3279–3284.

7. Trost, B. M.; Dirat, O.; Dudash Jr., J.; Hembre, E. J. *Angew. Chem. Int. Ed.* **2001**, *40*, 3658–3660.

8. Jiang, L.; Burke, S. D. *Org. Lett.* **2002**, *4*, 3411–3414.

9. Pineschi, M. *New J. Chem.* **2004**, *28*, 657–665.

10. Piarulli, U.; Daubos, P.; Claverie, C.; Roux, M.; Gennari, C. *Angew. Chem. Int. Ed.* **2003**, *42*, 234–236.

11. Piarulli, U.; Claverie, C.; Daubos, P.; Gennari, C.; Minnaard, A. J.; Feringa, B. L. *Org. Lett.* **2003**, *5*, 4493–4496.

12. Del Moro, G.; Crotti, P.; Di Bussolo, V.; Macchia, F.; Pineschi, M. *Org. Lett.* **2003**, *5*, 1971–1974.

13. Bertozzi, F.; Crotti, P.; Maccia, F.; Pineschi, M.; Arnold, A.; Feringa, B. L. *Org. Lett.* **2000**, *2*, 933–936.

14. Jacobsen, E. N.; Wu, M. H. In *Comprehensive Asymmetric Catalysis*, Jacobsen, E. N.; Pfaltz, A.; Yamamoto, H. (Eds.), Springer, New York, **1999**, p. 1309.

15. Schaus, S. E.; Jacobsen, E. N. *Org. Lett.* **2000**, *2*, 1001–1004.

16. Zhu, C.; Yuan, F.; Gu, W.; Pan, Y. *Chem. Commun.* **2003**, 692–693.

17. Matsunaga, S.; Das, J.; Roels, J.; Vogl, E. M.; Yamamoto, N.; Iida, T.; Yamaguchi, K.; Shibasaki, M. *J. Am. Chem. Soc.* **2000**, *122*, 2252–2260.

18. Zhu, C.; Yang, M.; Sun, J.; Zhu, Y.; Pan, Y. *Synlett* **2004**, 465–468.

19. Sagawa, S.; Abe, H.; Hase, Y.; Inaba, T. *J. Org. Chem.* **1999**, *64*, 4962–4965.

20. Jacobsen, E. N. *Acc. Chem. Res.* **2000**, *33*, 421–431.

21. Kassab, D. J.; Ganem, B. *J. Org. Chem.* **1999**, *64*, 1782–1783.

22. (a) Holland, J. M.; Lewis, M.; Nelson, A. *Angew. Chem. Int. Ed.* **2001**, *40*, 4082–4084; (b) Holland, J. M.; Lewis, M.; Nelson, A. *J. Org. Chem.* **2003**, *68*, 747–753.

23. Bartoli, G.; Bosco, M.; Carlone, A.; Locatelli, M.; Massaccesi, M.; Melchiorre, P.; Sambri, L. *Org. Lett.* **2004**, *6*, 2173–2176.

24. Haufe, G.; Bruns, S. *Adv. Synth. Catal.* **2002**, *344*, 165–171.

25. Denmark, S. E.; Barsanti, P. A.; Wong, K.-T.; Stavenger, R. A. *J. Org. Chem.* **1998**, *63*, 2428–2429.

26. Tao, B.; Lo, M. M.-C.; Fu, G. C. *J. Am. Chem. Soc.* **2001**, *123*, 353–354.

27. Nakajima, M.; Saito, M.; Uemura, M.; Hashimoto, S. *Tetrahedron Lett.* **2002**, *43*, 8827–8829.

28. Li, Z.; Fernández, M.; Jacobsen, E. N. *Org. Lett.* **1999**, *1*, 1611–1613.

29. (a) Müller, P.; Nury, P. *Org. Lett.* **1999**, *1*, 439–441; (b) Müller, P.; Nury, P. *Helv. Chim. Acta* **2001**, *84*, 662–677.

30. The use of oxabicycloalkenes as scaffolds for natural product synthesis, including desymmetrization approaches involving these molecules, has recently been reviewed; see: Hartung, I. V.; Hoffmann, H. M. R. *Angew. Chem. Int. Ed.* **2004**, *43*, 1934–1949.

31. Lautens, M.; Fagnou, K.; Rovis, T. *J. Am. Chem. Soc.* **2000**, *122*, 5650–5651.

32. Lautens, M.; Fagnou, K.; Taylor, M. *Org. Lett.* **2000**, *2*, 1677–1679.

33. Leong, P.; Lautens, M. *J. Org. Chem.* **2004**, *69*, 2194–2196.

34. Lautens, M.; Fagnou, K. *J. Am. Chem. Soc.* **2001**, *123*, 7170–7171.

35. Lautens, M.; Fagnou, K.; Zunic, V. *Org. Lett.* **2002**, *4*, 3465–3468.

36. Lautens, M.; Fagnou, K.; Yang, D. *J. Am. Chem. Soc.* **2003**, *125*, 14884–14892.

37. Lautens, M.; Fagnou, K. *Proc. Natl. Acad. Sci. USA* **2004**, *101*, 5455–5460.

38. Lautens, M.; Renaud, J.-L.; Hiebert, S. *J. Am. Chem. Soc.* **2000**, *122*, 1804–1805.

39. Lautens, M.; Hiebert, S.; Renaud, J.-L. *Org. Lett.* **2000**, *2*, 1971–1973.

40. Lautens, M.; Hiebert, S.; Renaud, J.-L. *J. Am. Chem. Soc.* **2001**, *123*, 6834–6839.

41. Bertozzi, F.; Pineschi, M.; Macchia, F.; Arnold, L. A.; Minnaard, A. J.; Feringa, B. L. *Org. Lett.* **2002**, *4*, 2703–2705.

42. Lautens, M.; Dockendorff, C.; Fagnou, K.; Malicki, A. *Org. Lett.* **2002**, *4*, 1311–1314.

43. Millward, D. B.; Sammis, G.; Waymouth, R. M. *J. Org. Chem.* **2000**, *65*, 3902–3909.

44. Lautens, M.; Rovis, T. *J. Am. Chem. Soc.* **1997**, *119*, 11090–11091.

45. Li, L.-P.; Rayabarapu, D. K.; Nandi, M.; Cheng, C.-H. *Org. Lett.* **2003**, *5*, 1621–1624.

46. Pérez Luna, A.; Cesario, M.; Bonin, M.; Micouin, L. *Org. Lett.* **2003**, *5*, 4771–4774.

47. La, D. S.; Alexander, J. B.; Cefalo, D. R.; Graf, D. D.; Hoveyda, A. H.; Schrock, R. R. *J. Am. Chem. Soc.* **1998**, *120*, 720–9721.

48. (a) Zhu, S. S.; Cefalo, D. R.; La, D. S.; Jamieson, J. Y.; Davis, W. M.; Hoveyda, A. H.; Schrock, R. R. *J. Am. Chem. Soc.* **1999**, *121*, 8251–8259; (b) Weatherhead, G. S.; Ford, J. G.; Alexanian, E. J.; Schrock, R. R.; Hoveyda, A. H. *J. Am. Chem. Soc.* **2000**, *122*, 1828–1829; (c) Tsang, W. C. P.; Jernelius, J. A.; Cortez, G. A.; Weatherhead, G. S.; Schrock, R. R.;

Hoveyda, A. H. *J. Am. Chem. Soc.* **2003**, *125*, 2591–2596; (d) Weatherhead, G. S.; Cortez, G. A.; Schrock, R. R.; Hoveyda, A. H. *Proc. Nat. Acad. Sci. USA* **2004**, *101*, 5805–5809.

49. Tsang, W. C. P.; Hultzsch, K. C.; Alexander, J. B.; Bonitatebus Jr., P. J.; Schrock, R. R.; Hoveyda, A. H. *J. Am. Chem. Soc.* **2003**, *125*, 2652–2666.

50. Dolman, S. J.; Sattely, E. S.; Hoveyda, A. H.; Schrock, R. R. *J. Am. Chem. Soc.* **2002**, *124*, 6991–6997.

51. Seiders, T. J.; Ward, D. W.; Grubbs, R. H. *Org. Lett.* **2001**, *3*, 3225–3228.

52. (a) Van Veldhuizen, J. J.; Garber, S. B.; Kingsbury, J. S.; Hoveyda, A. H. *J. Am. Chem. Soc.* **2002**, *124*, 4954–4955; (b) Van Veldhuizen, J. J.; Gillingham, D. G.; Garber, S. B.; Kataoka, O.; Hoveyda, A. H. *J. Am. Chem. Soc.* **2003**, *125*, 12502–12508.

53. Trost, B. M.; Mino, T. *J. Am. Chem. Soc.* **2003**, *125*, 2410–2411.

54. Oriyama, T.; Taguchi, H.; Terakado, D.; Sano, T. *Chem. Lett.* **2002**, 26–27.

55. Vedejs, E.; Daugulis, O.; Tuttle, N. *J. Org. Chem.* **2004**, *69*, 1389–1392.

56. Mizuta, S.; Sadamori, M.; Fujimoto, T.; Yamamoto, I. *Angew. Chem. Int. Ed.* **2003**, *42*, 3383–3385.

57. Matsumura, Y.; Maki, T.; Murakami, S.; Onomura, O. *J. Am. Chem. Soc.* **2003**, *125*, 2052–2053.

58. Reynolds, N. T.; Read de Alaniz, J.; Rovis, T. *J. Am. Chem. Soc.* **2004**, *126*, 9518–9519. A conceptually identical reaction was independently discovered by Bode and Chow; see Chow, K. Y.-K.; Bode, J. W. *J. Am. Chem. Soc.* **2004**, *126*, 8126–8127.

59. Ruble, J. C.; Tweddell, J.; Fu, G. C. *J. Org. Chem.* **1998**, *63*, 2794–2795.

60. Fu, G. C. *Acc. Chem. Res.* **2000**, *33*, 412–420.

61. Spivey, A. C.; Zhu, F.; Mitchell, M. B.; Davey, S. G.; Jarvest, R. L. *J. Org. Chem.* **2003**, *68*, 7379–7385.

62. (a) Sculimbrene, B. R.; Miller, S. J. *J. Am. Chem. Soc.* **2001**, *123*, 10125–10126; (b) Sculimbrene, B. R.; Morgan, A. J.; Miller, S. J. *J. Am. Chem. Soc.* **2002**, *124*, 11653–11656; (c) Sculimbrene, B. R.; Morgan, A. J.; Miller, S. J. *Chem. Commun.* **2003**, 1781–1785.

63. Hodgson, D. M.; Gibbs, A. R.; Lee, G. P. *Tetrahedron* **1996**, *52*, 14361–14384.

64. Seki, A.; Asami, M. *Tetrahedron* **2002**, *58*, 4655–4663.

65. Bertilsson, S. K.; Andersson, P. G. *Tetrahedron* **2002**, *58*, 4665–4668.

66. Pettersen, D.; Amedjkouh, M.; Ahlberg, P. *Tetrahedron* **2002**, *58*, 4669–4673.

67. Equey, O.; Alexakis, A. *Tetrahedron Asymm.* **2004**, *15*, 1069–1072.

68. (a) Asami, M.; Ishizaki, T.; Inoue, S. *Tetrahedron: Asymm.* **1994**, *5*, 793; (b) Asami, M.; Suga, T.; Honda, K.; Inoue, S. *Tetrahedron Lett.* **1997**, *38*, 6425.

69. Södergren, M. J.; Andersson, P. G. *J. Am. Chem. Soc.* **1998**, *120*, 10760–10761.

70. Liu, D.; Kozmin, S. A. *Angew. Chem. Int. Ed.* **2001**, *40*, 4757–4759.

71. (a) Jensen, D. R.; Pugsley, J. S.; Sigman, M. S. *J. Am. Chem. Soc.* **2001**, *123*, 7475–7476; (b) Ferreira, E. M.; Stoltz, B. M. *J. Am. Chem. Soc.* **2001**, *123*, 7725–7726; (c) Mandal, S. K.; Jensen, D. R.; Pugsley, J. S.; Sigman, M. S. *J. Org. Chem.* **2003**, *68*, 4600–4603.

72. Pérez Luna, A.; Ceschi, M.-A.; Bonin, M.; Micouin, L.; Husson, H.-P. *J. Org. Chem.* **2002**, *67*, 3522–3524.

73. Pérez Luna, A.; Bonin, M.; Micouin, L.; Husson, H.-P. *J. Am. Chem. Soc.* **2002**, *124*, 12098–12099.

74. Miyafuji, A.; Katsuki, T. *Tetrahedron* **1998**, *54*, 10339–10348.

75. Punniyamurthy, T.; Miyafuji, A.; Katsuki, T. *Tetrahedron Lett.* **1998**, *39*, 8295–8298.

76. Shimizu, H.; Nakata, K.; Katsuki, T. *Chem. Lett.* **2002**, 1080–1081.

77. (a) Bolm, C.; Beckmann, O. Baeyer-Villiger reaction, in *Comprehensive Asymmetric Catalysis*, Jacobsen, E. N.; Pfaltz, A.; Yamamoto, H. (Eds.), Springer-Verlag, Berlin, 1999, p. 803; (b) Strukul, G. *Angew. Chem. Int. Ed.* **1998**, *37*, 1198–1209.

78. (a) Uchida, T.; Katsuki, T.; Ito, K.; Akashi, S.; Ishii, A.; Kuroda, T. *Helv. Chim. Acta* **2002**, *85*, 3078–3089; (b) Watanabe, A.; Uchida, T.; Ito, K.; Katsuki, T. *Tetrahedron Lett.* **2002**, *43*, 4481–4485; (c) Bolm, C.; Beckmann, O.; Cosp, A.; Palazzi, C. *Synlett* **2001**, 1461–1463; (d) Bolm, C.; Beckmann, O.; Palazzi, C. *Can. J. Chem.* **2001**, *79*, 1593–1597; (e) Murahashi, S.-I.; Ono, S.; Imada, Y. *Angew. Chem. Int. Ed.* **2002**, *41*, 2366–2368; (f) Miyake, Y.; Nishibayashi, Y.; Uemura, S. *Bull. Chem. Soc. Jpn.* **2002**, *75*, 2233–2237.

79. Paneghetti, C.; Gavagnin, R.; Pinna, F.; Strukul, G. *Organometallics* **1999**, *18*, 5057–5065.

80. (a) Spivey, A. C.; Andrews, B. I. *Angew. Chem. Int. Ed.* **2001**, *40*, 3131–3134; (b) Chen, Y.; McDaid, P.; Deng, L. *Chem. Rev.* **2003**, *103*, 2965–2984.

81. Bolm, C.; Schiffers, I.; Dinter, C. L.; Gerlach, A. *J. Org. Chem.* **2000**, *65*, 6984–6991.

82. Chen, Y.; Tian, S.-K.; Deng, L. *J. Am. Chem. Soc.* **2000**, *122*, 9542–9543.

83. (a) Chen, Y.; Deng, L. *J. Am. Chem. Soc.* **2001**, *123*, 11302–11303; (b) Hang, J.; Tian, S.-K.; Tang, L.; Deng, L. *J. Am. Chem. Soc.* **2001**, *123*, 12696–12697.

84. Choi, C.; Tian, S.-K.; Deng, L. *Synthesis* **2001**, 1737–1741.

85. Song, Y.-M.; Choi, J. S.; Yang, J. W.; Han, H. *Tetrahedron Lett.* **2004**, *45*, 3301–3304.

86. Uozumi, Y.; Yasoshima, K.; Miyachi, T.; Nagai, S.-i. *Tetrahedron Lett.* **2001**, *42*, 411–414.

87. Bercot, E. A.; Rovis, T. *J. Am. Chem. Soc.* **2002**, *124*, 174–175.

88. Bercot, E. A.; Rovis, T. *J. Am. Chem. Soc.* **2004**, *126*, 10248–10249.

89. Shintani, R.; Fu, G. C. *Angew. Chem. Int. Ed.* **2002**, *41*, 1057–1059.

90. (a) Frauenrath, H.; Philipps, T. *Angew. Chem, Int. Ed. Engl.* **1986**, *25*, 274; (b) Frauenrath, H.; Reim, S.; Wiesner, A. *Tetrahedron Asymm.* **1998**, *9*, 1103–1106.

91. Tanaka, K.; Fu, G. C. *J. Am. Chem. Soc.* **2002**, *124*, 10296–10297.

92. Davies, H. M. L.; Beckwith, R. E. J. *Chem. Rev.* **2003**, *103*, 2861–2904.

93. Wardrop, D. J.; Forslund, R. E. *Tetrahedron Lett.* **2002**, *43*, 737–739.

94. (a) Bräse, S. *Synlett* **1999**, 1654–1656; (b) Willis, M. C.; Powell, L. H. W.; Claverie, C. K.; Watson, S. J. *Angew. Chem. Int. Ed.* **2004**, *43*, 1249–1251.

95. (a) Hayashi, T.; Niizuma, S.; Kamikawa, T.; Suzuki, N.; Uozumi, Y. *J. Am. Chem. Soc.* **1995**, *117*, 9101–9102; (b) Kamikawa, T.; Hayashi, T. *Tetrahedron* **1999**, *55*, 3455–3466.

96. Böttcher, A.; Schmalz, H.-G. *Synlett* **2003**, 1595–1597.

97. Jeong, N.; Kim, D. H.; Choi, J. H. *Chem. Commun.* **2004**, 1134–1135.

98. (a) Angelaud, R.; Babot, O.; Charvat, T.; Landais, Y. *J. Org. Chem.* **1999**, *64*, 9613–9624; (b) Landais, Y.; Zekri, E. *Tetrahedron Lett.* **2001**, *42*, 6547–6551; (c) Landais, Y.; Zekri, E. *Eur. J. Org. Chem.* **2002**, 4037–4053.

99. Spivey, A. C.; Andrews, B. I.; Brown, A. D.; Frampton, C. S. *Chem. Commun.* **1999**, 2523–2524.

100. Ohtsuka, Y.; Koyasu, K.; Ikeno, T.; Yamada, T. *Org. Lett.* **2001**, *3*, 2543–2546.

11

HISTORY AND PERSPECTIVE OF CHIRAL ORGANIC CATALYSTS

GÉRALD LELAIS AND DAVID W. C. MACMILLAN

Department of Chemistry, California Institute of Technology,
Pasadena, California, USA

11.1 INTRODUCTION

Since the late 1960s, advances in the development of metal-based chiral catalysts have provided a wealth of enantioselective oxidation, reduction, π-bond activation and Lewis acid-catalyzed reactions.[1–4] In contrast, the use of organic molecules as reaction catalysts has only gained widespread attention since the mid-1990s, a remarkable fact given the potential savings in cost, time, energy, operational complexity, and chemical waste.[5] During the last decade, organocatalysis has grown at an extraordinary pace in accord with the large number of researchers and ideas that continue to advance this burgeoning area. Moreover, this upstart field has grown from a small collection of chemically unique reactions to a thriving area of general catalysis concepts, atypical reactivity, and widely applicable reactions. The primary attractions of organic catalysis can be summarized by four central features of conceptual and operational significance:

1. This new field allows the development of novel modes of substrate activation, thereby enabling the invention of transformations that were previously unknown.

2. Organocatalytic processes provide the capacity to catalyze known transformations with chemo-, regio-, diastereo-, or enantioselectivities that are orthogonal or complementary to known metal-catalyzed systems.

New Frontiers in Asymmetric Catalysis, Edited by Koichi Mikami and Mark Lautens
Copyright © 2007 John Wiley & Sons, Inc.

3. Organic catalysts are generally insensitive to the oxygen and moisture in our natural atmosphere and, as such, do not require the use of air- and moisture-free reaction/storage vessels, inert-gas atmospheres, and moisture-free reagents/solvents.

4. A large variety of organic reagents are naturally available as single enantiomers (e.g., amino acids, carbohydrates, hydroxy acids). Therefore, organic catalysts are generally inexpensive to prepare, easy to handle, and readily accessible in industrial-scale quantities.

In light of these facts, organocatalysis has become a field of central importance to chemical synthesis and has been adopted as a major research area on a global scale. Moreover, widespread application of organocatalytic reactions has become prevalent in the pharmaceutical and biotechnology industries. Indeed, given the rapid uptake of such technologies within medicinal and process chemistry groups, it is clear that organocatalytic methods will come to be routinely exploited for the large-scale manufacturing of enantiopure drugs.

A literature survey of articles that focus on the use of organocatalytic concepts indicate that this area began exponential growth in 2000, and that interest in this field has continued to rapidly expand on a yearly basis ever since (Figure 11.1)![6] As a further example of this phenomenon, the total number of organocatalysis articles for year 2004 has been overtaken in the first 6 months of 2005.

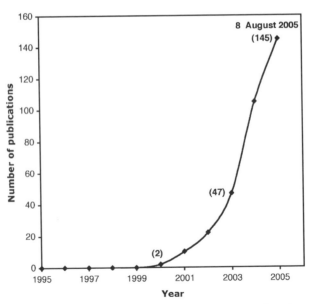

Figure 11.1. Diagram indicating the number of articles per year focusing on the use of organocatalytic concepts.

This chapter surveys the history of organocatalysis up to the present time and focuses on the creative insights that have triggered its currently rapid growth. To keep this chapter within a reasonable length, we have focused only on the major achievements in the areas of iminium, enamine, Brønsted acid, and phase transfer catalysis. For other important areas of organocatalysis, such as the use of hypervalent silicon in asymmetric transformations (i.e., Lewis base catalysis),[7–10] the chemistry of polymer-supported[11] and cinchona alkaloid[12–14] catalysts, or the application of nucleophilic[15–21] and acyl-anion[22–24] catalysis, we refer to their corresponding review articles. Section 11.2 describes the pioneering contributions that have set the foundations for this field. Emphasis is placed on the discoveries that outlined for the first time new organocatalytic activation concepts. The subsequent sections cover the four major subfields of asymmetric organocatalysis that have become the focus of broad attention in the chemical community since the late 1990s. Thus, Section 11.3 covers the new catalysis concept of iminium activation, wherein catalyst design has enabled the discovery of a broad range of enantioselective catalytic transformations. Section 11.4 provides an overview on enamine catalysis, and Section 11.5 focuses on reactions that rely on the breakthrough field of hydrogen-bonding activation (sometimes referred to as *Brønsted acid catalysis*). Section 11.6 covers the chemistry of phase transfer catalysis (PTC) and reactions involving chiral quaternary ammonium ion catalysts under homogeneous conditions. Finally, Section 11.7 includes remarks on the perspective of asymmetric organocatalysis.

11.2 HISTORICAL BACKGROUND

The field of asymmetric organocatalysis has a longstanding history that begins in the early twentieth century (see timeline in Figure 11.2).

Indeed, the first example of asymmetric organocatalytic reaction was reported by Bredig and Fiske in 1912.[25] In these studies, cinchona alkaloids catalyzed the addition of hydrogen cyanide to benzaldehyde (and later, the addition to other aldehydes)[26,27] affording optically active cyanohydrins with low enantiomeric excess (<10% ee).[25–27] Remarkably, it took almost 50 years before the pioneering work of Pracejus[28,29] described the use of *O*-benzoylquinine (**1**) as an organocatalyst to enable the enantioselective methanolysis of phenylmethylketene, providing the first example of an organocatalytic transformation with, at the time, notable levels of enantioselectivity (up to 76% ee) [Eq. (11.1)]:

Ketene methanolysis (Pracejus): 1960

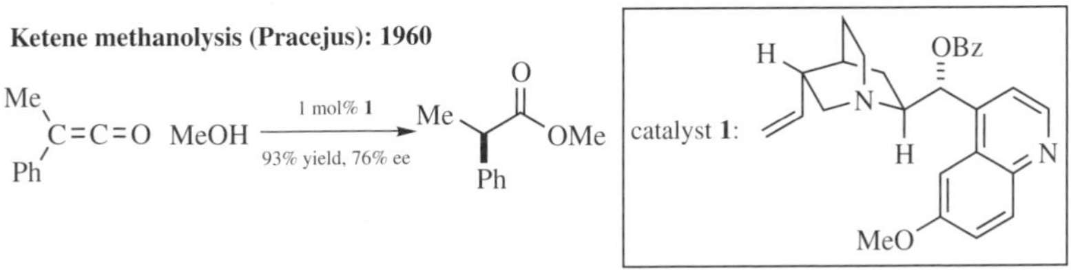

$$\begin{array}{c} Me \\ \diagdown \\ C=C=O \quad MeOH \xrightarrow[\substack{93\% \text{ yield, } 76\% \text{ ee}}]{1 \text{ mol}\% \textbf{1}} \\ \diagup \\ Ph \end{array} \qquad \begin{array}{c} O \\ \| \\ Me \diagdown\!\!\diagdown\!\!\diagup OMe \\ | \\ Ph \end{array} \qquad \text{catalyst } \textbf{1:} \qquad (11.1)$$

In the late 1960s and early 1970s, the groups of Hajos[5a] and Wiechert,[5b] respectively, performed the now-famous proline (**2**)-catalyzed intramolecular aldol

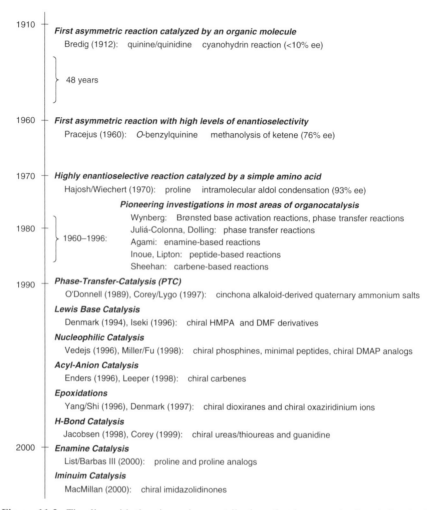

1910 — **First asymmetric reaction catalyzed by an organic molecule**
  Bredig (1912):   quinine/quinidine   cyanohydrin reaction (<10% ee)

  } 48 years

1960 — **First asymmetric reaction with high levels of enantioselectivity**
  Pracejus (1960):   *O*-benzylquinine   methanolysis of ketene (76% ee)

1970 — **Highly enantioselective reaction catalyzed by a simple amino acid**
  Hajosh/Wiechert (1970):   proline   intramolecular aldol condensation (93% ee)
    Pioneering investigations in most areas of organocatalysis
      Wynberg: Brønsted base activation reactions, phase transfer reactions
1980 —   1960–1996:   Juliá-Colonna, Dolling: phase transfer reactions
      Agami: enamine-based reactions
      Inoue, Lipton: peptide-based reactions
      Sheehan: carbene-based reactions
1990 — **Phase-Transfer-Catalysis (PTC)**
  O'Donnell (1989), Corey/Lygo (1997):   cinchona alkaloid-derived quaternary ammonium salts
  Lewis Base Catalysis
  Denmark (1994), Iseki (1996):   chiral HMPA and DMF derivatives
  Nucleophilic Catalysis
  Vedejs (1996), Miller/Fu (1998):   chiral phosphines, minimal peptides, chiral DMAP analogs
  Acyl-Anion Catalysis
  Enders (1996), Leeper (1998):   chiral carbenes
  Epoxidations
  Yang/Shi (1996), Denmark (1997):   chiral dioxiranes and chiral oxaziridinium ions
  H-Bond Catalysis
  Jacobsen (1998), Corey (1999):   chiral ureas/thioureas and guanidine
2000 — **Enamine Catalysis**
  List/Barbas III (2000):   proline and proline analogs
  Iminuim Catalysis
  MacMillan (2000):   chiral imidazolidinones

Figure 11.2. Timeline with the pioneering contributions that have set the foundation in the field of organocatalysis.

condensation of prochiral triketones (now commonly referred to as the *Hajos–Parrish reaction*) [Eq. (11.2)]. This desymmetrizing aldol provided enantioenriched bicyclic adducts in ≤93% ee for use in the total synthesis of steroids and other natural products.[30–39]

Intramolecular aldol reaction (Hajosh, Wiechert): 1971

$$\text{3 mol% 2} \qquad \text{quant., 93% ee} \qquad \text{catalyst 2:} \qquad (11.2)$$

Remarkably, this reaction and the underlying catalytic principle lay dormant for almost 25 years, before the pioneering work of Barbas et al.[5d] demonstrated the first intermolecular variant of this direct aldol protocol. Importantly, their studies inspired a large range of chemists to begin research in the field of (what became subsequently known as) enamine catalysis, thereby enabling the rebirth and timely growth of an extremely important subfield of organocatalysis.

In the 1970s and 1980s, many groups reported the use of cinchona alkaloids as chiral catalysts,[40–42] providing further evidence to their status as privileged catalysts for reactions such as hetero-[2+2] cycloadditions,[43–45] phase-transfer-catalyzed epoxidations,[46–53] alkylations,[54–57] conjugate additions,[58–73] and phosphonylation reactions of aldehydes.[74,75] It should be noted that other organic architectures have also been employed as organocatalysts for similar reactions,[51,76–88] in some cases with improved enantioselectivities. Among these examples, perhaps the most powerful transformation is the ammonium-salt (**3**) catalyzed methylation of an indanone derivative [Eq. (11.3)] as developed by the process research group at Merck, Rahway, NJ.[55–57] Indeed, it was this reaction that provided the foundation for the subfield of quaternary-ammonium-salts-mediated asymmetric phase transfer catalysis (PTC):

Phase-transfer methylation (Dolling, O'Donnell): 1984

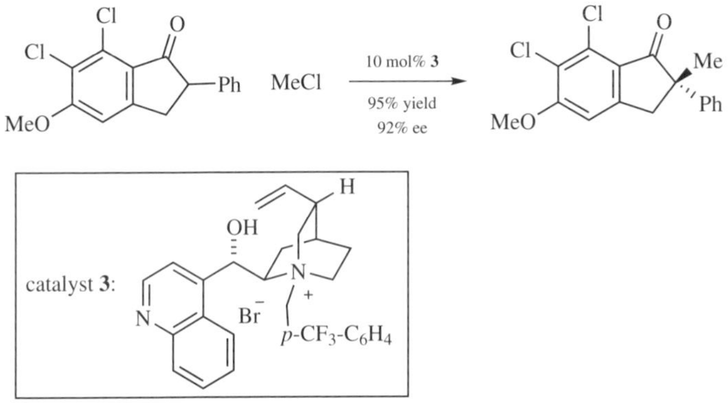

$$(11.3)$$

The next milestone in the realm of enantioselective organocatalysis was reached by Inoue and coworkers,[89–93] who elegantly modernized the cyanohydrin reaction, first outlined by Bredig and Fiske in 1912. In these studies, a cyclic histidine-containing dipeptide (**4**) catalyzed the HCN addition to aromatic aldehydes with high enantioselectivities (97% ee) [Eq. (11.4)]: a result that effectively paved the way for the field of peptide-catalyzed nucleophilic addition to aldehydes and imines:

Cynaohydrin reaction (Inoue): 1979

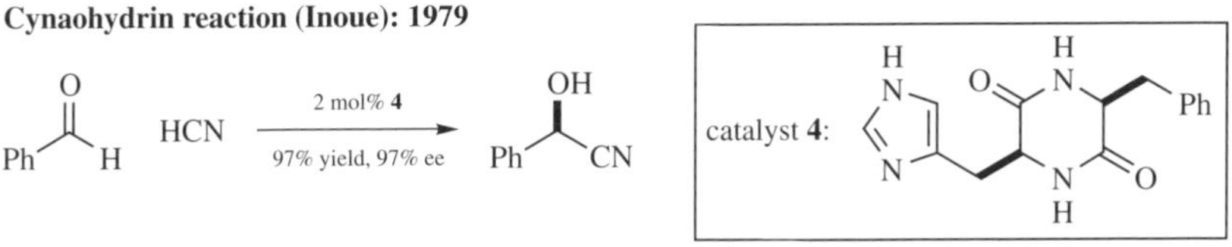

$$(11.4)$$

Perhaps more importantly, these studies provided the necessary impetus for the development of the hydrogen-bonding/Brønsted acid catalysis field. Indeed, shortly after Inoue's studies, the first reports of enantioenriched α-amino acid synthesis via enantioselective Strecker reaction (i.e., the addition of HCN to imines) appeared in the literature [Eq. (11.5)].[94,95] Further developments by Corey[5r] and Jacobsen[5q,s,t,96–98] outlined the use of a guanidine or acyclic peptide-based catalysts, respectively, to mediate this transformation. Notable in these studies are the divergent strategies of catalyst development, with Corey employing a C_2-symmetric guanidine (**5**) [Eq. (11.5)] and Jacobsen applying a high-throughput evaluation technique to identify useful peptide-like catalysts:

Hydrocyanation (Lipton, Corey, Jacobsen): 1996

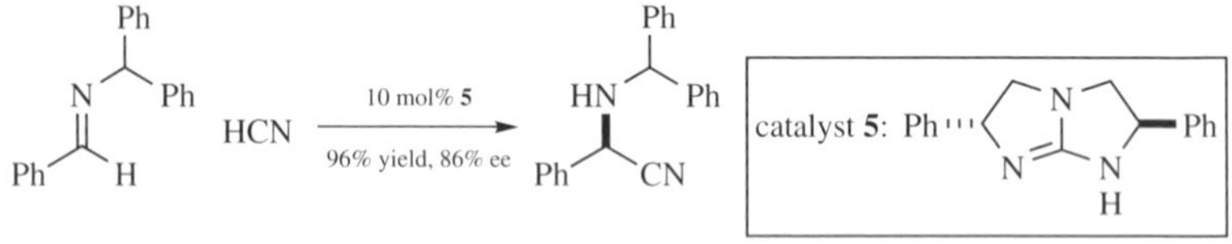

$$\text{(11.5)}$$

In the 1990s, short peptides,[5u,99] and other nucleophiles[15,16,100–107] were used as organocatalysts for a number of enantioselective acyl transfer processes; transformations that set the stage for the more recent research in the area of nucleophilic catalysis.[15–21] One of the most appealing approaches to enantioselective acyl transfer was outlined by Fu using an azaferrocene catalyst (**6**) [Eq. (11.6)].[15,103–107] While these pyridyl systems are not organic catalysts in the strictest sense, these azaferrocene compounds function as chiral dimethylaminopyridine equivalents for a broad range of acyl transfer processes:

Acyl transfer (Fu, Miller, Vedejs): 1996

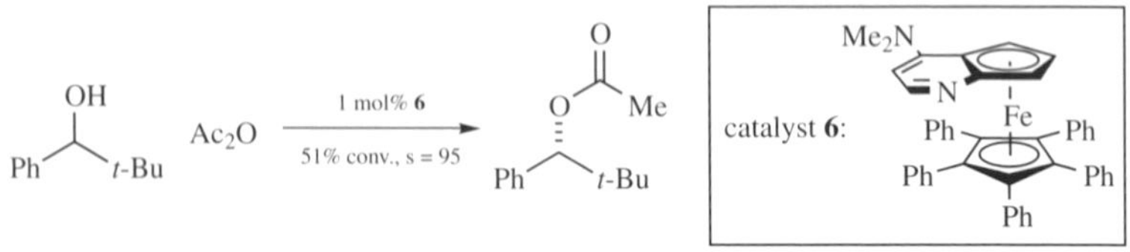

$$\text{(11.6)}$$

An important area of organocatalysis that was also initiated in the early 1990s has been the realm of Lewis base catalysis.[7–10] On the basis of the nonintuitive activation principle of silyl bonding rehybridization, Denmark and Iseki introduced chiral HMPA[108–114] and DMF variants[115,116] as effective catalysts for enantioselective allylation and aldol reactions of aldehydes:

Allylation (Iseki, Denmark): 1994

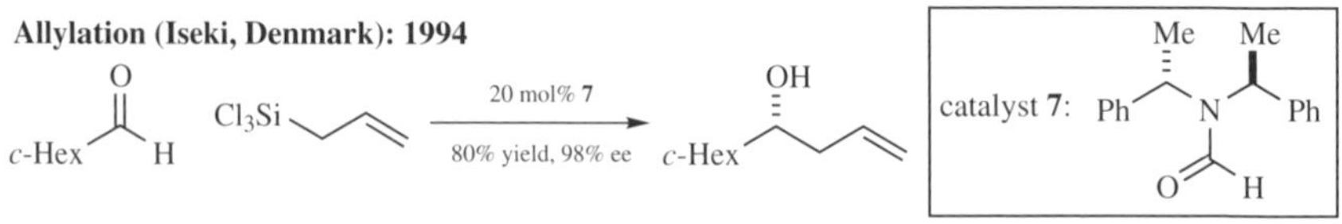

$$\text{(11.7)}$$

Among the most widely recognized areas of organic catalysis in the modern era have been the seminal studies of Yang,[5j,l,117,118] Shi,[5k,n,119–130] and Denmark[5m,131,132] toward the development of nonmetal enantioselective epoxidation reactions [Eq. (11.8)].[133–135] As outlined with the Shi protocol [Eq. (11.8)], these reactions employ chiral ketone catalysts to mediate the asymmetric transfer of oxygen from oxone to a range of unfunctionalized olefins with excellent levels of enantiocontrol:

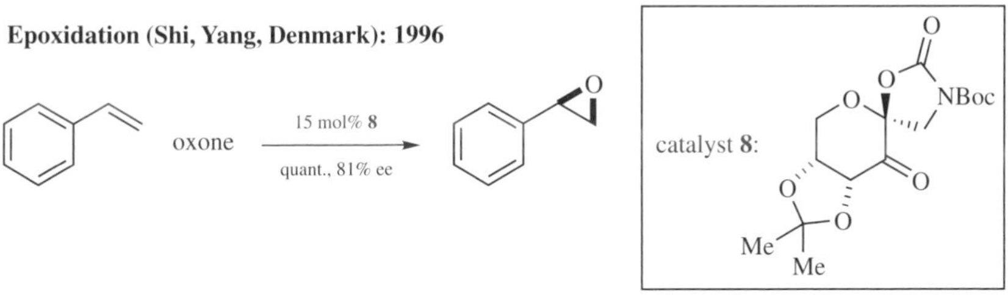

Epoxidation (Shi, Yang, Denmark): 1996

$$(11.8)$$

While the studies outlined in Eqs. (11.1)–(11.8) summarize the most important frontiers for metal-free synthesis, they also highlight a common limitation for the organocatalysis field. To date, there remain relatively few (less than 10) activation mechanisms that have been established to be amenable to organic catalysis. Accordingly, a primary objective for the advancement of the field of asymmetric organocatalysis has been the design and/or development of concepts that enable organic substrates to function as catalysts for a wide variety of new and established reactions.

11.3   IMINIUM CATALYSIS: A NEW CONCEPT IN ORGANOCATALYSIS

The Catalysis Concept of Iminium Activation   In 2000, the MacMillan laboratory[136] disclosed a new strategy for asymmetric synthesis based on the capacity of chiral amines to function as enantioselective catalysts for a range of transformations that traditionally use Lewis acids. This catalytic concept was founded on the mechanistic postulate that the reversible formation of iminium ions from α,β-unsaturated aldehydes and amines [Eq. (11.10)] might emulate the equilibrium dynamics and π-orbital electronics that are inherent to Lewis acid catalysis [i.e., lowest unoccupied molecular orbital (LUMO)-lowering activation] [Eq. (11.9)]:

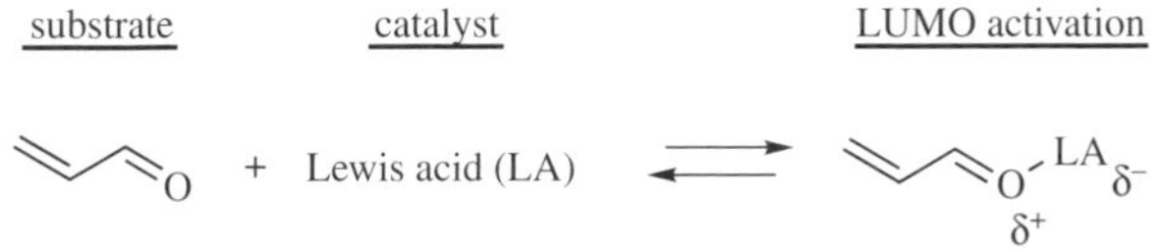

$$(11.9)$$

$$(11.10)$$

Catalyst Design   Preliminary experimental findings and computational studies demonstrated the importance of four objectives for the construction of a broadly useful iminium activation catalyst. Besides the importance of chiral amines to undergo reversible iminium ion formation with efficient reaction rates, it was proposed that high levels of iminium geometry control and selective π-facial discrimination of the olefin component were essential requirements for enantiocontrol. In addition, ease of catalyst preparation and implementation were considered crucial factors for widespread adoption of this new technology. Toward these goals, MacMillan identified a simple chiral amine, imidazolidinone **9**,[137] as a first-generation catalyst (Figure 11.3a).[136] As shown with computational modeling, the catalyst-activated iminium ion MM3-**10** was expected to form with (*E*)-isomer selectivity to avoid non bonding interactions between the substrate olefin and the *gem*-dimethyl substituents on the catalyst framework. In terms of enantiofacial discrimination, the calculated iminium structure MM3-**10** also reveals that the benzyl group on the catalyst framework will effectively shield the *Si*-face of the substrate, leaving the *Re*-face exposed for enantioselective bond formation. The effectiveness of imidazolidinone **9** was confirmed by its successful application to enantioselective Diels–Alder reactions,[136] nitrone cycloadditions,[138] and Friedel–Crafts alkylations of pyrroles.[139] However, diminished reactivity was observed when heteroaromatics, such as indoles and furans, were used as π-nucleophiles in lieu of pyrrole in similar Friedel–Crafts studies. To overcome such limitations, Mac-Millan introduced a second-generation imidazolidinone catalyst (**11**)[140] that realized improved reaction rates and reaction versatility. Catalyst improvements involved modification of **9** via removal of the inherent *trans*-methyl substituent to reduce the steric encumbrance around the reactive nitrogen lone pair. This was anticipated to provide increased rates of iminium formation and improved overall reaction rates

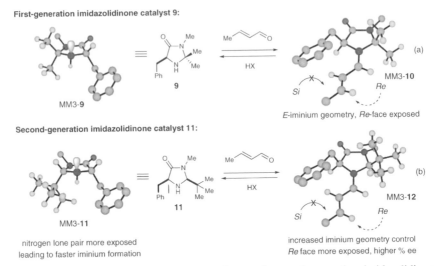

Figure 11.3. Computational models of the first- and second-generation imidazolidinone catalysts (**9** and **11**) and of the corresponding iminium ions.

thereafter. Replacement of the *cis*-methyl group with a *tert*-butyl functionality was proposed to increase iminium geometry control, while, at the same time, providing better coverage of the blocked *Si*-enantioface (Figure 11.3b).[141]

The effectiveness of imidazolidinone of type **11** was confirmed by successful application to a broad range of chemical transformations, including cycloadditions, conjugate additions, Friedel–Crafts alkylations, Mukaiyama–Michael additions, hydrogenations, cyclopropanations, and epoxidations. A summary of these enantioselective iminium catalyzed processes is provided by reaction subclass.

Cycloadditions   The original disclosure of the concept of iminium catalysis by the MacMillan group was made in the context of enantioselective catalytic Diels–Alder reactions. In these studies a range of α,β-unsaturated aldehydes were exposed to the corresponding cyclic and acyclic dienes in the presence of chiral imidazolidinone **9** to provide [4 + 2] cycloaddition adducts in useful yields and with high levels of enantioselectivity (Scheme 11.1a).[136] Importantly, these studies established the first known example of a highly enantioselective organocatalytic Diels–Alder reaction. A central tenet for the invention of iminium catalysis by MacMillan was the intent to develop an organocatalytic activation platform that could be expanded to a far-ranging series of transformations. Since 2000, iminium catalysis has already been successfully employed in a variety of new cycloaddition processes, including *exo*-Diels–Alder cycloadditions (Scheme 11.1b),[142] the first enantioselective catalytic Diels–Alder protocol for simple ketone dienophiles (Scheme 11.1c),[143] type I and type II intramolecular Diels–Alder reactions (Scheme 11.1d),[144] as well as nitrone cycloadditions (Scheme 11.1e)[138] using imidazolidinone catalysts **9**, **11**, **13**, and **14**. An elegant application of iminium catalysis by Harmata and coworkers using catalyst **11** has enabled the first organocatalytic [4 + 3]-cycloaddition

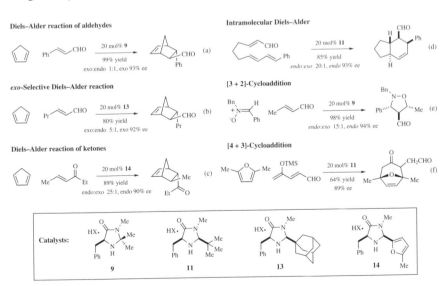

Scheme 11.1.

reaction to afford enantioenriched bicyclic ring systems (Scheme 11.1f),[145] products that are not readily available using Lewis acid catalysis.

Friedel–Crafts Alkylations and Mukaiyama–Michael Reactions   The metal-catalyzed addition of aromatic substrates to electron-deficient σ- and π-systems, commonly known as *Friedel–Crafts alkylation*, has long been established as a powerful strategy for C–C bond formation.[146–148] Surprisingly, however, relatively few asymmetric catalytic protocols have been reported that exploit this venerable reaction manifold, despite the widespread availability of electron-rich aromatics and the chemical utility of the resulting products.

In this context, MacMillan outlined the value of the iminium catalysis strategy via the development of asymmetric Friedel–Crafts variants that were not available using acid or metal catalysis. Indeed, it has been documented that α,β-unsaturated aldehydes are poor electrophiles for pyrrole, indole, or aryl conjugate addition because of the capacity of electron-rich aromatics to undergo acid catalyzed 1,2-carbonyl addition in lieu of 1,4-addition.[149,150] In 2001, the MacMillan laboratory disclosed that pyrroles undergo facile 1,4-addition to various α,β-unsaturated aldehydes in the presence of catalytic amounts of imidazolidinone **9** to provide the corresponding conjugate addition adducts in high yields and excellent enantioselectivity (Scheme 11.2a).[139]

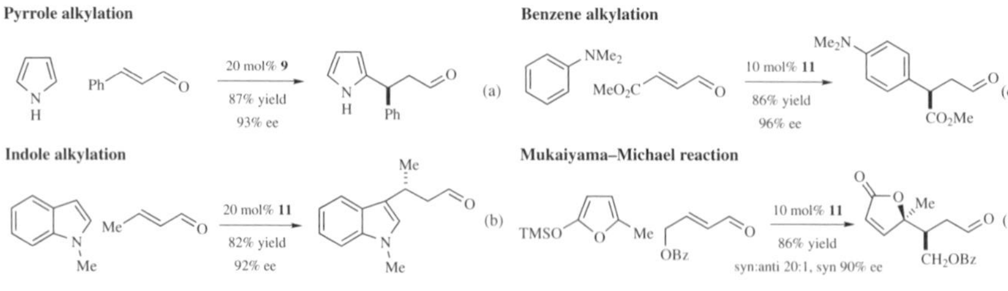

Scheme 11.2.

Alternatively, catalyst **11** affords highly selective alkylation adducts with a variety of π-nucelophiles, including indoles (Scheme 11.2b),[141] anilines (Scheme 11.2c),[151] and silyloxy furan derivatives (Scheme 11.2d).[152] It is important to note that only 1,4-addition products were formed in all of these reactions, thereby demonstrating the possibility of accessing complementary chemical selectivities using metal or organic catalysts.

Cascade Addition–Cyclization Reactions   Given the importance of cascade reactions in modern chemical synthesis,[153–156] the MacMillan group has proposed expansion of the realm of iminium catalysis to include the activation of tandem bond-forming processes, with a view toward the rapid construction of natural products. In this context, the addition–cyclization of tryptamines with α,β-unsaturated aldehydes in the presence of imidazolidinone catalysts **11** or **15** has been accomplished to provide pyrroloindoline adducts in high yields and with excellent enantioselectivities (Scheme 11.3a).[157] This transformation is successful

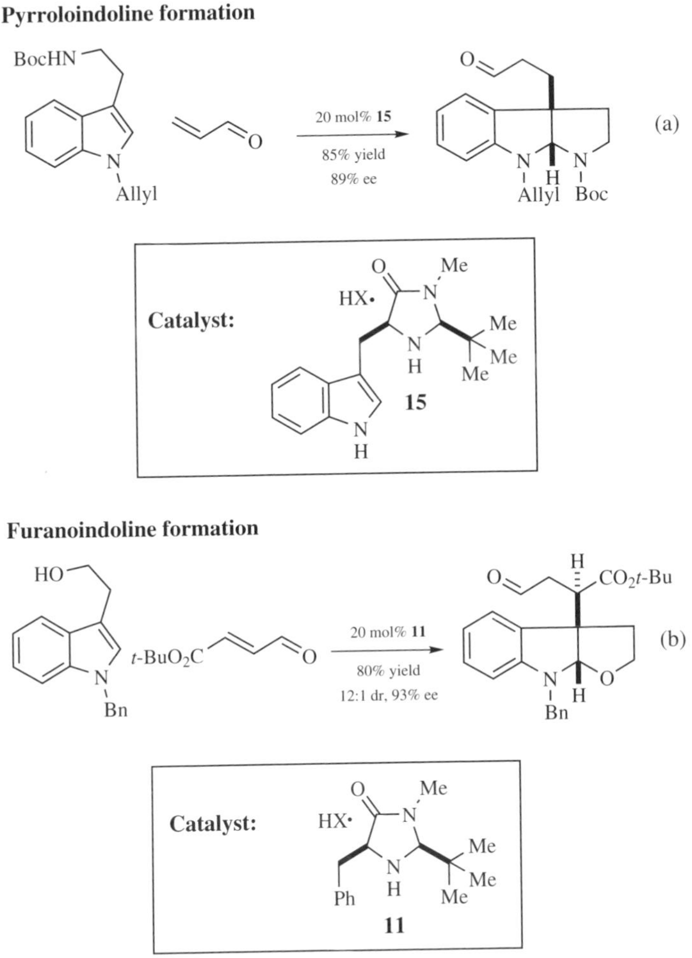

Pyrroloindoline formation

(a)

Furanoindoline formation

(b)

Scheme 11.3.

for a wide range of tryptamine and α,β-unsaturated aldehyde substrates. Moreover, this amine-catalyzed sequence has been extended to the enantioselective construction of furanoindoline frameworks (Scheme 11.3b).[157] Application of the pyrroloindoline-forming protocol to natural product synthesis has been accomplished in the form of the first enantioselective synthesis of (−)-flustramine B.

Transfer Hydrogenations   Within the vast architectural subclass known as *stereogenicity*, hydrogen is found to be the most common substituent. Not surprisingly, therefore, the field of enantioselective catalysis has focused great attention on the invention of hydrogenation technologies since the mid-1950s.[158–160] While these powerful transformations rely mainly on the use of organometallic catalysts and hydrogen gas, it is important to consider that the large majority of hydrogen-bearing stereocenters

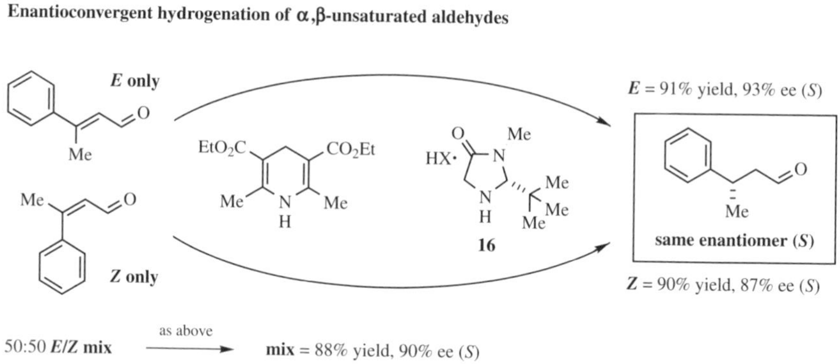

Enantioconvergent hydrogenation of α,β-unsaturated aldehydes

Scheme 11.4.

are created in biological cascade sequences involving enzymes and hydride reduction cofactors such as NADH or FADH$_2$.[161] On this basis, the use of small imidazolidinone catalysts, in combination with Hantzsch esters (NADH analogs), was proposed as an alternative metal-free approach to transfer hydrogenation. Indeed, independent studies from MacMillan and coworkers (using catalyst **16**)[162] and List and coworkers (using imidazolidinone **11**)[163,164] demonstrated that selective reduction of β,β-disubstituted-α,β-unsaturated aldehydes could be accomplished in good yield and excellent enantioselectivity via this biomimetic strategy (Scheme 11.4).[162] A notable feature of this transformation is that the sense of enantioinduction is not related to the olefin geometry of the starting materials. As a consequence, mixtures of E/Z-olefin isomers can be employed to provide enantiomerically pure hydrogenation adducts, a desirable yet rare feature for catalytic hydrogenations. This capacity to tolerate starting materials of low geometric purity will likely enhance the general utility of this operationally simple asymmetric reduction.

Iminium-Catalyzed Additions to α,β-Unsaturated Ketones   Given the inherent problems of forming tertrasubstituted iminium ions from ketones, along with the accordant issues associated with iminium ion geometry control, it is noteworthy that significant progress has been made in the development of iminium catalysts for enone substrates since 2001. Historically, the asymmetric Michael addition of carbogenic reagents to α,β-unsaturated ketones was first catalyzed by metalloprolinates in the early 1990s.[165–169] Approximately a decade later, Hanessian[170] reported the first organocatalytic variant, using L-proline (**2**) as the asymmetric catalyst in combination with *trans*-2,5-dimethylpiperazine to facilitate enantioselective nitroalkane additions to cyclic enones (Scheme 11.5a). In 2002, Jørgensen[171–175] and others[176–179] reported important expansions of the iminium catalysis paradigm to demonstrate the enantioselective conjugate addition of carbogenic nucleophiles such as nitroalkanes (Scheme 11.5b),[171] malonates (Scheme 11.5c),[172,177] cyclic (Scheme 11.5d), and acyclic 1,3-dicarbonyl compounds (Scheme 11.5e),[173–175,178] and β-ketosulfones (Scheme 11.5f)[175] to a number of acyclic α,β-unsaturated ketones. Among the amines employed in these studies, it is notable that iminium catalysts **17** and **18** furnished the desired Michael

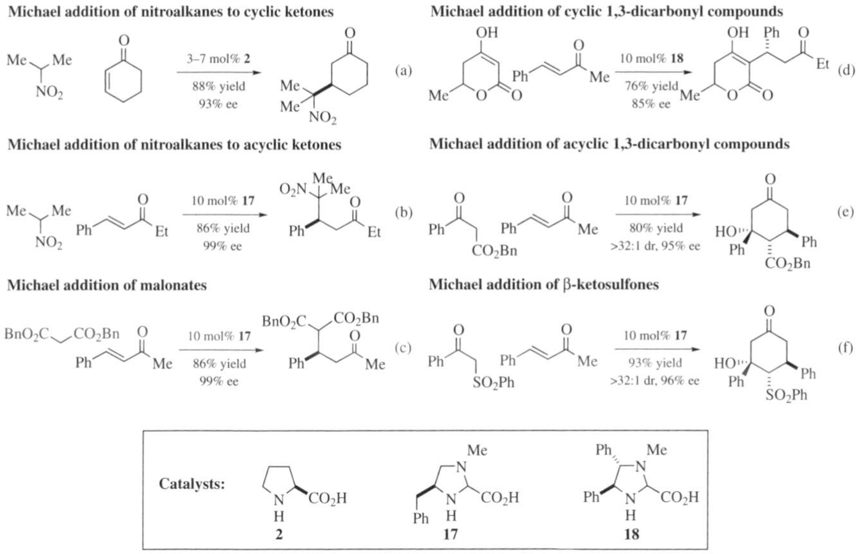
Michael addition of nitroalkanes to cyclic ketones
3–7 mol% 2
88% yield
93% ee
(a)
Michael addition of cyclic 1,3-dicarbonyl compounds
10 mol% 18
76% yield
85% ee
(d)
Michael addition of nitroalkanes to acyclic ketones
10 mol% 17
86% yield
99% ee
(b)
Michael addition of acyclic 1,3-dicarbonyl compounds
10 mol% 17
80% yield
>32:1 dr, 95% ee
(e)
Michael addition of malonates
10 mol% 17
86% yield
99% ee
(c)
Michael addition of β-ketosulfones
10 mol% 17
93% yield
>32:1 dr, 96% ee
(f)
Catalysts:
2
17
18

Scheme 11.5.

adducts with superior yields and enantioselectivities. Important examples from Jørgensen's work in the iminium catalysis area include the one-step synthesis of pharmaceutically relevant adducts related to warfarin.

Miscellaneous Iminium Catalyzed Transformations The enantioselective construction of three-membered hetero- or carbocyclic ring systems is an important objective for practitioners of chemical synthesis in academic and industrial settings. To date, important advances have been made in the iminium activation realm, which enable asymmetric entry to α-formyl cyclopropanes and epoxides. In terms of cyclopropane synthesis, a new class of iminium catalyst has been introduced, providing the enantioselective stepwise [2 + 1] union of sulfonium ylides and α,β-unsaturated aldehydes.[180] As shown in Scheme 11.6a, the zwitterionic hydro-indoline-derived catalyst (**19**) enables both iminium geometry control and directed electrostatic activation of sulfonium ylides in proximity to the incipient iminium reaction partner. This combination of geometric and stereoelectronic effects has been proposed as being essential for enantio- and diastereocontrol in forming two of the three cyclopropyl bonds.

The catalytic asymmetric epoxidation of α,β-unsaturated aldehydes has also been an important challenge in iminium catalysis and for chemical synthesis in general. More recently, Jørgensen and coworkers[181] have developed an asymmetric organocatalytic approach to α,β-epoxy aldehydes using pyrrolidine catalyst **20** and H_2O_2 as the stoichiometric oxidant. The reaction appears to be extremely general and will likely receive wide attention from the chemical synthesis community (Scheme 11.6b).

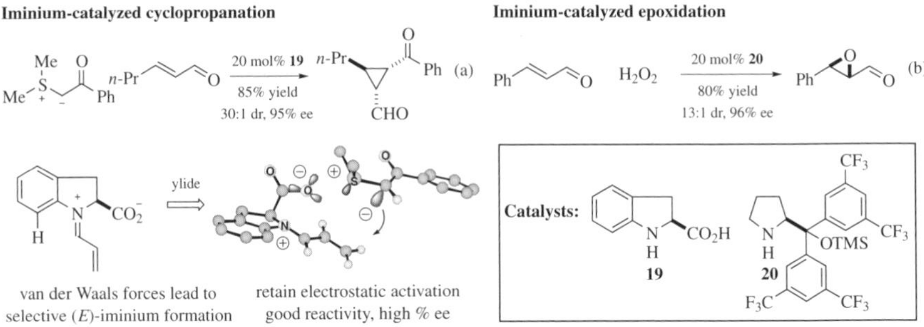

Iminium-catalyzed cyclopropanation

Iminium-catalyzed epoxidation

Scheme 11.6.

11.4 ENAMINE CATALYSIS: BIRTH, REBIRTH, AND RAPID GROWTH

The Catalysis Concept of Enamine Activation   Enamine catalysis is one of the most thoroughly investigated research areas within organocatalysis.[182–190] The underlying activation concept is based on the reversible condensation of an amine catalyst with a simple aldehyde or ketone to form an iminium intermediate that, on tautomerization, generates an enamine species that can subsequently be trapped by an appropriate electrophile [Eq. (11.11)]. Electronically orthogonal to iminium activation, enamine catalysis involves the highest occupied molecular orbital (HOMO)-raising activation of carbonyls and adjacent α-carbonyl carbons via formation of electron-rich amine-substituted olefins, which have sufficient π-electron density to participate in nucleophilic addition with a variety of carbon-, nitrogen-, oxygen-, sulfur- and halogen-centered electrophiles:

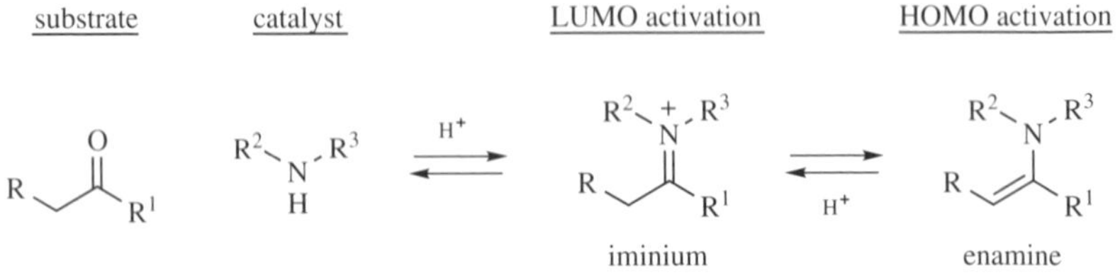

$$(11.11)$$

At present, one of the most successful catalysts for enamine activation has been proline (**2**). Proline is a cheap, widely and commercially available amino acid that can be found in both enantiomeric forms and, as such, represents a remarkable synthetic alternative to many established asymmetric catalysts. Given such attractive features, it has become the catalyst of choice for many enamine-catalyzed processes. However, various more recent studies have demonstrated that proline is not a universal catalyst for transformations that involve the α-functionalization of ketone or aldehyde carbonyls. Indeed, these studies have demonstrated that the iminium catalysts developed by MacMillan (imidazolidinones) and Jørgensen (pyrrolidines) are also highly effective for enamine activation with respect to enantioselectivity, reaction efficiency, and versatility across a wide spectrum of reactions. In a manner similar to the advent and development of the iminium catalysis field, enamine activation and the underpinning principles are finding widespread application in a number of vintage reactions, including aldol reactions, Mannich reactions, Michael additions, and Diels–Alder reactions, as well as in a number of unprecedented processes that involve enantioselective α-carbonyl functionalization.

Aldol Reactions[191–195]   The aldol transformation is a fundamental C–C bond-forming process in chemistry and biology. Since the mid-1970s, seminal research from the laboratories of Evans,[196–199] Heathcock,[200–204] Masamune,[205–208] and Mukaiyama[209–211] have established the aldol reaction as the principal chemical transformation for the stereoselective construction of complex polyol architecture. In the 1990s, Shibasaki[212–215] and Trost[216–218] outlined some of the first metal-catalyzed examples of enantioselective direct aldol reactions, an important class of aldol coupling transformations that do not require the pregeneration of enolates or enolate equivalents. With this in mind, it is noteworthy that the first example of a direct asymmetric catalytic aldol coupling was achieved in the organocatalysis arena in 1969. More specifically, the proline-catalyzed intramolecular aldol involving triketones, now termed the *Hajos–Parrish reaction* [Section 11.2, Eq. (11.2)],[5a,b] was indeed the first direct aldol of this type and the first catalytic enamine activation outside of biochemistry. Remarkably, the underlying catalytic principle of enamine activation remained undeveloped for almost 25 years before the pioneering work of Barbas and coworkers[5d] revealed that proline catalysis could be extended to the direct enantioselective intermolecular aldol reaction between ketones and aldehydes (Scheme 11.7a). Subsequent to this important publication, more than 50 manuscripts have been published on the subject of enamine-catalyzed aldol reactions by a large and diverse range of researchers.[5f,195,219–270]

With the development of organocatalytic intermolecular couplings between ketones and aldehydes in place, a new goal for aldol technology became the identification of catalytic methods that would allow the direct coupling of aldehyde substrates.[271] By the year 2002, this powerful yet elusive aldol variant had been accomplished only within the realm of enzymatic catalysis.[272,273] In the same year, MacMillan and coworkers[274–277] developed the first direct enantioselective aldehyde–aldehyde cross-aldol reaction via enamine activation using proline or imidazolidinone catalysts (Scheme 11.7b). The utility of this cross-aldol reaction of aldehydes was demonstrated in the rapid and highly selective construction of erythrose in one chemical step. This methodology became the foundation for a two-step synthesis of polyol-differentiated, highly enantioenriched carbohydrates (Scheme 11.7c),[278–281] key building blocks for saccharide and polysaccharide synthesis.

The proline-mediated intramolecular aldol condensation of dialdehyde substrates was also reported by List in 2003,[282] affording enantioselective synthesis of cyclic β-hydroxy aldehydes via a 6-*enolexo*-aldolization reaction (Scheme 11.7d).

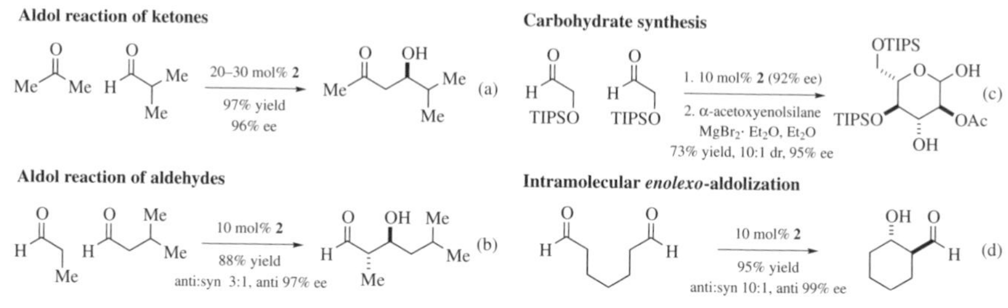

Aldol reaction of ketones

Carbohydrate synthesis

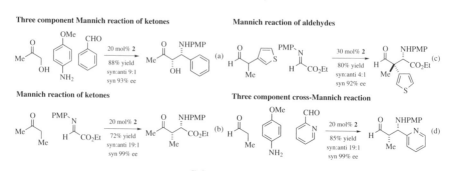

Scheme 11.7.

Mannich Reactions[283–285]   Expanding the concept of enamine catalysis outside the realm of aldol chemistry, Barbas[286] and List[5e,287] independently demonstrated the first highly selective amino-acid-catalyzed Mannich reaction. This three-component protocol, involving ketone donors and imine acceptors derived in situ from aldehydes and amines, provides β-amino ketones with excellent levels of stereoinduction (Scheme 11.8a). Similarly, the proline (**2**)-catalyzed reaction of ketones with preformed imines derived from ethyl glyoxylate furnished enantioenriched α-amino acid esters in a highly expeditious fashion (Scheme 11.8b).[288]

A major advancement for the subfield of enamine catalysis was achieved with the identification of aldehydes as useful donors for similar Mannich reactions.[289–295] In particular, the addition of mono- or disubstituted aldehydes to ketoimines or aldimines, respectively, represents an elegant and highly efficient approach to the enantioselective construction of quaternary α-amino acids (Scheme 11.8c).[294,295] A one-pot, three-component variant of the aldehyde Mannich reaction has also been recently disclosed (Scheme 11.8d).[292,293,296–300]

Three component Mannich reaction of ketones

Mannich reaction of aldehydes

Mannich reaction of ketones

Three component cross-Mannich reaction

Scheme 11.8.

Michael Reactions[301–304]   The Michael and analogous conjugate addition reactions represent a cornerstone of classical and modern chemical synthesis, and not surprisingly, therefore, catalytic variants of this venerable reaction have received increasing attention in recent years.[301–303,305] The first examples of enamine Michael additions using proline (**2**) as the organocatalyst have proven disappointing in terms of enantiocontrol,[87,88,176,220,306–310] stimulating the search for a more selective enamine catalyst system. In this context, imidazolidinones (initially

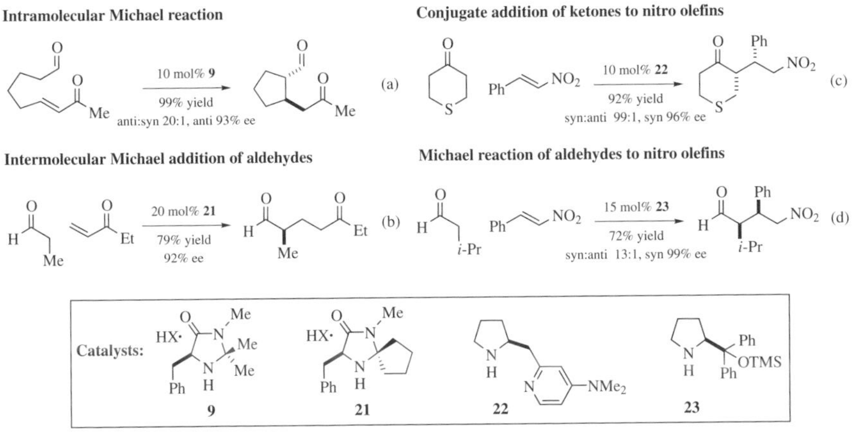

Intramolecular Michael reaction

Conjugate addition of ketones to nitro olefins

10 mol% **9**
99% yield
anti:syn 20:1, anti 93% ee   (a)

10 mol% **22**
92% yield
syn:anti 99:1, syn 96% ee   (c)

Intermolecular Michael addition of aldehydes

Michael reaction of aldehydes to nitro olefins

20 mol% **21**
79% yield
92% ee   (b)

15 mol% **23**
72% yield
syn:anti 13:1, syn 99% ee   (d)

Catalysts:

9   **21**   **22**   **23**

Scheme 11.9.

developed for iminium catalysis) have proved to be quite general for a number of enamine-activated Michael additions. For example, List[311] has elegantly demonstrated that imidazolidinone catalyst **9** can be successfully employed in the asymmetric intramolecular Michael reaction of formyl enones, affording cyclic *trans*-ketoaldehydes with high levels of selectivity (Scheme 11.9a). Moreover, MacMillan has employed a proline-catalyzed intramolecular Michael reaction to generate a *cis*-fused lactol as the key step in the total synthesis of (−)-littoralisone.[312] Gellman and coworkers[313] have outlined that an intermolecular variant of this process can be accomplished in lieu of aldehyde–aldehyde aldol couplings using imidazolidinone catalyst **21**, if catechol additives are employed (Scheme 11.9b). These additives are proposed to activate the Michael acceptor in the presence of the transiently generated enamines, thereby changing the dominant electrophile in this process from aldehyde to enone. The use of chiral pyrrolidine as enamine catalysts for Michael additions has also been achieved,[179,307,309,314–326] with pyrrolidines **22**[321] and **23**[179] providing superior results in the addition of unmodified ketones (Scheme 11.9c) or aldehydes (Scheme 11.9d), respectively, to nitroolefins.

Diels–Alder Reactions The organocatalytic Diels–Alder reaction of α,β-unsaturated carbonyl compounds can be performed either via iminium (see Section 11.3) or enamine catalysis. The first highly selective enamine-promoted cycloaddition reaction was reported by Jørgensen and coworkers,[327] who developed an amine-catalyzed inverse-electron-demand hetero-Diels–Alder (HDA) reaction (Scheme 11.10a).[328]

In extending this concept to transformations that formally deliver Diels–Alder products, a one-pot three-component Mannich/Michael reaction pathway was developed in which simple cyclic enones, formaldehyde, and aryl amines were treated with catalytic amounts of proline (**2**) to provide regio-, diastereo-, and enantioselective bicyclic compounds in high yields (Scheme 11.10b).[329] Multicomponent domino reactions involving a [4 + 2] cycloaddition step have also been reported.[177,330–333]

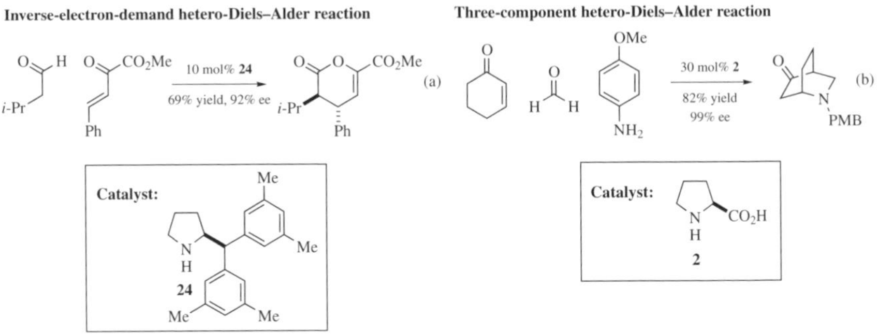

Inverse-electron-demand hetero-Diels–Alder reaction

Three-component hetero-Diels–Alder reaction

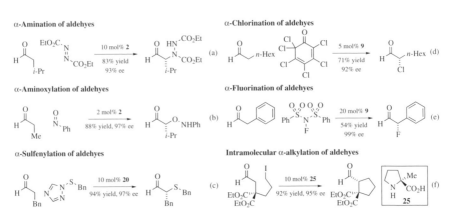

Scheme 11.10.

Enantioselective α-Functionalization of Aldehydes and Ketones The direct and enantiosective functionalization of enolates or enolate equivalents with carbon-, nitrogen-, oxygen-, sulfur- or halogen-centered electrophiles represents a powerful transformation of chemical synthesis and of fundamental importance to modern practitioners of asymmetric molecule construction. Independent studies from List, Jørgensen, Córdova, Hayashi, and MacMillan have demonstrated the power of enamine catalysis, developing catalytic enantioselective reactions such as direct α-aminations (Scheme 11.11a),[334–339] α-oxidations (Scheme 11.11b),[340–351] α-sulfenylations (Scheme 11.11c),[352] α-chlorinations (Scheme 11.11d),[353–356] and α-fluorinations (Scheme11.11e)[357–360] of aldehydes or ketones. These products have direct applications as synthetic building blocks for the preparation of α-amino acids, α-amino alcohols, 1,2-diols, terminal epoxides, and oxazolidines. A holy grail throughout the field of asymmetric catalysis (as of August 2005) has been the direct and enantioselective α-alkylation of carbonyls.[361] In this context, List and coworkers[362] have performed seminal research that has resulted in the first

α-Amination of aldehyes

α-Chlorination of aldehydes

α-Aminoxylation of aldehyes

α-Fluorination of aldehydes

α-Sulfenylation of aldehyes

Intramolecular α-alkylation of aldehyes

Scheme 11.11.

direct intramolecular aldehyde α-alkylation using enamine activation as the underpinning catalytic process (Scheme 11.11f). In these studies, List employs trialkylamine additives that, on in situ protonation, enable activation of the halogen leaving group via electrostatic interactions.[362] These studies will likely set the stage for the first intermolecular alkylation using enamine catalysis.

11.5 BRØNSTED ACID CATALYSIS: HYDROGEN-BONDING ACTIVATION

The Catalysis Concept of Enantioselective Brønsted Acid or Hydrogen-Bonding Activation A truly groundbreaking area of organocatalysis that has emerged since 1998 has been the enantioselective Brønsted acid or hydrogen-bonding catalysis.[363–366] While, in fact, there are two distinct modes of activation, the subfields of Brønsted acid and hydrogen bond catalysis are closely related in that both involve substrate LUMO-lowering activation via proton or hydrogen association. In a stricter sense, Brønsted acid activation involves substrate protonation with concomitant ion-pair formation between the activated substrate (cation) and the resulting catalyst conjugate base (anion). In contrast, hydrogen bond catalysis is defined as LUMO-lowering activation by the simultaneous sharing of a hydrogen atom by the substrate (hydrogen bond acceptor) and the catalyst (hydrogen bond donor). While such definitions allow theoretical differentiation of these two activation modes, at the present time it is not trivial to experimentally distinguish between these catalysis concepts. As such, in the following treatment of this specific organocatalysis area we will use the term *hydrogen-bonding activation* to represent both hydrogen bond catalysis and Brønsted acid activation.

In analogy to classical Lewis acid complexation of carbonyl compounds, a variety of chiral organic molecules such as alcohol, amidine, guanidine, urea, and thiourea derivatives undergo hydrogen bond interactions with C=O double bonds giving rise to LUMO-lowering activation (Figure 11.4a). Similarly, this catalytic concept can also be applied to imines, 1,2-dicarbonyls, and amides, providing a tremendous opportunity for vast range of enantioselective applications (Figure 11.4b). In many

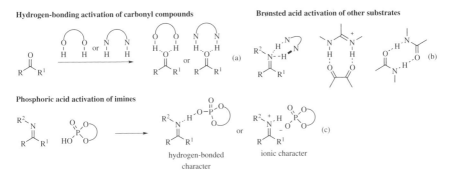

Figure 11.4. Hydrogen-bonding and Brønsted acid complexation modes for the LUMO-lowering activation of substrates inherent to the field of Brønsted acid catalysis.

cases, catalytic efficiency relies on the formation of strictly oriented hydrogen bonds via the bi- or multidentate nature of catalyst–substrate binding, which limits conformational degrees of freedom, thereby ensuring high levels of organizational control, a central requirement for enantioselective discrimination. Chiral BINOL-derived phosphoric acids have also been employed as hydrogen bond catalysts for the activation of imines (Figure 11.4c). The rigid biaryl framework of this phosphoric acid and the tight monodentate hydrogen-bonded complex of substrate and catalyst (with partial ionic character) are the major characteristics that enable a highly ordered asymmetric environment.

To date, hydrogen bond catalysis has been successfully utilized to facilitate enantioselective Michael additions, Baylis–Hillman reactions, Diels–Alder cycloadditions, and additions of π-nucleophiles to imines.

Nucleophilic Additions to Imines   The addition of carbon nucleophiles to imines using hydrogen bond catalysis provides a practical approach for the construction of a large subset of important organic architectures, including α- and β-amino acids. Among the first examples of asymmetric H-bond-catalyzed reactions have been the cyanohydrin and Strecker reactions described by the groups of Inoue,[89–93] Lipton,[94,95] and Corey[5r] in the 1980s and 1990s (see Section 11.2). However, it was the later improvements in scope and selectivity of the Strecker reaction by Jacobsen[5q,s,t,96–98] that effectively allowed the field of hydrogen bond catalysis to experience its real boom. Indeed, catalyst development and optimization via parallel high-throughput evaluations of peptide-like H-bond donors led to the discovery of chiral urea and thiourea catalysts for the highly selective hydrocyanation of both aldimines (Scheme 11.12a)[5q,t,96–98] and ketoimines (Scheme 11.12b).[5s,97,367,368] Noteworthy in these studies is the unique basis for substrate activation, with placement of the imine substrate in a bridging mode between both urea (or thiourea) hydrogens (cf. Figure 11.4b).[97]

Meanwhile, chiral (thio)urea catalysts have been employed for a variety of imine addition reactions consisting of Mannich,[98,369] aza-Henry,[370,371] Pictet–Spengler,[372] and hydrophosphonylation reactions.[373]

Interestingly, fundamentally different stereoinduction mechanisms have been proposed for the activation of a number of related imine substrates,[98] studies that resulted in the development of simple and highly effective new catalytic systems (**27**) for the addition of silyl ketene acetals to *N*-Boc-protected aldimines (Mannich reaction) (Scheme 11.12c).[98]

As a true testament to the potential long-term impact of H-bonding activation, a number of ureas, thioureas, and acid catalysts are now finding broad application in a large number of classical and modern carbon–carbon bond-forming processes. On one hand, Johnston's[374] chiral amidinium ion **28** was elegantly applied to the asymmetric aza-Henry reactions (Scheme 11.12d). On the other hand, chiral phosphoric acids (e.g., **29** and **30**), initially developed by Akiyama[375] and Terada,[376] have been successfully employed in Mannich reactions,[375,376] hydrophosphonylation reactions,[377] aza-Friedel–Crafts alkylations (Scheme 11.12e),[378] and in the first example of an enantioselective catalytic reductive amination coupling (Scheme 11.12f).[379–382]

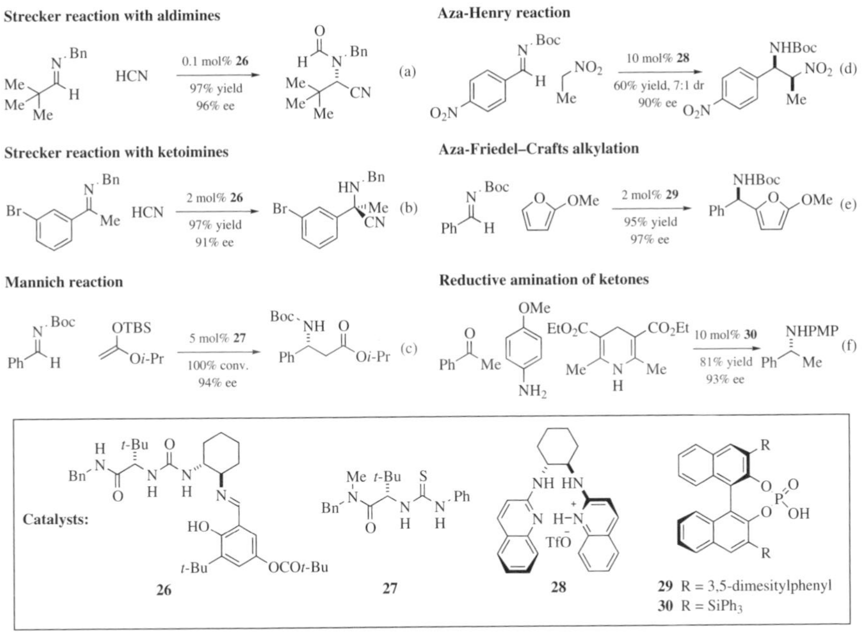

Scheme 11.12.

Conjugate Additions and Baylis–Hillman Reactions Peptide catalysts have reemerged as a viable approach to asymmetric catalysis.[383–385] In particular, Miller and coworkers[386,387] identified a series of peptide-based organocatalysts capable of catalyzing the enantioselective conjugate addition of azide to α,β-unsaturated imides (Scheme 11.13a).[388] Although the specific catalysis mechanism in this case is still under investigation, it is conceivable that substrate activation relies, at least partially, on hydrogen-bonding interactions.[389] Alternatively, Takemoto[390–392] and others[393–395] have developed chiral (thio)urea derivatives capable of catalyzing a range of conjugate additions such as Michael,[390–392] hetero-Michael[393,394] and Morita–Baylis–Hillman[395] reactions with moderate to good enantioselectivity (Scheme 11.13b).[396]

Other successful H-bond catalysis applications have been introduced by Schaus[397,398] and Sasai[399] involving asymmetric Morita–Baylis–Hillman (Scheme 11.13c) and aza-Morita–Baylis–Hillman reactions (Scheme 11.13d), respectively. Intriguingly, derivatized BINOL systems **33** and **34** provided optimal selectivities.

Miscellaneous Reactions Berkessel[400,401] has identified peptide-like urea-based bifunctional organocatalysts for the highly efficient dynamic kinetic resolution of azalactones (Scheme 11.14a). Another selective hydrogen-bonding activation mechanism that enables the addition of pyrroles to ketenes using catalytic quantities of azaferrocene **36** has been introduced by Fu and coworkers (Scheme 11.14b).[106]

Conjugate addition of azide

Morita–Baylis-Hillman reaction

Michael reaction

Aza-Morita–Baylis–Hillman reaction

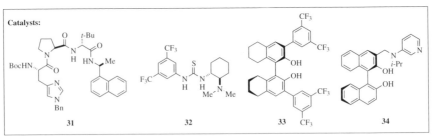

Scheme 11.13.

Dynamic kinetic resolution of azalactones

Hetero-Diels–Alder reaction

Pyrrole addition to ketenes

N-Selective nitroso aldol reaction

Diels–Alder reaction

O-Selective nitroso aldol reaction

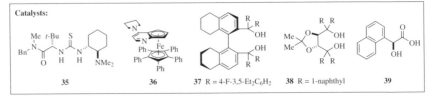

Scheme 11.14.

Interestingly, this catalyst and related systems typically function as chiral DMAP-type catalysts.[15,107] However, in this case, mechanistic studies suggest that the role of **36** is to serve as a conjugate base for the protic catalyst derivative.[106]

The most recent breakthroughs in hydrogen bond catalysis have been accomplished in the realm of the Diels–Alder reaction. While initial experiments in this area were conducted by Göbel and coworkers[402,403] using axially chiral amidinium ion catalysts,[404,405] impressive scope and selectivity advancements for the Diels–Alder (Scheme 11.14c),[406] and hetero-Diels–Alder variants (Scheme 11.14d)[407–410] were accomplished by Rawal and coworkers[406,407,409] using chiral diol **37** or TADDOL **38**-type catalysts. Importantly, these studies were the first to demonstrate that hydrogen bonding of a simple chiral alcohol to a carbonyl group could accomplish the catalytic activation and induction that was previously thought to be possible only in the domain of enzymes, catalytic antibodies, and chiral Lewis acids.[407]

Momiyama and Yamamoto[411] have further expanded the utility of H-bond-mediated reactions catalyzing nitrosobenzene addition to enamines using TADDOL **38** or hydroxy acid **39** as catalysts. Remarkably, the judicious selection of H-bond-catalyst/enamine combination resulted in the formation of only *N*-addition compounds with TADDOL **38** (Scheme 11.14e), while acid **39** furnished exclusively *O*-nitroso aldol products (Scheme 11.14f).

11.6 PHASE TRANSFER CATALYSIS (PTC)

The Catalysis Concept of Phase Transfer Activation Asymmetric phase transfer catalysis (PTC) represents one of the most powerful modes of substrate activation within the realm of organocatalysis.[12,412–420] In its most basic conceptual form, phase transfer activation is founded on the use of biphasic reaction systems along with catalysts (typically ammonium ions) that can accelerate the rate of ion transfer of a substrate from one phase (typically an aqueous solution) to a second phase (organic media) via the formation of catalyst–substrate ion pair. By this process, the activated substrate is transiently partitioned into a medium (or interface) in which a reagent or reaction partner already resides, thereby facilitating a bimolecular transformation. PTC has tremendous preparative advantages such as simple reaction procedures, mild conditions, and inexpensive and environmentally benign reagents, and, as a result, has been employed widely and in a variety of settings (including industrial applications). The most commonly employed asymmetric phase transfer catalysts are quaternary ammonium salts, typically derived from cinchona alkaloids. Such catalysts are inexpensive, readily modulated to a number of salts, and available in both pseudoenantiomeric forms (cinchonine vs. cinchonidine, and quinidine vs. quinine).[12] A variety of subtly distinct mechanistic scenarios have been advanced within the subfield of PTC; however, the enantioselective alkylation of methylene carbons represents the most thoroughly investigated application, and as such this mechanistic pathway is presented as a representative process in Figure 11.5.[421]

In the first step, deprotonation of the methylene-containing substrate generally takes place at the interface of two phases (liquid–liquid or solid–liquid).

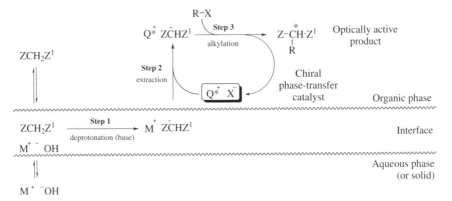

Figure 11.5. Representative example of the mechanistic pathway of phase transfer catalysis (PTC). (Z, $Z^{1\prime}$ = functional group; M = metal; Q* = chiral catalyst; R = alkyl or aryl reagent; X = halogen).

Subsequent ion exchange of the metal cation with the quaternary ammonium ion catalyst provides a lipophilic ion pair (step 2), which either reacts with the requisite alkyl electrophile at the interface (step 3) or is partitioned into the electrophile-containing organic phase, whereupon alkylation occurs and the catalyst is reconstituted. Enantioselective PTC has found application in a vast number of chemical transformations, including alkylations, conjugate additions, aldol reactions, oxidations, reductions, and C–X bond formations.[413]

Alkylations   Despite widespread demand for a direct and enantioselective variant, the catalytic alkylation reaction remains a challenging goal for methodological advancement. Within organocatalysis, PTC has historically been the most competent solution to the alkylation problem. Indeed, the first highly enantioselective PTC alkylation was reported in 1984 by the process research group at Merck[55-57] using cinchona-derived ammonium salt (**3**) as a phase transfer catalyst [Section 11.2, Eq. (11.3)]. Inspired by this work, and on the basis of similar achiral transformations,[422] O'Donnell and coworkers[5g,423–425] expanded the scope and defined the limitations of this methodology, developing a highly practical route to optically active α-amino acids (Scheme 11.15a). In this case, asymmetric PTC alkylation of a protected glycine derivative[422] with cinchona alkaloid-derived catalyst **40** furnished the corresponding α-amino acids in good yields, and with moderate enantioselectivity (Scheme 11.15a, entry 1). Independent studies by Corey[5h,5i,426] and Lygo[427–430] culminated in the development of a new class of cinchona alkaloid-derived phase transfer catalysts (e.g., **41** and **42**), which enabled the alkylation of glycine derivatives with extremely high levels of enantioselectivity (Scheme 11.15a, entries 2 and 3). Intriguingly, the adaptation of these reaction conditions to incorporate organic soluble, nonionic phosphazene (Schwesinger) bases,[431–433] led to the development of an efficient homogeneous catalytic alkylation protocol as a complementary method for the synthesis of α-amino acids.[434]

Since the mid-1990s, the phase transfer preparation of enantioenriched α-amino acids has received a great deal of attention,[417,418,435,436] and several types of

PTC alkylation of a protected glycine

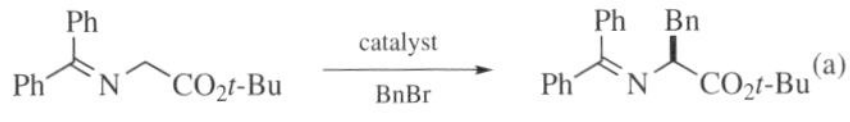

entry	cat. (mol%)	yield (%)	ee (%)
1	**40** (10)	87	81 (*S*)
2	**41** (10)	87	94 (*S*)
3	**42** (10)	68	91 (*S*)
4	**43** (1)	95	96 (*R*)

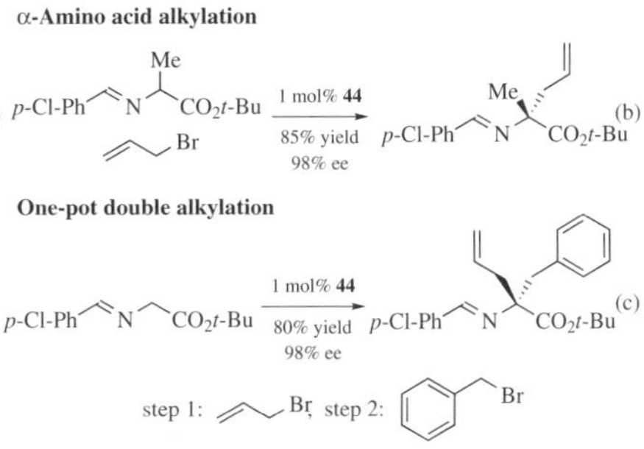

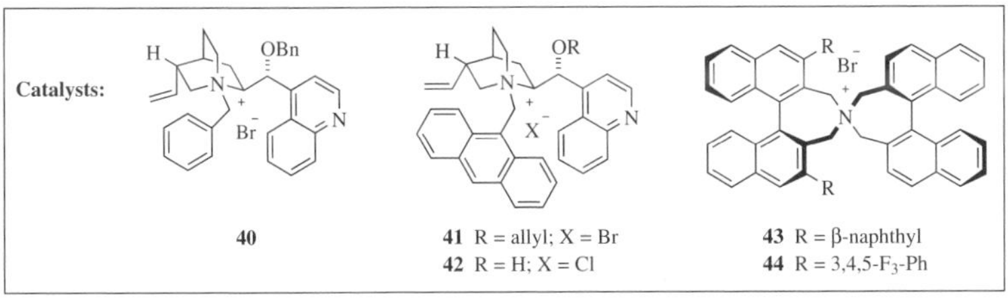

Scheme 11.15.

new catalyst structures have emerged.[419,420,437–449] Perhaps the most useful phase transfer catalysts prepared to date have been the C_2-symmetric chiral quaternary ammonium salts (e.g., **43** and **44**) invented by Maruoka and coworkers.[419,437–445] These remarkable systems mediate highly selective alkylation reactions using as little as 1 mol% catalyst. Moreover, Maruoka has undertaken substantial efforts toward the construction of α,α-dialkyl-α-amino acids (Scheme 11.15b), a structural entity that plays an important role in the design of peptides with nonnatural characteristics.[424,430,439,444,450–454] Notable in these studies is Maruoka's one-pot double alkylation of glycine derivatives to produce differentially substituted adducts with exceptionally high levels of enantiocontrol (Scheme 11.15c).[439]

Conjugate Additions Michael additions using PTC conditions have been investigated since the early 1980s.[69,81] Indeed, the first highly enantioselective phase transfer Michael addition was discovered by Cram and Sogah[81] using chiral crown ether complex **45** as the reaction catalyst (Scheme 11.16a). For the next 20 years sporadic examples of enantioselective PTC Michael reactions were reported.[413] However, it was only on the discovery of Corey's catalyst **41**[426,455–458] and similar cinchona alkaloid-derived quaternary ammonium salts[459–465] that major improvements in both selectivity and Michael reaction scope were accomplished. To date, a variety of new phase transfer catalyst structures have been developed that are highly amenable to such conjugate additions.[446,466–476]

Variants of the PTC Michael reaction have been successfully applied for a variety of chemical transformations, including the synthesis of functionalized α-amino acids (Scheme 11.16b),[426,446,457,458,468,470,472,475] the addition of malonates and 1,3-dicarbonyl compounds to α,β-unsaturated ketones (Scheme 11.16c),[460,462, 464,466,471,476] and the construction of γ-nitro carbonyl structures (Scheme 11.16d).[456,463,469,473,474] Again, Maruoka's quaternary

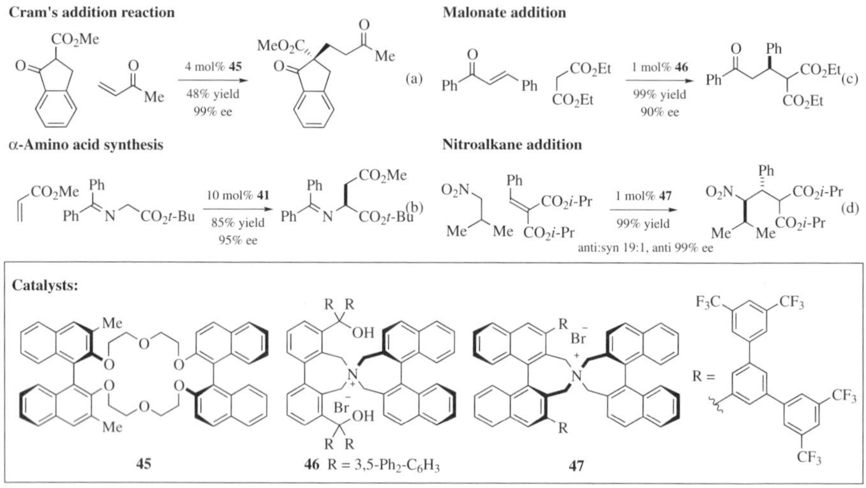

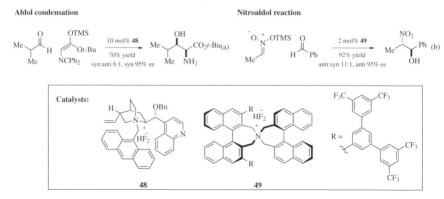

Scheme 11.16.

ammonium salts have proved extremely versatile for application to a vast range of Michael transformations.

Aldol and Related Condensations   As an elegant extension of the PTC–alkylation reaction, quaternary ammonium catalysts have been efficiently utilized in asymmetric aldol (Scheme 11.17a)[477–485] and nitroaldol reactions (Scheme 11.17b)[486,487] for the construction of optically active β-hydroxy-α-amino acids. In most cases, Mukaiyama–aldol-type reactions were performed, in which the coupling of silyl enol ethers with aldehydes was catalyzed by chiral ammonium fluoride salts, thus avoiding the need of additional bases, and allowing the reaction to be performed under homogeneous conditions.[477–482,485] It is important to note that salts derived from cinchona alkaloids provided preferentially *syn*-diastereomers, while Maruoka's catalysts afforded *anti*-diastereomers.

Scheme 11.17.

Epoxidations[488] and Darzens Condensations   The asymmetric catalytic epoxidation of α,β-unsaturated ketones using cinchona alkaloid-derived catalysts was
introduced in the 1970s.[46,48–50] However, high levels of enantioselectivity were
achieved only 20 years later, when Lygo,[489–491] Arai,[492–494] and others[5p,495–498]
prepared highly reactive cinchona alkaloid-based quaternary ammonium salts
(Scheme 11.18a). An alternative but equally impressive approach has involved
the use of poly(amino acid) catalysts, thus providing the enantioselective epoxidation of enones via the Juliá–Colonna mechanism.[51,83,85,499,500] Again, the versatility of Maruoka's quaternary ammonium salts has been demonstrated in the
epoxidation arena.[501]

A powerful approach to the synthesis of α,β-epoxy carbonyls and related compounds is found in the Darzens reaction (Scheme 11.18b). In this context, the
groups of Bakó[467,474,502] and Arai[503–508] have investigated the asymmetric PTC-
Darzens condensation using crown ethers and cinchona alkaloids-derived catalysts,
respectively, obtaining epoxides with moderate enantioselectivity.

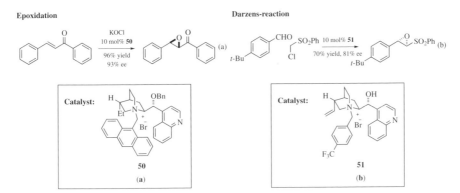

Scheme 11.18.

Miscellaneous PTC Reactions   The field of PTC is constantly expanding toward
the discovery of new enantioselective transformations. Indeed, more recent applications have demonstrated the capacity of chiral quaternary ammonium salts to catalyze
a number of transformations, including the Neber rearrangement (Scheme 11.19a),[509]
the trifluoromethylation of carbonyl compounds (Scheme 11.19b),[510,511] the Mannich
reaction (Scheme 11.19c),[512] and the nucleophilic aromatic substitution (S$_N$Ar)
reaction (Scheme 11.19d)[513] in good yield and, more importantly, with valuable levels
of enantioselectivity.

11.7   FUTURE PERSPECTIVE

Since 2000, the field of asymmetric catalysis has bloomed extensively (and perhaps
unexpectedly) with the introduction of a variety of metal-free catalysis concepts
that have collectively become known as *organocatalysis*. Perhaps more impressive

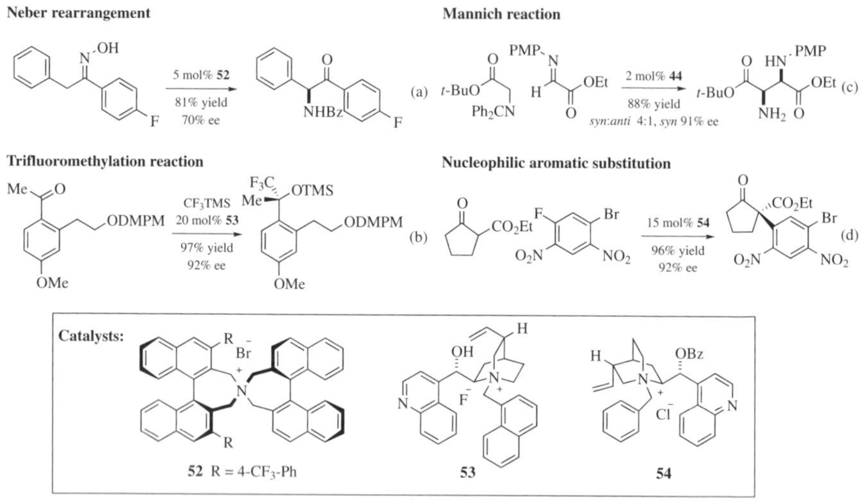

Scheme 11.19.

is the fact that the field of organocatalysis has quickly grown to become the third fundamental branch of catalysis that can be utilized for the construction of enantiopure organic structures, providing a valuable complement to organometallic and enzymatic catalysis. While substrate scope still remains an important issue for many organocatalytic reactions, an increasingly large number of transformations are now meeting the requisite high standards of "useful" enantioselective processes. At the present time, tremendous efforts are being focused on the discovery of catalysts with improved efficiency and turnover numbers, or reactions that can be performed with equimolar units of reagents.[179,241,245,269,313,324,326,514] Novel organocatalytic applications are constantly being described, stimulating this field to provide alternative catalysts for known processes and, more importantly, to provide chemical selectivities that cannot be accessed using other catalytic activation modes. Given the tremendous growth and impact of this catalytic field in less than 7 years, it will certainly be exciting to observe future organocatalytic developments over the next few decades. The utility of organo cascade catalysis and organocatalytic C–H bond activation are only a few potential areas of metal-free catalysis that will likely impact the world of asymmetric molecule construction in coming years.

ACKNOWLEDGMENTS

The authors would like to thank Dr. DeMichael Chung, Dr. Abbas M. Walji, Jamison B. Tuttle, and Robert K. Knowles for their thoughtful comments and helpful suggestions on various aspects of the manuscript. Financial support by the NIH National Institute of General Medical Sciences and gifts from Amgen, Lilly, Bristol-Myers Squibb, Johnson and Johnson, and Merck Research Laboratories are kindly acknowledged. G. L. is grateful to the Swiss National Foundation (Stefano Franscini Fond), the Roche Foundation, and the Novartis Foundation for postdoctoral fellowship support.

REFERENCES AND NOTES

1. Noyori R. (Ed.), *Asymmetric Catalysis in Organic Synthesis*, Wiley, New York, **1994**.

2. Jacobsen, E. N.; Pfaltz, A.; Yamamoto H. (Eds.), *Comprehensive Asymmetric Catalysis*, Springer, Heidelberg, **1999**.

3. Ojima I. (Ed.), *Catalytic Asymmetric Synthesis*, 2nd ed., Wiley-VCH, New York, **2000**.

4. Jacobsen, E. N.; Pfaltz, A.; Yamamoto H. (Eds.), *Comprehensive Asymmetric Catalysis. Supplement*, Springer, Heidelberg, **2004**.

5. For notable examples of organocatalytic reactions, see the following. Aldol reaction: (a) Hajos, Z. G.; Parrish, D. R. *J. Org. Chem.* **1974**, *39*, 1615–1621; (b) Eder, U.; Sauer, G.; Wiechert, R. *Angew. Chem. Int. Ed.* **1971**, *10*, 496–497; (c) Agami, C.; Meynier, F.; Puchot, C.; Guilhem, J.; Pascard, C. *Tetrahedron* **1984**, *40*, 1031–1038; (d) List, B.; Lerner, R. A.; Barbas C. F., III, *J. Am. Chem. Soc.* **2000**, *122*, 2395–2396; (e) List, B. *J. Am. Chem. Soc.* **2000**, *122*, 9336–9337; (f) List, B.; Pojarliev, P.; Castello, C. *Org. Lett.* **2001**, *3*, 573–575. Phase transfer catalysis: (g) O'Donnell, M. J.; Bennett, W. D.; Wu, S. D. *J. Am. Chem. Soc.* **1989**, *111*, 2353–2355; (h) Corey, E. J.; Xu, F.; Noe, M. C. *J. Am. Chem. Soc.* **1997**, *119*, 12414–12415; (i) Corey, E. J.; Bo, Y.; Busch-Petersen, J. *J. Am. Chem. Soc.* **1998**, *120*, 13000–13001. Epoxidation: (j) Yang, D.; Yip, Y.-C.; Tang, M.-W.; Wong, M.-K.; Zheng, J.-H.; Cheung, K.-K. *J. Am. Chem. Soc.* **1996**, *118*, 491–492; (k) Tu, Y.; Wang, Z.-X.; Shi, Y. *J. Am. Chem. Soc.* **1996**, *118*, 9806–9807; (l) Yang, D.; Wong, M.-K.; Yip, Y.-C.; Wang, X.-C.; Tang, M.-W.; Zheng, J.-H.; Cheung, K.-K. *J. Am. Chem. Soc.* **1998**, *120*, 5943–5952; (m) Denmark, S. E.; Wu, Z. *Synlett* **1999**, 847–859; (n) Tian, H.; She, X.; Shu, L.; Yu, H.; Shi, Y. *J. Am. Chem. Soc.* **2000**, *122*, 11551–11552. Baylis–Hillman reaction: (o) Iwabuchi, Y.; Nakatani, M.; Yokoyama, N.; Hatakeyama, S. *J. Am. Chem. Soc.* **1999**, *121*, 10219–10220; (p) Corey, E. J.; Zhang, F.-Y. *Org. Lett.* **1999**, *1*, 1287–1290. Asymmetric Strecker synthesis: (q) Sigman, M. S.; Jacobsen, E. N. *J. Am. Chem. Soc.* **1998**, *120*, 4901–4902; (r) Corey, E. J.; Grogan, M. J. *Org. Lett.* **1999**, *1*, 157–160; (s) Vachal, P.; Jacobsen, E. N. *Org. Lett.* **2000**, *2*, 867–870; (t) Sigman, M. S.; Vachal, P.; Jacobsen, E. N. *Angew. Chem. Int. Ed.* **2000**, *39*, 1279–1281. Acyl transfer: (u) Jarvo, E. R.; Copeland, G. T.; Papaioannou, N.; Bonitatebus, P. J.; Miller, S. J. *J. Am. Chem. Soc.* **1999**, *121*, 11638–11643.

6. The literature survey was obtained by searching in Web of Science under the topic *organocatal**, which comprises the words *organocatalysis, organocatalytic*, and *organocatalyst(s)*. This search is not meant to be complete in showing all publications that have an organocatalytic reaction, but mainly to indicate the origin of the concept of organocatalysis.

7. Denmark, S. E.; Stavenger, R. A. *Acc. Chem. Res.* **2000**, *33*, 432–440.

8. Denmark, S. E.; Beutner, G. L.; Wynn, T.; Eastgate, M. D. *J. Am. Chem. Soc.* **2005**, *127*, 3774–3789.

9. Shen, Y.-C.; Li, Y.-X.; Wen, X.-J.; Feng, X.-M. *Chin. J. Org. Chem.* **2005**, *25*, 272–281.

10. Rendler, S.; Oestreich, M. *Synthesis* **2005**, 1727–1747.

11. Benaglia, M.; Puglisi, A.; Cozzi, F. *Chem. Rev.* **2003**, *103*, 3401–3429.

12. Kacprzak, K.; Gawronski, J. *Synthesis* **2001**, 961–998.

13. Tian, S.-K.; Chen, Y.; Hang, J.; Tang, L.; McDaid, P.; Deng, L. *Acc. Chem. Res.* **2004**, *37*, 621–631.

14. Dálaigh, C. O. *Synlett* **2005**, 875–876.

15. Fu, G. C. *Acc. Chem. Res.* **2000**, *33*, 412–420.

16. Vedejs, E.; Daugulis, O.; MacKay, J. A.; Rozners, E. *Synlett* **2001**, 1499–1505.

17. Molt, O.; Schrader, T. *Synthesis* **2002**, 2633–2670.

18. France, S.; Guerin, D. J.; Miller, S. J.; Lectka, T. *Chem. Rev.* **2003**, *103*, 2985–3012.

19. Murugan, R.; Scriven, E. F. V. *Aldrichim. Acta* **2003**, *36*, 21–27.

20. Methot, J. L.; Roush, W. R. *Adv. Synth. Catal.* **2004**, *346*, 1035–1050.

21. Spivey, A. C.; Arseniyadis, S. *Angew. Chem. Int. Ed.* **2004**, *43*, 5436–5441.

22. Enders, D.; Balensiefer, T. *Acc. Chem. Res.* **2004**, *37*, 534–541.

23. Johnson, J. S. *Angew. Chem. Int. Ed.* **2005**, *43*, 1326–1328.

24. Christmann, M. *Angew. Chem. Int. Ed.* **2005**, *44*, 2632–2634.

25. Bredig, G.; Fiske, P. S. *Biochem. Z.* **1912**, *46*, 7–23.

26. Bredig, G.; Minaeff, M. *Biochem. Z.* **1932**, *249*, 241–244.

27. Prelog, V.; Wilhelm, M. *Helv. Chim. Acta* **1954**, *37*, 1634–1660.

28. Pracejus, H. *Justus Liebigs Ann. Chem.* **1960**, *634*, 9–22.

29. Pracejus, H.; Mätje, H. *J. Prakt. Chem.* **1964**, *24*, 195–205.

30. Ruppert, J.; Eder, U.; Wiechert, R. *Chem. Ber.* **1973**, *106*, 3636–3644.

31. Sauer, G.; Eder, U.; Haffer, G.; Neef, G.; Wiechert, R. *Angew. Chem. Int. Ed.* **1975**, *14*, 417–417.

32. Micheli, R. A.; Hajos, Z. G.; Cohen, N.; Parrish, D. R.; Portland, L. A.; Sciamanna, W.; Scott, M. A.; Wehrli, P. A. *J. Org. Chem.* **1975**, *40*, 675–681.

33. Danishefsky, S.; Cain, P. *J. Am. Chem. Soc.* **1975**, *97*, 5282–5284.

34. Danishefsky, S.; Cain, P. *J. Am. Chem. Soc.* **1976**, *98*, 4975–4983.

35. Takano, S.; Kasahara, C.; Ogasawara, K. *J. Chem. Soc. Chem. Commun.* **1981**, 635–637.

36. Woodward, R. B.; Logusch, E.; Nambiar, K. P.; Sakan, K.; Ward, D. E.; Au-Yeung, B.-W.; Balaram, P.; Browne, L. J.; Card, P. J.; Chen, C. H.; Chênevert, R. B.; Fliri, A.; Frobel, K.; Gais, H.-J.; Garratt, D. G.; Hayakawa, K.; Heggie, W.; Hesson, D. P.; Hoppe, D.; Hoppe, I.; Hyatt, J. A.; Ikeda, D.; Jacobi, P. A.; Kim, K. S.; Kobuke, Y.; Kojima, K.; Krowicki, K.; Lee, V. J.; Leutert, T.; Malchenko, S.; Martens, J.; Matthews, R. S.; Ong, B. S.; Press, J. B.; Babu, T. V. R.; Rousseau, G.; Sauter, H. M.; Suzuki, M.; Tatsuta, K.; Tolbert, L. M.; Truesdale, E. A.; Uchida, I.; Ueda, Y.; Uyehara, T.; Vasella, A. T.; Vladuchick, W. C.; Wade, P. A.; Williams, R. M.; Wong, H. N.-C. *J. Am. Chem. Soc.* **1981**, *103*, 3210–3213.

37. Baggiolini, E. G.; Iacobelli, J. A.; Hennessy, B. M.; Uskokovic, M. R. *J. Am. Chem. Soc.* **1982**, *104*, 2945–2948.

38. Wovkulich, P. M.; Baggiolini, E. G.; Hennessy, B. M.; Uskokovic, M. R.; Mayer, E.; Norman, A. W. *J. Org. Chem.* **1983**, *48*, 4433–4436.

39. Danishefsky, S. J.; Masters, J. J.; Young, W. B.; Link, J. T.; Snyder, L. B.; Magee, T. V.; Jung, D. K.; Isaacs, R. C. A.; Bornmann, W. G.; Alaimo, C. A.; Coburn, C. A.; Di Grandi, M. J. *J. Am. Chem. Soc.* **1996**, *118*, 2843–2859.

40. Pracejus, H. *Fortschr. chem. Forsch.* **1967**, *8*, 493–553.

41. Morrison, J. D.; Mosher H. S. (Eds.), *Asymmetric Organic Reactions*, Prentice-Hall, Englewood Cliffs., NJ, **1971**.

42. Wynberg, H. In *Topics in Stereochemistry*, Vol. 16, Eliel, E. L.; Wilen, S. H.; Allinger N.L. (Eds.), Wiley, New York, **1986**, pp. 87–129.

43. Borrmann, D.; Wegler, R. *Chem. Ber.* **1967**, *100*, 1575–1579.

44. Wynberg, H.; Staring, E. G. J. *J. Am. Chem. Soc.* **1982**, *104*, 166–168.

45. Wynberg, H.; Staring, E. G. J. *J. Chem. Soc. Chem. Commun.* **1984**, 1181–1182.

46. Helder, R.; Hummelen, J. C.; Laane, R. W. P. M.; Wiering, J. S.; Wynberg, H. *Tetrahedron Lett.* **1976**, *17*, 1831–1834.

47. Wynberg, H. *Chimia* **1976**, *30*, 445–451.

48. Hummelen, J. C.; Wynberg, H. *Tetrahedron Lett.* **1978**, *19*, 1089–1092.

49. Marsman, B.; Wynberg, H. *J. Org. Chem.* **1979**, *44*, 2312–2314.

50. Wynberg, H.; Marsman, B. *J. Org. Chem.* **1980**, *45*, 158–161.

51. Juliá, S.; Masana, J.; Vega, J. C. *Angew. Chem. Int. Ed.* **1980**, *19*, 929–931.

52. Snatzke, G.; Wynberg, H.; Feringa, B.; Marsman, B. G.; Greydanus, B.; Pluim, H. *J. Org. Chem.* **1980**, *45*, 4094–4096.

53. Harigaya, Y.; Yamaguchi, H.; Onda, M. *Heterocycles* **1981**, *15*, 183–185.

54. Juliá, S.; Ginebreda, A.; Guixer, J.; Masana, J.; A. Tomás, Colonna, S. *J. Chem. Soc. Perkin Trans. 1* **1981**, 574–577.

55. Dolling, U.-H.; Davis, P.; Grabowski, E. J. J. *J. Am. Chem. Soc.* **1984**, *106*, 446–447.

56. Battacharya, A.; Dolling, U.-H.; Grabowski, E. J. J.; Karady, S.; Ryan, K. M.; Weinstock, L. M. *Angew. Chem. Int. Ed.* **1986**, *25*, 476–477.

57. Hughes, D. L.; Dolling, U.-H.; Ryan, K. M.; Schoenewaldt, E. F.; Grabowski, E. J. J. *J. Org. Chem.* **1987**, *52*, 4745–4752.

58. Långström, B.; Bergson, G. *Acta Chem. Scand.* **1973**, *27*, 3118–3119.

59. Wynberg, H.; Helder, R. *Tetrahedron Lett.* **1975**, *16*, 4057–4060.

60. Helder, R.; Arends, R.; Bolt, W.; Hiemstra, H.; Wynberg, H. *Tetrahedron Lett.* **1977**, *18*, 2181–2182.

61. Pracejus, H.; Wilcke, F. W.; Hanemann, K. *J. Prakt. Chem.* **1977**, *319*, 219–229.

62. Hermann, K.; Wynberg, H. *Helv. Chim. Acta* **1977**, *60*, 2208–2212.

63. Kobayashi, N.; Iwai, K. *J. Am. Chem. Soc.* **1978**, *100*, 7071–7072.

64. Colonna, S.; Hiemstra, H.; Wynberg, H. *J. Chem. Soc. Chem. Commun.* **1978**, 238–239.

65. Trost, B. M.; Shuey, C. D.; DiNinno, F. Jr.; McElvain, S. S. *J. Am. Chem. Soc.* **1979**, *101*, 1284–1285.

66. Hermann, K.; Wynberg, H. *J. Org. Chem.* **1979**, *44*, 2238–2244.

67. Hiemstra, H.; Wynberg, H. *J. Am. Chem. Soc.* **1981**, *103*, 417–430.

68. Gawronski, J.; Gawronska, K.; Wynberg, H. *J. Chem. Soc. Chem. Commun.* **1981**, 307–308.

69. Colonna, S.; Re, A.; Wynberg, H. *J. Chem. Soc. Perkin Trans. 1* **1981**, 547–552.

70. Shibasaki, M.; Nishida, A.; Ikegami, S. *J. Chem. Soc. Chem. Commun.* **1982**, 1324–1325.

71. Hodge, P.; Khoshdel, E.; Waterhouse, J. *J. Chem. Soc. Perkin Trans. 1* **1983**, 2205–2209.

72. Hodge, P.; Khoshdel, E.; Waterhouse, J.; Fréchet, J. M. J. *J. Chem. Soc. Perkin Trans. 1* **1985**, 2327–2331.

73. Latvala, A.; Stanchev, S.; Linden, A.; Hesse, M. *Tetrahedron Asym.* **1993**, *4*, 173–176.

74. Wynberg, H.; Smaardijk, A. A. *Tetrahedron Lett.* **1983**, *24*, 5899–5900.

75. Smaardijk, A. A.; Noorda, S.; van Bolhuis, F.; Wynberg, H. *Tetrahedron Lett.* **1985**, *26*, 493–496.

76. Yamaguchi, K.; Minoura, Y. *Chem. Ind.* **1975**, 478–480.

77. Fukushima, H.; Ohashi, S.; Inoue, S. *Makromol. Chem.* **1975**, *176*, 2751–2753.

78. Fukushima, H.; Inoue, S. *Makromol. Chem.* **1975**, *176*, 3609–3611.

79. Fukushima, H.; Inoue, S. *Makromol. Chem.* **1976**, *177*, 2617–2626.

80. Fiaud, J.-C. *Tetrahedron Lett.* **1975**, 3495–3496.

81. Cram, D. J.; Sogah, G. D. Y. *J. Chem. Soc. Chem. Commun.* **1981**, 625–628.

82. Banfi, S.; Colonna, S.; Juliá, S.; *Synth. Commun.* **1983**, *13*, 1049–1052.

83. Colonna, S.; Molinari, H.; Banfi, S.; Juliá, S.; Masana, J.; Alvarez, A. *Tetrahedron* **1983**, *39*, 1635–1641.

84. Banfi, S.; Colonna, S.; Molinari, H.; Juliá, S. *Synth. Commun.* **1983**, *13*, 901–904.

85. Banfi, S.; Colonna, S.; Molinari, H.; Juliá, S.; Guixer, J. *Tetrahedron* **1984**, *40*, 5207–5211.

86. Colonna, S.; Juliá, S.; Molinari, H.; Banfi, S. *Heterocycles* **1984**, *21*, 548–548.

87. Kozikowski, A. P.; Mugrage, B. B. *J. Org. Chem.* **1989**, *54*, 2274–2275.

88. Hirai, Y.; Terada, T.; Yamazaki, T.; Momose, T. *J. Chem. Soc. Perkin Trans. 1* **1992**, 509–516.

89. Oku, J.-i.; Ito, N.; Inoue, S. *Makromol. Chem.* **1979**, *180*, 1089–1091.

90. Oku, J.-i.; Inoue, S. *J. Chem. Soc. Chem. Commun.* **1981**, 229–230.

91. Oku, J.-i.; Ito, N.; Inoue, S. *Makromol. Chem.* **1982**, *183*, 579–586.

92. Asada, S.; Kobayashi, Y.; Inoue, S. *Makromol. Chem.* **1985**, *186*, 1755–1762.

93. Tanaka, K.; Mori, A.; Inoue, S. *J. Org. Chem.* **1990**, *55*, 181–185.

94. Iyer, M. S.; Gigstad, K. M.; Namdev, N. D.; Lipton, M. *J. Am. Chem. Soc.* **1996**, *118*, 4910–4911.

95. Iyer, M. S.; Gigstad, K. M.; Namdev, N. D.; Lipton, M. *Amino Acids* **1996**, *11*, 259–268.

96. Su, J. T.; Vachal, P.; Jacobsen, E. N. *Adv. Synth. Catal.* **2001**, *343*, 197–200.

97. Vachal, P.; Jacobsen, E. N. *J. Am. Chem. Soc.* **2002**, *124*, 10012–10014.

98. Wenzel, A. G.; Lalonde, M. P.; Jacobsen, E. N. *Synlett* **2003**, 1919–1922.

99. Copeland, G. T.; Jarvo, E. R.; Miller, S. J. *J. Org. Chem.* **1998**, *63*, 6784–6785.

100. Vedejs, E.; Daugulis, O.; Diver, S. T. *J. Org. Chem.* **1996**, *61*, 430–431.

101. Vedejs, E.; Daugulis, O. *J. Am. Chem. Soc.* **1999**, *121*, 5813–5814.

102. Vedejs, E.; MacKay, J. A. *Org. Lett.* **2001**, *3*, 535–536.

103. Ruble, J. C.; Latham, H. A.; Fu, G. C. *J. Am. Chem. Soc.* **1997**, *119*, 1492–1493.

104. Fu, G. C. *Pure Appl. Chem.* **2001**, *73*, 347–349.

105. Fu, G. C. *Pure Appl. Chem.* **2001**, *73*, 1113–1116.

106. Hodous, B. L.; Fu, G. C. *J. Am. Chem. Soc.* **2002**, *124*, 10006–10007.

107. Fu, G. C. *Acc. Chem. Res.* **2004**, *37*, 542–547.

108. Denmark, S. E.; Coe, D. M.; Pratt, N. E.; Griedel, B. D. *J. Org. Chem.* **1994**, *59*, 6161–6163.

109. Denmark, S. E.; Winter, S. B. D.; Su, X.; Wong, K.-T. *J. Am. Chem. Soc.* **1996**, *118*, 7404–7405.

110. Iseki, K.; Kuroki, Y.; Takahashi, M.; Kobayashi, Y. *Tetrahedron Lett.* **1996**, *37*, 5149–5150.

111. Denmark, S. E.; Wong, K.-T.; Stavenger, R. A. *J. Am. Chem. Soc.* **1997**, *119*, 2333–2334.

112. Denmark, S. E.; Winter, S. B. D. *Synlett* **1997**, 1087–1089.

113. Iseki, K.; Kuroki, Y.; Takahashi, M.; Kishimoto, S.; Kobayashi, Y. *Tetrahedron* **1997**, *53*, 3513–3526.

114. Denmark, S. E.; Stavenger, R. A.; Wong, K.-T. *J. Org. Chem.* **1998**, *63*, 918–919.

115. Iseki, K.; Mizuno, S.; Kuroki, Y.; Kobayashi, Y. *Tetrahedron Lett.* **1998**, *39*, 2767–2770.

116. Iseki, K.; Mizuno, S.; Kuroki, Y.; Kobayashi, Y. *Tetrahedron* **1999**, *55*, 977–988.

117. Yang, D.; Wang, X.-C.; Wong, M.-K.; Yip, Y.-C.; Tang, M.-W. *J. Am. Chem. Soc.* **1996**, *118*, 11311–11312.

118. Yang, D.; Yip, Y.-C.; Chen, J.; Cheung, K.-K. *J. Am. Chem. Soc.* **1998**, *120*, 7659–7660.

119. Wang, Z.-X.; Tu, Y.; Frohn, M.; Shi, Y. *J. Org. Chem.* **1997**, *62*, 2328–2329.

120. Wang, Z.-X.; Tu, Y.; Frohn, M.; Zhang, J.-R.; Shi, Y. *J. Am. Chem. Soc.* **1997**, *119*, 11224–11235.

121. Frohn, M.; Dalkiewicz, M.; Tu, Y.; Wang, Z.-X.; Shi, Y. *J. Org. Chem.* **1998**, *63*, 2948–2953.

122. Wang, Z.-X.; Shi, Y. *J. Org. Chem.* **1998**, *63*, 3099–3104.

123. Tu, Y.; Wang, Z.-X.; Frohn, M.; He, M.; Yu, H.; Tang, Y.; Shi, Y. *J. Org. Chem.* **1998**, *63*, 8475–8485.

124. Cao, G.-A.; Wang, Z.-X.; Tu, Y.; Shi, Y. *Tetrahedron Lett.* **1998**, *39*, 4425–4428.

125. Zhu, Y.; Tu, Y.; Yu, H.; Shi, Y. *Tetrahedron Lett.* **1998**, *39*, 7819–7822.

126. Frohn, M.; Zhou, X.; Zhang, J.-R.; Tang, Y.; Shi, Y. *J. Am. Chem. Soc.* **1999**, *121*, 7718–7719.

127. Wang, Z.-X.; Miller, S. M.; Anderson, O. P.; Shi, Y. *J. Org. Chem.* **1999**, *64*, 6443–6458.

128. Wang, Z.-X.; Cao, G.-A.; Shi, Y. *J. Org. Chem.* **1999**, *64*, 7646–7650.

129. Warren, J. D.; Shi, Y. *J. Org. Chem.* **1999**, *64*, 7675–7677.

130. Shu, L.; Shi, Y. *Tetrahedron Lett.* **1999**, *40*, 8721–8724.

131. Denmark, S. E.; Wu, Z.; Crudden, C. M.; Matsuhashi, H. *J. Org. Chem.* **1997**, *62*, 8288–8289.

132. Denmark, S. E.; Matsuhashi, H. *J. Org. Chem.* **2002**, *67*, 3479–3486.

133. Worth mentioning are also the studies by Aggarwal (Aggarwal, V. K.; Wang, M. F. *Chem. Commun.* **1996**, 191–192) and Page (Page, P. C. B.; Rassias, G. A.; Bethell, D.; Schilling, M. B. *J. Org. Chem.* **1998**, *63*, 2774–2777), in which chiral iminium salts were used as catalysts for asymmetric epoxidation.

134. For reviews of asymmetric catalytic epoxidation with chiral ketones, see: (a) Ref. 5m; (b) Frohn, M.; Shi, Y. *Synlett* **2000**, 1979–2000; (c) Shi, Y. *J. Synth. Org. Chem. Jpn.* **2002**, *60*, 342–349; (d) Shi, Y. *Acc. Chem. Res.* **2004**, *37*, 488–496; (e) Yang, D. *Acc. Chem. Res.* **2004**, *37*, 497–505.

135. For additional reviews of asymmetric catalytic epoxidation, see: (a) Li, A.-H.; Dai, L.-X.; Aggarwal, V. K. *Chem. Rev.* **1997**, *97*, 2341–2372; (b) Aggarwal, V. K.; Richardson, J. *Chem. Commun.* **2003**, 2644–2651; (c) Zhang, Z.-G.; Wang, X.-Y.; Sun, C.; Shi, H.-C. *Chin. J. Org. Chem.* **2004**, *24*, 7–14; (d) Aggarwal, V. K.; Winn, C. *Acc. Chem. Res.* **2004**, *37*, 611–620; (e) Xia, Q.-H.; Ge, H.-Q.; Ye, C.-P.; Liu, Z.-M.; Su, K.-X. *Chem. Rev.* **2005**, *105*, 1603–1662.

136. Ahrendt, K. A.; Borths, C. J.; MacMillan, D. W. C. *J. Am. Chem. Soc.* **2000**, *122*, 4243–4244.

137. Solodin, I.; Goldberg, Y.; Zelcans, G.; Lukevics, E. *J. Chem. Soc. Chem. Commun.* **1990**, 1321–1322.

138. Jen, W. S.; Wiener, J. J. M.; MacMillan, D. W. C. *J. Am. Chem. Soc.* **2000**, *122*, 9874–9875.

139. Paras, N. A.; MacMillan, D. W. C. *J. Am. Chem. Soc.* **2001**, *123*, 4370–4371.

140. Naef, R.; Seebach, D. *Helv. Chim. Acta* **1985**, *68*, 135–143.

141. Austin, J. F.; MacMillan, D. W. C. *J. Am. Chem. Soc.* **2002**, *124*, 1172–1173.

142. Huang, Y.; MacMillan, D. W. C.; unpublished results.

143. Northrup, A. B.; MacMillan, D. W. C. *J. Am. Chem. Soc.* **2002**, *124*, 2458–2460.

144. Wilson, R. M.; Jen, W. S.; MacMillan, D. W. C. *J. Am. Chem. Soc.* **2005**, *127*, 11616–11617.

145. Harmata, M.; Ghosh, S. K.; Hong, X.; Wacharasindhu, S.; Kirchhoefer, P. *J. Am. Chem. Soc.* **2003**, *125*, 2058–2059.

146. Olah G. A. (Ed.), *Friedel-Crafts and Related Reactions*, Vols. 1–4, Wiley-Interscience, New York, **1963–1965**.

147. Olah G. A. (Ed.), *Friedel-Crafts Chemistry*, Wiley, New York, **1973**.

148. Roberts, R. M.; Khalaf A. A. (Eds.), *Friedel-Crafts Alkylation Chemistry: A Century of Discovery*, Marcel Dekker, New York, **1984**.

149. Strell, M.; Kalojanoff, A. *Chem. Ber.* **1954**, *87*, 1025–1032.

150. Gupta, R. R.; Kumar, M.; Gupta, V. *Heterocyclic Chemistry*, Vols. 2–3, Springer-Verlag, Heidelberg, **1999**.

151. Paras, N. A.; MacMillan, D. W. C. *J. Am. Chem. Soc.* **2002**, *124*, 7894–7895.

152. Brown, S. P.; Goodwin, N. C.; MacMillan, D. W. C. *J. Am. Chem. Soc.* **2003**, *125*, 1192–1194.

153. Tietze, L. F. *Chem. Rev.* **1996**, *96*, 115–136.

154. Nicolaou, K. C.; Montagnon, T.; Snyder, S. A. *Chem. Commun.* **2003**, 551–564.

155. Ramón, D. J.; Yus, M. *Angew. Chem. Int. Ed.* **2005**, *44*, 1602–1634.

156. Wasilke, J.-C.; Obrey, S. J.; Baker, R. T.; Bazan, G. C. *Chem. Rev.* **2005**, *105*, 1001–1020.

157. Austin, J. F.; Kim, S.-G.; Sinz, C. J.; Xiao, W.-J.; MacMillan, D. W. C. *Proc. Natl. Acad. Sci. USA* **2004**, *101*, 5482–5487.

158. Akabori, S.; Sakurai, S.; Izumi, Y.; Fujii, Y. *Nature* **1956**, *178*, 323–324.

159. Ohkuma, T.; Kitamura, M.; Noyori, R. In *Catalytic Asymmetric Synthesis*, 2nd ed., Ojima, I. (Ed.) Wiley-VCH, New York, **2000**.

160. Noyori, R. *Angew. Chem. Int. Ed.* **2002**, *41*, 2008–2022.

161. Alberts, B.; Johnson, A.; Lewis, J.; Raff, M.; Roberts, K.; Walter P. (Eds.), *Molecular Biology of the Cell*, 4th ed., Garland, New York, **2002**.

162. Ouellet, S. G.; Tuttle, J. B.; MacMillan, D. W. C. *J. Am. Chem. Soc.* **2005**, *127*, 32–33.

163. Yang, J. W.; Hechavarria Fonseca, M. T.; List, B. *Angew. Chem. Int. Ed.* **2004**, *43*, 6660–6662.

164. Yang, J. W.; Hechavarria Fonseca, M. T.; Vignola, N.; List, B. *Angew. Chem. Int. Ed.* **2005**, *44*, 108–110.

165. Yamaguchi, M.; Yokota, N.; Minami, T. *J. Chem. Soc. Chem. Commun.* **1991**, 1088–1089.

166. Yamaguchi, M.; Shiraishi, T.; Hirama, M. *Angew. Chem. Int. Ed.* **1993**, *32*, 1176–1178.

167. Yamaguchi, M.; Shiraishi, T.; Igarashi, Y.; Hirama, M. *Tetrahedron Lett.* **1994**, *35*, 8233–8236.

168. Yamaguchi, M.; Shiraishi, T.; Hirama, M. *J. Org. Chem.* **1996**, *61*, 3520–3530.

169. Yamaguchi, M.; Igarashi, Y.; Reddy, R. S.; Shiraishi, T.; Hirama, M. *Tetrahedron* **1997**, *53*, 11223–11236.

170. Hanessian, S.; Pham, V. *Org. Lett.* **2000**, *2*, 2975–2978.

171. Halland, N.; Hazell, R. G.; Jørgensen, K. A. *J. Org. Chem.* **2002**, *67*, 8331–8338.

172. Halland, N.; Aburel, P. S.; Jørgensen, K. A. *Angew. Chem. Int. Ed.* **2003**, *42*, 661–665.

173. Halland, N.; Hansen, T.; Jørgensen, K. A. *Angew. Chem. Int. Ed.* **2003**, *42*, 4955–4957.

174. Halland, N.; Aburel, P. S.; Jorgensen, K. A. *Angew. Chem. Int. Ed.* **2004**, *43*, 1272–1277.

175. Pulkkinen, J.; Aburel, P. S.; Halland, N.; Jørgensen, K. A. *Adv. Synth. Catal.* **2004**, *346*, 1077–1080.

176. Benaglia, M.; Cinquini, M.; Cozzi, F.; Puglisi, A.; Celentano, G. *J. Mol. Catal. A: Chemical* **2003**, *204–205*, 157–163.

177. Dhevalapally, B.; Barbas III, C. F. *Chem. Eur. J.* **2004**, *10*, 5323–5331.

178. Gryko, D. *Tetrahedron Asym.* **2005**, *16*, 1377–1383.

179. Hayashi, Y.; Gotoh, H.; Hayashi, T.; Shoji, M. *Angew. Chem. Int. Ed.* **2005**, *44*, 4212–4215.

180. Kunz, R. K.; MacMillan, D. W. C. *J. Am. Chem. Soc.* **2005**, *127*, 3240–3241.

181. Marigo, M.; Franzén, J.; Poulsen, T. B.; Zhuang, W.; Jørgensen, K. A. *J. Am. Chem. Soc.* **2005**, *127*, 6964–6965.

182. Gröger, H.; Wilken, J. *Angew. J.; Chem. Int. Ed.* **2001**, *40*, 529–532.

183. List, B. *Synlett* **2001**, 1675–1686.

184. List, B. *Tetrahedron* **2002**, *58*, 5573–5590.

185. Gathergood, N. *Aust. J. Chem.* **2002**, *55*, 615–615.

186. Movassaghi, M.; Jacobsen, E. N. *Science* **2002**, *298*, 1904–1905.

187. Li, J.-W.; Xu, L.-W.; Xia, C.-G. *Chin. J. Org. Chem.* **2004**, *24*, 23–28.

188. List, B. *Acc. Chem. Res.* **2004**, *37*, 548–557.

189. Notz, W.; Tanaka, F.; Barbas III, C. F. *Acc. Chem. Res.* **2004**, *37*, 580–591.

190. Allemann, C.; Gordillo, R.; Clemente, F. R.; Cheong, P. H.-Y.; Houk, K. N. *Acc. Chem. Res.* **2004**, *37*, 558–569.

191. Alcaide, B.; Almendros, P. *Eur. J. Org. Chem.* **2002**, 1595–1601.

192. Palomo, C.; Oiarbide, M.; García, J. M. *Chem. Eur. J.* **2002**, *8*, 37–44.

193. Alcaide, B.; Almendros, P. *Angew. Chem. Int. Ed.* **2003**, *42*, 858–860.

194. Palomo, C.; Oiarbide, M.; García, J. M. *Chem. Soc. Rev.* **2004**, *33*, 65–75.

195. Saito, S.; Yamamoto, H. *Acc. Chem. Res.* **2004**, *37*, 570–579.

196. Evans, D. A.; Vogel, E.; Nelson, J. V. *J. Am. Chem. Soc.* **1979**, *101*, 6120–6123.

197. Evans, D. A.; Bartroli, J.; Shih, T. L. *J. Am. Chem. Soc.* **1981**, *103*, 2127–2129.

198. Evans, D. A.; Nelson, J. V.; Vogel, E.; Taber, T. R. *J. Am. Chem. Soc.* **1981**, *103*, 3099–3111.

199. Evans, D. A.; Nelson, J. V.; Taber, T. R. In *Topics in Stereochemistry*, Vol. 13, Allinger, N. L.; Eliel, E. L.; Wilen, S. H. (Eds.), Wiley, New York, **1982**, pp. 1–115.

200. Kleschick, W. A.; Buse, C. T.; Heathcock, C. H. *J. Am. Chem. Soc.* **1977**, *99*, 247–248.

201. Heathcock, C. H.; White, C. T. *J. Am. Chem. Soc.* **1979**, *101*, 7076–7077.

202. Heathcock, C. H. *Science* **1981**, *214*, 395–400.

203. Heathcock, C. H. In *Asymmetric Synthesis*, Vol. 3, part B, Morrison, J. D. (Ed.), Academic Press, Orlando, FL, **1984**, pp. 111–212.

204. Danda, H.; Hansen, M. M.; Heathcock, C. H. *J. Org. Chem.* **1990**, *55*, 173–181.

205. Masamune, S.; Ali, S. A.; Snitman, D. L.; Garvey, D. S. *Angew. Chem. Int. Ed.* **1980**, *19*, 557–558.

206. Masamune, S.; Choy, W.; Kerdesky, F. A. J.; Imperiali, B. *J. Am. Chem. Soc.* **1981**, *103*, 1566–1568.

207. Masamune, S.; Sato, T.; Kim, B. M.; Wollmann, T. A. *J. Am. Chem. Soc.* **1986**, *108*, 8279–8281.

208. Kim, B. M.; Williams, S. F.; Masamune, S. In *Comprehensive Organic Synthesis*, Vol. 2, Trost, B. M.; Fleming, I. (Eds.) Pergamon, Oxford, **1991**, Chap. 1.7, pp. 239–275.

209. Mukaiyama, T.; Narasaka, K.; Banno, K. *Chem. Lett.* **1973**, 1011–1014.

210. Mukaiyama, T.; Banno, K.; Narasaka, K. *J. Am. Chem. Soc.* **1974**, *96*, 7503–7509.

211. Kobayashi, S.; Uchiro, H.; Shiina, I.; Mukaiyama, T. *Tetrahedron* **1993**, *49*, 1761–1772.

212. Yamada, Y. M. A.; Yoshikawa, N.; Sasai, H.; Shibasaki, M. *Angew. Chem. Int. Ed.* **1997**, *36*, 1871–1873.

213. Yoshikawa, N.; Yamada, Y. M. A.; Das, J.; Sasai, H.; Shibasaki, M. *J. Am. Chem. Soc.* **1999**, *121*, 4168–4178.

214. Yoshikawa, N.; Kumagai, N.; Matsunaga, S.; Moll, G.; Ohshima, T.; Suzuki, T.; Shibasaki, M. *J. Am. Chem. Soc.* **2001**, *123*, 2466–2467.

215. Yoshikawa, N.; Shibasaki, M. *Tetrahedron* **2001**, *57*, 2569–2579.

216. Trost, B. M.; Ito, H. *J. Am. Chem. Soc.* **2000**, *122*, 12003–12004.

217. Trost, B. M.; Ito, H.; Silcoff, E. R. *J. Am. Chem. Soc.* **2001**, *123*, 3367–3368.

218. Trost, B. M.; Silcoff, E. R.; Ito, H. *Org. Lett.* **2001**, *3*, 2497–2500.

219. Notz, W.; List, B. *J. Am. Chem. Soc.* **2000**, *122*, 7386–7387.

220. Sakthivel, K.; Notz, W.; Bui, T.; Barbas III, C. F. *J. Am. Chem. Soc.* **2001**, *123*, 5260–5267.

221. Saito, S.; Nakadai, M.; Yamamoto, H. *Synlett* **2001**, 1245–1248.

222. Bahmanyar, S.; Houk, K. N. *J. Am. Chem. Soc.* **2001**, *123*, 11273–11283.

223. Benaglia, M.; Celentano, G.; Cozzi, F. *Adv. Synth. Catal.* **2001**, *343*, 171–173.

224. Dickerson, T. J.; Janda, K. D. *J. Am. Chem. Soc.* **2002**, *124*, 3220–3221.

225. Nakadai, M.; Saito, S.; Yamamoto, H. *Tetrahedron* **2002**, *58*, 8167–8177.

226. Rankin, K. N.; Gauld, J. W.; Boyd, R. J. *J. Phys. Chem. A* **2002**, *106*, 5155–5159.

227. Arnó, M.; Domingo, L. R. *Theor. Chem. Acc.* **2002**, *108*, 232–239.

228. Benaglia, M.; Cinquini, M.; Cozzi, F.; Puglisi, A.; Celentano, G. *Adv. Synth. Catal.* **2002**, *344*, 533–542.

229. Córdova, A.; Notz, W.; Barbas III, C. F. *Chem. Commun.* **2002**, 3024–3025.

230. Loh, T.-P.; Feng, L.-C.; Yang, H.-Y.; Yang, J.-Y. *Tetrahedron Lett.* **2002**, *43*, 8741–8743.

231. Kotrusz, P.; Kmentová, I.; Gotov, B.; Toma, S.; Solcániová, E.; *Chem. Commun.* **2002**, 2510–2511.

232. Liu, H.; Peng, L.; Zhang, T.; Li, Y. *New J. Chem.* **2003**, *27*, 1159–1160.

233. Dhar, D.; Beadham, I.; Chandrasekaran, S. *Proc. Indian Acad. Sci. (Chem. Sci.)* **2003**, *115*, 365–372.

234. Darbre, T.; Machuqueiro, M. *Chem. Commun.* **2003**, 1090–1091.

235. Bahmanyar, S.; Houk, K. N.; Martin, H. J.; List, B. *J. Am. Chem. Soc.* **2003**, *125*, 2475–2479.

236. Peng, Y.-Y.; Ding, Q.-P.; Li, Z.; Wang, P. G.; Cheng, J.-P. *Tetrahedron Lett.* **2003**, *44*, 3871–3875.

237. Sekiguchi, Y.; Sasaoka, A.; Shimomoto, A.; Fujioka, S.; Kotsuki, H. *Synlett* **2003**, 1655–1658.

238. Kofoed, J.; Nielsen, J.; Reymond, J.-L. *Bioorg. Med. Chem. Lett.* **2003**, *13*, 2445–2447.

239. Martin, H. J.; List, B. *Synlett* **2003**, 1901–1902.

240. Fache, F.; Piva, O. *Tetrahedron Asym.* **2003**, *14*, 139–143.

241. Tang, Z.; Jiang, F.; Yu, L.-T.; Cui, X.; Gong, L.-Z.; Mi, A.-Q.; Jiang, Y.-Z.; Wu, Y.-D. *J. Am. Chem. Soc.* **2003**, *125*, 5262–5263.

242. Szöllosi, G; London, G.; Baláspiri, L.; Somlai, C.; Bartók, M.; *Chirality* **2003**, *15*, S90–96.

243. Pan, Q.; Zou, B.; Wang, Y.; Ma, D. *Org. Lett.* **2004**, *6*, 1009–1012.

244. Torii, H.; Nakadai, M.; Ishihara, K.; Saito, S.; Yamamoto, H. *Angew. Chem. Int. Ed.* **2004**, *43*, 1983–1986.

245. Tang, Z.; Jiang, F.; Cui, X.; Gong, L.-Z.; Mi, A.-Q.; Jiang, Y.-Z.; Wu, Y.-D. *Proc. Natl. Acad. Sci. USA* **2004**, *101*, 5755–5760.

246. Shen, Z.-X.; Zhou, H.; Ma, J.-M.; Liu, Y. H.; Zhang, Y.-W. *Chin. J. Org. Chem.* **2004**, *24*, 1213–1216.

247. Hartikka, A.; Arvidsson, P. I. *Tetrahedron Asym.* **2004**, *15*, 1831–1834.

248. Zhong, G.; Fan, J.; Barbas III, C. F. *Tetrahedron Lett.* **2004**, *45*, 5681–5684.

249. Berkessel, A.; Koch, B.; Lex, J. *Adv. Synth. Catal.* **2004**, *346*, 1141–1146.

250. Tanimori, S.; Naka, T.; Kirihata, M. *Synth. Commun.* **2004**, *34*, 4043–4048.

251. Lacoste, E.; Landais, Y.; Schenk, K.; Verlhac, J.-B.; Vincent, J.-M. *Tetrahedron Lett.* **2004**, *45*, 8035–8038.

252. Shi, L.-X.; Sun, Q.; Ge, Z.-M.; Zhu, Y.-Q.; Cheng, T.-M.; Li, R.-T.; *Synlett* **2004**, 2215–2217.

253. Casas, J.; Sundén, H.; Córdova, A.; *Tetrahedron Lett.* **2004**, *45*, 6117–6119.

254. Edin, M.; Bäckvall, J.-E.; A. Córdova, *Tetrahedron Lett.* **2004**, *45*, 7697–7701.

255. Ward, D. E.; Jheengut, V. *Tetrahedron Lett.* **2004**, *45*, 8347–8350.

256. Chandrasekhar, S.; Narsihmulu, C.; Ramakrishna Reddy, N.; Shameem Sultana, S. *Tetrahedron Lett.* **2004**, *45*, 4581–4582.

257. Wu, Y.-S.; Shao, W.-Y.; Zheng, C.-Q.; Huang, Z.-L.; Cai, J.; Deng, Q.-Y. *Helv. Chim. Acta* **2004**, *87*, 1377–1384.

258. Nyberg, A. I.; Usano, A.; Pihko, P. M. *Synlett* **2004**, 1891–1896.

259. Hayashi, Y.; Tsuboi, W.; Shoji, M.; Suzuki, N. *Tetrahedron Lett.* **2004**, *45*, 4353–4356.

260. Córdova, A. *Tetrahedron Lett.* **2004**, *45*, 3949–3952.

261. Gruttadauria, M.; Riela, S.; Lo Meo, P.; D'Anna, F.; Noto, R. *Tetrahedron Lett.* **2004**, *45*, 6113–6116.

262. Ibrahem, I.; Córdova, A. *Tetrahedron Lett.* **2005**, *46*, 3363–3367.

263. Enders, D.; Grondal, C. *Angew. Chem. Int. Ed.* **2005**, *44*, 1210–1212.

264. Suri, J. T.; Ramachary, D. B.; Barbas III, C. F. *Org. Lett.* **2005**, *7*, 1383–1385.

265. Ding, Y.; Zhang, Y.-W.; Liu, Y.-H.; Luo, X.-Q.; Ma, J.-M.; Shen, Z.-X. *Chin. J. Org. Chem.* **2005**, *25*, 567–569.

266. Cobb, A. J. A.; Shaw, D. M.; Longbottom, D. A.; Gold, J. B.; Ley, S. V. *Org. Biomol. Chem.* **2005**, *3*, 84–96.

267. Rogers, C. J.; Dickerson, T. J.; Brogan, A. P.; Janda, K. D. *J. Org. Chem.* **2005**, *70*, 3705–3708.

268. Krattiger, P.; Kovasy, R.; Revell, J. D.; Ivan, S.; Wennemers, H. *Org. Lett.* **2005**, *7*, 1101–1103.

269. Guo, H.-M.; Cun, L.-F.; Gong, L.-Z.; Mi, A.-Q.; Jiang, Y.-Z. *Chem. Commun.* **2005**, 1450–1452.

270. Liu, Y.-H.; Zhang, Y.-W.; Ding, Y.-P.; Shen, Z.-X.; Luo, X.-Q. *Chin. J. Chem.* **2005**, *23*, 634–636.

271. A nondirect enantioselective cross-aldol reaction between two discrete aldehyde components has been achieved; see Denmark, S. E.; Ghosh, S. K. *Angew. Chem. Int. Ed.* **2001**, *40*, 4759–4762.

272. Gijsen, H. J. M.; Wong, C.-H. *J. Am. Chem. Soc.* **1994**, *116*, 8422–8423.

273. Exposure of acetaldehyde to proline leads to the corresponding trimer dehydration adduct 5-hydroxy-(2*E*)-hexenal in 10% yield and 90% ee. For reference, see Córdova, A.; Notz, W.; Barbas III, C. F. *J. Org. Chem.* **2002**, *67*, 301–303.

274. Northrup, A. B.; MacMillan, D. W. C. *J. Am. Chem. Soc.* **2002**, *124*, 6798–6799.

275. Northrup, A. B.; Mangion, I. K.; Hettche, F.; MacMillan, D. W. C. *Angew. Chem. Int. Ed.* **2004**, *43*, 2152–2154.

276. Storer, R. I.; MacMillan, D. W. C. *Tetrahedron* **2004**, *60*, 7705–7714.

277. Mangion, I. K.; Northrup, A. B.; MacMillan, D. W. C. *Angew. Chem. Int. Ed.* **2004**, *43*, 6722–6724.

278. Northrup, A. B.; MacMillan, D. W. C. *Science* **2004**, *305*, 1752–1755.

279. Casas, J.; Engqvist, M.; Ibrahem, I.; Kaynak, B.; Córdova, A. *Angew. Chem. Int. Ed.* **2005**, *44*, 1343–1345.

280. For a recent review on the organocatalytic synthesis of carbohydrates, see Kazmaier, U. *Angew. Chem. Int. Ed.* **2005**, *44*, 2186–2188.

281. For a recent review on the organocatalytic synthesis of carbohydrates, see Limbach, M. *Chem. Biodiv.* **2005**, *2*, 825–836.

282. Pidathala, C.; Hoang, L.; Vignola, N.; List, B. *Angew. Chem. Int. Ed.* **2003**, *42*, 2785–2788.

283. Córdova, A. *Acc. Chem. Res.* **2004**, *37*, 102–112.

284. Tanaka, F.; Barbas III, C. F. In *Enantioselective Synthesis of β-Amino Acids*, 2nd ed. Juaristi, E.; Soloshonok, V. (Eds.) Wiley, Hoboken, NJ **2005**, Chap. 9.

285. Kobayashi, S.; Ishitani, H. *Chem. Rev.* **1999**, *99*, 1069–1094.

286. Notz, W.; Sakthivel, K.; Bui, T.; Zhong, G.; Barbas III, C. F. *Tetrahedron Lett.* **2001**, *42*, 199–201.

287. List, B.; Pojarliev, P.; Biller, W. T.; Martin, H. J. *J. Am. Chem. Soc.* **2002**, *124*, 827–833.

288. Córdova, A.; Notz, W.; Zhong, G.; Betancort, J. M.; Barbas III, C. F. *J. Am. Chem. Soc.* **2002**, *124*, 1842–1843.

289. Córdova, A.; Watanabe, S.-i.; Tanaka, F.; Notz, W.; Barbas III, C. F. *J. Am. Chem. Soc.* **2002**, *124*, 1866–1867.

290. Córdova, A.; Barbas III, C. F. *Tetrahedron Lett.* **2002**, *43*, 7749–7752.

291. Watanabe, S.-i.; Córdova, A.; Tanaka, F.; Barbas III, C. F. *Org. Lett.* **2002**, *4*, 4519–4522.

292. A. Córdova, Barbas III, C. F. *Tetrahedron Lett.* **2003**, *44*, 1923–1926.

293. Notz, W.; Tanaka, F.; Watanabe, S.-i.; Chowdari, N. S.; Turner, J. M.; Thayumanavan, R.; Barbas III, C. F. *J. Org. Chem.* **2003**, *68*, 9624–9634.

294. Zhuang, W.; Saaby, S.; Jørgensen, K. A. *Angew. Chem. Int. Ed.* **2004**, *43*, 4476–4478.

295. Chowdari, N. S.; Suri, J. T.; Barbas III, C. F. *Org. Lett.* **2004**, *6*, 2507–2510.

296. Hayashi, Y.; Tsuboi, W.; Ashimine, I.; Urushima, T.; Shoji, M.; Sakai, K. *Angew. Chem. Int. Ed.* **2003**, *42*, 3677–3680.

297. Córdova, A. *Synlett* **2003**, 1651–1654.

298. Córdova, A. *Chem. Eur. J.* **2004**, *10*, 1987–1997.

299. Münch, A.; Wendt, B.; Christmann, M. *Synlett* **2004**, 2751–2755.

300. Ibrahem, I.; Córdova, A. *Tetrahedron Lett.* **2005**, *46*, 2839–2843.

301. Sibi, M. P.; Manyem, S. *Tetrahedron* **2000**, *56*, 8033–8061.

302. Krause, N.; Hoffmann-Röder, A. *Synthesis* **2001**, 171–196.

303. Berner, O. M.; Tedeschi, L.; Enders, D. *Eur. J. Org. Chem.* **2002**, 1877–1894.

304. Christoffers, J.; Baro, A. *Angew. Chem. Int. Ed.* **2003**, *42*, 1688–1690.

305. Yamaguchi, M. In *Comprehensive Asymmetric Catalysis*, Jacobsen, E.N.; Pfaltz, A.; Yamamoto, H. (Eds.), Springer, Heidelberg, **1999**, Chap. 31.2.

306. List, B.; Pojarliev, P.; Martin, H. J. *Org. Lett.* **2001**, *3*, 2423–2425.

307. Betancort, J. M.; Sakthivel, K.; Thayumanavan, R.; Barbas III, C. F. *Tetrahedron Lett.* **2001**, *42*, 4441–4444.

308. Enders, D.; Seki, A. *Synlett* **2002**, 26–28.

309. Kotrusz, P.; Toma, S.; Schmalz, H.-G.; Adler, A. *Eur. J. Org. Chem.* **2004**, 1577–1583.

310. Salaheldin, A. M.; Yi, Z.; Kitazume, T. *J. Fluor. Chem.* **2004**, *125*, 1105–1110.

311. Hechavarria Fonseca, M. T.; List, B. *Angew. Chem. Int. Ed.* **2004**, *43*, 3958–3960.

312. Mangion, I. K.; MacMillan, D. W. C. *J. Am. Chem. Soc.* **2005**, *127*, 3696–3697.

313. Peelen, T. J.; Chi, Y.; Gellman, S. H. *J. Am. Chem. Soc.* **2005**, *127*, 11598–11599.

314. Betancort, J. M.; Barbas III, C. F. *Org. Lett.* **2001**, *3*, 3737–3740.

315. Alexakis, A.; Andrey, O. *Org. Lett.* **2002**, *4*, 3611–3614.

316. Andrey, O.; Vidonne, A.; Alexakis, A. *Tetrahedron Lett.* **2003**, *44*, 7901–7904.

317. Andrey, O.; Alexakis, A.; Bernardinelli, G. *Org. Lett.* **2003**, *5*, 2559–2561.

318. Andrey, O.; Alexakis, A.; Tomassini, A.; Bernardinelli, G. *Adv. Synth. Catal.* **2004**, *346*, 1147–1168.

319. Melchiorre, P.; Jørgensen, K. A. *J. Org. Chem.* **2003**, *68*, 4151–4157.

320. Betancort, J. M.; Sakthivel, K.; Thayumanavan, R.; Tanaka, F.; Barbas III, C. F. *Synthesis* **2004**, 1509–1521.

321. Ishii, T.; Fujioka, S.; Sekiguchi, Y.; Kotsuki, H. *J. Am. Chem. Soc.* **2004**, *126*, 9558–9559.

322. Cobb, A. J. A.; Longbottom, D. A.; Shaw, D. M.; Ley, S. V. *Chem. Commun.* **2004**, 1808–1809.

323. Mase, N.; Thayumanavan, R.; Tanaka, F.; Barbas III, C. F. *Org. Lett.* **2004**, *6*, 2527–2530.

324. Wang, W.; Wang, J.; Li, H. *Angew. Chem. Int. Ed.* **2005**, *44*, 1369–1371.

325. Mitchell, C. E. T.; Cobb, A. J. A.; Ley, S. V. *Synlett* **2005**, 611–614.

326. Chi, Y.; Gellman, S. H. *Org. Lett.* **2005**, *7*, 4253–4256.

327. Juhl, K.; Jørgensen, K. A. *Angew. Chem. Int. Ed.* **2003**, *42*, 1498–1501.

328. For other examples, see: (a) Wabnitz, T. C.; Saaby, S.; Jørgensen, K. A. *Org. Biomol. Chem.* **2004**, *2*, 828–834; (b) Yamamoto, Y.; Momiyama, N.; Yamamoto, H. *J. Am. Chem. Soc.* **2004**, *126*, 5962–5963; (c) Hayashi, Y.; Yamaguchi, J.; Hibino, K.; Sumiya, T.; Urushima, T.; Shoji, M.; Hashizume, D.; Koshino, H. *Adv. Synth. Catal.* **2004**, *346*, 1435–1439.

329. Sundén, H.; Ibrahem, I.; Eriksson, L.; Córdova, A. *Angew. Chem. Int. Ed.* **2005**, *44*, 4877–4880.

330. List, B.; Castello, C. *Synlett* **2001**, 1687–1689.

331. Ramachary, D. B.; Chowdari, N. S.; Barbas III, C. F. *Synlett* **2003**, 1910–1914.

332. Ramachary, D. B.; Chowdari, N. S.; Barbas III, C. F. *Angew. Chem. Int. Ed.* **2003**, *42*, 4233–4237.

333. Ramachary, D. B.; Anebouselvy, K.; Chowdari, N. S.; Barbas III, C. F. *J. Org. Chem.* **2004**, *69*, 5838–5849.

334. Bøgevig, A.; Juhl, K.; Kumaragurubaran, N.; Zhuang, W.; Jørgensen, K. A. *Angew. Chem. Int. Ed.* **2002**, *41*, 1790–1793.

335. List, B. *J. Am. Chem. Soc.* **2002**, *124*, 5656–5657.

336. Kumaragurubaran, N.; Juhl, K.; Zhuang, W.; Bøgevig, A.; Jørgensen, K. A. *J. Am. Chem. Soc.* **2002**, *124*, 6254–6255.

337. Vogt, H.; Vanderheiden, S.; Bräse, S. *Chem. Commun.* **2003**, 2448–2449.

338. Chowdari, N. S.; Barbas III, C. F. *Org. Lett.* **2005**, *7*, 867–870.

339. For recent reviews on α-amination, see: (a) Duthaler, R. O. *Angew. Chem. Int. Ed.* **2003**, *42*, 975–978; (b) Janey, J. M. *Angew. Chem. Int. Ed.* **2005**, *44*, 4292–4300.

340. Zhong, G. *Angew. Chem. Int. Ed.* **2003**, *42*, 4247–4250.

341. Brown, S. P.; Brochu, M. P.; Sinz, C. J.; MacMillan, D. W. C. *J. Am. Chem. Soc.* **2003**, *125*, 10808–10809.

342. Hayashi, Y.; Yamaguchi, J.; Hibino, K.; Shoji, M. *Tetrahedron Lett.* **2003**, *44*, 8293–8296.

343. Bøgevig, A.; Sundén, H.; Córdova, A. *Angew. Chem. Int. Ed.* **2004**, *43*, 1109–1112.

344. Hayashi, Y.; Yamaguchi, J.; Sumiya, T.; Shoji, M. *Angew. Chem. Int. Ed.* **2004**, *43*, 1112–1115.

345. Momiyama, N.; Torii, H.; Saito, S.; Yamamoto, H. *Proc. Natl. Acad. Sci. USA* **2004**, *101*, 5374–5378.

346. Córdova, A.; Sundén, H.; Bøgevig, A.; Johansson, M.; Himo, F. *Chem. Eur. J.* **2004**, *10*, 3673–3684.

347. Hayashi, Y.; Yamaguchi, J.; Sumiya, T.; Hibino, K.; Shoji, M. *J. Org. Chem.* **2004**, *69*, 5966–5973.

348. For recent reviews on α-aminoxylation, see: (a) Merino, P.; Tejero, T. *Angew. Chem. Int. Ed.* **2004**, *43*, 2995–2997; (b) Ref. 339b.

349. Córdova, A.; Sundén, H.; Engqvist, M.; Ibrahem, I.; Casas, J. *J. Am. Chem. Soc.* **2004**, *126*, 8914–8915.

350. H. Sundén, Engqvist, M.; Casas, J.; Ibrahem, I.; Córdova, A.; *Angew. Chem. Int. Ed.* **2004**, *43*, 6532–6535.

351. Engqvist, M.; Casas, J.; Sundén, H.; Ibrahem, I.; Córdova, A. *Tetrahedron Lett.* **2005**, *46*, 2053–2057.

352. Marigo, M.; Wabnitz, T. C.; Fielenbach, D.; Jørgensen, K. A. *Angew. Chem. Int. Ed.* **2005**, *44*, 794–797.

353. Brochu, M. P.; Brown, S. P.; MacMillan, D. W. C. *J. Am. Chem. Soc.* **2004**, *126*, 4108–4109.

354. Halland, N.; Braunton, A.; Bachmann, S.; Marigo, M.; Jørgensen, K. A. *J. Am. Chem. Soc.* **2004**, *126*, 4790–4791.

355. Marigo, M.; Bachmann, S.; Halland, N.; Braunton, A.; Jørgensen, K. A. *Angew. Chem. Int. Ed.* **2004**, *43*, 5507–5510.

356. For a recent review on α-halogenations, see Oestreich, M. *Angew. Chem. Int. Ed.* **2005**, *44*, 2324–2327.

357. Enders, D.; Hüttl, M. R. M. *Synlett* **2005**, 991–993.

358. Marigo, M.; Fielenbach, D.; Braunton, A.; Kjœrsgaard, A.; Jørgensen, K. A. *Angew. Chem. Int. Ed.* **2005**, *44*, 3703–3706.

359. Steiner, D. D.; Mase, N.; Barbas III, C. F. *Angew. Chem. Int. Ed.* **2005**, *44*, 3706–3710.

360. Beeson, T. D.; MacMillan, D. W. C. *J. Am. Chem. Soc.* **2005**, *127*, 8826–8828.

361. For examples of catalytic asymmetric indirect alkylations, see Doyle, A. G.; Jacobsen, E. N. *J. Am. Chem. Soc.* **2005**, *127*, 62–63 and references cited therein.

362. Vignola, N.; List, B. *J. Am. Chem. Soc.* **2004**, *126*, 450–451.

363. Schreiner, P. R. *Chem. Soc. Rev.* **2003**, *32*, 289–296.

364. Oestreich, M. *Nachr. Chem.* **2004**, *52*, 35–38.

365. Pihko, P. M. *Angew. Chem. Int. Ed.* **2004**, *43*, 2062–2064.

366. Bolm, C.; Rantanen, T.; Schiffers, I.; Zani, L. *Angew. Chem. Int. Ed.* **2005**, *44*, 1758–1763.

367. For a review on catalytic asymmetric cyanation of ketoimines, see Spino, C. *Angew. Chem. Int. Ed.* **2004**, *43*, 1764–1766.

368. A new cinchona alkaloid-derived catalyst has been developed for the enantioselective Strecker reaction of aryl aldimines via hydrogen-bonding activation. For reference, see Huang, J.; Corey, E. J. *Org. Lett.* **2004**, *6*, 5027–5029.

369. Wenzel, A. G.; Jacobsen, E. N. *J. Am. Chem. Soc.* **2002**, *124*, 12964–12965.

370. Okino, T.; Nakamura, S.; Furukawa, T.; Takemoto, Y. *Org. Lett.* **2004**, *6*, 625–627.

371. Yoon, T. P.; Jacobsen, E. N. *Angew. Chem. Int. Ed.* **2005**, *44*, 466–468.

372. Taylor, M. S.; Jacobsen, E. N. *J. Am. Chem. Soc.* **2004**, *126*, 10558–10559.

373. Joly, G. D.; Jacobsen, E. N. *J. Am. Chem. Soc.* **2004**, *126*, 4102–4103.

374. Nugent, B. M.; Yoder, R. A.; Johnston, J. N. *J. Am. Chem. Soc.* **2004**, *126*, 3418–3419.

375. Akiyama, T.; Itoh, J.; Yokota, K.; Fuchibe, K. *Angew. Chem. Int. Ed.* **2004**, *43*, 1566–1568.

376. Uraguchi, D.; Terada, M. *J. Am. Chem. Soc.* **2004**, *126*, 5356–5357.

377. Akiyama, T.; Morita, H.; Itoh, J.; Fuchibe, K. *Org. Lett.* **2005**, *7*, 2583–2585.

378. Uraguchi, D.; Sorimachi, K.; Terada, M. *J. Am. Chem. Soc.* **2004**, *126*, 11804–11805.

379. Rueping, M.; Sugiono, E.; Azap, C.; Theissmann, T.; Bolte, M. *Org. Lett.* **2005**, *7*, 3781–3783.

380. Hoffmann, S.; Seayad, A. M.; List, B. *Angew. Chem. Int. Ed.* **2005**, *44*, 7424–7427.

381. Storer, R. I.; Carrera, D. E.; Ni, Y.; MacMillan, D. W. C. *J. Am. Chem. Soc.* **2006**, *128*, 84–86.

382. For the reduction of ketimines with trichlorosylane via a possible hydrogen-bonding activation, see Malkov, A. V.; Mariani, A.; MacDougall, K. N.; Kocovsky, P. *Org. Lett.* **2004**, *6*, 2253–2256.

383. Jarvo, E. R.; Miller, S. J. *Tetrahedron* **2002**, *58*, 2481–2495.

384. Berkessel, A. *Curr. Opin. Chem. Biol.* **2003**, *7*, 409–419.

385. Miller, S. J. *Acc. Chem. Res.* **2004**, *37*, 601–610.

386. Horstmann, T. E.; Guerin, D. J.; Miller, S. J. *Angew. Chem. Int. Ed.* **2000**, *39*, 3635–3638.

387. Tsuboike, K.; Guerin, D. J.; Mennen, S. M.; Miller, S. J. *Tetrahedron* **2004**, *60*, 7367–7374.

388. For an example of asymmetric Michael addition using a chiral diamine–dipeptide catalyst, see Tsogoeva, S. B.; Jagtap, S. B. *Synlett* **2004**, 2624–2626.

389. Harriman, D. J.; Deslongehamps, G. *J. Comput. Aided Mol. Design.* **2004**, *18*, 303–308.

390. Okino, T.; Hoashi, Y.; Takemoto, Y. *J. Am. Chem. Soc.* **2003**, *125*, 12672–12673.

391. Hoashi, Y.; Yabuta, T.; Takemoto, Y. *Tetrahedron Lett.* **2004**, *45*, 9185–9188.

392. Hoashi, Y.; Okino, T.; Takemoto, Y. *Angew. Chem. Int. Ed.* **2005**, *44*, 4032–4035.

393. Sohtome, Y.; Tanatani, A.; Hashimoto, Y.; Nagasawa, K. *Chem. Pharm. Bull.* **2004**, *52*, 477–480.

394. Li, B.-J.; Jiang, L.; Liu, M.; Chen, Y.-C.; Ding, L.-S.; Wu, Y. *Synlett* **2005**, 603–606.

395. Sohtome, Y.; Tanatani, A.; Hashimoto, Y.; Nagasawa, K. *Tetrahedron Lett.* **2004**, *45*, 5589–5592.

396. For an example of asymmetric Michael reaction using guanidine-based catalysts, see: Ma, D.; Cheng, K. *Tetrahedron Asym.* **1999**, *10*, 713–719.

397. McDougal, N. T.; Schaus, S. E. *J. Am. Chem. Soc.* **2003**, *125*, 12094–12095.

398. McDougal, N. T.; Trevellini, W. L.; Rodgen, S. A.; Kliman, L. T.; Schaus, S. E. *Adv. Synth. Catal.* **2004**, *346*, 1231–1240.

399. Matsui, K.; Takizawa, S.; Sasai, H. *J. Am. Chem. Soc.* **2005**, *127*, 3680–3681.

400. Berkessel, A.; Cleemann, F.; Mukherjee, S.; Müller, T. N; Lex, J. *Angew. Chem. Int. Ed.* **2005**, *44*, 807–811.

401. Berkessel, A.; Mukherjee, S.; Cleemann, F.; Müller, T. N.; Lex, J. *Chem. Commun.* **2005**, 1898–1900.

402. Schuster, T.; Bauch, M.; Dürner, G.; Göbel, M. W.; *Org. Lett.* **2000**, *2*, 179–181.

403. Tsogoeva, S. B.; Dürner, G.; Bolte, M.; Göbel, M. W. *Eur. J. Org. Chem.* **2003**, 1661–1664.

404. For other pioneering examples of hydrogen-bond-mediated Diels–Alder reactions using chiral PHANOLs as organocatalysts affording almost racemic addition products, see: (a) Braddock, D. C.; MacGilp, I. D.; Perry, B. G. *Synlett* **2003**, 1121–1124; (b) Braddock, D. C.; MacGilp, I. D.; Perry, B. G. *Adv. Synth. Catal.* **2004**, *346*, 1117–1130.

405. For highly enantioselective [2 + 2] photocycloadditions via hydrogen-bonding activation using excess chiral Kemp triacid derivatives as host molecules, see: (a) Bach, T.; Bergmann, H. *J. Am. Chem. Soc.* **2000**, *122*, 11525–11526; (b) Bach, T.; Bergmann, H.; Grosch, B.; Harms, K. *J. Am. Chem. Soc.* **2002**, *124*, 7982–7990.

406. Thadani, A. N.; Stankovic, A. R.; Rawal, V. H. *Proc. Natl. Acad. Sci. USA* **2004**, *101*, 5846–5850.

407. Huang, Y.; Unni, A. K.; Thadani, A. N.; Rawal, V. H. *Nature* **2003**, *424*, 146–146.

408. Du, H.; Zhao, D.; Ding, K. *Chem. Eur. J.* **2004**, *10*, 5964–5970.

409. Unni, A. K.; Takenaka, N.; Yamamoto, H.; Rawal, V. H. *J. Am. Chem. Soc.* **2005**, *127*, 1336–1337.

410. Tanoi, T.; Mikami, K. *Tetrahedron Lett.* **2005**, *46*, 6355–6358.

411. Momiyama, N.; Yamamoto, H. *J. Am. Chem. Soc.* **2005**, *127*, 1080–1081.

412. Nelson, A. *Angew. Chem. Int. Ed.* **1999**, *38*, 1583–1585.

413. O'Donnell, M. J. In *Catalytic Asymmetric Synthesis*, 2nd ed. Ojima, I. (Ed.), Wiley-VCH, New York, **2000**.

414. Jones, R. A. *Quaternary Ammonium Salts: Their Use in Phase-Transfer Catalysis*, Academic Press, London, **2001**.

415. Lygo, B. In *Rodd's Chemistry of Carbon Compounds: Asymmetric Catalysis, Vol. 5*, 2nd ed., Sainsbury, M. (Ed.), Elsevier Science, Oxford, **2001**.

416. Shen, Z.-X.; Kong, A.-D.; Chen, W.-Y.; Zhang, Y.-W. *Chin. J. Org. Chem.* **2003**, *23*, 10–21.

417. O'Donnell, M. J. *Acc. Chem. Res.* **2004**, *37*, 506–517.

418. Lygo, B.; Andrew, B. I. *Acc. Chem. Res.* **2004**, *37*, 518–525.

419. Ooi, T.; Maruoka, K. *Acc. Chem. Res.* **2004**, *37*, 526–533.

420. Ohshima, T. *Chem. Pharm. Bull.* **2004**, *52*, 1031–1052.

421. Rabinovitz, M.; Cohen, Y.; Halpern, M. *Angew. Chem. Int. Ed.* **1986**, *25*.

422. O'Donnell, M. J.; Eckrich, T. M. *Tetrahedron Lett.* **1978**, *19*, 4625–4628.

423. Lipkowitz, K. B.; Cavanaugh, M. W.; Baker, B.; O'Donnell, M. J. *J. Org. Chem.* **1991**, *56*, 5181–5192.

424. O'Donnell, M. J.; Wu, S. *Tetrahedron Asym.* **1992**, *3*, 591–594.

425. O'Donnell, M. J.; Wu, S.; Huffman, J. C. *Tetrahedron* **1994**, *50*, 4507–4518.

426. Corey, E. J.; Noe, M. C.; Xu, F. *Tetrahedron Lett.* **1998**, *39*, 5347–5350.

427. Lygo, B.; Wainwright, P. G. *Tetrahedron Lett.* **1997**, *38*, 8595–8598.

428. Lygo, B.; Crosby, J.; Peterson, J. A. *Tetrahedron Lett.* **1999**, *40*, 1385–1388.

429. Lygo, B. *Tetrahedron Lett.* **1999**, *40*, 1389–1392.

430. Lygo, B.; Crosby, J.; Peterson, J. A. *Tetrahedron Lett.* **1999**, *40*, 8671–8674.

431. Schwesinger, R. *Chimia* **1985**, *39*, 269–272.

432. Schwesinger, R.; Willaredt, J.; Schlemper, H.; Keller, M.; Schmitt, D.; Fritz, H. *Chem. Ber.* **1994**, *127*, 2435–2454.

433. Schwesinger, R.; Schlemper, H.; Hasenfratz, C.; Willaredt, J.; Dambacher, T.; Breuer, T.; Ottaway, C.; Fletschinger, M.; Boele, J.; Fritz, H.; Putzas, D.; Rotter, H. W.; Bordwell, F. G.; Satish, A. V.; Ji, G.-Z.; Peters, E.-M.; Peters, K.; von Schnering, H. G.; Walz, L. *Liebigs Ann.* **1996**, *7*, 1055–1081.

434. O'Donnell, M. J.; Delgado, F.; Hostettler, C.; Schwesinger, R. *Tetrahedron Lett.* **1998**, *39*, 8775–8778.

435. O'Donnell, M. J. *Aldrichim. Acta* **2001**, *34*, 3–15.

436. Maruoka, K.; Ooi, T. *Chem. Rev.* **2003**, *103*, 3013–3028.

437. Ooi, T.; Kameda, M.; Maruoka, K. *J. Am. Chem. Soc.* **1999**, *121*, 6519–6520.

438. Ooi, T.; Kameda, M.; Tannai, H.; Maruoka, K. *Tetrahedron Lett.* **2000**, *41*, 8339–8342.

439. Ooi, T.; Takeuchi, M.; Kameda, M.; Maruoka, K. *J. Am. Chem. Soc.* **2000**, *122*, 5228–5229.

440. Ooi, T.; Uematsu, Y.; Kameda, M.; Maruoka, K. *Angew. Chem. Int. Ed.* **2002**, *41*, 1551–1554.

441. Ooi, T.; Uematsu, Y.; Maruoka, K. *Adv. Synth. Catal.* **2002**, *344*, 288–291.

442. Ooi, T.; Tayama, E.; Maruoka, K. *Angew. Chem. Int. Ed.* **2003**, *42*, 579–582.

443. Ooi, T.; Kameda, M.; Maruoka, K. *J. Am. Chem. Soc.* **2003**, *125*, 5139–5151.

444. Ooi, T.; Kubota, Y.; Maruoka, K. *Synlett* **2003**, 1931–1933.

445. Maruoka, K. *Pure Appl. Chem.* **2005**, *77*, 1285–1296.

446. Shibuguchi, T.; Fukuta, Y.; Akachi, Y.; Sekine, A.; Ohshima, T.; Shibasaki, M. *Tetrahedron Lett.* **2002**, *43*, 9539–9543.

447. Ohshima, T.; Shibuguchi, T.; Fukuta, Y.; Shibasaki, M. *Tetrahedron* **2004**, *60*, 7743–7754.

448. Kita, T.; Georgieva, A.; Hashimoto, Y.; Nakata, T.; Nagasawa, K. *Angew. Chem. Int. Ed.* **2002**, *41*, 2832–2834.

449. Mase, N.; Ohno, T.; Hoshikawa, N.; Ohishi, K.; Morimoto, H.; Yoda, H.; Takabe, K. *Tetrahedron Lett.* **2003**, *44*, 4073–4075.

450. Belokon', Y. N.; Kochetkov, K. A.; Churkina, T. D.; Ikonnikov, N. S.; Chesnokov, A. A.; Larionov, O. V.; Parmár, V. S.; Kumar, R.; Kagan, H. B. *Tetrahedron Asym.* **1998**, *9*, 851–857.

451. Belokon', Y. N.; Kochetkov, K. A.; Churkina, T. D.; Ikonnikov, N. S.; Vyskocil, S.; Kagan, H. B. *Tetrahedron Asym.* **1999**, *10*, 1723–1728.

452. Ooi, T.; Takeuchi, M.; Ohara, D.; Maruoka, K. *Synlett* **2001**, 1185–1187.

453. Jew, S.-s.; Jeong, B.-S.; Lee, J.-H.; Yoo, M.-S.; Lee, Y.-J.; Park, B.-s.; Kim, M. G.; Park, H.-g. *J. Org. Chem.* **2003**, *68*, 4514–4516.

454. Kitamura, M.; Shirakawa, S.; Maruoka, K. *Angew. Chem. Int. Ed.* **2005**, *44*, 1549–1551.

455. Zhang, F.-Y.; Corey, E. J. *Org. Lett.* **2000**, *2*, 1097–1100.

456. Corey, E. J.; Zhang, F.-Y. *Org. Lett.* **2000**, *2*, 4257–4259.

457. O'Donnell, M. J.; Delgado, F.; Domínguez, E.; de Blas, J.; Scott, W. L. *Tetrahedron Asym.* **2001**, *12*, 821–828.

458. Corey, E. J.; Noe, M. C. *Org. Synth.* **2003**, *80*, 38–45.

459. Arai, S.; Nakayama, K.; Ishida, T.; Shioiri, T. *Tetrahedron Lett.* **1999**, *40*, 4215–4218.

460. Perrard, T.; Plaquevent, J.-C.; Desmurs, J.-R.; Hébrault, D. *Org. Lett.* **2000**, *2*, 2959–2962.

461. Zhang, F.-Y.; Corey, E. J. *Org. Lett.* **2001**, *3*, 639–641.

462. Kim, D. Y.; Huh, S. C.; Kim, S. M. *Tetrahedron Lett.* **2001**, *42*, 6299–6301.

463. Kim, D. Y.; Huh, S. C. *Tetrahedron* **2001**, *57*, 8933–8938.

464. Thierry, B.; Perrard, T.; Audouard, C.; Plaquevent, J.-C.; Cahard, D. *Synthesis* **2001**, 1742–1746.

465. Zhang, F.-Y.; Corey, E. J. *Org. Lett.* **2004**, *6*, 3397–3399.

466. Loupy, A.; Zaparucha, A. *Tetrahedron Lett.* **1993**, *34*, 473–476.

467. P. Bakó, Czinege, E.; Bakó, T; Czugler, M.; Töke, L. *Tetrahedron Asym.* **1999**, *10*, 4539–4551.

468. Arai, S.; Tsuji, R.; Nishida, A. *Tetrahedron Lett.* **2002**, *43*, 9535–9537.

469. Ooi, T.; Doda, K.; Maruoka, K. *J. Am. Chem. Soc.* **2003**, *125*, 9022–9023.

470. Akiyama, T.; Hara, M.; Fuchibe, K.; Sakamoto, S.; Yamaguchi, K. *Chem. Commun.* **2003**, 1734–1735.

471. Ooi, T.; Miki, T.; Taniguchi, M.; Shiraishi, M.; Takeuchi, M.; Maruoka, K. *Angew. Chem. Int. Ed.* **2003**, *42*, 3796–3798.

472. Arai, S.; Tokumaru, K.; Aoyama, T. *Chem. Pharm. Bull.* **2004**, *52*, 646–648.

473. Ooi, T.; Fujioka, S.; Maruoka, K. *J. Am. Chem. Soc.* **2004**, *126*, 11790–11791.

474. Bakó, P.; Makó, A.; Keglevich, G.; Kubinyi, M.; Pál, K. *Tetrahedron Asym.* **2005**, *16*, 1861–1871.

475. Lygo, B.; Allbutt, B.; Kitrton, E. H. M. *Tetrahedron Asym.* **2005**, *46*, 4461–4464.

476. Ooi, T.; Ohara, D.; Fukumoto, K.; Maruoka, K. *Org. Lett.* **2005**, *7*, 3195–3197.

477. Gasparski, C. M.; Miller, M. J. *Tetrahedron* **1991**, *47*, 5367–5378.

478. Ando, A.; Miura, T.; Tatematsu, T.; Shioiri, T. *Tetrahedron Lett.* **1993**, *34*, 1507–1510.

479. Shioiri, T.; Bohsako, A.; Ando, A. *Heterocycles* **1996**, *42*, 93–97.

480. Horikawa, M.; Busch-Petersen, J.; Corey, E. J. *Tetrahedron Lett.* **1999**, *40*, 3843–3846.

481. Ooi, T.; Doda, K.; Maruoka, K. *Org. Lett.* **2001**, *3*, 1273–1276.

482. Bluet, G.; Campagne, J.-M. *J. Org. Chem.* **2001**, *66*, 4293–4298.

483. Ooi, T.; Taniguchi, M.; Kameda, M.; Maruoka, K. *Angew. Chem. Int. Ed.* **2002**, *41*, 4542–4544.

484. Ooi, T.; Kameda, M.; Taniguchi, M.; Maruoka, K. *J. Am. Chem. Soc.* **2004**, *126*, 9685–9694.

485. Ooi, T.; Taniguchi, M.; Doda, K.; Maruoka, K. *Adv. Synth. Catal.* **2004**, *346*, 1073–1076.

486. Corey, E. J.; Zhang, F.-Y. *Angew. Chem. Int. Ed.* **1999**, *38*, 1931–1934.

487. Ooi, T.; Doda, K.; Maruoka, K. *J. Am. Chem. Soc.* **2003**, *125*, 2054–2055.

488. Nemoto, T.; Ohshima, T.; Shibasaki, M. *J. Synth. Org. Chem. Jpn.* **2002**, *60*, 94–105.

489. Lygo, B.; Wainwright, P. G. *Tetrahedron Lett.* **1998**, *39*, 1599–1602.

490. Lygo, B.; Wainwright, P. G. *Tetrahedron* **1999**, *55*, 6289–6300.

491. Lygo, B.; To, D. C. M. *Tetrahedron Lett.* **2001**, *42*, 1343–1346.

492. Arai, S.; Tsuge, H.; Shioiri, T. *Tetrahedron Lett.* **1998**, *39*, 7563–7566.

493. Arai, S.; Oku, M.; Miura, M.; Shioiri, T. *Synlett* **1998**, 1201–1202.

494. Arai, S.; Tsuge, H.; Oku, M.; Miura, M.; Shioiri, T. *Tetrahedron* **2002**, *58*, 1623–1630.

495. Adam, W.; Rao, P. B.; Degen, H.-G.; Saha-Möller, C. R. *Tetrahedron Asym.* **2001**, *12*, 121–125.

496. Adam, W.; Rao, P. B.; Degen, H.-G.; Levai, A.; Patonay, T.; Saha-Möller, C. R. *J. Org. Chem.* **2002**, *67*, 259–264.

497. Ye, J.; Wang, Y.; Liu, R.; Zhang, G.; Zhang, Q.; Chen, J.; Liang, X. *Chem. Commun.* **2003**, 2714–2715.

498. Ye, J.; Wang, Y.; Chen, J.; Liang, X. *Adv. Synth. Catal.* **2004**, *346*, 691–696.

499. Juliá, S.; Guixer, J.; Masana, J.; Rocas, J.; Colonna, S.; Annuziata, R.; Molinari, H. *J. Chem. Soc., Perkin Trans. 1* **1982**, 1317–1324.

500. For a review on catalytic asymmetric epoxidations using polyamino acids as catalysts, see Porter, M. J.; Roberts, S. M.; Skidmore, J. *Bioorg. Med. Chem.* **1999**, *7*, 2145–2156.

501. Ooi, T.; Ohara, D.; Tamura, M.; Maruoka, K. *J. Am. Chem. Soc.* **2004**, *126*, 6844–6845.

502. Bakó, P.; Szöllösi, A.; Bombicz, P.; Töke, L. *Synlett* **1997**, 291–292.

503. Arai, S.; Shioiri, T. *Tetrahedron Lett.* **1998**, *39*, 2145–2148.

504. Arai, S.; Ishida, T.; Shioiri, T. *Tetrahedron Lett.* **1998**, *39*, 8299–8302.

505. Arai, S.; Shirai, Y.; Ishida, T.; Shioiri, T. *Chem. Commun.* **1999**, 49–50.

506. Arai, S.; Shirai, Y.; Ishida, T.; Shioiri, T. *Tetrahedron* **1999**, *55*, 6375–6386.

507. Arai, S.; Shioiri, T. *Tetrahedron* **2002**, *58*, 1407–1413.

508. Arai, S.; Tokumaru, K.; Aoyama, T. *Tetrahedron Lett.* **2004**, *45*, 1845–1848.

509. Ooi, T.; Takahashi, M.; Doda, K.; Maruoka, K. *J. Am. Chem. Soc.* **2002**, *124*, 7640–7641.

510. Iseki, K.; Nagai, T.; Kobayashi, Y. *Tetrahedron Lett.* **1994**, *35*, 3137–3138.

511. Caron, S.; Do, N. M.; Arpin, P.; Larivée, A. *Synthesis* **2003**, 1693–1698.

512. Ooi, T.; Kameda, M.; Fujii, J.-i.; Maruoka, K. *Org. Lett.* **2004**, *6*, 2397–2399.

513. Bella, M.; Kobbelgaard, S.; Jørgensen, K. A. *J. Am. Chem. Soc.* **2005**, *127*, 3670–3671.

514. Ishihara, K.; Nakano, K. *J. Am. Chem. Soc.* **2005**, *127*, 10504–10505.

12

CHIRAL BRØNSTED/LEWIS ACID CATALYSTS

Kazuaki Ishihara

Graduate School of Engineering, Nagoya University, Furo-cho, Chikusa, Nagoya, Japan

Hisashi Yamamoto

Department of Chemistry, The University of Chicago, Chicago, Illinois USA

12.1 INTRODUCTION

There has been great interest in the area of chiral acid catalysts in organic synthesis over the past few decades. This topic has been the subject of several previous reviews.[1] For example, the book *Lewis Acids in Organic Synthesis* (edited by Hisashi Yamamoto) was published by Wiley-VCH in 2000.[1e] In this chapter, successful and significant chiral Brønsted acid catalysts, chiral Lewis acid catalysts [typical Lewis acidic elements: main group elements, B(III) and Al(III), and early transition metal, Ti(IV)], and Lewis acid-assisted chiral Brønsted acid catalysts developed after 2000 are discussed. Chiral acid/base catalysts will be discussed in Chapter 13 by Shibasaki and Kanai.

12.2 CHIRAL BRØNSTED ACID CATALYSTS

Since Curran and Kuo[2a,b] and Schreiner and coworkers[2c,d] reported that urea and thiourea derivatives act like Lewis acid catalysts, several chiral urea and thiourea catalysts have been designed by Jacobsen et al.[3,4] and Takemoto et al.[5]

New Frontiers in Asymmetric Catalysis, Edited by Koichi Mikami and Mark Lautens
Copyright © 2007 John Wiley & Sons, Inc.

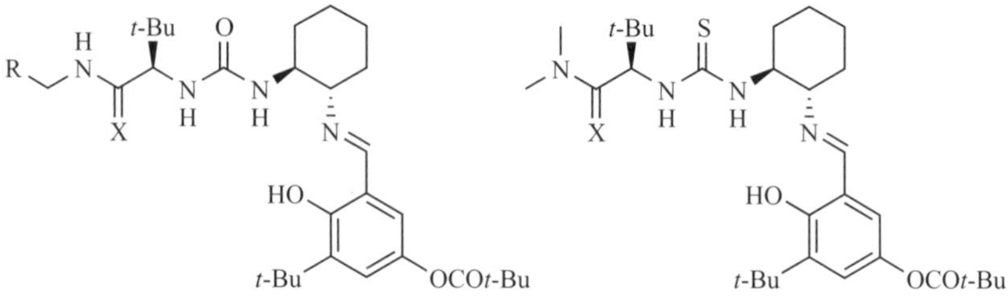

1 (R = Ph; X = O)
2 (R = polystyrene; X = O or S)

3

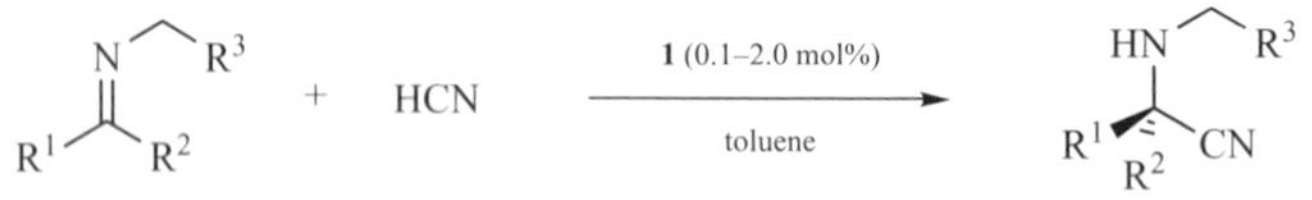

R^1 = aryl or alkyl, R^2 = H or Me
R^3 = aryl, alkyl, heteroatom
>50 examples

85–99% yield
70–99% ee

Scheme 12.1

The highly enantioselective addition of HCN not only to aldimines[3a,b] but also to methylketoimines[3c] has been achieved using readily accessible and recyclable chiral urea catalysts. High yields and high enantioselectivities of up to 99% ee have been obtained for more than 50 examples. Catalyst **1** has been identified and optimized from a parallel library of Schiff base derivatives that has been synthesized and screened on solid-phase substrates. Resin-bound catalyst **2** can be used in the enantioselective synthesis of Strecker adducts on a preparative scale. According to mechanistic studies, the two urea protons of **1** are essential for its catalyst activity, and the Strecker reaction with **1** proceeds through binding of the imine substrate as the Z-isomer.[3d] Based on these mechanistic studies, catalyst **3** has been rationally designed as the most enantioselective Strecker catalyst prepared to date (Scheme 12.1).[3d]

Furthermore, a highly efficient route to N-tert-butoxycarbonyl (Boc)-protected β-amino acids via the enantioselective addition of silyl ketene acetals to N-Boc-aldimines catalyzed by thiourea catalyst **4** has been reported (Scheme 12.2).[4] From a steric and electronic standpoint, the N-Boc imine substrates used in this reaction are fundamentally different from the N-alkyl derivatives used in the Strecker reaction.

The Michael reaction of malonates to nitroolefins and the aza-Henry reaction[5a] of nitroalkanes to N-phosphinoylimines[5b] are catalyzed by thiourea derivative **5a** to provide the respective products in good and moderate enantioselectivities. Thiourea

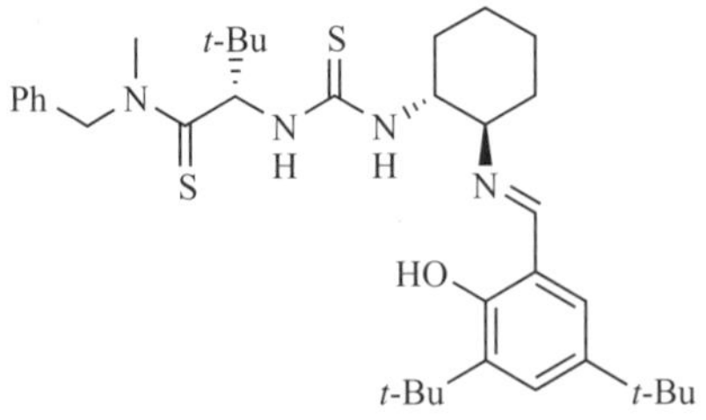

4

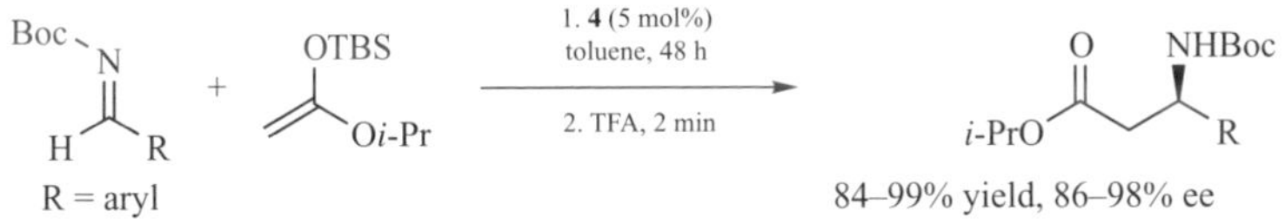

Boc–N = CH–R
R = aryl

+ OTBS / Oi-Pr

1. **4** (5 mol%)
toluene, 48 h

2. TFA, 2 min

i-PrO–C(O)–CH2–CH(NHBoc)–R

84–99% yield, 86–98% ee

Scheme 12.2

catalyst acts as a bifunctional organocatalyst that has two acidic urea hydrogens that activate nitroolefins and a basic dimethylamino group that activate malonates. More recently, Berkessel et al. reported a broadly applicable organocatalytic method for dynamic kinetic resolution of azlactones with good enantioselectivity.[5c] The method is based on a bifunctional urea–amine catalyst such as **5b**, which activates the substrate azlactones to nucleophilic attack though the formation of a hydrogen-bonded supramolecular aggregate (Scheme 12.3).

The acyl-Pictet–Spengler reaction is also catalyzed by chiral thiourea derivative **6** to provide N-acetyl β-carbolines in high enantioselectivities.[6] Notably, thiourea derivatives can activate not only electronically distinct imine derivatives such as N-alkyl and N-Boc imines but also a weakly Lewis basic N-acyliminium ion with high enantioselectivity using a chiral hydrogen bond donor (Scheme 12.4).

A simple, commercially available chiral alcohol, α,α,α'α'-tetraaryl-1,3-dioxolane-4,5-dimethanol (TADDOL, **7a**), catalyzes the hetero- and carbo-Diels–Alder reactions of aminosiloxydienes with aldehydes and α-substituted acroleins to afford the dihydropyrones and cyclohexenones, respectively, in good yields and high enantioselectivities.[7a,b] More recently, it was reported that axially chiral biaryl diols **7b** and **7c** were more highly effective catalysts for enantioselective hetero-Diels–Alder reactions (Scheme 12.5).[7c]

The highly enantioselective Morita–Baylis–Hillman reaction of cyclohexenone with aldehydes is catalyzed by a chiral BINOL-derived Brønsted acid **8** in the presence of triethylphosphine as the nucleophilic promoter (Scheme 12.6).[8]

Highly enantioselective Mannich-type reactions of N-(2-hydroxyphenyl) aldimines with ketene trimethylsilyl acetals[9a] and of N-Boc-aldimines with acetyl acetone[9b] or furan[9c] are catalyzed by chiral phosphonic acids **9** derived from 3,3'-diaryl-(R)-BINOL and POCl3 (Scheme 12.7).[9]

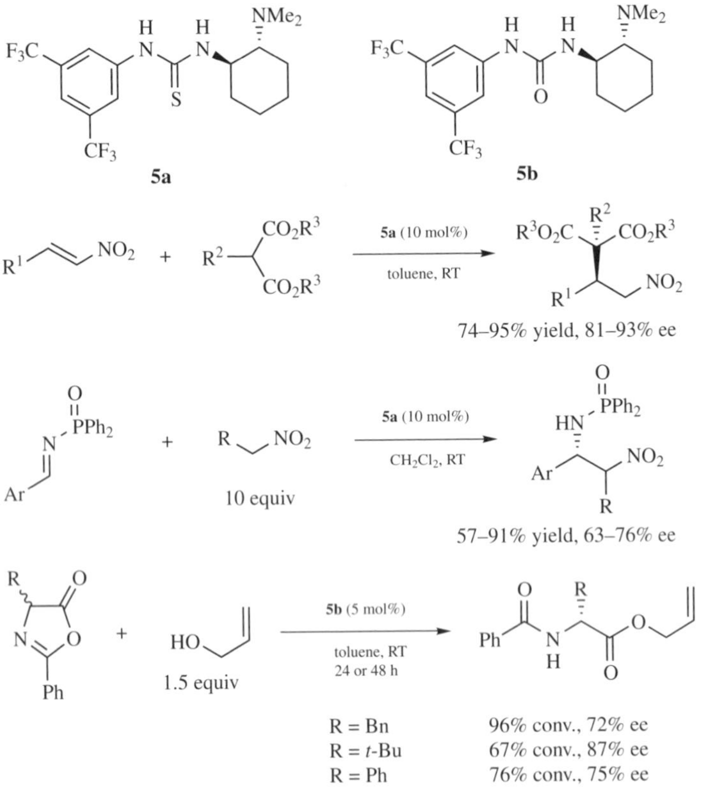

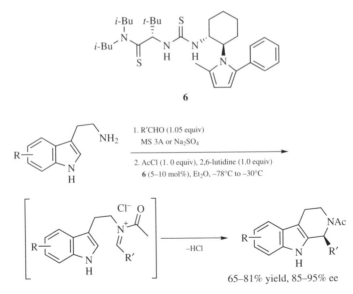

5a

5b

R^1 ⌒ NO_2 + R^2 with CO_2R^3 / CO_2R^3 → (5a (10 mol%), toluene, RT) → R^3O_2C R^2 CO_2R^3 ... R^1 NO_2

74–95% yield, 81–93% ee

Ar ⌒ N—$\overset{O}{\overset{\|}{P}}Ph_2$ + R ⌒ NO_2 (10 equiv) → (5a (10 mol%), CH_2Cl_2, RT) → HN—$\overset{O}{\overset{\|}{P}}Ph_2$ ... Ar / NO_2 / R

57–91% yield, 63–76% ee

R — oxazolone ring, N / O / Ph + HO ⌒ (1.5 equiv) → (5b (5 mol%), toluene, RT, 24 or 48 h) → Ph—$\overset{O}{\|}$—N(H)—CH(R)—C(=O)—O ⌒

R = Bn — 96% conv., 72% ee
R = t-Bu — 67% conv., 87% ee
R = Ph — 76% conv., 75% ee

Scheme 12.3

6

R — indole — ethyl — NH_2 →
1. R′CHO (1.05 equiv) MS 3A or Na_2SO_4
2. AcCl (1.0 equiv), 2,6-lutidine (1.0 equiv) **6** (5–10 mol%), Et_2O, –78°C to –30°C

[Cl^- O / R — indole — ethyl — N^+ — C(=O) — ... — R′] →(–HCl)→ R — tetrahydro-β-carboline — NAc ... R′

65–81% yield, 85–95% ee

Scheme 12.4

362

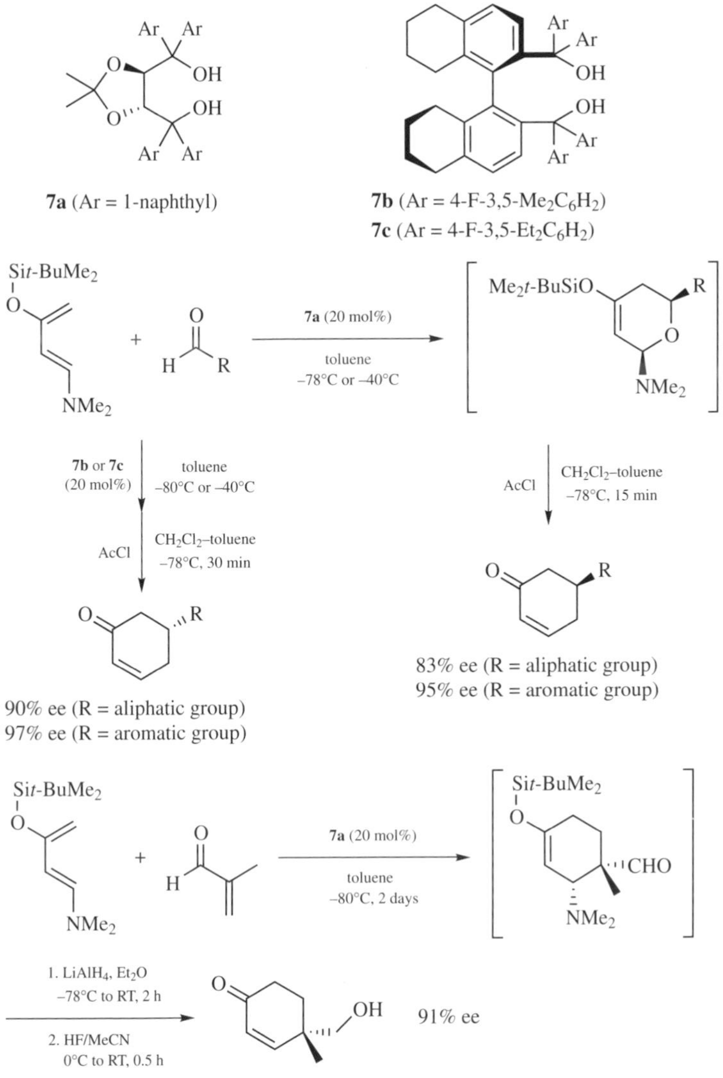

7a (Ar = 1-naphthyl)

7b (Ar = 4-F-3,5-Me$_2$C$_6$H$_2$)
7c (Ar = 4-F-3,5-Et$_2$C$_6$H$_2$)

90% ee (R = aliphatic group)
97% ee (R = aromatic group)

83% ee (R = aliphatic group)
95% ee (R = aromatic group)

91% ee

Scheme 12.5

12.3 CHIRAL LEWIS ACID CATALYSTS

12.3.1 B(III)

The cationic chiral Lewis acids **10**, generated from the corresponding oxazaboroli-dines by protonation by trifluoromethanesulfonic acid, are excellent catalysts for the enantioselective reaction of 2-substituted acroleins, α-unsaturated α,β-enones, α-unsaturated acrylic acid esters, and α-unsaturated acrylic acids with a variety

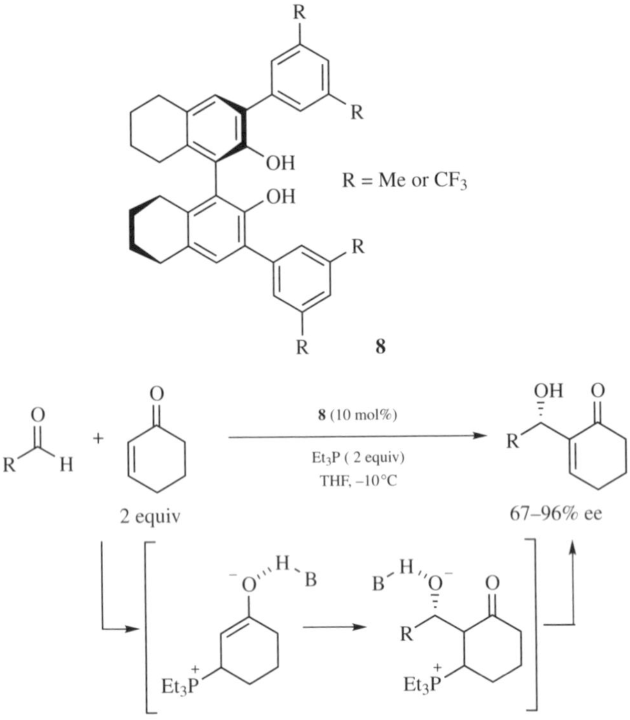

R = Me or CF$_3$

8

Scheme 12.6

of dienes.[10a,b] Interestingly, the face selectivities of the enantioselective Diels–Alder reactions of α-unsaturated α,β-enones, acrylic acid esters, and acrylic acids are opposite to those observed for 2-substituted acroleins. Corey et al. proposed two pretransition-state assemblies that include attractive H–O interactions between an α-olefinic hydrogen or a formyl hydrogen and oxaborolidinyl oxygen (Scheme 12.8).

Later, Corey reported that the bulky superacid triflylimide (Tf$_2$NH) protonates chiral oxazaborolidines to form superactive, stable, chiral acids **11a** and **11b**, which are highly effective catalysts for a wide variety of enantioselective Diels–Alder reactions that were beyond the reach of synthetic chemists (Scheme 12.9).[10c,d]

Futatsugi and Yamamoto have demonstrated that the combination of a chiral oxazaborolidine **12** and SnCl$_4$ leads to an extremely active Lewis acid-assisted chiral Lewis acid (LLA) as a promising catalyst for enantioselective Diels–Alder reactions (Scheme 12.10).[11] The enantioselectivity can be preserved even in the presence of a large amount of achiral Lewis acid activator (SnCl$_4$), which implicates the generation of highly reactive species. Only 0.5

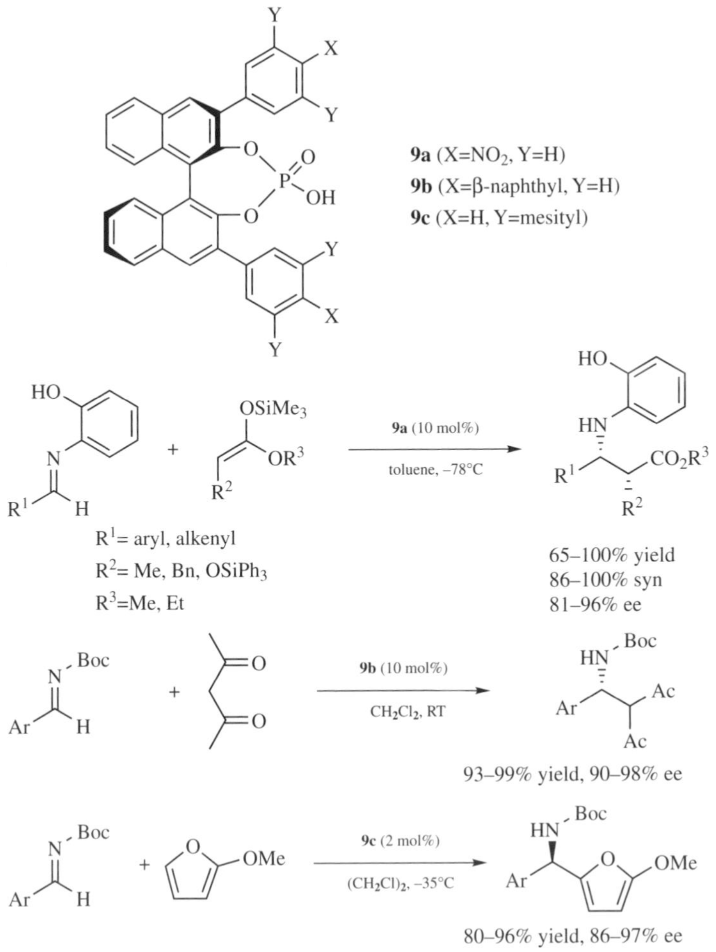

9a (X=NO$_2$, Y=H)
9b (X=β-naphthyl, Y=H)
9c (X=H, Y=mesityl)

R^1= aryl, alkenyl
R^2= Me, Bn, OSiPh$_3$
R^3=Me, Et

65–100% yield
86–100% syn
81–96% ee

9a (10 mol%)
toluene, –78°C

93–99% yield, 90–98% ee

9b (10 mol%)
CH$_2$Cl$_2$, RT

80–96% yield, 86–97% ee

9c (2 mol%)
(CH$_2$Cl)$_2$, –35°C

Scheme 12.7

mol% of SnCl$_4$ is needed to promote the reaction. Its inherent reactivity and asymmetric induction ability can be maintained even in the presence of a small amount of water as well as other Lewis bases by adding a slightly large amount of SnCl$_4$.

The enantioselective addition of an allylsilane to an aldehyde catalyzed by chiral acyloxyborane (CAB) **13** is an excellent method for obtaining optically active homoallyl alcohols.[12] Itsuno and Kumagai reported that the synthesis of a new optically active polymer with chirality on the mainchain is possible by applying this reaction to the asymmetric polymerization of bis(allylsilane) and dialdehyde (Scheme 12.11).[13]

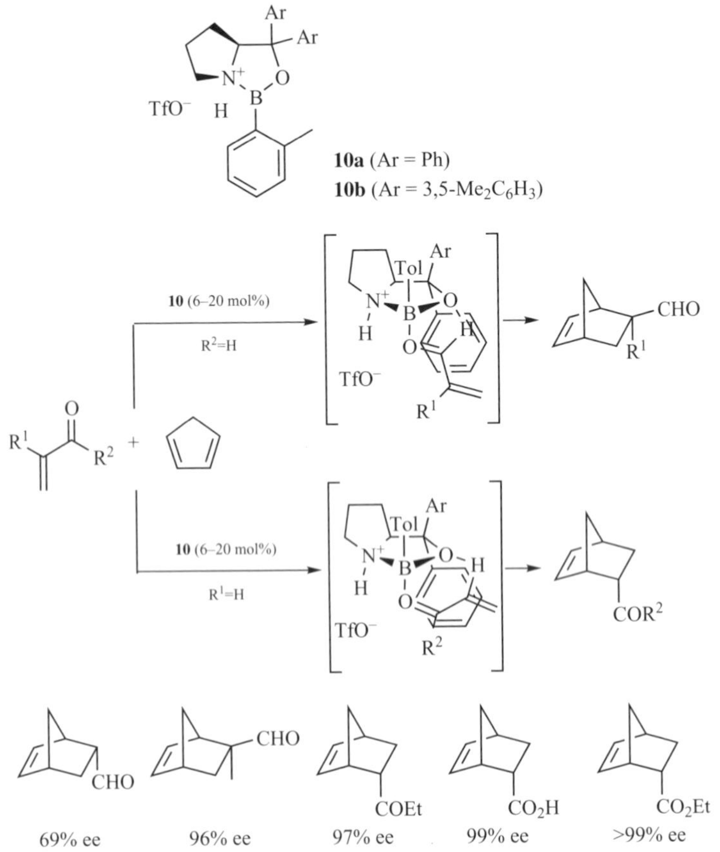
TfO⁻ H
10a (Ar = Ph)
10b (Ar = 3,5-Me2C6H3)
10 (6–20 mol%)
R2=H
R1=H
CHO
COR2
69% ee
96% ee
97% ee
99% ee
>99% ee

Scheme 12.8

12.3.2 Al(III)

The chiral dialuminum Lewis acid **14**, which is effective as an asymmetric Diels–Alder catalyst, has been prepared from DIBAH and BINOL derivatives (Scheme 12.12).[14] The catalytic activity of **14** is significantly greater than that of monoaluminum reagents. The catalyst achieves high reactivity and selectivity by an intramolecular interaction of two aluminum Lewis acids. Similarly, the chiral trialuminum Lewis acid **15** is quantitatively formed from optically pure 3-(2,4,6-triisopropylphenyl)binaphthol (2 equiv) and Me_3Al (3 equiv) in CH_2Cl_2 at room temperature (Scheme 12.12).[14] The novel structure of **15** has been ascertained by ^1H NMR spectroscopic analysis and measurement of the methane gas evolved. Trinuclear aluminum catalyst **15** is effective for the Diels–Alder reaction of methacrolein with cyclopentadiene. Diels–Alder adducts have been obtained in 99% yield with 92% exo selectivity. Under optimum reaction conditions, the

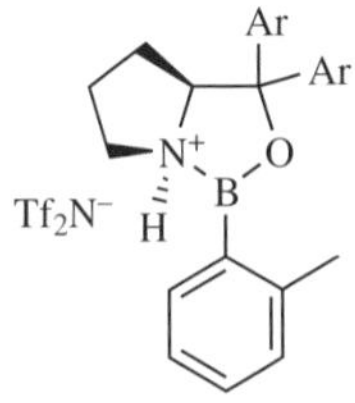

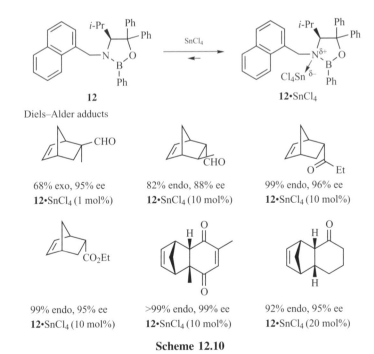

11a (Ar=Ph)
11b (Ar=3,5-Me$_2$C$_6$H$_3$)

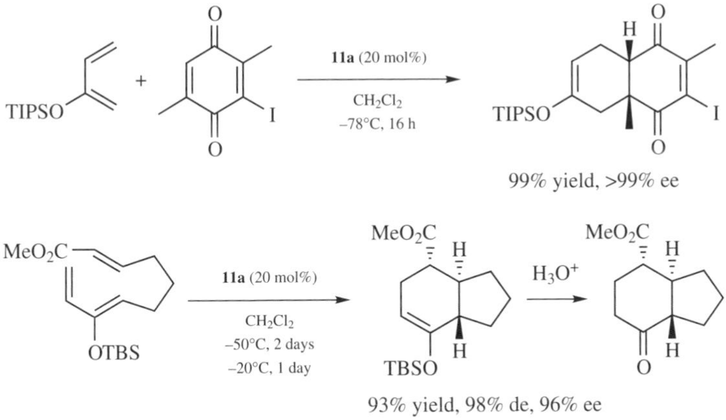

99% yield, >99% ee

93% yield, 98% de, 96% ee

Scheme 12.9

12

Diels–Alder adducts

12·SnCl$_4$

68% exo, 95% ee
12·SnCl$_4$ (1 mol%)

82% endo, 88% ee
12·SnCl$_4$ (10 mol%)

99% endo, 96% ee
12·SnCl$_4$ (10 mol%)

99% endo, 95% ee
12·SnCl$_4$ (10 mol%)

>99% endo, 99% ee
12·SnCl$_4$ (10 mol%)

92% endo, 95% ee
12·SnCl$_4$ (20 mol%)

Scheme 12.10

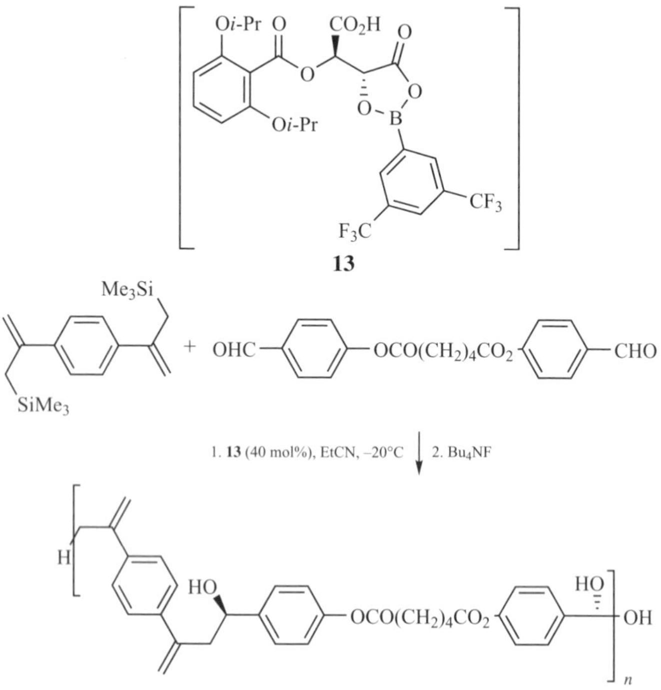

99% yield, Mn = 14,000, $M_w/M_n = 4.62$, $[\Phi]_{405} = -1144.1$

Scheme 12.11

exo adduct has been obtained with 75% ee. The intramolecular association of **14** as shown in Scheme 12.12 might lead to enhanced Lewis acidity of the aluminum atom located in the center of the chiral scaffold. The chiral heterogeneous Lewis acid **16** prepared from trimethylaluminum, menthol, and tetrachlorobisphenol A has been found to promote the asymmetric Diels–Alder reaction of methacrolein with cyclopentadiene and can be reused more than 3 times. However, the catalytic asymmetric Diels–Alder reaction has not yet been achieved (Scheme 12.13).[15]

A very simple chiral Lewis acid, prepared by mixing optically pure BINOL with 3 equiv of Me_3Al, catalyzes the [4+2] cycloaddition of *N*-hydroxy-*N*-phenylacrylamine with cyclopentadiene at 0°C in high yield (>96%) and a fairly good level of enantioselectivity (91% ee). Facile conversion of the products to the corresponding alcohols or aldehydes makes the hydroxamic acid intermediates particularly useful (Scheme 12.14).[16]

12.3.3 Ti(IV)

(*R*,*R*)-3-Aza-3-benzyl-1,5-diphenylpentane ligated Ti(IV) complex **17** is quite effective for the asymmetric inverse-electron-demand Diels–Alder reaction of

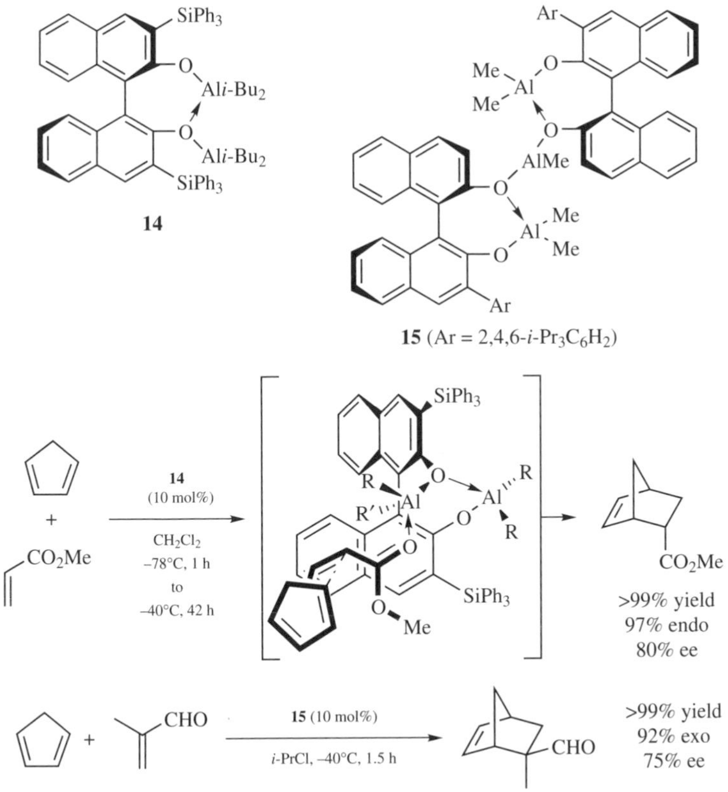

14

15 (Ar = 2,4,6-*i*-Pr₃C₆H₂)

Scheme 12.12

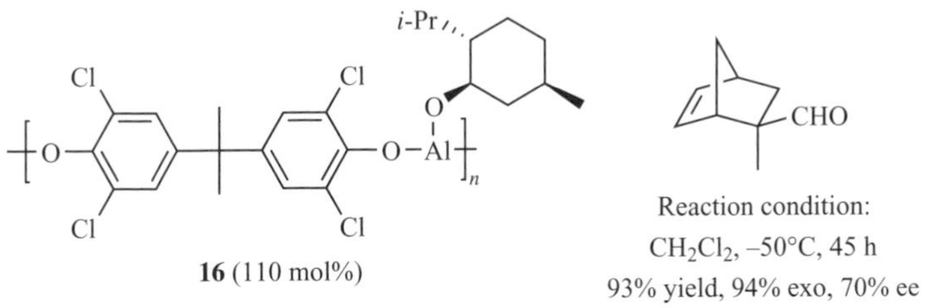

16 (110 mol%)

Reaction condition:
CH₂Cl₂, −50°C, 45 h
93% yield, 94% exo, 70% ee

Scheme 12.13

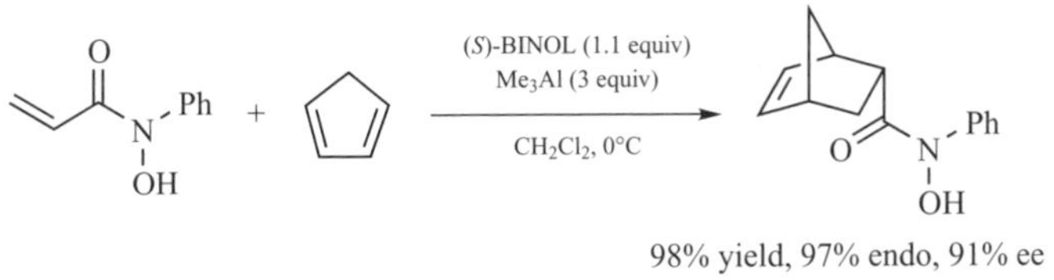

98% yield, 97% endo, 91% ee

Scheme 12.14

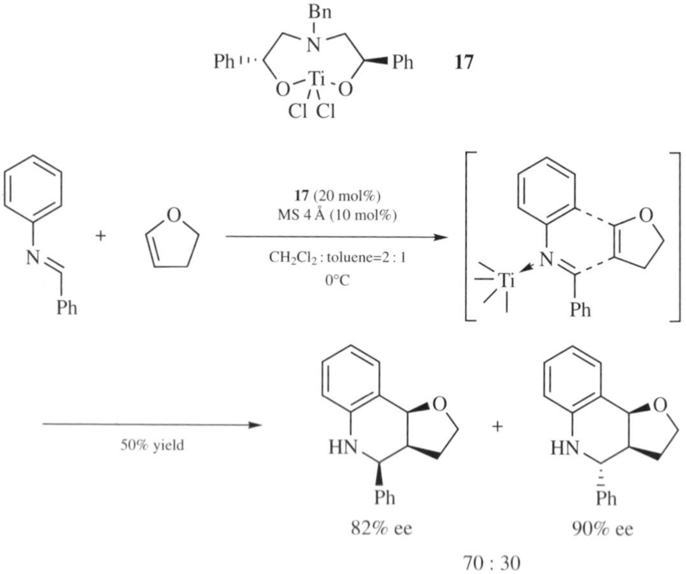

Scheme 12.15

benzylidene aniline with electron-rich dienophiles. Some increase in enantioselectivity has been observed with the addition of molecular sieves to the reaction mixture (Scheme 12.15).[17]

A chiral dinuclear Ti(IV) Lewis acid catalyst **18** can be prepared in situ from a 1 : 2 molar mixture of (R)-3,3′-di(2-mesitylethynyl)binaphthol and Ti(Oi-Pr)$_4$ at ambient temperature. The 3- and 3′-substituents on the chiral ligand are effective for preventing undesired aggregation between Ti(IV) complexes and increasing the enantioselectivity (up to 82% ee) in the Diels–Alder reaction of methacrolein with cyclopentadiene (Scheme 12.16).[18]

Chiral tetranuclear Ti(IV) cluster **19**, a cubic structure that consists of four Ti atoms and OHs, and six (R)-BINOL ligands bridging two Ti atoms as ligands, has been shown to be a novel chiral Lewis acid catalyst for the [2 + 3] cycloaddition reaction with nitrones. Chiral Ti(IV) clusters with 7,7′-substituted (R)-BINOL ligands have been synthesized to give enhanced enantiomeric excesses up to 78% ee (Scheme 12.17).[19]

A chiral dinuclear Ti(IV) oxide **20** has been successfully designed by Maruoka and coworkers and can be used for the strong activation of aldehydes, thereby allowing a new catalytic enantioselective allylation of aldehydes with allyltributyltin (Scheme 12.18).[20] The chiral catalyst **20** can be readily prepared either by treatment of bis(triisopropoxy)titanium oxide [(i-PrO)$_3$Ti-O-Ti(Oi-Pr)$_3$] with (S)-BINOL or by the reaction of ((S)-binaphthoxy)isopropoxytitanium chloride with silver(I) oxide. The reaction of 3-phenylpropanal with allyltributyltin (1.1 equiv) under the influence of **20** (10 mol%) gives 1-phenyl-5-hexen-3-ol

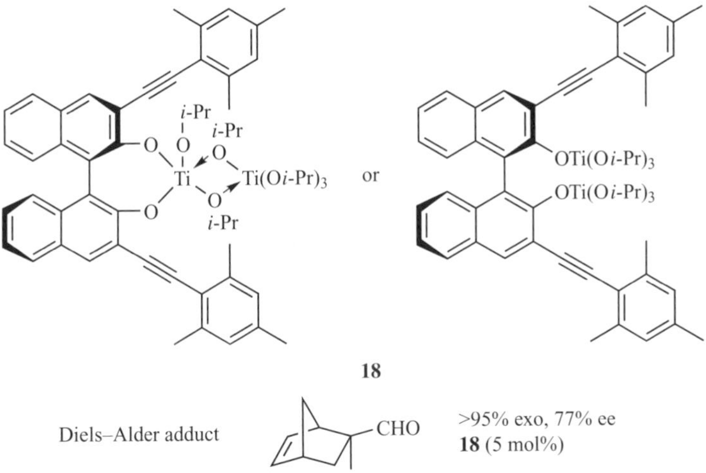

Scheme 12.16

in 84% yield with 99% ee. This asymmetric approach provides a very useful method for obtaining high reactivity and selectivity by the simple introduction of the M–O–M unit in the design of chiral Lewis acid catalysts. The authors proposed that the high reactivity of this chiral dinuclear Ti(IV) oxide might be ascribed to the intramolecular coordination of one isopropoxy oxygen atom to the other titanium center, thereby enhancing the otherwise weak Lewis acidity of the original Ti(IV) center for carbonyl activation. This mode of activation

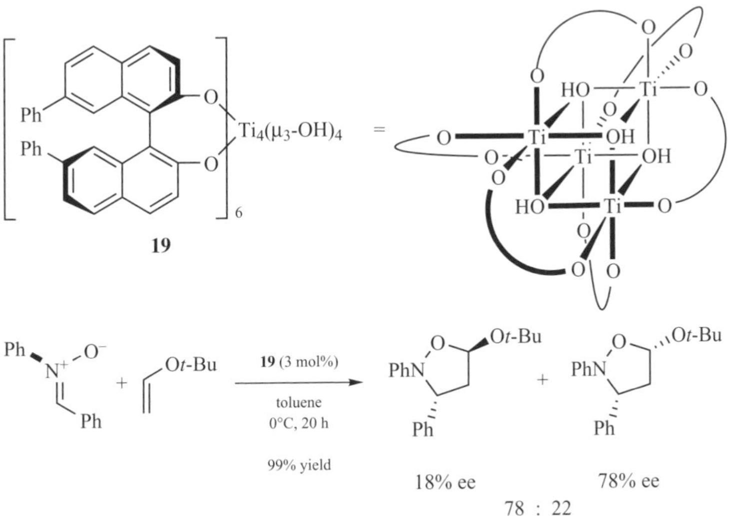

Scheme 12.17

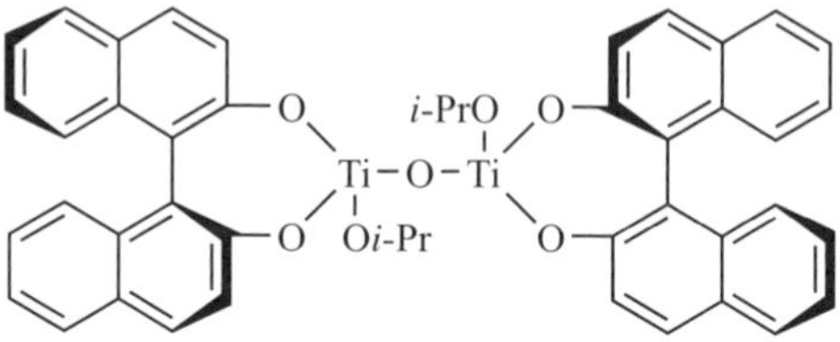
i-PrO
Ti
O
Ti
Oi-Pr
O
O
O
O

20

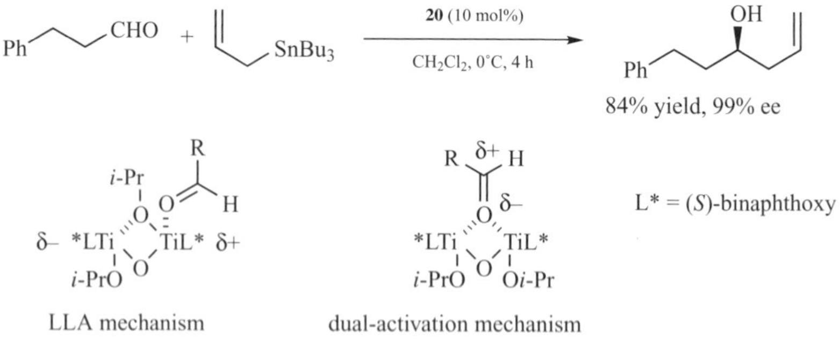
Ph
CHO
SnBu3
20 (10 mol%)
CH2Cl2, 0°C, 4 h
OH
Ph
84% yield, 99% ee
L* = (S)-binaphthoxy
LLA mechanism
dual-activation mechanism

Scheme 12.18

is a typical example of the LLA mechanism. Alternatively, a dual activation of the carbonyl group by the simultaneous coordination of two Ti centers has also been proposed as the origin of the high reactivity.

Marhwald reported that ligand exchange of Ti(*rac*-BINOLate)(O*t*-Bu)$_2$ with optically active α-hydroxy acids presents an unexpected and novel approach to enantioselective direct aldol reactions of aldehydes and ketones (Scheme 12.19).[21] The aldol products have been isolated with a high degree of syn diastereoselectivity. High enantioselectivities have been observed when using simple optically pure α-hydroxy acids.

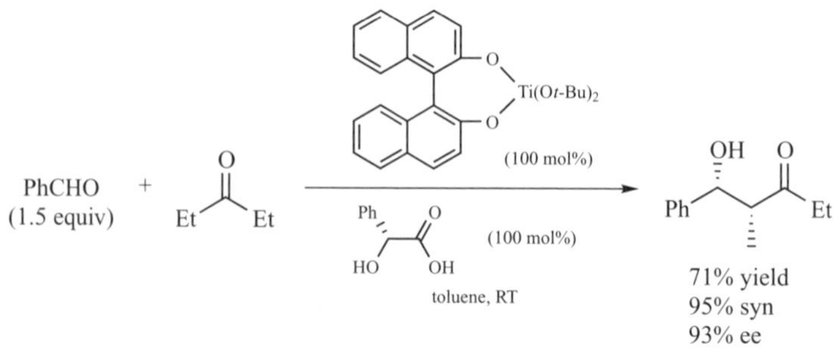
Ti(Ot-Bu)2
(100 mol%)
PhCHO
(1.5 equiv)
Et
Et
(100 mol%)
toluene, RT
OH
Ph
Et
71% yield
95% syn
93% ee

Scheme 12.19

12.4   LEWIS ACID–ASSISTED CHIRAL BRØNSTED ACID CATALYSTS

An enantio- and diastereoselective stepwise cyclization of polyprenoids induced by Lewis acid-assisted chiral Brønsted acids (chiral LBAs) and achiral LBAs has been developed by Ishihara, Yamamoto, and coworkers.[22] (−)-Ambrox® can be synthesized via the enantioselective cyclization of (E,E)-homofarnesyl triethylsilyl ether with SnCl$_4$-coordinated (R)-(o-fluorobenzyloxy)-2′-hydroxy-1,1′-binaphthyl (**22**) and subsequent diastereoselective cyclization with CF$_3$CO$_2$H•SnCl$_4$ as key steps.[23] Protection of (E,E)-homofarnesol by a triethylsilyl group increases the enantioselectivity of chiral LBA-induced cyclization and both the chemical yield and diastereoselectivity in the subsequent cyclization. The enantioselective cyclization of homo(polyprenyl)arenes possessing an aryl group is also induced by **22**•SnCl$_4$. Several optically active podocarpa-8,11,13-triene diterpenoids and a (−)-tetracyclic polyprenoid of sedimentary origin are synthesized (75–80% ee) by the enantioselective cyclization of homo(polyprenyl) benzene derivatives induced by **22**•SnCl$_4$ and subsequent diastereoselective cyclization induced by BF$_3$•Et$_2$O/EtNO$_2$ or F$_3$CCO$_2$H•SnCl$_4$ (Scheme 12.20).

BINOL derivative•SnCl$_4$ complexes are useful not only as artificial cyclases but also as enantioselective protonation reagents for silyl enol ethers.[24] However, their exact structures have not been determined. SnCl$_4$-free BINOL derivatives are

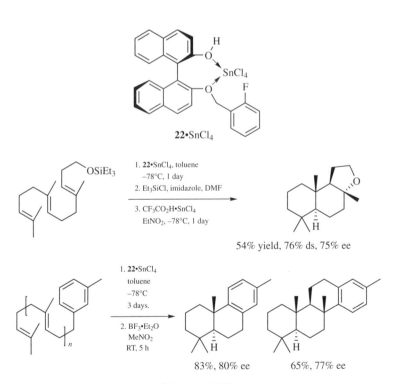

Scheme 12.20

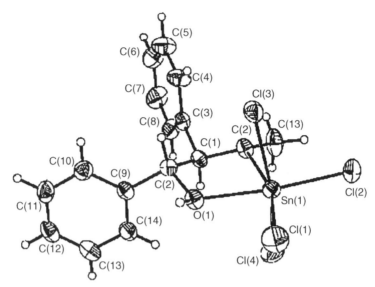

Figure 12.1. ORTEP drawing of **23**•SnCl₄.

preferentially crystallized in the presence of SnCl₄, because BINOL derivatives•SnCl₄ are less likely to undergo crystallization than SnCl₄-free BINOL derivatives, and the complexation is reversible. Fortunately, a colorless crystal of (*R,R*)-1,2-diphenylethane-2-methoxyethanol (**23**)•SnCl₄ has been obtained from a 1:1 molar mixture of **23** and SnCl₄.[25] The X-ray structure of **23**•SnCl₄ is shown in Figure 12.1. Although the activated proton (H$_{act}$) in **23**•SnCl₄ cannot be located exactly, the high electron density is certainly distributed in the pseudoequatorial direction.

Monoalkyl ethers of (*R,R*)-1,2-bis[3,5-bis(trifluoromethyl)phenyl]ethanediol, **24**, have been examined for the enantioselective protonation of silyl enol ethers and ketene disilyl acetals in the presence of SnCl₄ (Scheme 12.21) [25]. The corresponding ketones and carboxylic acids have been isolated in quantitative yield. High enantioselectivities have been observed for the protonation of trimethylsilyl enol ethers derived from aromatic ketones and ketene bis(trimethylsilyl)acetals derived from 2-arylalkanoic acids.

2,6-Di[(1′*R*,2′*R*)-*trans*-2′-(3″,5″-xylyl)cyclohexanoxy]phenol (**25**)•SnCl₄ and **24**•SnCl₄ show a tighter chelation structure than does **22**•SnCl₄.[26] **25**•SnCl₄ is highly effective for the enantioselective cyclization of (homogeranyl)arenes. For example, 4-(homogeranyl)toluene transforms to a trans-fused tricyclic compound with 85% ee in the presence of **25**•SnCl₄ (Scheme 12.22). These results suggest that five-membered chelation structures of 2-alkoxyalchohols and SnCl₄ are suitable for use as LBAs.

26•SnCl₄ is also superior to **22**•SnCl₄, and is effective for the enantioselective cyclization of 2-(polyprenyl)phenol derivatives to give polycyclic terpenoids bearing a chroman skeleton.[27] The synthetic utility of **26**•SnCl₄ is demonstrated by very

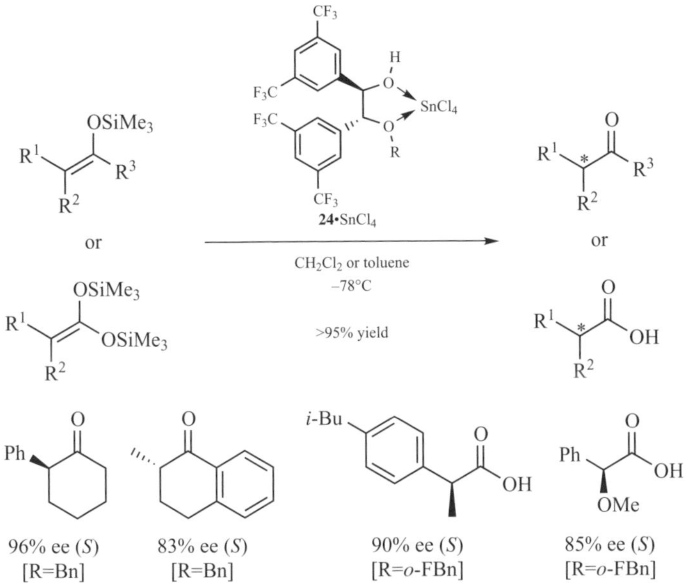

Scheme 12.21

efficient routes to (−)-chromazonarol, (+)-8-*epi*-puupehedione, a key synthetic intermediate of (+)-widendiol, and (−)-11′-deoxytaondiol methyl ether (Scheme 12.23).

The observed absolute stereopreference can be understood in terms of two proposed transition-state assemblies, **27** and **28** (Figure 12.2). The direction of the H—O bond of (*R*)-**26** might be fixed in the naphthoxy plane by bidentate chelation

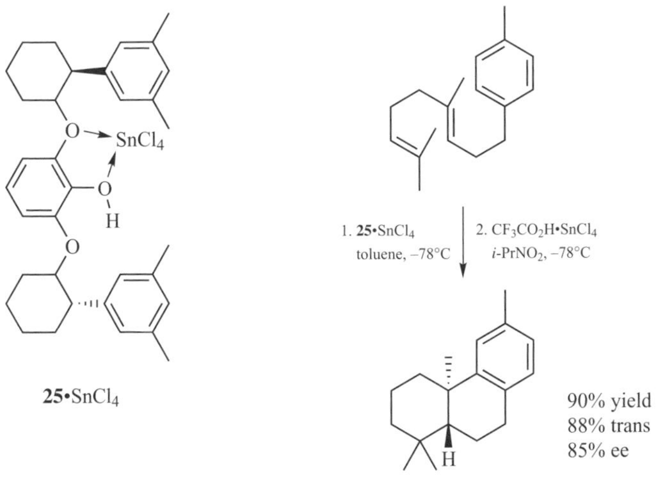

Scheme 12.22

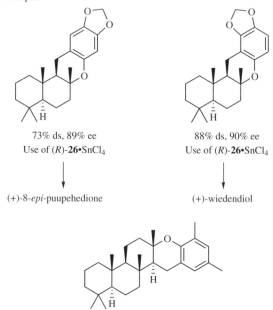

1. (S)-**26**•SnCl₄
2. CF₃CO₂H•SnCl₄

40%

69% ds, 88% ee
cf. (R)-**22**•SnCl₄: 55% ds, 40% ee

Other examples:

73% ds, 89% ee
Use of (R)-**26**•SnCl₄

88% ds, 90% ee
Use of (R)-**26**•SnCl₄

(+)-8-*epi*-puupehedione

(+)-wiedendiol

(−)-11′-deoxytaondiol methyl ether
48% ds, 90% ee

Scheme 12.23

of SnCl₄. As in their previous report,[25] the stereochemical course in the enantiose-lective cyclization would be controlled by a linear OH/π interaction with an initial protonation step. The *Re*-face of the terminal isoprenyl group of polyprenoids preferentially approaches the activated proton of LBA perpendicular to its H—O

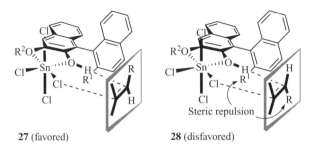

27 (favored)　　　　　　28 (disfavored)

Figure 12.2. Possible explanation for the absolute stereochemistry.

bond. While **27** is favored due to minimum steric repulsion, **28** is disfavored because of severe steric repulsion between R and R^1.

Ishihara, Yamamoto, and coworkers demonstrated that **26•**;SnCl$_4$ is an artificial cyclase that is useful for not only achiral but also chiral substrates.[28] The diastereoselective cyclization of (±)-nerolidol (**29**) has been examined with 1 equiv of the achiral LBA, 2-methoxyphenol (**32**)•SnCl$_4$, in dichloromethane at −78°C (entry 1, Table 12.1). Cyclization of (±)-**29** bearing an acid-sensitive allylic hydroxy group gives a complex reaction mixture, and the desired trans-fused 2-oxabicyclo[4.4.0]

TABLE 12.1. Double Asymmetric Induction in Cyclization of (±)-29 with LBAs

Entry	ArOH	Yield (%) 30 + 31	Ratio 30 : 31	Ee (%) (Rotation) 30, 31	From (S)-29 (−)-30 : (+)-31	From (R)-29 (+)-30 : (−)-31
1^a	32^c	<10	37 : 63	—	37 : 63	37 : 63
2^b	32^c	0	—	—	—	—
3^a	(R)-**26**	32	9 : 91	91 (−), 78 (−)	45 : 55	<1 : >99
4^b	(R)-**26**	13	28 : 72	97(−), 90(−)	88 : 12	<1 : >99

^{a}CH$_2$Cl$_2$ was used as solvent.
^{b}Toluene was used as solvent.
^{c}2-Methoxyphenol (**32**).

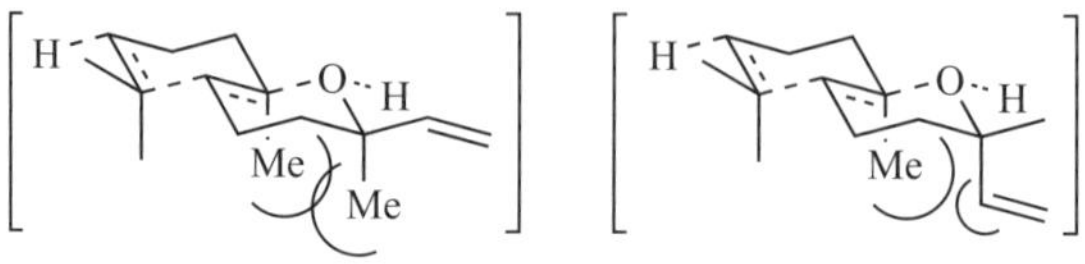

33 (favored)   34 (disfavored)

Figure 12.3. Two transition-state assemblies **33** and **34** in the proton-induced cyclization of (±)-**29**.

decanes have been obtained in less than 10% yield as a 37 : 63 mixture of (±)-caparrapi oxide (**30**) and (±)-8-epicaparrapi oxide (**31**). This diastereomeric ratio is due to substrate control; transition-state assembly **34** is more favorable than transition-state assembly **33** because of the steric difference between 3-vinyl group and the 3-methyl group of (±)-**29** (Figure 12.3). When (R)-**26** instead of **32** is used as a Brønsted acid, a 9 : 91 mixture of (−)-**30** (91% ee) and (−)-**31** (78% ee) is obtained in 32% yield (entry 3). This result indicates that (+)-**30** and (−)-**31** are obtained from (S)-**29** and (R)-**29** with 10% and >99% de, respectively. In the former case, low diastereoselectivity is due to the mismatch in asymmetric induction between substrate control and reagent control. In the latter case, high diastereoselectivity is due to the double asymmetric induction of substrate control and reagent control. The use of toluene in place of CH$_2$Cl$_2$ lowers the chemical yields of **30** and **31**, but raises their enantioselectivities to 97% ee and 90% ee. Notably, (−)-**30** is obtained from (S)-**29** with 76% de due to reagent control, which overcomes substrate control.

To improve the chemical yield of **30** or **31**, (±)-(E)-3-hydroxy-3,7,11-trimethyl-6,10-dodecadienyl 1-phenylacetate (**35**), which is less acid-sensitive than (±)-**29**, has been used as a substrate for cyclization with (R)-**26**·SnCl$_4$ (Scheme 12.24). The enantioselectivity is higher in the order CH$_2$Cl$_2$ toluene < chloropropane, while the chemical yield of **36** and **37** increases in the order toluene < chloropropane ≪ CH$_2$Cl$_2$. Chloropropane is superior to toluene with respect to both enantioselectivity and reactivity. When a 1 : 1 mixed solvent of chloropropane and CH$_2$Cl$_2$ is used, a 44 : 56 mixture of (−)-**36** (82% ee) and (−)-**37** (82% ee) is

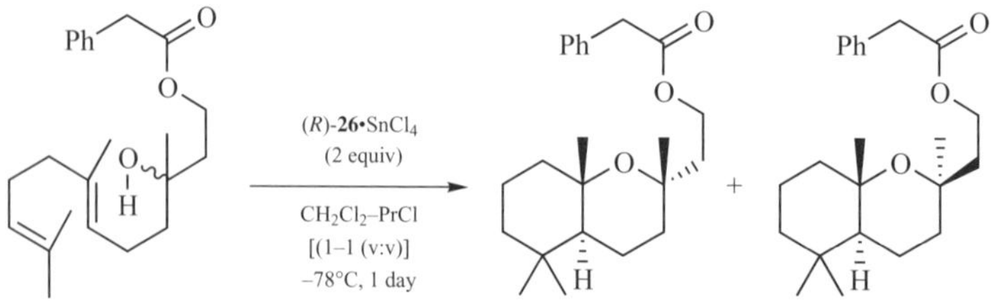

(±)-**35**

(R)-**26**·SnCl$_4$
(2 equiv)

CH$_2$Cl$_2$–PrCl
[(1–1 (v:v)]
−78°C, 1 day

65% yield

(−)-**36**   (−)-**37**

82% ee   82% ee

44 : 56

Scheme 12.24

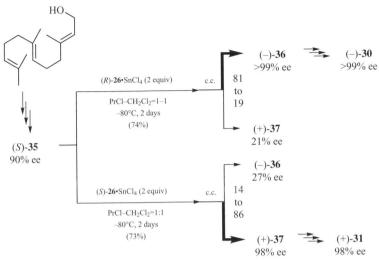

Scheme 12.25

obtained in 65% yield. The substrate control of **35** is lower than that of **29** because of a slight difference in the thermodynamic stabilities of **36** and **37**. Compounds **36** and **37** are easily separable by column chromatography on silicagel. (*S*)-**35** can be easily prepared with 90% ee in 91% overall yield from (*E,E*)-farnesol in three steps.[28] The asymmetric cyclization of (*S*)-**35** induced by 2 equiv of (*R*)-**5**·SnCl$_4$ gives an 81 : 19 mixture of (−)-**36** (>99% ee) and (+)-**37** (21% ee) in 74% yield. On the other hand, the asymmetric cyclization of (*S*)-**35** induced by 2 equiv of (*S*)-**26**·SnCl$_4$ gives a 14 : 86 mixture of (−)-**36** (27% ee) and (+)-**37** (98% ee) in 73% yield. The substrate control of **35** is much lower than the reagent control by **26**·SnCl$_4$. Optically pure (−)-**36** and (+)-**37** are easily separated by column chromatography on silicagel. Optically pure (−)-caparrapi oxide **30** has been obtained in 92% overall yield from (−)-**36** in three steps.[28] In the same manner, (+)-8-epicaparrapi oxide **31** (98% ee) has been obtained in 91% overall yield from (+)-**37** (Scheme 12.25).[28]

12.5 CONCLUSIONS AND OUTLOOK

Recently, the power of designer acid catalysts has generally increased as a result of the development of the catalytic enantioselective versions described here. In particular, combined acid catalysis is still very much in a state of infancy, and there is still much more to learn with regard to new reactivity.[1f] The ultimate goal is a more reactive, more selective, and more versatile catalyst. We believe that the realization of such an objective would be a tremendous benefit for the further development of organic synthesis, including green chemistry.

REFERENCES

1. (a) Yamamoto, H. (Ed.), *Lewis Acid Reagents, A Practical Approach*, Oxford Univ. Press, **1999**; (b) Santelli, M.; Pons, J.-M. *Lewis Acids and Selectivity in Organic Synthesis*, CRC Press, Boca Raton, FL, **1996**; (c) *Methods of Organic Chemistry* (Houben–Weyl), *Additional and Supplementary Volume to the 4th Edition*, Vol. E 21b; *Stereoselective Synthesis*, Helmchen, G.; Hoffmann, R. W.; Mulzer, J.; Schaumann, E. (eds.), Thieme, Stuttgart, **1995**; (d) Schinzer, D. (Ed.), *Selectivities in Lewis Acid Promoted Reaction*, Kluwer Academic Publishers, Dordrecht, Boston, London, **1988**; (e) Yamamoto, H. (Ed.), *Lewis Acids in Organic Synthesis*, Vols. 1, 2, Wiley-VCH, Weinheim, **2000**; (f) Yamamoto, H.; Futatsugi, K. *Angew. Chem. Int. Ed.* **2005**, *44*, 1924–1942.

2. (a) Curran, D. P.; Kuo, L. H. *J. Org. Chem.* **1994**, *59*, 3259–3261; (b) Curran, D. P.; Kuo, L. H. *Tetrahedron Lett.* **1995**, *37*, 6647–6650; (c) Schreiner, P. R.; Wittkopp, A. *Org. Lett.* **2002**, *4*, 217–220; (d) Wittkopp, A.; Schreiner, P. R. *Chem. Eur. J.* **2003**, *9*, 407–411.

3. (a) Sigman, M. S.; Vachal, P.; Jacobsen, E. N. *Angew. Chem. Int. Ed.* **2000**, *39*, 1279–1281; (b) Vachal, P.; Jacobsen, E. N. *Org. Lett.* **2000**, *2*, 867–870; (c) Su, J. T.; Vachal, P.; Jacobsen, E. N. *Adv. Synth. Catal.* **2001**, *343*, 197–200; (d) Vachal, P.; Jacobsen, E. N. *J. Am. Chem. Soc.* **2002**, *124*, 10012–10014.

4. Wenzel, A. G.; Jacobsen, E. N. *J. Am. Chem. Soc.* **2002**, *124*, 12964–12965.

5. (a) Okino, T.; Hoashi, Y.; Takemoto, Y. *J. Am. Chem. Soc.* **2003**, *125*, 12672–12673; (b) Okino, T.; Nakamura, S.; Furukawa, T.; Takemoto, Y. *Org. Lett.* **2003**, *6*, 625–628; (c) Berkessek, A.; Cleemann, F.; Mukherjee, S.; Müller, T. N.; Lex, J. *Angew. Chem. Int. Ed.* **2005**, *44*, 807–811.

6. Taylor, M. S.; Jacobsen, E. N. *J. Am. Chem. Soc.* **2004**, *126*, 10558–10559.

7. (a) Huang, Y.; Unni, A. K.; Thadani, A. N.; Rawal, V. H. *Nature* **2003**, *424*, 146; (b) Thadani, A. N.; Stankovic, A. R.; Rawal, V. H. *Proc. Nat. Acad. Sci. USA* **2004**, *101*, 5846–5850; (c) Unni, A. K.; Takenaka, N.; Yamamoto, H.; Rawal, V. H. *J. Am. Chem. Soc.* **2005**, *127*, 1336–1337.

8. McDougal, N. T.; Schaus, S. E. *J. Am. Chem. Soc.* **2003**, *125*, 12094–12095.

9. (a) Akiyama, T.; Itoh, J.; Yokota, K.; Fuchibe, K. *Angew. Chem. Int. Ed.* **2003**, *43*, 1566–1568; (b) Uraguchi, D.; Terada, M. *J. Am. Chem. Soc.* **2004**, *126*, 5356–5357; (c) Uraguchi, D.; Sorimachi, K.; Terada, M. *J. Am. Chem. Soc.* **2004**, *126*, 11804–11805.

10. (a) Corey, E. J.; Shibata, T.; Lee, T. W. *J. Am. Chem. Soc.* **2002**, *124*, 3808–3809; (b) Ryu, D. H.; Lee, T. W.; Corey, E. J. *J. Am. Chem. Soc.* **2002**, *124*, 9992–9993; (c) Ryu, D. H.; Corey, E. J. *J. Am. Chem. Soc.* **2003**, *125*, 6388–6390; (d) Zhou, G.; Hu, Q-Y.; Corey, E. J. *Org. Lett.* **2003**, *5*, 3979–3982.

11. Futatsugi, K.; Yamamoto, H. *Angew. Chem. Int. Ed.* **2005**, *44*, 1484–1487.

12. 12 Ishihara, K.; Mouri, M.; Gao, Q.; Maruyama, T.; Furuta, K.; Yamamoto, H. *J. Am. Chem. Soc.* **1993**, *115*, 11490–11495.

13. Kumagai, T.; Itsuno, S. *Macromol. Rapid Commun.* **2001**, *22*, 741–745.

14. Ishihara, K.; Kobayashi, J.; Inanaga, K.; Yamamoto, H. *Synlett* **2001**, *3*, 394–396.

15. Nagasawa, T.; Kudo, N.; Hida, Y.; Ito, K.; Ohba, Y. *Bull. Chem. Soc. Jpn.* **2001**, *74*, 989–990.

16. Corminboeuf, O.; Renaud, P. *Org. Lett.* **2002**, *4*, 1731–1733.

17. Sundararajan, G.; Prabagaran, N.; Varghese, B. *Org. Lett.* **2001**, *3*, 1973–1976.

18. Ishihara, K.; Kobayashi, J.; Nakano, K.; Ishibashi, H.; Yammaoto, H. *Chirality* **2003**, *15*, 135–138.

19. Mikami, K.; Ueki, M.; Matsumoto, Y.; Terada, M. *Chirality* **2001**, *13*, 541–544.

20. Hanawa, H.; Hashimoto, T.; Maruoka, K. *J. Am. Chem. Soc.* **2003**, *125*, 1708–1709.

21. Mahrwald, R. *Org. Lett.* **2000**, *2*, 4011–4012.

22. (a) Ishihara, K.; Nakamura, S.; Yamamoto, H. *J. Am. Chem. Soc.* **1999**, *121*, 4906–4907; (b) Nakamura, S.; Ishihara, K.; Yamamoto, H. *J. Am. Chem. Soc.* **2000**, *122*, 8131–8140.

23. (a) Ishihara, K.; Ishibashi, H.; Yamamoto, H. *J. Am. Chem. Soc.* **2001**, *123*, 1505–1506; (b) Ishihara, K.; Ishibashi, H.; Yamamoto, H. *J. Am. Chem. Soc.* **2002**, *124*, 3647–3655.

24. (a) Ishihara, K.; Kaneeda, M.; Yamamoto, H. *J. Am. Chem. Soc.* **1994**, *116*, 11179–11180; (b) Ishihara, K.; Nakamura, S.; Kaneeda, M.; Yamamoto, H. *J. Am. Chem. Soc.* **1996**, *118*, 12854–12855; (c) Ishihara, K.; Nakamura, S.; Yamamoto, H. *Croat. Chem. Acta* **1996**, *69*, 513–517; (d) Yanagisawa, A.; Ishihara, K.; Yamamoto, H. *Synlett* **1997**, 411–420. (e) Ishihara, K.; Ishida, Y.; Nakamura, S.; Yamamoto, H. *Synlett* **1997**, 758–760. (f) Ishihara, K.; Nakamura, S.; Kaneeda, M.; Ishihara, K.; Yamamoto, H. *J. Am. Chem. Soc.* **2000**, *122*, 8120–8130. (g) Ishibashi, H.; Ishihara, K.; Yamamoto, H. *Chem. Rec.* **2002**, *2*, 177–188.

25. Ishihara, K.; Nakashima, D.; Yamamoto, H. *J. Am. Chem. Soc.* **2003**, *125*, 24–25.

26. Kumazawa, K.; Ishihara, K.; Yamamoto, H. *Org. Lett.* **2004**, *6*, 2551–2554.

27. Ishibashi, H.; Ishihara, K.; Yamamoto, H. *J. Am. Chem. Soc.* **2004**, *126*, 11122–11123.

28. (a) Muhammet, U.; Ishibashi, H.; Ishihara, K.; Yamamoto, H. *Org. Lett.* **2005**, *7*, 1601–1604; (b) Muhammet, U.; Ishihara, K.; Yamamoto, H. *Bio. Med. Chem.* **2005** *13*, 5055–5065.

13

CHIRAL BIFUNCTIONAL ACID/BASE CATALYSTS

MASAKATSU SHIBASAKI AND MOTOMU KANAI

Graduate School of Pharmaceutical Sciences, The University of Tokyo Hongo, Bunkyo-ku, Tokyo, Japan

13.1 INTRODUCTION

Successful asymmetric catalysis generally requires two concepts: reactivity enhancement and enantiocontrol. Bases can enhance the reactivity mainly through activation of the nucleophile side. Two methods are possible for the activation of nucleophiles; working as a Brönsted base to generate metal-conjugated nucleophiles through deprotonation, or as a Lewis base through coordinating to metals and thus polarizing nucleophilic center–metal bonds. Thus, chiral base catalysis can be defined as catalytic enantioselective reactions using chirally modified nucleophiles by asymmetric catalysts.

In this chapter, we discuss recent (reported mainly during 2000–2005) asymmetric reactions catalyzed by chiral bases. Because practicality is an important factor in the present asymmetric catalysis, we restricted our discussion mainly to the reactions giving over 90% ee unless the conversion is novel. We notice, however, that there are many potentially useful and scientifically interesting reactions, in which enantioselectivity does not exceed the practical range at this moment. Chiral organic base (proline and cinchona alkaloids)-catalyzed reactions were discussed in Chapter 11 by Lelais and MacMillan.

We divided the topic into five main sections. There are no purely acid- and base-catalyzed reactions, however, at least when metals are involved in the reaction. Metals should activate electrophiles as a Lewis acid. Thus, base catalysis discussed

New Frontiers in Asymmetric Catalysis, Edited by Koichi Mikami and Mark Lautens
Copyright © 2007 John Wiley & Sons, Inc.

in this chapter always includes Lewis acidic (or Brönsted acidic) function. We use the term "bifunctional catalysis" for the reactions in which two functions (Brönsted base–Lewis acid, Brönsted base–Brönsted acid, or Lewis base–Lewis acid) work at the distinct parts (or molecules) of the catalyst(s).

13.2   CHIRAL BRÖNSTED BASE CATALYSIS

There are not many highly enantioselective reactions catalyzed by one metal-centered chiral Brönsted bases. One plausible reason is the necessity to use weakly positive (thus, not very Lewis acidic) metals to produce enough Brönsted basicity for nucleophile generation. As a result, it is difficult to construct efficient chiral environment around prochiral electrophiles (such as aldehydes and enones). On the other hand, prochiral carbons on the nucleophile side exist close to the chirality of asymmetric catalysts. Therefore, higher enantioinduction is possible on the nucleophile side than on the electrophile side in the chiral Brönsted base catalysis. A seminal example, the catalytic enantioselective Michael addition of stabilized carbanions to enones using KO^tBu–chiral crown ether complex **1**, can be categorized into this classification [Eq. (13.1)].[1] High enantioselectivity was produced on the nucleophile side:

$$(13.1)$$

Despite the generally conceivable drawback in the one metal-center Brönsted base asymmetric catalyst, there are examples that produce high enantioselectivity on the electrophile side. Our group reported a catalytic enantioselective Michael reaction of malonates to cyclic enones using La-linked-BINOL complex **2** [Eq. (13.2)].[2] Excellent enantioselectivity was obtained from a wide range of cyclic enones with five- to nine-membered rings. This reaction takes advantage of the ambiphilic character of lanthanide naphthoxide; the lanthanum naphthoxide acts as a Brönsted base to deprotonate malonates, and the nucleophile attacks an activated enone coordinating to the chirally modified lanthanum working as a Lewis acid. The chiral catalyst is stable to air, and reusable at least for 4 times without significantly changing the enantioselectivity. Through a modification of the linker heteroatom from oxygen to nitrogen (**3**), it became possible to use α-ketoesters as a nucleophile:[3]

$$(13.2)$$

61–98%
82–>99% ee

Marks' group developed novel catalytic intramolecular hydroamination reactions utilizing the high Brönsted basicity of organolanthanides, in conjunction with the characteristic facile insertion of a carbon–carbon unsaturated bond into lanthanide–nitrogen σ bonds [Eq. (13.3)].[4] In this reaction, organolanthanide **4** deprotonates amine **5** to generate lanthanide amide **6** in the first step. Then, this intermediate undergoes irreversible, turnover-limiting intramolecular olefin insertion to produce new organolanthanide **8** presumably via a four-centered transition state **7**. The rapid protonolysis of the lanthanide–carbon bond by substrate amine yields the azacyclic product **9** and regenerates the catalytically active species **6**. This reaction was extended to catalytic enantioselective reactions to afford chiral nitrogen-containing heterocyclic compounds. A more recent example using chiral bis(oxazoline) complex is shown in Eq. (13.4).[5] Although enantioselectivity is in the moderate range at this moment, further studies should improve the efficiency of this important atom-economical reaction.

$$(13.3)$$

$$(13.4)$$

> 98% conv.
40–67% ee

13.3 CHIRAL BRÖNSTED BASE–LEWIS ACID BIFUNCTIONAL CATALYSIS

Our group developed enantioselective catalysts containing both Lewis acid and Brönsted base functions in one molecule. According to this concept, the Brönsted base moiety deprotonates nitroalkanes, ketones, or malonates to generate a nucleophile, and the Lewis acid moiety activates an electrophile [see **11** in Eq. (13.5)]. This type of dual activation of two reacting substrates occurs simultaneously at the defined positions by the asymmetric catalyst, which leads to high reactivity and enantioselectivity with broad substrate generality. Because the initial stage, which works using catalysts **12** and **13**, was reviewed previously,[6] we focus our discussion here on the recent progress using zinc-linked-BINOL complex and the direct catalytic asymmetric aldol-Tishchenko reaction:

$$(13.5)$$

Our group recently reported that a chiral polymetallic complex prepared from Et_2Zn and linked-BINOL in a 4:1 ratio is an excellent catalyst for direct aldol reaction [Eq. (13.6)],[7] Mannich reaction [Eq. (13.7)],[8] and Michael reaction [Eq. (13.8)][9] of α-hydroxy ketones. Under the optimized conditions in the presence of MS 3A, catalyst amount was reduced to as low as 0.01 mol%. Precatalyst crystal structure was determined as a 3:2 complex of zinc and linked-BINOL (**14**). On the basis of CSI-MS analysis, this complex is converted to oligomeric zinc-rich species containing the donor α-hydroxy ketone in the reaction mixture, and this polyzinc complex should be the active catalyst. One of the zinc naphthoxides should work as a Brönsted base to generate a zinc enolate through deprotonation of α-hydroxy ketone, and another zinc atom should work as a Lewis acid to activate substrates. This type of bifunctional catalysis should define the entry of the enolate to the substrate, thus yielding products with high enantioselectivity. In all the reactions using (S,S)-linked BINOL, both syn and anti diastereomers have (R)-configuration at the α-carbons; this fact suggests that the asymmetric catalyst differentiates the enantioface of the enolate with high selectivity:

$$(13.6)$$

$$(13.7)$$

$$(13.8)$$

14

Another more recent contribution from our laboratories in this area is the development of the direct catalytic asymmetric aldol-Tishchenko reaction.[10] Our group previously developed the first catalytic enantioselective intermolecular aldol reaction of unmodified methyl ketones using a chiral heteropolymetallic complex.[11] This topic is currently one of the most actively investigated areas in synthetic organic chemistry.[12] Direct aldol reaction of ethyl ketones is, however, viewed as a formidable challenge, due to the fast retroaldol reactions.[13] Our group has reported a preliminary solution to this problem by coupling the reversible aldol reaction with the irreversible Tishchenko reaction. Using the catalyst prepared from La(OTf)$_3$, BINOL, and BuLi in a 1:3:5.6 ratio, reactions between aromatic aldehydes and electron-deficient aromatic ketones proceeded at room temperature, giving the product diol with high enantio- and diastereo-selectivities [Eq. (13.9)]. The reaction mechanism was proposed involving nonstereoselective aldolization followed by stereodetermining Tishchenko reaction through a cyclic transition state **17**, which was initially proposed by Evans et al.[14] Because the Tishchenko reaction from *syn*-aldolate **16** through the unfavorable transition state **18** is slow, the retroaldol reaction predominates from *syn*-**16**. Thus, the Tishchenko product from *anti*-aldolate **15** was the sole product. It was also confirmed that enantioselectivity was determined in the Tishchenko step:

M* = La(OTf)$_3$ + (*R*)–BINOL + BuLi (1:3:5.6) (10 mol %)

(13.9)

Trost's group reported direct catalytic enantioselective aldol reaction of unmodified ketones using dinuclear Zn complex **21** [Eq. (13.10)].[15] This reaction is noteworthy because products from linear aliphatic aldehydes were also obtained in reasonable chemical yields and enantioselectivity, in addition to secondary and tertiary alkyl-substituted aldehydes. Primary alkyl-substituted aldehydes are normally problematic substrates for direct aldol reaction because self-aldol condensation of the aldehydes complicates the reaction. Bifunctional Zn catalysis **22** was proposed, in which one Zn atom acts as a Lewis acid to activate an aldehyde and the other Zn-alkoxide acts as a Brönsted base to generate a Zn-enolate. The same catalyst was also applied to an enantioselective nitroaldol reaction,[16] a syn-selective direct asymmetric aldol reaction of α-hydroxy ketones,[17] and Mannich reactions.[18]

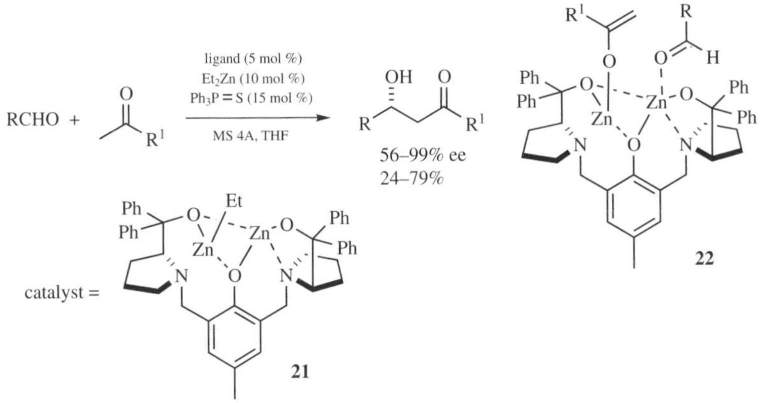

(13.10)

Snapper and Hoveyda reported a catalytic enantioselective Strecker reaction of aldimines using peptide-based chiral titanium complex [Eq. (13.11)].[19] Rapid and combinatorial tuning of the catalyst structure is possible in their approach. Based on kinetic studies, bifunctional transition state model **24** was proposed, in which titanium acts as a Lewis acid to activate an imine and an amide carbonyl oxygen acts as a Brönsted base to deprotonate HCN. Related catalyst is also effective in an enantioselective epoxide opening by cyanide:[20]

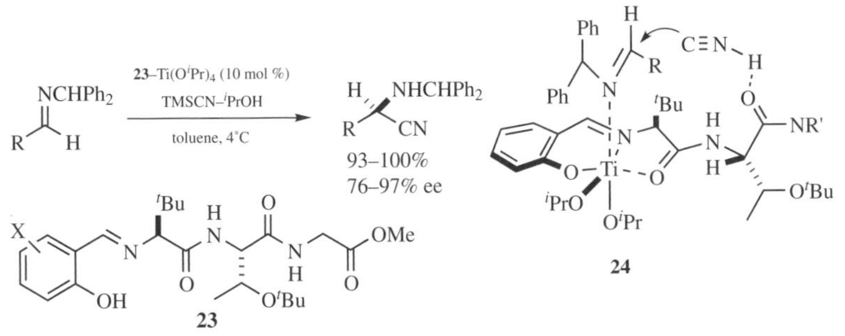

(13.11)

After the famous achievement of catalytic hydrolytic kinetic resolution of terminal epoxide using (salen)–Co(III)–OAc complex,[21] Jacobsen's group elucidated that this reaction proceeds via a cooperative bimetallic mechanism [Eq. (13.12)]; one Co complex acts as a Brönsted base to generate a cobalt hydroxide, and another Co complex acts as a Lewis acid to activate an epoxide. With this reaction mechanism, the reaction rate is expected to be the fastest when the ratio of the nucleophile Co—OH and the Lewis acid Co—X is close to 1:1. As expected, significantly improved reaction rate and catalyst turnover were produced when a 1:1 mixture of (salen)Co—Cl (which is readily converted to Co—OH through Cl transfer during the reaction) and (salen)Co—SbF$_6$ was used as a catalyst.[22] Use of (salen)Co—OTs catalyst, which generates a 1:1 mixture of Co—OH and Co—OTs during the reaction, is practically more convenient because only one catalytic species is required. The homobimetallic system also produced remarkable results in the (salen)Co-catalyzed phenolic kinetic resolution and the (salen)Co-catalyzed asymmetric ring-opening reaction of epoxides.[23] Cooperative homobimetallic catalysis is also applied to the (salen)Al complex-catalyzed enantioselective addition reactions of weakly acidic species, such as cyanide, hydrazoic acid, thiols, nitrogen heterocycles, and oximes:[24]

e.g., R = Bu catalyst (X = OAc) 0.5 mol %: 16 h 43%, >99% ee epoxide
catalyst (X = OTs) 0.15 mol %: 3 h 43%, >99% ee epoxide

A more recent, more advanced concept termed *cooperative dual catalysis* was postulated by Jacobsen's group in catalytic enantioselective conjugate addition of cyanide to α,β-unsaturated imides.[25] In the presence of 2 mol% μ-oxobimetallic (salen)Al complex (**25**) and 3 mol% (pybox)Er complex (**26**), the reaction proceeded with excellent enantioselectivity [Eq. (13.13)]. Only a trace amount (<3%) of product was obtained with a separate use of these two chiral metal complexes. Kinetic studies indicated that both metal complexes are involved in the rate-determining step. In addition, matched combination was found between two chiral ligands coordinating to the metals. Therefore, these two metal complexes function cooperatively in the asymmetric induction.

93–97% ee
80–94%

25: X = OAl(salen)   **26**   cooperative dual-metal catalysis

Although bases are not chirally modified (a stoichiometric amount of base was used in Evans' reaction), three important contributions in catalytic enantioselective carbon–carbon bond-forming reactions are described in this section:

1. Evans' group reported the Ni(II)-catalyzed enantioselective aldol reaction of in situ generated chiral metal enolate derived from an ester surrogate, N-propionylthiazolidinethione.[26] Excellent enantio- and syn-selectivity were produced from a range of aldehydes, including aromatic and easily enolizable primary alkyl-substituted aldehydes [Eq. (13.14)]. The reaction was proposed to proceed through a nickel enolate, which is generated via deprotonation of the propionate coordinating to the Lewis acidic cationic nickel by the Brönsted base (lutidine). TMSOTf traps the resulting nickel aldolate. Stoichiometric amounts of TMSOTf and lutidine were required in this reaction possibly to prevent the retroaldol reaction and/or to facilitate the catalytic cycle by enhancing the product dissociation rate from the catalyst; no reaction occurred in the absence of TMSOTf:

$$(13.14)$$

90–97% ee
88:12–97:3 syn-selectivity
46–86%

2. Shair's group reported a catalytic enantioselective aldol-type reaction using $Cu(OTf)_2$ (10 mol%)-PhBox (13 mol%) complex as a catalyst and methyl malonic acid half-thioester as a nucleophile.[27] A slight (3 mol%) excess of ligand to copper is proposed to function as a base to deprotonate the half-thioester. Mild reaction conditions are noteworthy for this reaction. Lack of strong Lewis acids or strongly basic intermediates enables it to be compatible with hydroxyl groups, phenols, enolizable aldehydes, and even carboxylic acid. The enantio- and diastereoselectivity are generally high (89–96% ee, syn:anti = 2.2:1-36:1). A typical example that demonstrates the characteristics of this reaction is shown in Eq (13.15):

$$(13.15)$$

3. Carreira et al. reported an enantioselective addition of terminal acetylenes to aldehydes using catalytic amounts of N-methylephedrin (**27**)–Zn(OTf)$_2$ complex and a base (Et$_3$N), [Eq. (13.16)].[28] Excellent enantioselectivity was produced from a range of alkynes and aldehydes, including enolizable aldehydes. Results using aromatic aldehydes as substrates were not reported.[29] The authors proposed that the reaction should proceed through zinc alkynilides generated via deprotonation of an activated alkynes coordinating to Zn(OTf)$_2$ by a catalytic Brönsted base. This study was extended to a catalytic enantioselective conjugate addition of phenylacetylene using a chiral copper complex with developing a novel phosphine ligand **28** [Eq. (13.17)].[30]

$$(13.16)$$

$$(13.17)$$

13.4   CHIRAL BRÖNSTED BASE–BRÖNSTED ACID BIFUNCTIONAL CATALYSIS

Sodeoka and coworkers reported a chiral Pd-catalyzed enantioselective Michael reaction of 1,3-diketones and β-ketoesters to enones, [Eq. (13.18)].[31] It was proposed that the catalyst **29** exists in equilibrium with **30** through liberating H$_2$O and TfOH, and the palladium hydroxide part of **30** would act as a Brönsted base to deprotonate 1,3-dicarbonyl compounds, while TfOH activates enones. This type of synergistic nucleophile generation and electrophile activation makes the highly enantioselective catalytic Michael reaction possible. Related catalysts also promote enantioselective fluorination of β-ketoesters[32] and conjugate addition of amines.[33]

$$(13.18)$$

Inspired by the reaction mechanism of Noyori's catalytic enantioslective transfer hydrogenation of ketones (**32**) using a chiral Ru-amido complex **31**,[34] Ikariya et al. reported that **31** can also function as a unique Brönsted base–Brönsted acid catalyst promoting enantioselective Michael reaction of malonates,[35] β-ketoesters,[36] and α-nitro esters[37] to cyclic enones and nitroalkenes.[38] A typical example is shown in Eq (13.19). In this reaction, a ruthenium amide acts as a Brönsted base, and generates a C-bound Ru-malonato complex through deprotonation. Michael addition of malonate should proceed to an enone coordinating to, and thus activated by, the relatively acidic NH proton in the catalyst (**33**). The electronic characteristic of the catalyst could be tuned by the sulfonamide and arene moieties, which allowed for application to a range of C—C bond-forming reactions:

$$(13.19)$$

13.5   CHIRAL LEWIS BASE CATALYSIS

Two patterns are possible in the activation mechanism by simple chiral Lewis base catalysts. One is through the activation of nucleophiles such as allyltrichlorosilanes or ketene trichlorosilyl acetals via hypervalent silicate formation using organic Lewis bases such as chiral phosphoramides[39] or N-oxides.[40] In this case, catalysts are pure organic compounds (see Chapter 11). The other is through the activation of nucleophiles by anionic Lewis base conjugated to metals. In this case, transmetallation is the key for the nucleophile activation. This type of asymmetric catalysis is the main focus of this section.

Catalytic asymmetric hydrosilylation of prochiral ketones has a rich history, and chiral complexes involving such metals as Rh, Zn, and Ti have been reported.[41] Efficient catalytic enantioselective hydrosilylation reactions using chiral copper hydride complexes as a catalyst and PMHS (polymethylhydrosiloxane) as a stoichiometric reducing reagent have been developed by the Lipshutz group. Excellent enantioelectivity was produced in hydrosilylation of ketones,[42] imines,[43] and α,β-unsaturated carbonyl substrates[44] [Eq. (13.20)]. The substrate/ligand ratio (S/L) reached up to 275,000. Use of chiral bisphosphines with bulky substituents on the phosphorus atoms (such as DTBM-SEGPHOS) is the key for the high catalyst turnover:

$$(13.20)$$

Aldol reactions of silyl enolates are promoted by a catalytic amount of transition metals through transmetallation generating transition metal enolates.[45] In 1995, Shibasaki and Sodeoka reported an enantioselective aldol reaction of enol silyl ethers to aldehydes using a Pd-BINAP complex in wet DMF.[46] Later, this finding was extended to a catalytic enantioselective Mannich-type reaction to α-imino esters by Sodeoka's group [Eq. (13.21)].[47] Detailed mechanistic studies revealed that the binuclear μ-hydroxo complex **34** is the active catalyst, and the reaction proceeds through a palladium enolate. The transmetallation step would be facilitated by the hydroxo ligand transfer onto the silicon atom of enol silyl ethers:

$$(13.21)$$

In 1998, Carreira reported a copper fluoride-catalyzed enantioselective aldol reaction of aldehydes using a silyl dienolate **35** as nucleophile [Eq. (13.22)].[48] High enantioselectivity was produced from aromatic and α,β-unsaturated aldehydes. Mechanistic studies using IR spectroscopy suggested that the reaction proceeds through a chiral copper dienolate **36**, which is generated through transmetallation from silyl dienolate. Once **36** reacts with an aldehyde, copper aldolate **37** is generated, which is silylated by the silyl dienolate **35** to produce silyl aldolate **38** with regenerating the copper enolate **36**. This dienolate aldol reaction was extended to a catalytic enantioselective addition to ketones by Campagne:[49]

$$(13.22)$$

Y. Yamamoto and coworkers reported that bis-π-allylpalladium complex can act as a nucleophile, in sharp contrast to the ordinary behavior of π-allylpalladium complexes acting as electrophiles.[50] Bis-π-allylpalladiums are generated through transmetallation from allyltin compounds or allylsilanes. Because the sterically bulky allyl group is nontransferable to aldehydes or imines, this finding was extended to a catalytic enantioselective allylation of aldimines using chiral bis-π-allylpalladium catalyst **39** containing a bulky allyl group [Eq. (13.23)].[51] High enantioselectivity was produced mainly from aromatic imines:

$$(13.23)$$

H. Yamamoto reported an enantioselective allylation of aldehydes catalyzed by AgF–p-tol-BINAP complex, [Eq. (13.24)].[52] High enantioselectivity was obtained

from aromatic and α,β-unsaturated aldehydes. Both *E*- and *Z*-crotylsilanes produced the corresponding anti-diastereoisomer selectively (anti/syn = ~ 94/6).
Because the peak ascribed to $(MeO)_3SiF$ was observed and no peaks of crotylsilane
were observed by ^{13}C NMR when the crotylsilane was mixed with AgF in a 1:1
ratio, the authors proposed that crotylsilver is generated from crotylsilane through
transmetallation. Reaction of the thus-generated crotylsilver with an aldehyde
might proceed via a six-membered transition state. The anti-selectivity might be
explained if the geometry isomerization of the crotylsilver is faster than the addition
to aldehydes. The related catalyst can promote an enantioselective aldol reaction of
trimethoxysilyl enol ethers to aldehydes[53] and an enantioselective protonation of
trimethylsilyl enol ethers:[54]

$$
RCHO + R' \underset{}{\diagdown}\diagup Si(OMe)_3 \xrightarrow[\substack{(R)\text{-tol-BINAP (6 mol\%)}\\ \\ CH_3OH}]{\text{AgF (10 mol\%)}} R \diagup\!\!\diagdown_{R'}^{OH} \diagdown\!\!\diagup \qquad (13.24)
$$

78–96% ee
67–99%

Ketones are generally more difficult substrates than aldehydes for catalytic enantioselective reactions. For steric and electronic reasons, ketones are much less reactive than aldehydes. In addition, differentiation between the two lone pairs of the
carbonyl oxygen atom, which is required for enantioselectivity, is also difficult
in ketones. Therefore, the catalyst should be more active and enantioselective
when targeting ketones than aldehydes. Developing an enantioselective catalyst
that can promote reactions targeting ketones is at the frontier in organic synthesis.
Our group developed a catalytic enantioselective allylation of ketones using
CuF–^{i}Pr-DuPHOS as a catalyst, $La(O^iPr)_3$ as a cocatalyst, and allylboronate or cro-
tylboronate as an allylation reagent [Eq. (13.25)].[55,56] A catalytic amount of
$La(O^iPr)_3$ greatly accelerates the reaction rate without affecting the enantioselectivity. Because the same enantioselectivity was produced using either allylboronate,
allyltrimethoxysilane, or allyltributyltin, the actual nucleophile is an allylcopper
generated through transmetallation.

$$ (13.25) $$

This chemistry was extended to a catalytic enantioselective alkenylation and phenylation of aldehydes and α-ketoesters.[57] Using CuF–DTBM-SEGPHOS complex, products were obtained with excellent enantioselectivity from a wide range of aldehydes including aromatic and aliphatic aldehdyes, [Eq. (13.26)]. Previously catalytic enantioselective vinylation and phenylation are restricted using the corresponding zinc reagents.[58] The active nucleophile is proposed to be an alkenyl or phenyl copper, based on NMR studies. The chiral CuF catalyst can also be applied to a catalytic enantioselective aldol reaction to ketones:[59]

$$(13.26)$$

13.6   CHIRAL LEWIS BASE–LEWIS ACID BIFUNCTIONAL CATALYSIS

If nucleophile activation by a Lewis base is combined with electrophile activation by a Lewis acid (**40**), higher enantioselectivity and catalyst activity can be expected. This type of reaction mechanism was proposed for catalytic enantioselective reduction of ketones using CBS catalyst (**41**)[60] and dialkylzinc addition to aldehydes (**42**).[61] These catalysts utilize heteroatoms directly conjugated to Lewis acid metals as Lewis bases. Recent successful examples in this category have been developed mainly by our group. Our concept for more flexible bifunctional asymmetric catalyst design is separating these two functionality with maintaining the cooperativity:[62]

$$(13.27)$$

In 1999, we reported a catalytic enantioselective cyanosilylation of aldehydes with a broad substrate generality using a newly designed bifunctional catalyst **43** [Eq. (13.28)].[63] This catalyst promotes the reaction with a dual activation of aldehydes and TMSCN by the Lewis acid Al and the Lewis base phosphine oxide. Support for the bifunctional mechanism was based on IR studies and kinetic data. This reaction was applied to our catalytic asymmetric total synthesis of epothilones[64] and a chiral building block of an antihyperglycemic agent.[65] Catalyst **43** was also effective for catalytic enantioselective Strecker reaction of aldimines.[66] A catalyst related to **43** containing diethylamino group as a Lewis base was developed by Nájera and Saá et al.:[67]

$$(13.28)$$

We reported the first catalytic enantioselective Reissert reaction of quinolines and isoquinolines.[68] Specifically, the isoquinoline Reissert reaction produced chiral tetrasubstituted carbons with excellent enantioselectivity [Eq. (13.29)].[69] A dual-activation mechanism **45** was proposed with the Lewis acid activating substrate acyl isoquinoline intermediate and the Lewis base activating TMSCN. More recently, we succeeded in developing the catalytic enantioselective Reissert reaction of pyridine derivatives with a new bifunctional catalyst derived from Et_2AlCl and C1 symmetric chiral ligand **46** containing sulfoxides as a Lewis base in a 1:2 ratio [Eq. (13.30)].[70] Catalysts containing a phosphine oxide as a Lewis base such as **43** produced only less than 10% ee. Because the acyl nicotinamide intermediate is more electrophilic than acyl (iso)quinolinium, strongly Lewis basic phosphine oxide could promote a monoactivation reaction pathway by the Lewis base, which should produce low enantioselectivity. Because a sulfoxide is a weaker Lewis base than a phosphine oxide, a sulfoxide-containing catalyst should facilitate the desired dual-activation pathway of an acyl pyridinium and TMSCN by the Lewis acid and Lewis base, respectively. In addition, both the ratio Al:**46** and the chirality of the sulfoxides are very important in this reaction. On the basis of ESI-MS studies of the catalyst, we proposed that a 2:3 complex of Al and the ligand is the highly enantioselective catalyst. C_1-symmetric chirality of the sulfoxides should be essential for stabilizing the 2:3 complex through bridging the two Al atoms. Thus, the chiral solfoxide might have dual roles in this reaction: a Lewis base to activate TMSCN and a stabilizer of the enantioselective 2:3 complex. Because of the importance of heterocycles as a building block of pharmaceuticals, these reactions are useful for the synthesis of biologically active molecules:

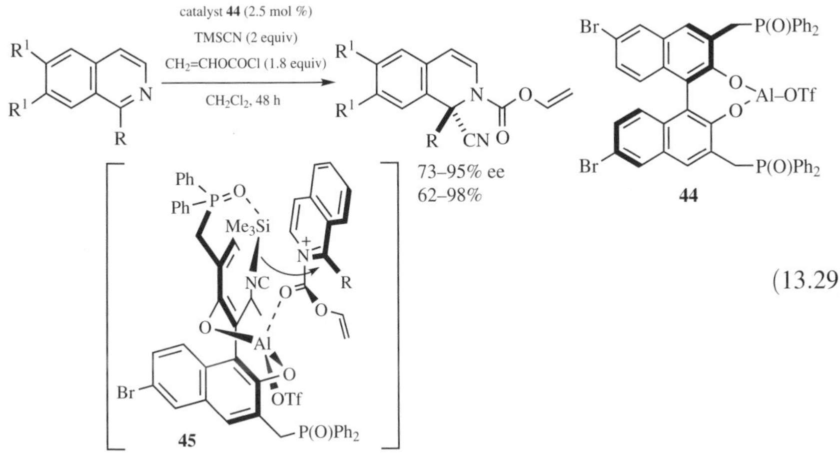

(13.29)

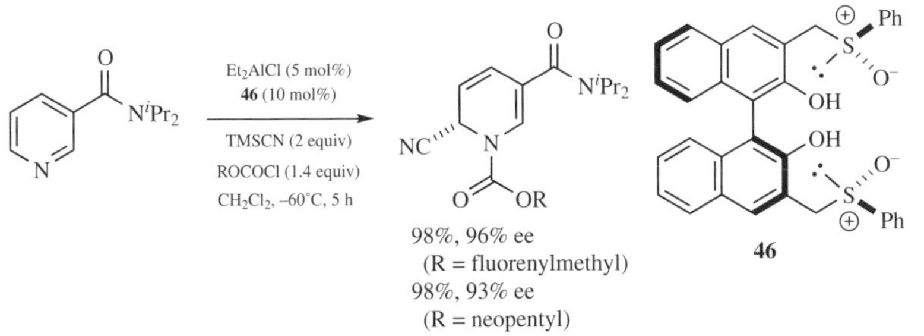

(13.30)

98%, 96% ee
(R = fluorenylmethyl)
98%, 93% ee
(R = neopentyl)

We reported a catalytic enantioselective cyanosilylation of ketones that produces chiral tetrasubstituted carbons from a wide range of substrate ketones [Eq. (13.31)].[71,72] The catalyst is a titanium complex of a D-glucose-derived ligand **47**. It was proposed that the reaction proceeds through a dual activation of substrate ketone by the titanium and TMSCN by the phosphine oxide (**51**), thus producing (R)-ketone cyanohydrins:

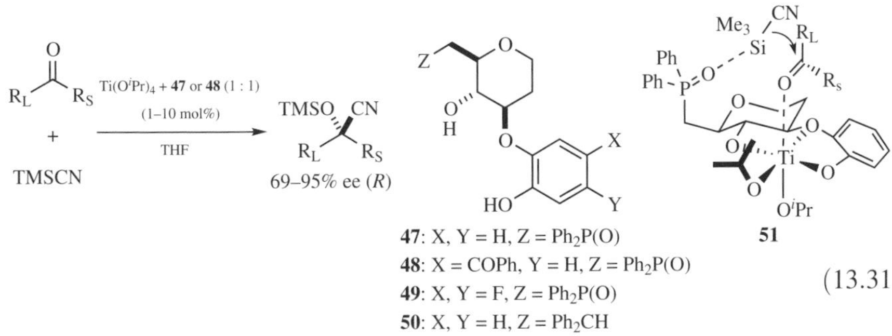

69–95% ee (R)

47: X, Y = H, Z = Ph$_2$P(O)
48: X = COPh, Y = H, Z = Ph$_2$P(O)
49: X, Y = F, Z = Ph$_2$P(O)
50: X, Y = H, Z = Ph$_2$CH

(13.31)

The enantioselectivity was dramatically reversed to (S)-selective when an asymmetric catalyst was prepared from the same ligand **47** and Gd(O^{i}Pr)$_3$ in a 1:2 ratio

[Eq. (13.22)].[73] The switch in enantioselectivity was attributed to the change in the reaction mechanism. On the basis of ESI-MS studies, the active Gd catalyst was proposed to be a 2:3 complex of Gd and **47**, and a gadolinium cyanide generated through transmetalation from TMSCN works as the active nucleophile. The cyanide should be transferred to an activated ketone coordinating to the Lewis acidic Gd metal in an intramolecular fashion (**52**). Lewis base phosphine oxide is again playing an essential role in this reaction. The reaction proceeded very slowly when a control catalyst prepared from ligand **50** without a Lewis base was used. Because no meaningful ESI-MS peak was observed in the catalyst prepared from **50**, it is postulated that the phosphine oxide stabilizes the active 2:3 complex **52** while bridging two Gd metals. These reactions are useful for the synthesis of biologically active compounds such as camptothecins, oxybutynin, and fostriecin.[74] More recently, an asymmetric catalyst generated from Gd(HMDS)$_3$ and ligand **49** in a 2:3 ratio was demonstrated as a better catalyst than that prepared from Gd(O^{i}Pr)$_3$, especially for electron-deficient reactive ketones.[75] This catalyst was applied to an asymmetric synthesis of a versatile synthetic intermediate for triazole antifungal agents:

(13.32)

Corey reported a catalytic enantioselective cyanosilylation of methyl ketones using combination of a chiral oxazaborolidinium and an achiral phosphine oxide, [Eq. (13.23)].[76] An intermolecular dual activation of a substrate by boron and TMSCN by the achiral phosphine oxide (MePh$_2$PO) is proposed as a transition-state model (**54**). The same catalyst was also used for cyanosilylation of aldehydes:[77]

(13.33)

Our group reported that the Gd complex derived from ligand **49** is also an excellent catalyst for Strecker reaction of phosphinoyl ketoimines [Eq. (13.34)].[78,79] Enantioselectivity, substrate generality, and catalyst turnover efficiency are all improved greatly in the presence of a stoichiometric amount of protic additive, 2,6-dimethylphenol. According to the ESI-MS studies of the catalyst in the presence of 2,6-dimethylphenol, the active catalyst structure changed from TMS containing **56** to protonated **57**, and **57** should be more active and enantioselective than **56** in Strecker reaction of ketoimines. Moreover, catalyst activity was further improved when using HCN as a protic additive. Only 0.1 mol% catalyst loading was enough for excellent conversion and enantioselectivity. Because HCN can also act as a cyanide source, it is possible to reduce the amount of TMSCN to 2.5 mol%. A proposed catalytic cycle is shown in Eq. (13.34). After the enantioselective intramolecular cyanide transfer to an activated ketoimine (**58**), the generated zwitterionic intermediate **59** should collapse via an intramolecular proton transfer to liberate the product. The active catalyst **57** should be regenerated only through a silylated 2:3 complex **56** from the resulting alkoxide complex **55**, because a catalytic amount of TMSCN was essential for the reaction to proceed:

$$(13.34)$$

The same catalyst is also effective for an enantioselective conjugate addition of cyanide to α,β-unsaturated N-acylpyrroles, [Eq. (13.35)].[80,81] The product is a precursor for chiral γ-amino acids:

$$(13.35)$$

83–98% ee
78–99%

In 1996, Carreira reported a catalytic enantioselective allylation of aldehydes using BINOL-modified TiF$_4$.[82] More recently, Kobayashi reported a catalytic enantioselective allylation of hydrazono esters in aqueous media using ZnF$_2$–chiral diamine complex, [Eq. (13.36)].[83] In both reactions, reaction mechanism via dual activation of the substrate by Lewis acid metals and allylsilanes by fluoride is proposed:

$$(13.36)$$

65–86% ee
61–92%

Since a dual-activation transition-state model **42** was proposed for catalytic enantioselective dialkylzinc addition to aldehydes by Noyori et al.,[61] numerous chiral catalysts were studied to evaluate their enantioselectivity using this reaction.[84] Here, we focus our discussion on more recent reports of catalytic enantioselective organozinc addition to ketones and α-ketoesters.[85] The first example of catalytic enantioselective addition of an organozinc (phenylzinc reagent prepared from 3.5 equiv of diphenylzinc and 1.5 equiv of MeOH) to ketones was reported by Fu et al. using Noyori's catalyst (DAIB), taking advantage of the higher reactivity of phenylzinc compared to alkylzinc reagents.[86] Extension of Carreira's alkynylation of aldehydes[28] to a catalytic enantioselective alkynylation of α-ketoesters has been reported by Jian et al., [Eq. (13.37)].[87,88] Using catalytic Zn(OTf)$_2$–chiral amino alcohol (**61**) complex (20 mol%) in the presence of a catalytic amount of Et$_3$N (30 mol%), high enantioselectivity was produced from aromatic- or t-alkyl-substituted α-ketoesters that contain no acidic α-protons:

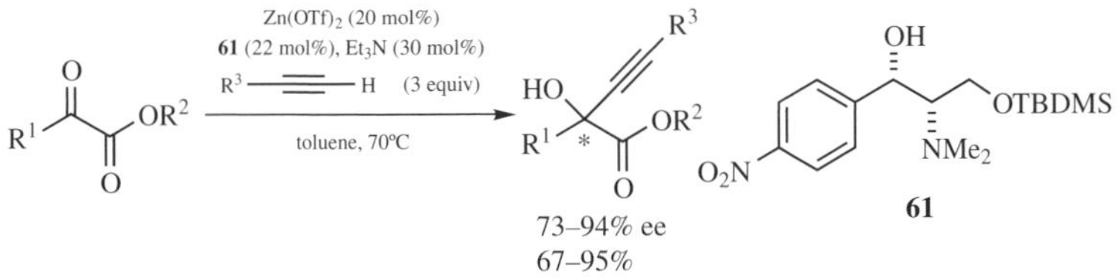

$$
\text{(13.37)}
$$

Dialkylzincs are much less reactive than phenyl or alkynylzincs. In 2002, Kozlowski et al. developed a chiral salen-based catalyst **62** that can promote the diethylzinc addition to α-ketoesters in high yield, [Eq. (13.38)].[89] In their catalysis, titanium acts as a Lewis acid, and amine nitrogen acts as a Lewis base (**63**). The enantioselectivity was up to 78% ee:

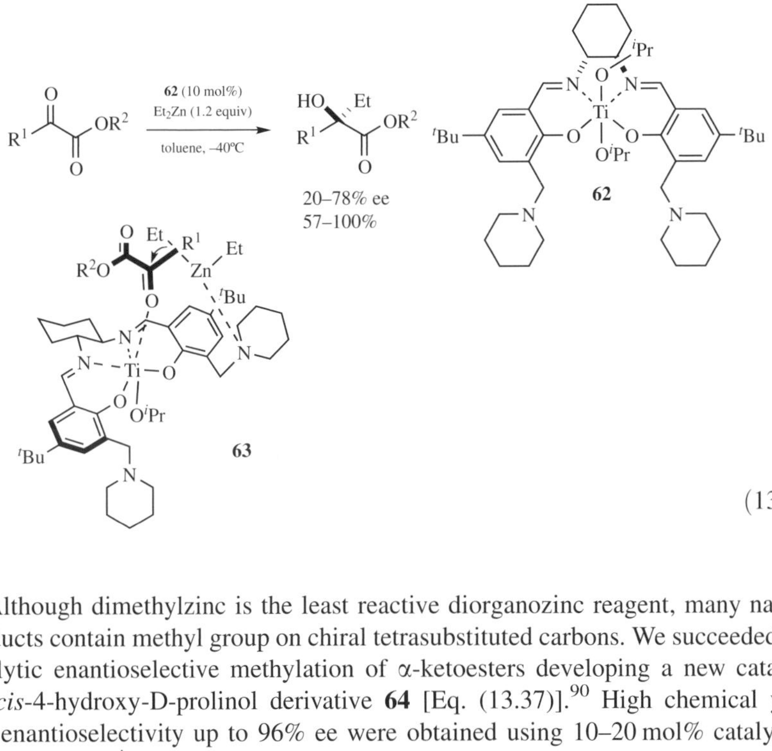

$$
\text{(13.38)}
$$

Although dimethylzinc is the least reactive diorganozinc reagent, many natural products contain methyl group on chiral tetrasubstituted carbons. We succeeded in a catalytic enantioselective methylation of α-ketoesters developing a new catalyst, 2,4-*cis*-4-hydroxy-D-prolinol derivative **64** [Eq. (13.37)].[90] High chemical yield and enantioselectivity up to 96% ee were obtained using 10–20 mol% catalyst in the presence of ^{i}PrOH as a catalytic additive (27 mol%). Slow addition of dimethylzinc was necessary for high enantioselectivity. We speculated that two zinc alkoxides (anionic Lewis bases) cooperatively activate the nucleophile, which should result in stronger activation of dimethylzinc. Catalysts containing *trans*-hydroxyl functions at the 2- and 4-positions of the pyrrolidine ring produced only low chemical yield and enantioselectivity:

$$\begin{array}{c}\text{72–96\% ee}\\\text{42–95\%}\end{array} \qquad (13.39)$$

An interesting bifunctional system with a combination of In(OTf)$_3$ and benzoyl-quinine **65** was developed in β-lactam formation reaction from ketenes and an imino ester by Lectka [Eq. (13.40)].[91] High diastrereo- and enantioselectivity as well as high chemical yield were produced with the bifunctional catalysis. In the absence of the Lewis acid, polymerization of the acid chloride and imino ester occurred, and product yield was moderate. It was proposed that quinine activates ketenes (generated from acyl chloride in the presence of proton sponge) as a nucleophile to generate an enolate, while indium activates the imino ester, which favors the desired addition reaction (**66**):

$$(13.40)$$

13.7 CONCLUSION

Chiral base catalysis was classified into five sections and reviewed. Although the reactions described herein are promoted by Brönsted or Lewis bases, the Lewis acidic characteristics of metals play important roles in both substrate activation and enantioselection. Compared with chiral Lewis acid-catalyzed reactions, however, an advantage of the base-catalyzed reactions may be the possibility of using more stable nucleophiles or prenucleophiles. In addition to the reactions mentioned in this chapter, purely organic Lewis base-catalyzed reactions and Lewis base–Lewis acid-catalyzed reactions are also important. Asymmetric catalysis is progressing toward becoming a more efficient and practical process with increased user- and environmental friendliness. In this sense, chiral acid/base bifunctional catalysis continues to be a main concept.

REFERENCES AND NOTES

1. Cram, D. J.; Sogah, D. Y. *J. Chem. Soc., Chem. Commun.* **1981**, 625–626.

2. (a) Kim, Y. S.; Matsunaga, S.; Das, J.; Sekine, A.; Ohshima, T.; Shibasaki, M. *J. Am. Chem. Soc.* **2000**, *122*, 6506–6507. For a review on *O*-linked-BINOL: (b) Matsunaga, S.; Ohshima, T.; Shibasaki, M. *Adv. Synth. Catal.* **2002**, *344*, 3–15.

3. Majima, K.; Takita, R.; Okada, A.; Ohshima, T.; Shibasaki, M. *J. Am. Chem. Soc.* **2003**, *125*, 15837–15845.

4. For a more recent review: Hong, S.; Marks, T. J. *Acc. Chem. Res.* **2004**, *37*, 673–686.

5. Hong, S.; Tian, S.; Metz, T.; Marks, T. J. *J. Am. Chem. Soc.* **2003**, *125*, 14768–14783. For other examples, see Ref. 4.

6. (a) Shibasaki, M.; Sasai, H.; Arai T. *Angew. Chem. Int. Ed. Engl.* **1997**, *36*, 1236–1256; (b) Shibasaki, M.; Kanai, M. *Chem. Pharm. Bull.* **2001**, *49*, 511–524; (c) Shibasaki, M.; Kanai, M.; Funabashi, K. *Chem. Commun.* **2002**, 1989–1999; (d) Shibasaki, M.; Yamamoto, Y. (Eds.), *Multimetallic Catalysts in Organic Synthesis*, Wiley-VCH; Weinheim, **2004**; (e) Ma, J.-A.; Cahard, D. *Angew. Chem. Int. Ed.* **2004**, *43*, 4566–4583.

7. Kumagai, N.; Matsunaga, S.; Kinoshita, T.; Harada, S.; Okada, S.; Sakamoto, S.; Yamaguchi, K.; Shibasaki, M. *J. Am. Chem. Soc.* **2003**, *125*, 2169–2178.

8. (a) Matsunaga, S.; Kumagai, N.; Harada, S.; Shibasaki, M. *J. Am. Chem. Soc.* **2003**, *125*, 4712–4713; (b) Matsunaga, S.; Yoshida, T.; Morimoto, H.; Kumagai, N.; Shibasaki, M. *J. Am. Chem. Soc.* **2004**, *126*, 8777–8785.

9. Harada, S.; Kumagai, N.; Kinoshita, T.; Matsunaga, S.; Shibasaki, M. *J. Am. Chem. Soc.* **2003**, *125*, 2582–2590.

10. Gnanadesikan, V.; Horiuchi, Y.; Ohshima, T.; Shibasaki, M. *J. Am. Chem. Soc.* **2004**, *126*, 7782–7783.

11. (a) Yamada, Y. M. A.; Yoshikawa, N.; Sasai, H.; Shibasaki, M. *Angew. Chem., Int. Ed. Engl.* **1997**, *36*, 1871–1873; (b) Yoshikawa, N.; Yamada, Y. M. A.; Das, J.; Sasai, H.; Shibasaki, M. *J. Am. Chem. Soc.* **1999**, *121*, 4168–4178.

12. For reviews on direct catalytic enantioselective aldol reactions, see: (a) Shibasaki, M.; Matsunaga, S.; Kumagai, N. In *Modern Aldol Reactions*, Mahrwald, R. (Ed.), Wiley-VCH, Weinheim, **2004**; Vol. 2, pp. 197–227; (b) Saito, S.; Yamamoto, H. *Acc. Chem. Res.* **2004**, *37*, 570–579; (c) Notz, W.; Tanaka, F.; Barbas, C. F., III. *Acc. Chem. Res.* **2004**, *37*, 580–591.

13. For the first example of direct catalytic asymmetric aldol reaction of ethyl ketones, see Mahrwald, R.; Ziemer, B. *Tetrahedron Lett.* **2002**, *43*, 4459–4461.

14. Evans, D. A.; Hoveyda, A. H. *J. Am. Chem. Soc.* **1990**, *112*, 6447–6449.

15. (a) Trost, B. M.; Ito, H. *J. Am. Chem. Soc.* **2000**, *122*, 12003–12004; (b) Trost, B. M.; Silcoff, E. R.; Ito, H. *Org. Lett.* **2001**, *3*, 2497–2500.

16. (a) Trost, B. M.; Yeh, V. S. C. *Angew. Chem. Int. Ed.* **2002**, *41*, 861–863; (b) Trost, B. M.; Yeh, V. S. C.; Ito, H.; Bremeyer, N. *Org. Lett.* **2002**, *4*, 2621–2623.

17. Trost, B. M.; Ito, H.; Silcoff, E. R. *J. Am. Chem. Soc.* **2001**, *123*, 3367–3368.

18. Trost, B. M.; Terrell, L. R. *J. Am. Chem. Soc.* **2003**, *125*, 338–339.

19. Josephsohn, N. S.; Kuntz, K. W.; Snapper, M. L.; Hoveyda, A. H. *J. Am. Chem. Soc.* **2001**, *123*, 11594–11599.

20. (a) Shimizu, K. D.; Cole, B. M.; Krueger, C. A.; Kuntz, K.; Snapper, M. L.; Hoveyda, A. H. *Angew. Chem. Int. Ed. Engl.* **1997**, *36*, 1704–1707; (b) for up-to-date extension of this concept, see Wieland, L. C.; Deng, H.; Snapper, M. L.; Hoveyda, A. H. *J. Am. Chem. Soc.* **2005**, *127*, 15453–15456.

21. (a) Tokunaga, M.; Larrow, J. F.; Kakiuchi, F.; Jacobsen, E. N. *Science* **1997**, *277*, 936–938; (b) Schaus, S. E.; Brandes, B. D.; Larrow, J. F.; Tokunaga, M.; Hansen, K. B.; Gould, A. E.; Furrow, M. E.; Jacobsen, E. N. *J. Am. Chem. Soc.* **2002**, *124*, 1307–1315.

22. Nielsen, L. P. C.; Stevenson, C. P.; Blackmond, D. G.; Jacobsen, E. N. *J. Am. Chem. Soc.* **2004**, *126*, 1360–1352.

23. For reviews, see: (a) Jacobsen, E. N.; Wu, M. H. In *Comprehensive Asymmetric Catalysis*; Jacobsen, E. N., Pfaltz, A., Yamamoto, H. (Eds.), Springer–Verlag, Heidelberg, **1999**, Vol. III, Chap. 35; (b) Jacobsen, E. N. *Acc. Chem. Res.* **2000**, *33*, 421–431.

24. Taylor, M. S.; Zalatan, D. N.; Lerchner, A. M.; Jacobsen, E. M. *J. Am. Chem. Soc.* **2005**, *127*, 1313–1317 and references cited therein.

25. Sammis, G. M.; Danjo, H.; Jacobsen, E. N. *J. Am. Chem. Soc.* **2004**, *126*, 9928–9929.

26. Evans, D. A.; Downey, C. W.; Hubbs, J. L. *J. Am. Chem. Soc.* **2003**, *125*, 8706–8707.

27. Magdziak, D.; Lalic, G.; Lee, H. M.; Fortner, K. C.; Aloise, A. D.; Shair, M. D. *J. Am. Chem. Soc.* **2005**, *127*, 7284–7285.

28. Anand, N. K.; Carreira, E. M. *J. Am. Chem. Soc.* **2001**, *123*, 9687–9688.

29. A related reaction using an In–BINOL catalyst that can be applied to aromatic aldehydes has been reported: Takita, R.; Yakura, K.; Ohshima, T.; Shibasaki, M. *J. Am. Chem. Soc.* **2005**, *127*, 13760–13761.

30. (a) Knöpfel, T. F.; Zarotti, P.; Ichikawa, T.; Carreira, E. M. *J. Am. Chem. Soc.* **2005**, *127*, 9682–9683; (b) Knöpfel, T. F.; Carreira, E. M. *J. Am. Chem. Soc.* **2003**, *125*, 6054–6055.

31. (a) Hamashima, Y.; Hotta, D.; Sodeoka, M. *J. Am. Chem. Soc.* **2002**, *124*, 11240–11241; (b) Hamashima, Y.; Takano, H.; Hotta, D.; Sodeoka, M. *Org. Lett.* **2003**, *5*, 3225–3228.

32. Hamashima, Y.; Yagi, K; Takano, H.; Tamás, L.; Sodeoka, M. *J. Am. Chem. Soc.* **2002**, *124*, 14530–14531.

33. Hamashima, Y.; Somei, H.; Shimura, Y.; Tamura, T.; Sodeoka, M. *Org. Lett.* **2004**, *6*, 1861–1864.

34. (a) Hashiguchi, S.; Fujii, A.; Takehara, J.; Ikariya, T.; Noyori, R. *J. Am. Chem. Soc.* **1995**, *117*, 7562–7563; (b) Noyori, R.; Yamakawa, M.; Hashiguchi, S. *J. Org. Chem.* **2001**, *66*, 7931–7944.

35. Watanabe, M.; Murata, K.; Ikariya, T. *J. Am. Chem. Soc.* **2003**, *125*, 7508–7509.

36. Wang, H.; Watanabe, M.; Ikariya, T. *Tetrahedron Lett.* **2005**, *46*, 963–966.

37. Ikariya, T.; Wang, H.; Watanabe, M.; Murata, K. *J. Organomet. Chem.* **2004**, *689*, 1377–1381.

38. Watanabe, M.; Ikagawa, A.; Wang, H.; Murata, K.; Ikariya, T. *J. Am. Chem. Soc.* **2004**, *126*, 11148–11149.

39. For a review, see Denmark, S E.; Fu, J. *Chem. Commun.* **2003**, 167-170.

40. Nakajima, M.; Saito, M.; Shiro, M.; Hashimoto, S.-i.; *J. Am. Chem. Soc.* **1998**, *120*, 6419–6420.

41. Nishiyama, H.; Itoh, K. Asymmetric hydrosilylation and related reactions. In *Catalytic Asymmetric Syntehsis*; Ojima, I. (Ed.), Wiley VCH, New York, **2000**; Chap. 2.

42. (a) Lipshutz, B. H.; Noson, K.; Chrisman, W.; Lower, A. *J. Am. Chem. Soc.* **2003**, *125*, 8779–8789; (b) Lipshutz, B. H.; Caires, C. C.; Kuipers, P.; Chrisman, W. *Org. Lett.* **2003**, *5*, 3085–3088; (c) Lipshutz, B. H.; Lower, A.; Noson, K. *Org. Lett.* **2002**, *4*, 4045–4048.

43. Lipshutz, B. H.; Shimizu, H. *Angew. Chem. Int. Ed.* **2004**, *43*, 2228–2230.

44. (a) Lipshutz, B. H.; Servesko, J. M. *Angew. Chem. Int. Ed.* **2003**, *42*, 4789–4782; (b) Lipshutz, B. H.; Servesko, J. M.; Petersen, T. B.; Papa, P. P.; Lover, A. A. *Org. Lett.* **2004**, *6*, 1273–1275; (c) Lipshutz, B. H.; Servesko, J. M.; Taft, B. R. *J. Am. Chem. Soc.* **2004**, *126*, 8352–8353.

45. (a) Sato, S.; Matsuda, I.; Izumi, Y. *Tetrahedron Lett.* **1986**, *27*, 5517–5520; (b) Slough, G. A.; Bergman, R. G.; Heathcock, C. H. *J. Am. Chem. Soc.* **1989**, *111*, 938–949.

46. Sodeoka, M.; Ohrai, K.; Shibasaki, M. *J. Org. Chem.* **1995**, *60*, 2648–2649.

47. (a) Hagiwara, E.; Fujii, A.; Sodeoka, M. *J. Am. Chem. Soc.* **1998**, *120*, 2474–2475; (b) Fujii, A.; Hagiwara, E.; Sodeoka, M. *J. Am. Chem. Soc.* **1999**, *121*, 5450–5458.

48. (a) Krüger, J.; Carreira, E. M. *J. Am. Chem. Soc.* **1998**, *120*, 837–838; (b) Pagenkopf, B. L.; Krüger, J.; Stojanovic, A.; Carreira, E. M. *Angew. Chem. Int. Ed.* **1998**, *37*, 3124–3126; (c) Pagenkopf, B. L.; Carreira, E. M. *Chem. Eur. J.* **1999**, *5*, 3437–3442.

49. Moreau, X.; Bazán-Tejeda, B.; Campagne, J.-M. *J. Am. Chem. Soc.* **2005**, *127*, 7288–7289.

50. (a) Nakamura, H.; Iwama, H.; Yamamoto, Y. *J. Am. Chem. Soc.* **1996**, *118*, 6641–6647; (b) Nakamura, H.; Shim, J.-G.; Yamamoto, Y. *J. Am. Chem. Soc.* **1997**, *119*, 8113–8114.

51. (a) Nakamura, H.; Nakamura, K.; Yamamoto, Y. *J. Am. Chem. Soc.* **1998**, *120*, 4242–4243; (b) Nakamura, H.; Nakamura, K.; Yamamoto, Y. *J. Org. Chem.* **1999**, *64*, 2614–2615; (c) Fernandes, R. A.; Stimac, A.; Yamamoto, Y. *J. Am. Chem. Soc.* **2003**, *125*, 14133–14139; (d) Fernandes, R. A.; Yamamoto, Y. *J. Org. Chem.* **2004**, *69*, 735–738.

52. (a) Yanagisawa, A.; Kageyama, H.; Nakatsuka, Y.; Asakawa, K.; Matsumoto, Y.; Yamamoto, H. *Angew. Chem. Int. Ed.* **1999**, *38*, 3701–3703; (b) Wadamoto, M.; Ozasa, N.; Yanagisawa, A.; Yamamoto, H. *J. Org. Chem.* **2003**, *68*, 5593-5601; (c) this catalyst was extended to enantioselective allylsilylation of ketones: Wadamoto, M.; Yamamoto, H. *J. Am. Chem. Soc.* **2005**, *127*, 14556–14557.

53. Yanagisawa, A.; Nakatsuka, Y.; Asakawa, K.; Kageyama, H.; Yamamoto, H. *Synlett* **2001**, 69–72.

54. Yanagisawa, A.; Touge, T.; Arai, T. *Angew. Chem. Int. Ed.* **2005**, *44*, 1546–1548.

55. (a) Wada, R.; Oisaki, K.; Kanai, M.; Shibasaki, M. *J. Am. Chem. Soc.* **2004**, *126*, 8910–8911; (b) prior to the development of the catalytic enantioselective allyboration of ketones, we reported a catalytic enantioselective allylsilylation of ketones using CuF–tol-BINAP catalyst—in this allylsilylation, however, enantioselectivity was moderate: Yamasaki, S.; Fujii, K.; Wada, R.; Kanai, M.; Shibasaki, M. *J. Am. Chem. Soc.* **2002**, *124*, 6536–6537.

56. Other examples of catalytic enantioselective allylation reactions utilize chiral Lewis acid catalyst and allyltin reagents: (a) Casolari, S.; D'Addario, D.; Tagliavini, E. *Org. Lett.* **1999**, *1*, 1061–1063; (b) Cunningham, A.; Woodward, S. *Synthesis* **2002**, 43–44; (c) Waltz, K. M.; Gavenonis, J.; Walsh, P. J. *Angew. Chem., Int. Ed.* **2002**, *41*, 3697–3699; (d) Kii, S.; Maruoka, K. *Chirality* **2003**, *15*, 68–70.

57. Tomita, D.; Wada, R.; Kanai, M.; Shibasaki, M. *J. Am. Chem. Soc.* **2005**, *127*, 4138–4139.

58. For selected examples, see: (a) Oppolzer, W.; Radinov, R. N. *Helv. Chim. Acta* **1992**, *75*, 170–173; (b) Wipf, P.; Ribe, S. *J. Org. Chem.* **1998**, *63*, 6454–6455; (c) Chen, Y. K.; Lurain, A. E.; Walsh, P. J. *J. Am. Chem. Soc.* **2002**, *124*, 12225–12231; (d) Li, H.; Walsh, P. J. *J. Am. Chem. Soc.* **2004**, *126*, 6538–6539.

59. (a) Oisaki, K.; Zhao, D.; Suto, Y.; Kanai, M.; Shibasaki, M. *Tetrahedron Lett.* **2005**, *46*, 4325–4329; (b) Oisaki, K.; Suto, Y.; Kanai, M.; Shibasaki, M. *J. Am. Chem. Soc.* **2003**, *125*, 5644–5645.

60. Corey, E. J.; Helal, C. J. *Angew. Chem., Int. Ed.* **1998**, *37*, 1986–2012.

61. Noyori, R; Kitamura, M. *Angew. Chem., Int. Ed. Engl.* **1991**, *30*, 49–69.

62. Kanai, M.; Kato, N.; Ichikawa, E.; Shibasaki, M. *Synlett* **2005**, 1491–1508.

63. (a) Hamashima, Y.; Sawada, D.; Kanai, M.; Shibasaki M. *J. Am. Chem. Soc.* **1999**, *121*, 2641–2642; (b) Hamashima, Y., Sawada, D.; Nogami, H.; Kanai, M.; Shibasaki, M. *Tetrahedron* **2001**, *57*, 805–814.

64. Sawada, D.; Kanai, M.; Shibasaki, M. *J. Am. Chem. Soc.* **2000**, *122*, 10521–10532.

65. Takamura, M.; Yanagisawa, H.; Kanai, M.; Shibasaki, M. *Chem. Pharm. Bull.* **2002**, *50*, 1118–1121.

66. (a) Takamura, M.; Hamashima, Y.; Usuda, H.; Kanai, M.; Shibasaki, M. *Angew. Chem., Int. Ed.* **2000**, *39*, 1650–1652; (b) Takamura, M.; Hamashima, Y.; Usuda, H.; Kanai, M.; Shibasaki, M. *Chem. Pharm. Bull.*, **2000**, *48*, 1586–1592.

67. (a) J. Casas, C. Nájera, J. M. Sansano, J. M. Saá, *Org. Lett.* **2002**, *4*, 2589–2592; (b) A. Baeza, J. Casas, C. Nájera, J. M. Sansano, J. M. Saá, *Angew. Chem. Int. Ed.* **2003**, *42*, 3143–3146; (c) J. Casas, A. Baeza, J. M. Sansano, C. Nájera, J. Saá, *Tetrahedron: Asymmetry* **2003**, *14*, 197–200; (d) J. Casas, C. Nájera, J. M. Sansano, J. M. Saá, *Tetrahedron* **2004**, *60*, 10487–10496.

68. (a) Takamura, M.; Funabashi, K.; Kanai, M.; Shibasaki, M. *J. Am. Chem. Soc.* **2000**, *122*, 6327–6328; (b) Takamura, M.; Funabashi, K.; Kanai, M.; Shibasaki, M. *J. Am. Chem. Soc.* **2001**, *123*, 6801–6808.

69. Funabashi, K.; Ratni, H.; Kanai, M.; Shibasaki, M. *J. Am. Chem. Soc.* **2001**, *123*, 10784–10785.

70. Ichikawa, E.; Suzuki, M.; Yabu, K.; Albert, M.; Kanai, M.; Shibasaki, M. *J. Am. Chem. Soc.* **2004**, *126*, 11808–11809.

71. (a) Hamashima, Y.; Kanai, M.; Shibasaki, M. *J. Am. Chem. Soc.* **2000**, *122*, 7412–7413; (b) Hamashima, Y.; Kanai, M.; Shibasaki, M. *Tetrahedron Lett.* **2001**, *42*, 691–694.

72. For examples of catalytic enantioselective cyanation of ketones from other groups, see: (a) Belokon', Y. N.; Green, B.; Ikonnikov, N. S.; North, M.; Tararov, V. I. *Tetrahedron Lett.* **1999**, *38*, 6669–6672; (b) Belokon', Y. N.;Caveda-Cepas, S.; Green, B.; Ikonnikov, N. S.; Khrustalev, V. N.; Larichev, V. S.; Moscalenko, M. A.; North, M.; Orizu, C.; Tararov, V. I.; Tasinazzo, M.; Timofeeva, G. I.; Yashkina, L. V. *J. Am. Chem. Soc.* **1999**, *121*, 3968–3973; (c) Belokon', Y. N.; Green, B.; Ikonnikov, N. S.; Larichev, V. S.; Lokshin, B. V.; Moscalenko, M. A.; North, M.; Orizu, C.; Peregudov, A. S.; Timofeeva, G. I. *Eur. J. Org. Chem.*

2000, 2655–2661; (d) Belokon', Y. N.; Green, B.; Ikonnikov, N. S.; North, M.; Parsons, T.; Tararov, V. I. *Tetrahedron*, **2001**, *57*, 771–779; (e) Tian, S.-K.; Deng, L. *J. Am. Chem. Soc.* **2001**, *123*, 6195–6196; (f) Tian, S.-K.; Deng, L. *J. Am. Chem. Soc.* **2003**, *125*, 9900–9901; (g) Deng, H.; Isler, M. P.; Snapper, M. L.; Hoveyda, A. H. *Angew. Chem. Int. Ed.* **2002**, *41*, 1009–1012; (h) Chen, F.; Feng, Z.; Qin, B.; Zhang, G.; Jiang, Y. *Org. Lett.* **2003**, *5*, 949–952; (i) Shen, Y.; Feng, X.; Li, Y.; Zhang, G.; Jiang, Y. *Eur. J. Org. Chem.* **2004**, 129–137; (j) Fuerst, D. E.; Jacobsen, E. N. *J. Am. Chem. Soc.* **2005**, *127*, 8964–8965; (k) Liu, X.; Qin, B.; Zhou, X.; He, B.; Feng, X. *J. Am. Chem. Soc.* **2005**, *127*, 12224–12225.

73. Yabu, K.; Masumoto, S.; Yamasaki, S.; Hamashima, Y.; Kanai, M.; Du, W.; Curran, D. P.; Shibasaki, M. *J. Am. Chem. Soc.* **2001**, *123*, 9908–9909.

74. For catalytic enantioselective synthesis of camptothecin intermediate, see: (a) Yabu, K.; Masumoto, S.; Kanai, M.; Curran, D. P.; Shibasaki, M. *Tetrahedron Lett.* **2002**. *43*, 2923–2926; (b) Yabu, K.; Masumoto, S.; Kanai, M.; Shibasaki, M. *Heterocycles* **2003**, *59*, 369–385. For oxybutynin synthesis, see; (c) Masumoto, S.; Suzuki, M.; Kanai, M.; Shibasaki, M. *Tetrahedron Lett.* **2002**. *43*, 8647–8651; (d) Masumoto, S.; Suzuki, M.; Kanai, M.; Shibasaki, M. *Tetrahedron* **2004**, *60*, 10497–10504. For fostriecin and its analog synthesis, see: (e) Fujii, K.; Maki, K.; Kanai, M.; Shibasaki, M. *Org. Lett.* **2003**, *5*, 733–736; (f) Maki, K.; Motoki, R.; Fujii, K.; Kanai, M.; Kobayashi, T.; Tamura, S.; Shibasaki, M. *J. Am. Chem. Soc.* **2005**, *127*, 17111–17117.

75. Suzuki, M.; Kato, N.; Kanai, M.; Shibasaki, M. *Org. Lett.* **2005**, *7*, 2527–2530.

76. Ryu, D. H.; Corey, E. J. *J. Am. Chem. Soc.* **2005**, *127*, 5384–5387.

77. Ryu, D. H.; Corey, E. J. *J. Am. Chem. Soc.* **2004**, *126*, 8106–8107.

78. (a) Masumoto, S.; Usuda, H.; Suzuki, M.; Kanai, M.; Shibasaki, M. *J. Am. Chem. Soc.* **2003**, *125*, 5634–5635; (b) Kato, N.; Suzuki, M.; Kanai, M.; Shibasaki, M. *Tetrahedron Lett.* **2004**, *45*, 3147–3151; (c) Kato, N.; Suzuki, M.; Kanai, M.; Shibasaki, M. *Tetrahedron Lett.* **2004**, *45*, 3153–3155.

79. For catalytic enantioselective Strecker reaction of ketoimines from other groups, see: (a) Vachal, P.; Jacobsen, E. N. *Org. Lett.* **2000**, *2*, 867–870; (b) Vachal, P.; Jacobsen, E. N. *J. Am. Chem. Soc.* **2002**, *124*, 10012–10014; (c) Chavarot, M.; Byrne, J. J.; Chavant, P. Y.; Vallée, Y. *Tetrahedron: Asymmetry* **2001**, *12*, 1147–1150.

80. (a) Mita, T.; Sasaki, K.; Kanai, M.; Shibasaki, M. *J. Am. Chem. Soc.* **2005**, *127*, 514–515; (b) for an extension of this catalyst to the first enantioselective *meso*-aziridine opening by cyanide, see Mita, T.; Fujimori, I.; Wada, R.; Wen, J.; Kanai, M.; Shibasaki, M. *J. Am. Chem. Soc.* **2005**, *127*, 11252–11253.

81. For catalytic enantioselective conjugate addition of cyanide to α,β-unsaturated carbonyl compounds, see: Sammis, G. M.; Jacobsen, E. N. *J. Am. Chem. Soc.* **2003**, *125*, 4442–4443. See also Ref. 25.

82. Gauthier Jr., D. R.; Carreira, E. M. *Angew. Chem. Int. Ed. Engl.* **1996**, *35*, 2363–2365.

83. Hamada, T.; Manabe, K.; Kobayashi, S. *Angew. Chem. Int. Ed.* **2003**, *42*, 3927–3930.

84. Pu, L.; Yu, H.-B. *Chem. Rev.* **2001**, *101*, 757–824.

85. For a review, see: (a) Ramón, D. J.; Yus, M. *Angew. Chem. Int. Ed.* **2004**, *43*, 284–287. There are significant advances in catalytic enantioselective introduction of alkyl, vinyl, and phenyl groups to simple ketones using alkyltitanium reagents generated via in situ transmetalation from zinc to titanium. We did not mention these reactions in detail in this review, because the reaction mechanism still remains ambiguous if those reactions proceed through one-center catalysis or two-center catalysis. For selected examples,

see: (b) Ramón, D. J.; Yus, M. *Tetrahedron Lett.* **1998**, *39*, 1239–1242; (c) Yus, M.; Ramón, D. J.; Prieto, O. *Tetrahedron Asym.* **2002**, *13*, 2291–2293; (d) GarcÙa, C.; LaRochelle, L. K.; Walsh, P. J. *J. Am. Chem. Soc.* **2002**, *124*, 10970–10971; (e) Jeon, S.-J.; Walsh, P. J. *J. Am. Chem. Soc.* **2003**, *125*, 9544–9545; (f) Li, H.; Walsh, P. J. *J. Am. Chem. Soc.* **2004**, *126*, 6538–6539; (g) Betancort, J. M.; GarcÙa, C.; Walsh, P. J. *Synlett* **2004**, 749–760.

86. Dosa, P. I.; Fu, G. C. *J. Am. Chem. Soc.* **1998**, *120*, 445–446.

87. Jiang, B.; Chen, Z.; Tang, X. *Org. Lett.* **2002**, *4*, 3451–3453.

88. For other examples of catalytic enantioselective alkynylation of α-ketoesters using a stoichiometric Zn metal, see: (a) Cozzi, P. G. *Angew. Chem. Int. Ed.* **2003**, *42*, 2895–2898; (b) Lu, G.; Li, X.; Jia, X.; Chan, W. L.; Chan, A. S. *Angew. Chem. Int. Ed.* **2003**, *42*, 5057–5058.

89. (a) Dimauro, E. F.; Kozlowski, M. C. *Org. Lett.* **2002**, *4*, 3781–3784; (b) Dimauro, E. F.; Kozlowski, M. C. *J. Am. Chem. Soc.* **2002**, *124*, 12668–12669. For a recent example, see Ref. 20b.

90. Funabashi, K.; Jachmann, M.; Kanai, M.; Shibasaki, M. *Angew. Chem. Int. Ed.* **2003**, *42*, 5489–5492.

91. France, S.; Shah, M. H.; Weatherwax, A.; Wack, H.; Roth, J. P.; Lectka, T. *J. Am. Chem. Soc.* **2005**, *127*, 1206–1215.

INDEX

New Frontiers in Asymmetric Catalysis, Edited by Koichi Mikami and Mark Lautens
Copyright © 2007 John Wiley & Sons, Inc.